REGIONAL HUMAN ANATOMY: A Laboratory Workbook for Use with Models and Prosections

Frederick E. Grine

State University of New York at Stony Brook

Boston Burr Ridge, IL Dubuque, IA Madison, WI New York San Francisco St. Louis
Bangkok Bogotá Caracas Kuala Lumpur Lisbon London Madrid Mexico City
Milan Montreal New Delhi Santiago Seoul Singapore Sydney Taipei Toronto

McGraw-Hill Higher Education

A Division of The ***McGraw-Hill*** *Companies*

REGIONAL HUMAN ANATOMY: A LABORATORY WORKBOOK FOR USE WITH MODELS AND PROSECTIONS

Published by McGraw-Hill, a business unit of The McGraw-Hill Companies, Inc., 1221 Avenue of the Americas, New York, NY 10020.

This book is printed on recycled, acid-free paper containing 10% postconsumer waste.

1 2 3 4 5 6 7 8 9 0 QPD/QPD 0 9 8 7 6 5 4 3 2 1

ISBN 0-07-240223-7

Sponsoring editor: *Martin J. Lange*
Developmental editor: *Kristine A. Queck*
Senior development manager: *Kristine Tibbetts*
Marketing manager: *Michelle Watnick*
Project manager: *Joyce Watters*
Production supervisor: *Kara Kudronowicz*
Designer: *K. Wayne Harms*
Cover designer: *Kristy Goddard*
Interior designer: *Rokusek Design*
Cover image: *Phototake*
Media technology senior producer: *Barbara R. Block*
Compositor: *Carlisle Communications, Ltd.*
Typeface: *10/12 Goudy*
Printer: *Quebecor World Dubuque, IA*

www.mhhe.com

CONTENTS

Preface x

Laboratory 1 Anatomical Terminology, General Osteology, and General Arthrology 1

1.1 ANATOMICAL TERMINOLOGY 2
The Anatomical Position 2
Planes of Section and Reference 2
Terms of Direction 2
Special Terms of Direction 4, Some Other Terms of Direction 4
Terms of Movement 6
1.2 GENERAL OSTEOLOGY 10
Features That Are Common to All Bones 10
Overview of the Skeleton 12
Primary Division of the Skeleton 13
Common Osteological Terms 14
Epiphyses and Bone Growth 14
1.3 GENERAL ARTHROLOGY 16
Fibrous Joints 16
Cartilaginous 16
Synovial Joints 18
Structure of a Typical Synovial Joint 18

Laboratory 2 The Back 21

2.1 THE INTEGUMENT 22
2.2 THE VERTEBRAL COLUMN 24
Structure of a Typical Vertebra 24
Characteristics of Different Vertebrae 25
Cervical 25, Thoracic 27, Lumbar 27, Sacral 28, Coccygeal 28
Curvatures of the Vertebral Column 29
Stabilizing Ligaments of the Vertebral Column 29
2.3 SPINAL CORD AND SPINAL NERVES 31
Spinal Cord and Meninges 31
Spinal Nerves 33
2.4 MUSCLES OF THE BACK 36
Superficial Muscles of the Back 36
Intrinsic Muscles of the Back 38
Superficial Intrinsic Muscles 38, Deep Intrinsic Muscles 39

Laboratory 3 The Upper Limb 41

3.1 BONES OF THE UPPER LIMB 43
Pectoral Girdle 43
Arm 45
Forearm 45
Hand 48
3.2 JOINTS OF THE UPPER LIMB 49
Shoulder Joint 49
Elbow Joint 51
Wrist Joint 53
The Flexor Retinaculum 53
3.3 INNERVATION OF THE UPPER LIMB 55
The Brachial Plexus 56
Roots 56, *Trunks* 56, *Cords* 56, *Peripheral Nerves* 56
Nerves to Muscles That Move the Arm 57
Nerves to Muscles That Move the Forearm, Wrist, and Digits 57
Courses of the Peripheral Nerves 59
Ventral Compartment Nerves 59, *Dorsal Compartment Nerves* 59
3.4 MUSCLES OF THE UPPER LIMB 60
Muscles That Move the Arm 60
Thoracohumeral Muscles 60, *Scapulohumeral Muscles* 61, *Rotator Cuff Muscles* 61
Muscles That Move the Forearm 65
Elbow Flexors 65, *Elbow Extensors* 65, *Pronators of the Forearm* 67, *Supinators of the Forearm* 67
Muscles That Move the Wrist 68
Flexors of the Wrist 68, *Extensors of the Wrist* 68
Muscles That Move the Fingers 70
Extrinsic Flexors of the Fingers 70, *Extrinsic Extensors of the Fingers* 70, *Intrinsic Muscles of the Fingers* 72
Muscles That Move the Thumb 74
Extrinsic Muscles of the Thumb 74, *Intrinsic Muscles of the Thumb* 76
3.5 BLOOD VESSELS OF THE UPPER LIMB 78
Arteries 78
Veins 80
Deep Veins 80, *Superficial Veins* 80

Laboratory 4 The Lower Limb 83

4.1 BONES OF THE LOWER LIMB 85
Pelvic Girdle 85
Thigh 87
Leg 88
Foot 90
The Arch of the Foot 91
4.2 JOINTS OF THE LOWER LIMB 93
Hip Joint 93
Knee Joint 95
Ankle Joint 97

4.3 INNERVATION OF THE LOWER LIMB 99
The Lumbosacral Plexus 102
Peripheral Nerves 102
Nerves to Muscles That Move the Thigh 103
Nerves to Muscles That Move the Leg, Ankle, and Digits 103
Courses of the Peripheral Nerves 105
Dorsal Compartment Nerves 105, *Ventral Compartment Nerves* 105
4.4 MUSCLES OF THE LOWER LIMB 107
Muscles That Move the Thigh 107
Iliopsoas Group Muscles 107, *Gluteal Group Muscles* 107, *Lateral Rotator Group Muscles* 110, *Adductor Group Muscles* 112
Muscles That Move the Leg 114
Leg Flexors 114, *Leg Extensors* 118
Muscles That Move the Foot 120
Plantarflexors 120, *Dorsiflexors* 122
Muscles That Move the Toes 123
Extrinsic Flexors 123, *Extrinsic Extensors* 123, *Intrinsic Muscles* 125
4.5 BLOOD VESSELS OF THE LOWER LIMB 127
Arteries 127
Veins 129
Deep Veins 129, *Superficial Veins* 129

Laboratory 5 The Neck 131
5.1 BONES AND CARTILAGES OF THE NECK 133
Bones 133
Cervical Vertebrae 133, *Hyoid Bone* 133
Cartilages 133
Tracheal Cartilages 133, *Larynx* 133
5.2 NERVES OF THE NECK 138
Spinal Nerves of the Neck 138
The Cervical Plexus 138
Cranial Nerves in the Neck 140
Autonomic Nerves in the Neck 142
Parasympathetic Fibers 142, *The Sympathetic Chain* 142
5.3 MUSCLES OF THE NECK 145
Muscles That Move the Scapula 145
Muscles That Move the Head 145
Muscles That Move the Neck 146
A Muscle of "Facial" Expression 149
Muscles That Move the Hyoid and Larynx 150
Hyolaryngeal Muscles 150, *Intrinsic Laryngeal Muscles* 154
The Pharynx and Esophagus 156
5.4 BLOOD VESSELS OF THE NECK 158
Arteries 158
Branches of the Internal Carotid Artery 158, *Branches of the External Carotid Artery* 158, *Branches of the Subclavian Artery* 158
Veins 160
Relationships of the Great Vessels in the Root of the Neck 162

5.5 THYROID AND PARATHYROID GLANDS 164
Anatomy and Position of the Glands 164
Blood Supply of the Glands 164

Laboratory 6 The Head 167
6.1 THE SKULL 168
Bones of the Cranium 168
Frontal 168, *Parietal* 168, *Occipital* 168, *Temporal* 168, *Sphenoid* 169
Bones of the Face 169
Maxilla 169, *Nasal* 169, *Lacrimal* 169, *Zygomatic* 169, *Ethmoid* 169, *Inferior Nasal Concha* 169, *Vomer* 170, *Palatine* 170, *Mandible* 170
6.2 THE DENTITION 175
Permanent Dentition 175
Deciduous Dentition 175
Dental Development and Eruption 176
Dental Anatomy 177
6.3 MUSCLES OF THE HEAD 178
Muscles of Facial Expression 178
Muscles of Mastication 180
Muscles That Move the Soft Palate 182
Muscles That Move the Tongue 184
Intrinsic Tongue Muscles 184, *Extrinsic Tongue Muscles* 184
6.4 NASAL AND ORAL CAVITIES 186
Nasal Cavity and Nasopharynx 186
Oral Cavity and Oropharynx 186
6.5 BLOOD VESSELS OF THE HEAD 188
Arteries 188
External Carotid Artery 188, *Vertebral and Internal Carotid Arteries: Blood Supply of the Brain* 188
Veins and Venous Sinuses 190
External Jugular Vein 190, *Endocranial Venous Sinuses: Blood Drainage from the Brain* 190

Laboratory 7 The Brain and Cranial Nerves 193
7.1 THE BRAIN 194
Rhombencephalon 196
Medulla Oblongata 196, *Pons* 196, *Cerebellum* 196
Mesencephalon 198
Midbrain 198
Prosencephalon 200
Diencephalon 200, *Cerebrum* 202
Ventricles and Cerebrospinal Fluid 207
Ventricular System 207, *CSF Circulation and Resorption* 207
7.2 CRANIAL NERVES 209
Olfactory (I) 210
Optic (II) 211

Oculomotor (III) 211
Trochlear (IV) 211
Abducens (VI) 211
Trigeminal (V) 213
Facial (VII) 215
Vestibulocochlear (VIII) 217
Glossopharyngeal (IX) 218
Vagus (X) 220
Accessory (XI) 222
Hypoglossal (XII) 223

Laboratory 8 The Eye and The Ear 225

8.1 THE EYE 226
External Features 226
The Bony Orbit 228
The Eyelids 229
The Lacrimal Apparatus 231
Extrinsic Eye Muscles 232
A Muscle That Moves the Eyelid 232, Muscles That Move the Eyeball 232
The Eyeball 235
8.2 THE EAR 237
External Ear 237, Middle Ear 237, Inner Ear 237
The External Ear 239
The Middle Ear 241
Bones of the Middle Ear 241, Muscles of the Middle Ear 241
The Inner Ear 242
Structures Related to Equilibrium 244, Structures Related to Hearing 245

Laboratory 9 The Thorax 247

9.1 THE BREAST 249
9.2 THE THORACIC SKELETON 251
Vertebrae 251
Sternum 251
Ribs and Costal Cartilages 251
Respiratory Movement of the Thoracic Skeleton 252
9.3 SKELETAL MUSCLES OF THE THORAX 254
Muscles of the Back 254
Muscles of the Upper Limb 254
Costoscapular Muscles 254
Sternocostal Muscle 255
Intercostal Muscles 256
Abdominal Diaphragm 257
9.4 THE THORACIC CAVITY 259
9.5 THE MEDIASTINUM 260
9.6 THE RESPIRATORY APPARATUS 262
Trachea and Bronchial Tree 262
The Lungs 264

9.7 THE HEART 266
Location of the Heart 266
The Pericardium 267
The Heart as a Pump 269
The Heart and Great Vessels 271
Anterior Aspect of the Heart 271, *Posterior Aspect of the Heart* 271, *The Great Vessels* 271
Internal Aspect of the Heart (Chambers and Valves) 273
Surface Projection and Auscultation of Heart Valves 274
Conducting System and Innervation of the Heart 276
Blood Supply of the Heart 277
9.8 LYMPHATICS IN THE THORAX 279
9.9 BLOOD VESSELS OF THE THORAX 281
Arteries 281
Veins 283
9.10 NERVES OF THE THORAX 284
Spinal Nerves in the Thorax 284
Cranial Nerve in the Thorax 285
Sympathetic Chain in the Thorax 286

Laboratory 10 The Abdomen 291

10.1 THE ABDOMINAL SKELETON 293
Relationship of Abdominal Viscera to the Rib Cage 294
10.2 SKELETAL MUSCLES OF THE ABDOMEN 295
Roof of the Abdominal Cavity 295
Anterolateral Walls of the Abdominal Cavity 295
Anterior Muscle 295, *Anterolateral Muscles* 297,
Posterior Wall of the Abdominal Cavity 298
10.3 ABDOMINAL CAVITY AND PERITONEUM 300
10.4 DIGESTIVE CANAL AND ORGANS 302
The Stomach 304
The Small Intestine 306
The Large Intestine 308
The Liver 310
The Gallbladder and Biliary Tree 312
The Pancreas 312
10.5 BLOOD VESSELS OF THE GUT 314
Arterial Supply 314
Celiac Artery 314, *Superior Mesenteric Artery* 316, *Inferior Mesenteric Artery* 317
Venous Drainage 318
10.6 LYMPHATIC ORGANS AND LYMPH DRAINAGE 320
Lymph Drainage 320
The Spleen 320
10.7 THE KIDNEYS AND ADRENAL GLANDS 322
Blood Vessels of the Kidneys and Adrenal Glands 322
Arteries 322, *Veins* 322
Structure of the Kidney 324

The Nephron 326
Tubular Component 326 *Vascular Component* 326,
10.8 GONADAL BLOOD VESSELS 328
Gonadal Arteries 328
Gonadal Veins 328
10.9 NERVES IN THE ABDOMEN 329
Spinal Nerves in the Abdomen 329
Cranial Nerve in the Abdomen 329
Sympathetic Chain in the Abdomen 329

Laboratory 11 The Pelvis 331
11.1 THE PELVIC SKELETON 333
Bony Pelvis 333
Pelvic Ligaments 333
Sex Differences in the Bony Pelvis 335
11.2 MUSCLES OF THE PELVIS 337
The Pelvic Diaphragm 337
Muscles of the Perineum 340
Muscles of the Urogenital Triangle 342
Muscles of the Anal Triangle 345
11.3 PERITONEUM IN THE PELVIC CAVITY 346
11.4 COMMON PELVIC VISCERA 347
Urinary Bladder and Ureters 347
11.5 MALE GENITALIA 349
Scrotum 349
Testis 352
Epididymis 352
Ductus Deferens 352
Seminal Vesicle and Ejaculatory Duct 354
Prostate Gland 354
Bulbourethral Glands 354
Penis 356
11.6 FEMALE GENITALIA 358
Labia Majora and Minora (Vulva and Vestibule) 358
Clitoris 361
Vagina 362
Uterus 362
Uterine Tubes 363
Ovary 365
Ligaments of the Uterus and Ovaries 366
11.7 BLOOD VESSELS OF THE PELVIS 368
Arteries 368
Veins 369
11.8 NERVES OF THE PELVIS 371
Spinal Nerves in the Pelvis 371
Sympathetic Nerves in the Pelvis 373
Parasympathetic Nerves in the Pelvis 373

Index *375*

PREFACE

The study of human anatomy provides a basic foundation of knowledge that is indispensable for students wishing to pursue careers in fields such as medicine, dentistry, nursing, physical therapy, the allied health professions, and athletic training. A knowledge of human anatomy is likely to be useful to artists, and it will provide insight for students who are simply curious about themselves.

There are two approaches to the study of human anatomy. One is systemic. As its name implies, it deals with anatomical systems, such as the nervous system, the skeletal system, the cardiovascular system, etc. The second approach is regional. As its name implies, it deals with anatomical regions, such as the head, the upper extremity, the thorax, the abdomen, etc. Both approaches have their advantages and disadvantages.

A systemic approach allows one to conceive anatomical detail within the context of the structural and/or functional system of which it is a part. It does not, however, facilitate understanding of important anatomical relationships. This is the strength of the regional approach. I believe that both approaches should be employed in a course on human anatomy. This is the pedagogy that I employ in the upper division undergraduate course on human anatomy that I teach at Stony Brook. This introductory course has a lecture component, in which I approach the body from a systemic point of view. It also has a laboratory component that is based on regional anatomy. Through the use of these complimentary approaches in the lectures and laboratories, the student is exposed to anatomical terminology and detail from different perspectives. I believe that reiteration such as this is key to learning anatomy.

Numerous textbooks and laboratory manuals are available for use in introductory courses in human anatomy. So, why write this one?

Undergraduate level textbooks almost always employ a systemic perspective. I use such a textbook in my course. However, the laboratory manuals that accompany these textbooks also take a systemic approach.

On the other hand, graduate level textbooks are universally written from a regional point of view, and graduate level laboratory manuals take a regional approach because they are guides for cadaveric dissection.

However, most undergraduate courses in human anatomy do not entail the dissection of a human cadaver. Rather, they employ models, prosected cadavers, and computer animated dissection. Increasingly, in fact, graduate level courses in medical, dental, osteopathic and other professional schools are coming to rely less on dissection and more on prosections and models.

The raison d'être of this particular book was my own need for an introductory (undergraduate) level regional human anatomy laboratory workbook that did not entail the dissection of a human (or cat) cadaver.

HOW TO USE THIS WORKBOOK

This is not a regional dissection manual. Rather, it is a workbook to be used in conjunction with human anatomy laboratory courses that entail the use of anatomical models, prosections, computer animated dissection, and the like. Such lab sessions will be of limited duration (usually three to four hours per week). It is, therefore, necessary to engage the student in as much interaction as possible *in preparation* for the laboratory.

The goal of *Regional Human Anatomy* is to provide students with hands-on exercises that will expose them to material sufficient to make each lab session more productive. Drawing an anatomical structure and writing out its name are excellent ways of reinforcing both the memorization of terms and the visualization of anatomical relationships.

In this book, each section of text is accompanied by one or more related illustrations. Anatomical structures identified in the text are highlighted with **boldface** type and assigned a number: **1**. In the accompanying illustration, the structure is identified by the same **1**. The student should write the name of the structure on the line adjacent to the number, and then color in the structure to which the line points. In this way, the student writes down the name of an anatomical structure, and closely approximates drawing it. After completing these prescribed exercises, the student is left with a set of colored and labeled drawings that will serve as a handy reference throughout their course of study.

At the end of each section of text, a direction box signified by a pencil identifies the accompanying illustrations that are to labeled and colored. With few exceptions, the choice of color is left up to the student. Anatomy books generally employ red for arteries, blue for veins, pink for muscles, yellow for nerves, and green for lymphatics. Where possible, the student should try and stick to these conventions, but in many cases this will not be desirable. For example, some illustrations identify different arteries—it would be folly to color them all red! When necessary, specific coloring instructions are noted in the direction boxes.

In addition, two special boxes placed throughout the text provide helpful insights to students as they complete the labeling and coloring exercises.

The occasional box with a question mark signifies a mental or physical exercise that the student is asked to undertake that will help clarify anatomical relationships. Thus, for example, the student might be asked to rotate a limb to better understand its innervation. Alternatively, the box may contain a pertinent question that will help clarify a possibly confusing anatomical entity. The answer to each question is apparent in the section of text that precedes it.

The box signified by a stethoscope and books indicates a discussion containing additional background information, or information of clinical relevance that will help clarify a particular relationship or emphasize the potential significance of an anatomical structure. I have chosen to keep clinical references to a bare minimum rather than pepper the text with extraneous information.

ORGANIZATION OF THIS WORKBOOK

Even though *Regional Human Anatomy* is not a dissection manual, it is organized along the lines of one.

Because this workbook might best be used in conjunction with a (systemic) textbook, the order of the labs reflects the order in which the most relevant anatomical systems are covered in such books. These are usually organized from anatomical terminology → integumentary system → skeletal system → muscular system → nervous system → circulatory system → respiratory system → digestive system → urinary system → reproductive system.

Thus, Laboratory 1 entails an introduction to anatomical terminology. Laboratory 2, *The Back,* deals with the integumentary system, which would be encountered if one were to begin dissection there. Laboratories 3 and 4, *The Upper Limb* and *The Lower Limb,* are largely devoted to the muscular and skeletal systems. The nervous system is largely the focus of Laboratories 7 and 8, and the circulatory and respiratory systems are dealt with largely in Laboratory 9, *The Thorax.* In large measure, Laboratory 10, *The Abdomen,* is devoted to the digestive and urinary systems, and the reproductive systems are covered in Laboratory 11, *The Pelvis.* Of course, a one-to-one correspondence between systemic and regional anatomy is not possible—perhaps it is not even desirable—but there can be substantial overlap.

In many places in this workbook, anatomical structures are covered in considerable detail. This level of scholarly devotion may be considered unnecessary, depending upon the nature of the course that is being taught and the student audience for whom it is intended. I have chosen to include such detail because it is a simple matter for an instructor to highlight those features that are relevant (or indicate those that are not relevant) for their students, and because I hope that *Regional Human Anatomy* will be found useful not only in undergraduate, but also in graduate level introductory classes.

ACKNOWLEDGEMENTS

I would like to express my gratitude to my mentors, from whom I first learned human anatomy—Phillip V. Tobias, Jack Allan and Beverly Kramer. They instilled in me an appreciation of anatomical design while I was a graduate student in the Department of Anatomical Sciences at the University of the Witwatersrand Medical School.

I owe considerable thanks to my colleagues in the Department of Anatomical Sciences, The State University of New York Medical School at Stony Brook. Brigitte Demes, John Fleagle, Catherine Forster, William Jungers, David Krause, Susan Larson, Maureen O'Leary, Callum Ross, and Randall Susman have contributed substantially to my knowledge of the structure of the human body. Their constant personal friendship, encouragement, and support are greatly appreciated.

I am particularly indebted to Jack T. Stern, Jr., Chair of the Department of Anatomical Sciences at Stony Brook University. Not only has he been a font of authoritative information about the finer details of anatomical structure, he has provided innumerable insights into the elegant relationship between morphology and function. He has graciously permitted me to borrow freely from his own books—*Essentials of Gross Anatomy* and *Core Concepts in Anatomy*—for the better explanations of anatomical form that are presented here.

In addition to my colleagues in the Department of Anatomical Sciences at Stony Brook, who have provided invaluable comments and suggestions on the manuscript, a number of very talented individuals have made further improvements to it while pursuing their graduate degrees here. I thank Roshna Wunderlich, Kamla Ahluwalia, Robert Asher, Robert Fajardo, Christopher Heesy, Michael Lague, John Polk, Brian Richmond and Nancy Stevens for substantially delaying their own research work to contribute comments to various chapters.

McGraw-Hill assembled an outstanding panel of anatomists to review the manuscript. Alfred Kwasi Boateng (*Florida Community College at Jacksonville*), Bradley S. Bowden (*Alfred University*), Charles Bursey (*Penn State University, Shenango*), Lisa K. Conley (*Carroll College*), and Leah Dvorak (*Concordia University Wisconsin*) contributed substantially to the improvement of this workbook with their suggestions, comments, and criticisms.

I am grateful to the members of McGraw-Hill's Applied Biology editorial team. Sponsoring Editor Marty Lange and Senior Development Manager Kristine Tibbetts aided in the development of this book. Developmental Editor Kristine Queck provided innumerable suggestions that substantially improved the final appearance of the book. Her patience, helpfulness, and co-operative spirit have been inspirational. My gratitude is extended to Project Manager Joyce Watters for attending to the myriad of details that go into the production of a book such as this.

My heartfelt appreciation is extended to Jack Haley and his very talented staff of illustrators at Imagineering Scientific and Technical Artworks, Inc. They expertly rendered my unimaginably crude drawings into beautiful artwork that will be a joy to color. I am truly grateful for their patience and ability.

ABOUT THE AUTHOR

Frederick E. Grine

Fred Grine is Professor of Anthropology and Anatomical Sciences at the State University of New York, Stony Brook, where he teaches human anatomy, forensic osteology, and human evolution. He received his bachelor's degree from Washington and Jefferson College, and his Ph.D. through the Department of Anatomical Sciences at the University of the Witwatersrand Medical School, Johannesburg, South Africa.

He edited *The Evolutionary History of the "Robust" Australopithecines*, and has written over 95 scientific research articles. His principal research focus relates to the study of human evolution. His research work has also explored the functional implications of primate tooth enamel structure, the elucidation of prehistoric diets from tooth wear, and the paleobiology of "mammal-like reptiles."

LABORATORY

1

Anatomical Terminology, General Osteology, and General Arthrology

1.1 ANATOMICAL TERMINOLOGY

The Anatomical Position

To avoid confusion, anatomical terms are applied to the body in a single anatomical position. These terms are, therefore, constant regardless of whether the subject is standing, sitting or lying down, whether the subject's head is turned, or whether the subject's limbs are in any of a number of different positions.

In the Anatomical Position, the body stands erect, the eyes look straight to the front, the upper limbs hang at the sides with the palms facing forward, and the lower limbs are parallel with the toes pointing directly forward.

The body shown in figure 1.1 is the anatomical position as viewed from the front and the side.

Planes of Section and Reference

Median Sagittal 1 A vertical plane that passes through the midline of the body, dividing it into equal left and right halves.

Sagittal 2 Any vertical plane parallel to the median sagittal plane.

Coronal 3 Any vertical plane perpendicular to the sagittal that divides the body into front (anterior) and back (posterior) portions.

Transverse 4 With reference to the head, neck, and trunk: any horizontal plane that divides the body into upper (superior) and lower (inferior) portions. With reference to the limbs, any plane perpendicular to the long axis of the element.

Longitudinal Any plane that sections an element parallel to its long axis.

Terms of Direction

Superior or **Cranial** 5 Refers to one structure being above, or closer to the head than another.

Inferior or **Caudal** 6 Refers to one structure being below, or closer to the feet or tail bone (coccyx) than another.

Anterior or **Ventral** 7 Refers to one structure being in front of another.

Posterior or **Dorsal** 8 Refers to one structure being behind another.

Medial 9 Refers to one structure being closer to the median sagittal plane than another.

Lateral 10 Refers to one structure being further away from the median sagittal plane than another.

Proximal 11 Refers to a structure of the upper limb or lower limb that is closer to the root (attachment) of the limb than another.

Distal 12 Refers to a structure of the upper or lower limb that is further away from the root of the limb than another.

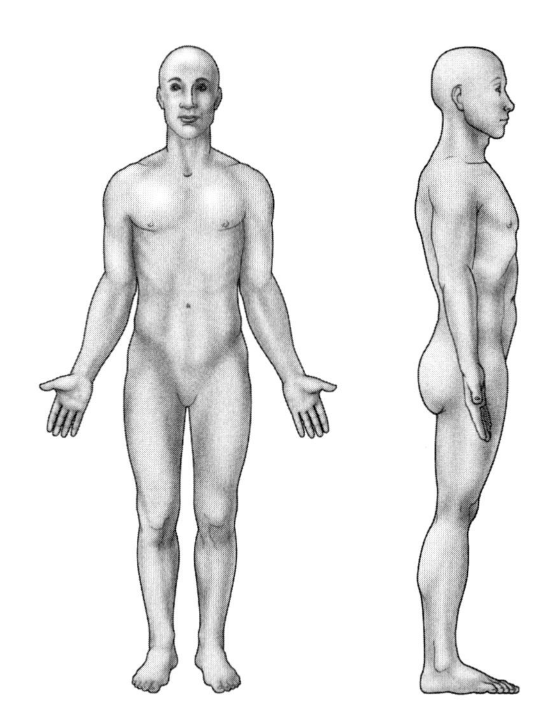

Figure 1.1 The Anatomical Position

In figure 1.2, identify these planes of reference and terms of direction. It is recommended that you color the planes in light colors and the directional arrows in brighter colors.

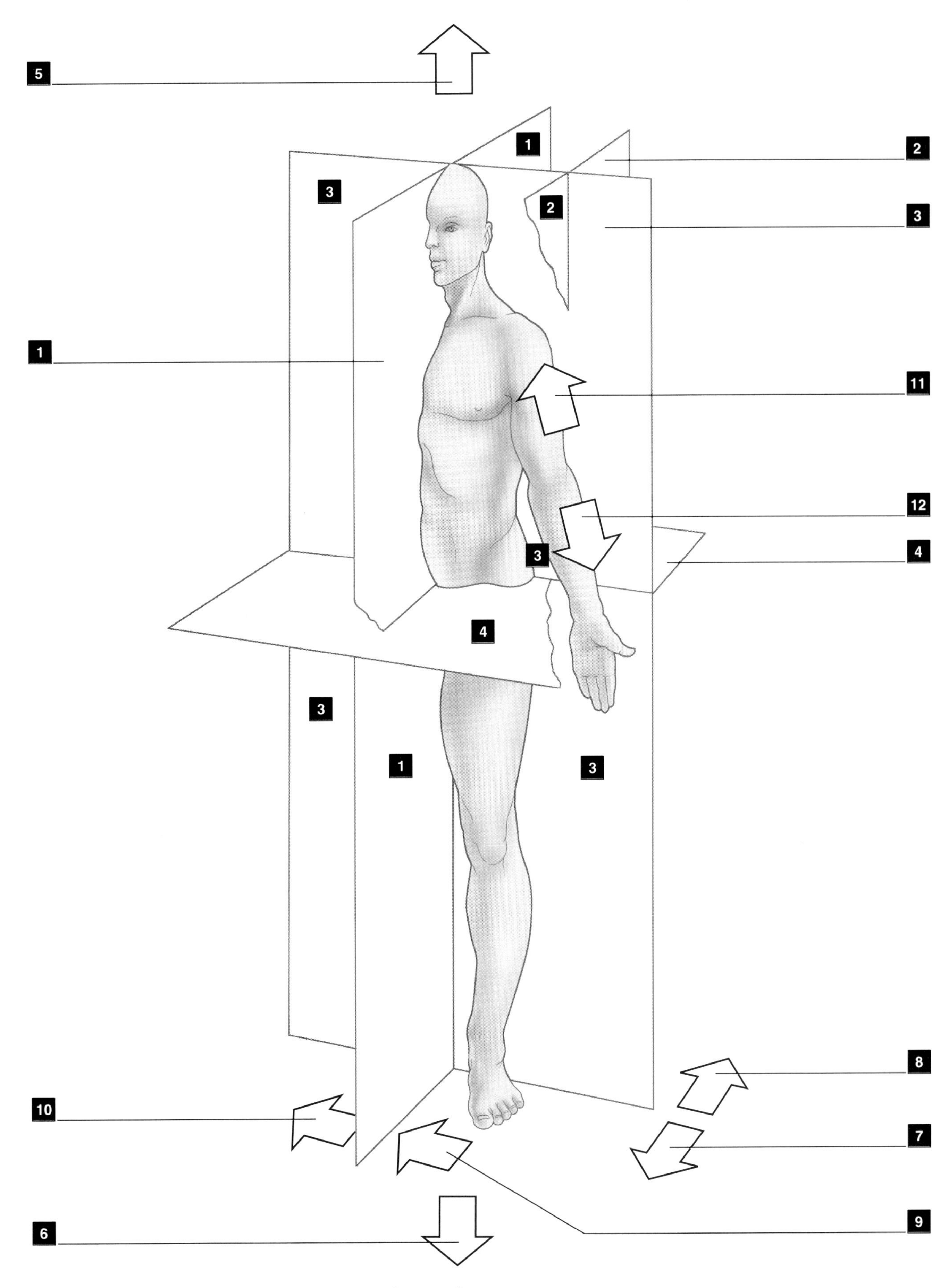

Figure 1.2 Anatomical Planes of Reference and Terms of Direction

Special Terms of Direction

Hands and Feet

Palmar 1 The ventral surface (palm) of the hand or fingers.
Dorsal 2 The dorsal surface (back) of the hand or fingers.
Plantar 3 The bottom, or ventral surface (sole), of the foot or toes.
Dorsal 4 The top, or dorsal surface, of the foot or toes.

Teeth

Mesial 5 Toward the anterior midline (i.e., the midpoint between the two central incisor teeth) of the dental arch.
Distal 6 Away from the anterior midline of the dental arch.
Lingual 7 Toward the tongue. Used in reference to all teeth.
Labial 8 Toward the lips. Used in reference to the incisors and canines.
Buccal 9 Toward the cheek. Used in reference to the premolars and molars.
Occlusal 10 The chewing surface that contacts the teeth in the other jaw. Used in reference to premolars and molars.

Some Other Terms of Direction

Superficial or **External** 11 Refers to one structure being closer to the outside or exterior surface than another. The closer a structure is to the external environment, the more superficial it is said to be.
Deep or **Internal** 12 Refers to one structure being closer to the center, or middle of the head, trunk, or limbs than another. The closer a structure is to the center, the deeper it is said to be.

In figure 1.3, identify these terms of direction.

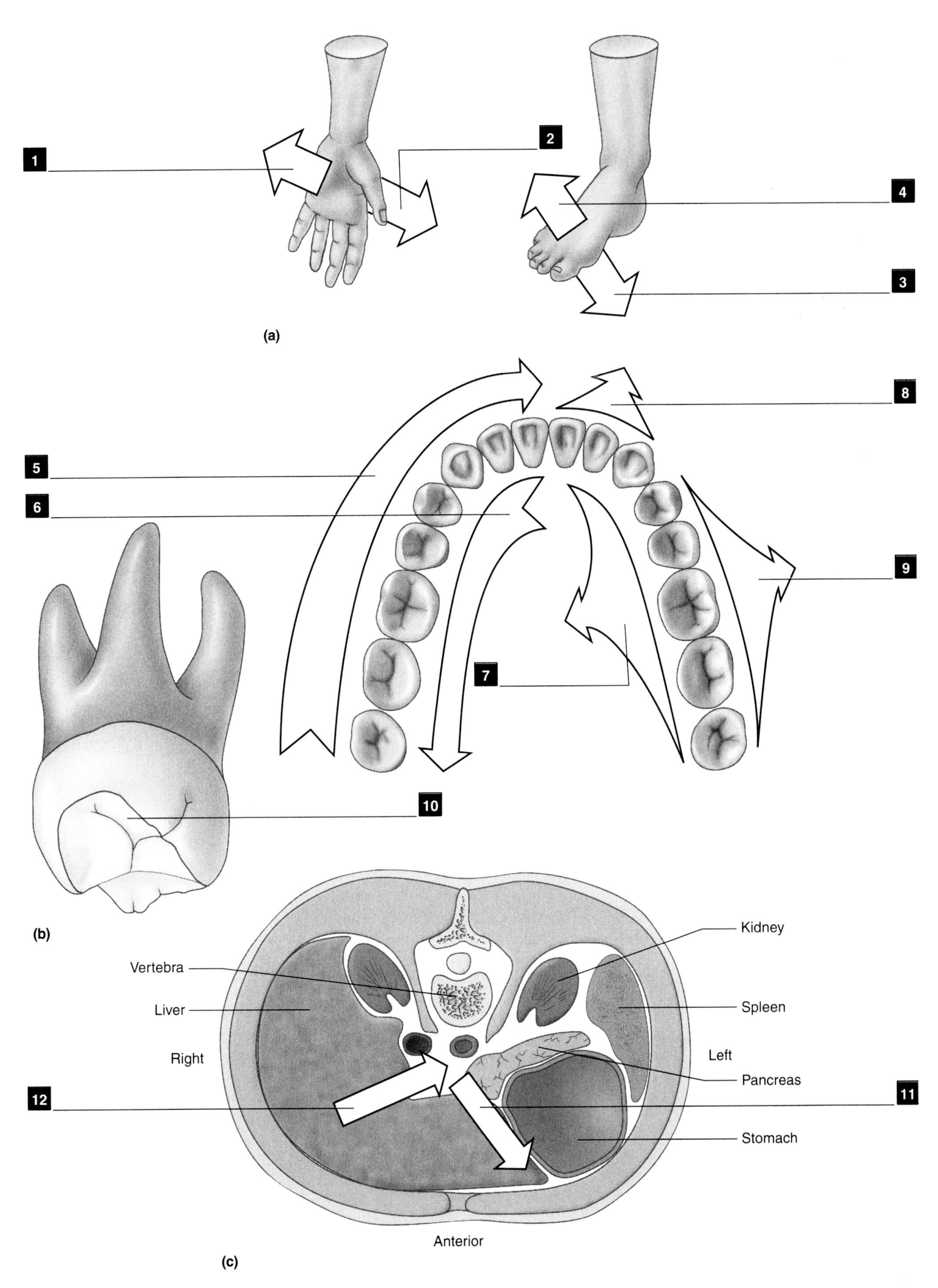

Figure 1.3 **Terms of Direction**
(a) hands and feet; (b) teeth; (c) transverse section.

Terms of Movement

Flexion 1 Movement that decreases the angle between two parts. Usually a movement to a more anterior location from anatomical position.

Extension 2 Movement that increases the angle between two parts. Usually, a movement to a more posterior location from anatomical position.

Abduction 3 Movement of an appendage away from the median sagittal plane.

Adduction 4 Movement of an appendage toward the median sagittal plane.

Rotation 5 Movement that occurs around a central axis of a body part.

Pronation 6 Rotary movement of the forearm so that the palm of the hand faces posteriorly (in extension) or inferiorly (in flexion).

Supination 7 Rotary movement of the forearm so that the palm to face anteriorly (in extension) or superiorly (in flexion).

Eversion 8 Turning the sole of the foot outward to face away from the midline.

Inversion 9 Turning the sole of the foot inward to face toward the midline.

Plantarflexion 10 Movement of the foot at the ankle so that the ball of the foot moves inferiorly, as when you stand on your toes.

Dorsiflexion 11 Movement of the foot at the ankle so that the ball of the foot moves superiorly, as when you stand on your heels.

Circumduction 12 Movement of an appendage in a circular or cone-shaped path. This involves abduction, adduction, flexion, and extension.

Opposition 13 Movement of the thumb to touch a fingertip. This involves abduction, flexion, and medial rotation of the thumb.

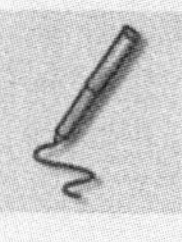

In figure 1.4, identify these terms of movement.

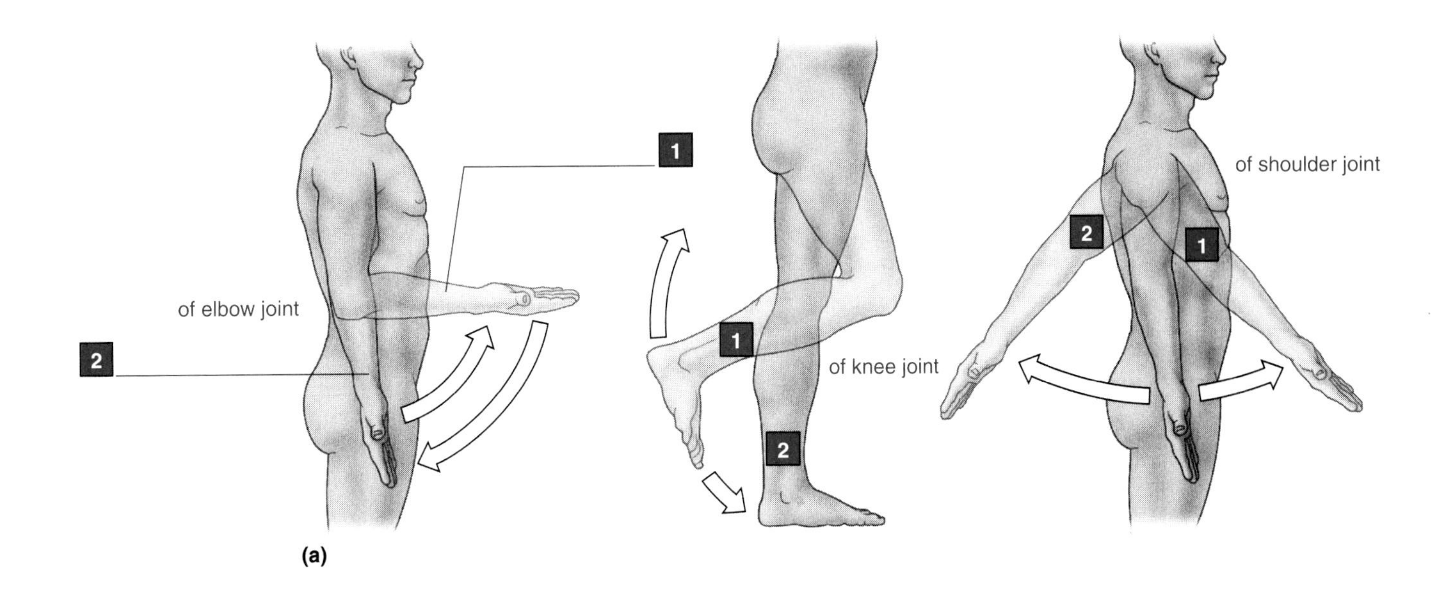

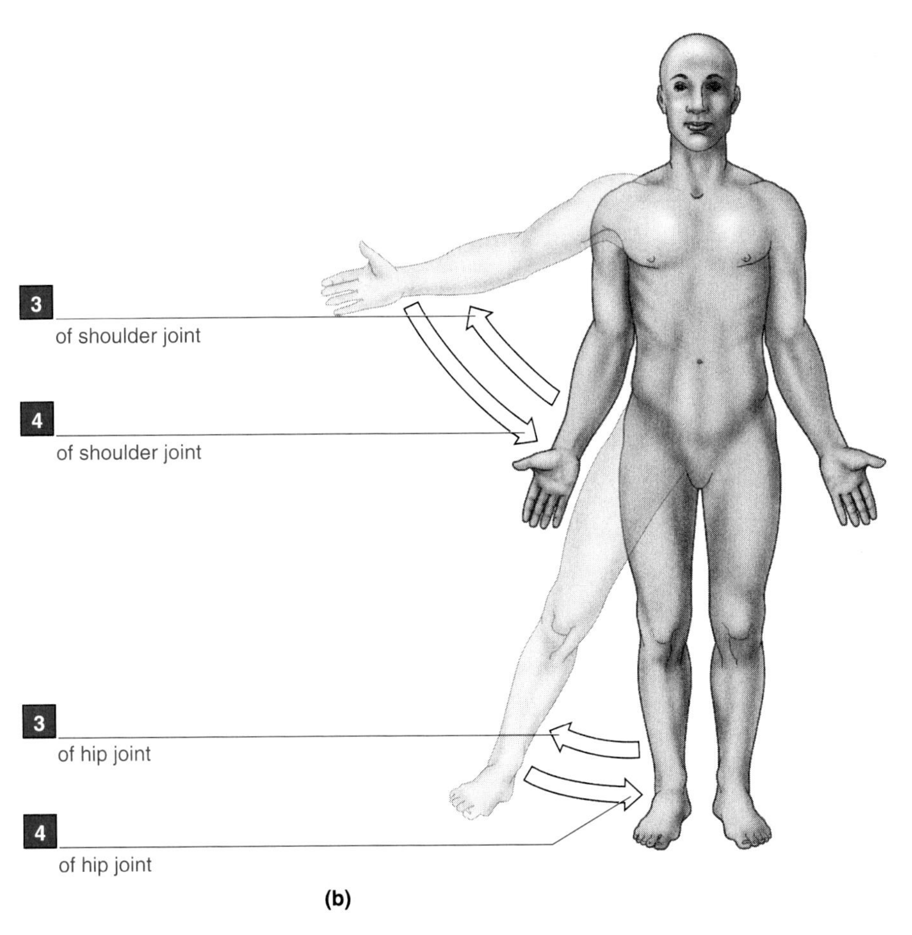

Figure 1.4 **Terms of Movement**
(a) flexion and extension; (b) abduction and adduction;

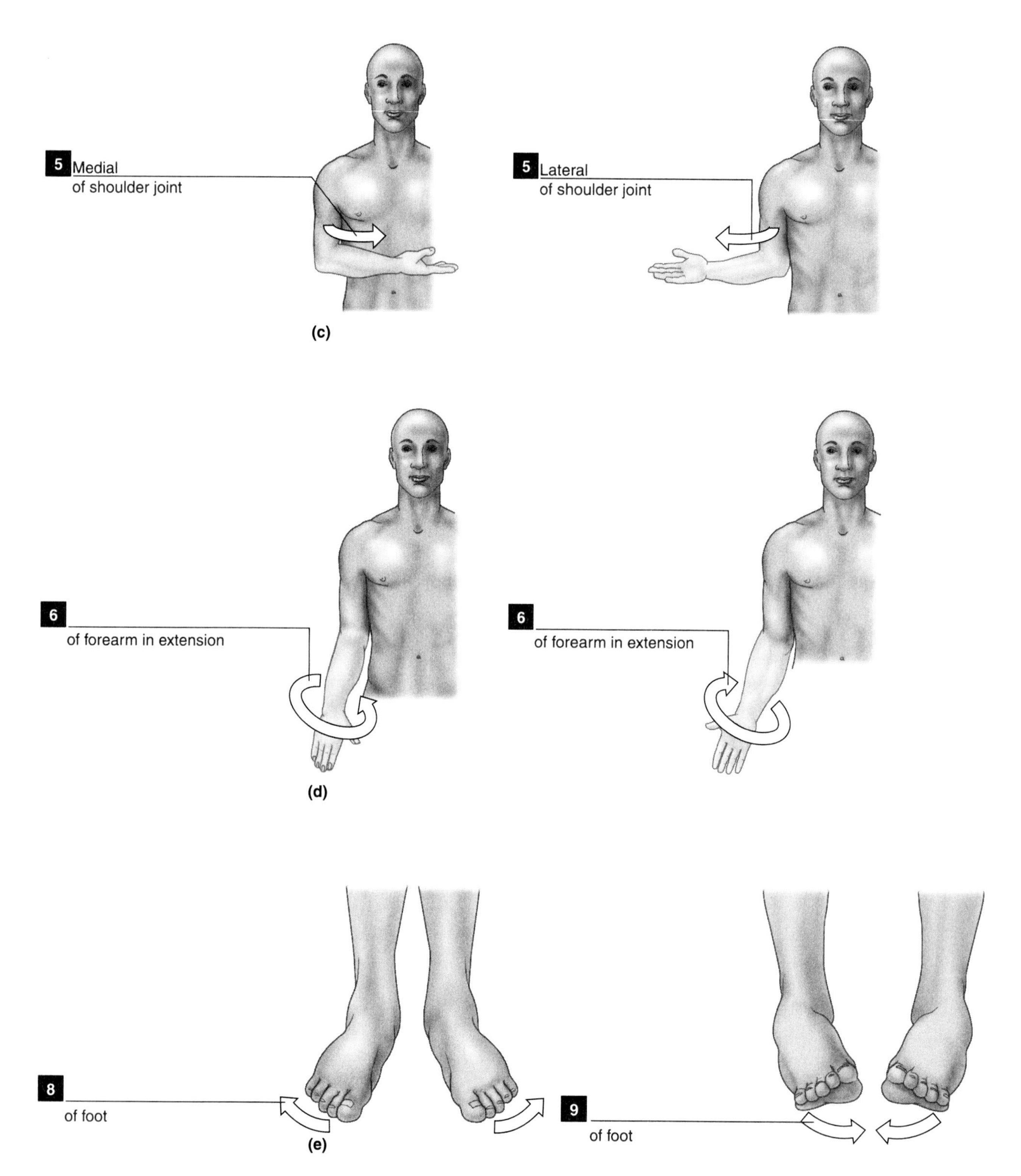

Figure 1.4 **Terms of Movement (continued)**
(c) rotation; (d) pronation and supination; (e) eversion and inversion;

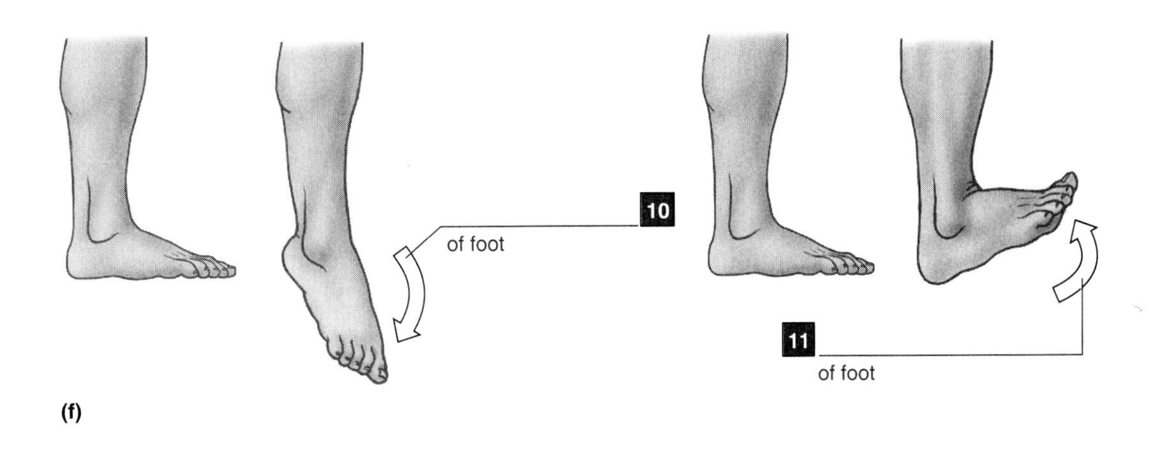

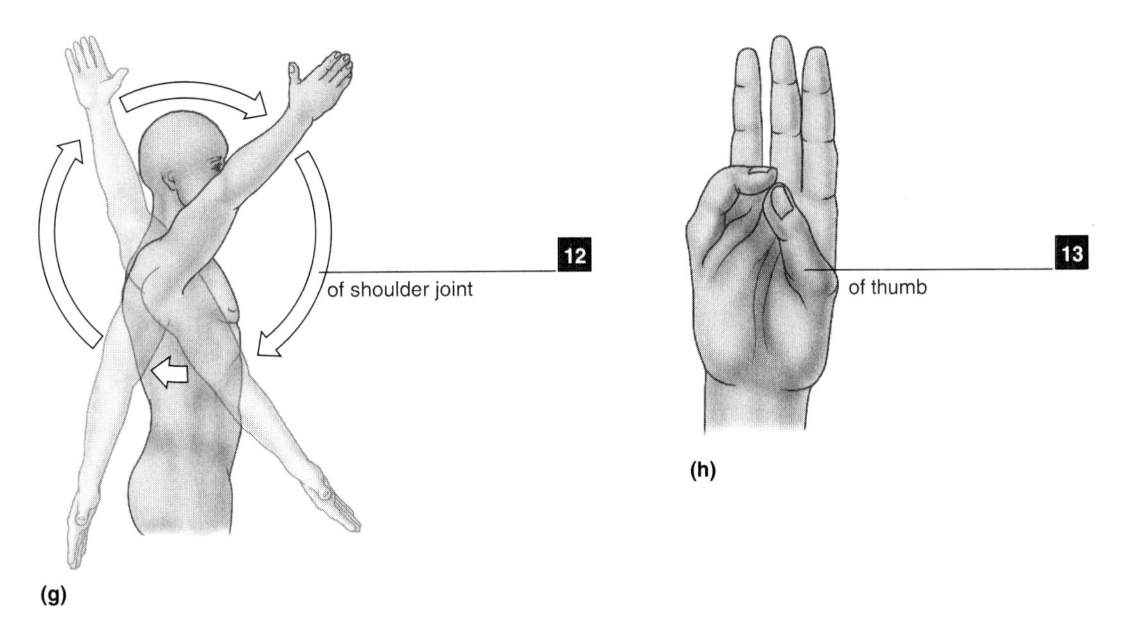

Figure 1.4 **Terms of Movement (continued)**
(f) plantarflexion and dorsiflexion; (g) circumduction; (h) opposition.

1.2 GENERAL OSTEOLOGY

Features That Are Common to All Bones

There are two types of bone tissue: **Compact (Ivory) Bone** 1 and **Cancellous (Spongy) Bone** 2**.** Cancellous bone is located internal to compact bone.

Struts of bone that project into the marrow cavity from the compact bone are known as **Trabeculae** 3**.** Trabeculae increase in density toward the proximal and distal ends of long bones. The spaces between the trabeculae are filled with red marrow, in which red blood cells are produced.

Bone is surrounded externally by layers of connective tissue called the **Periosteum** 4**.** The tendons of muscles attach to the periosteum. There is no periosteum on articular joint surfaces; they are covered with hyaline cartilage. The surface of the bone that borders the marrow cavity is lined with a connective tissue called *Endosteum.*

The bulk of compact bone in an adult consists of Osteonal or Haversian Bone. Osteonal bone is comprised of individual **Osteons** 5**.**

At the center of each osteon is the **Haversian Canal** 6**,** which carries nerves and blood vessels. Haversian canals, also known as *Central Canals* tend to run longitudinally.

Bones are highly vascular. Many bones, especially those of the arm, forearm, thigh, and leg are supplied by a large vessel known as the **Nutrient Artery** 7**.** The nutrient artery enters the bone through the *Nutrient Foramen.*

The long bones have a tubular shaft known as the **Diaphysis** 8 made up of compact bone that surrounds the **Medullary (Marrow) Cavity** 9**.** The medullary cavity is filled with fatty yellow marrow. At each end of the bone is the **Epiphysis** 10**.** Some bones have several epiphyses, whereas the long bones of the fingers and toes have only one epiphysis each.

Separating the epiphysis and diaphysis is a trabecular-filled expansion known as the **Metaphysis** 11**.**

In subadult individuals there is a cartilaginous **Epiphyseal (Growth) Plate** 12 between the metaphysis and epiphysis.

In figure 1.5, identify and color these features of bones.

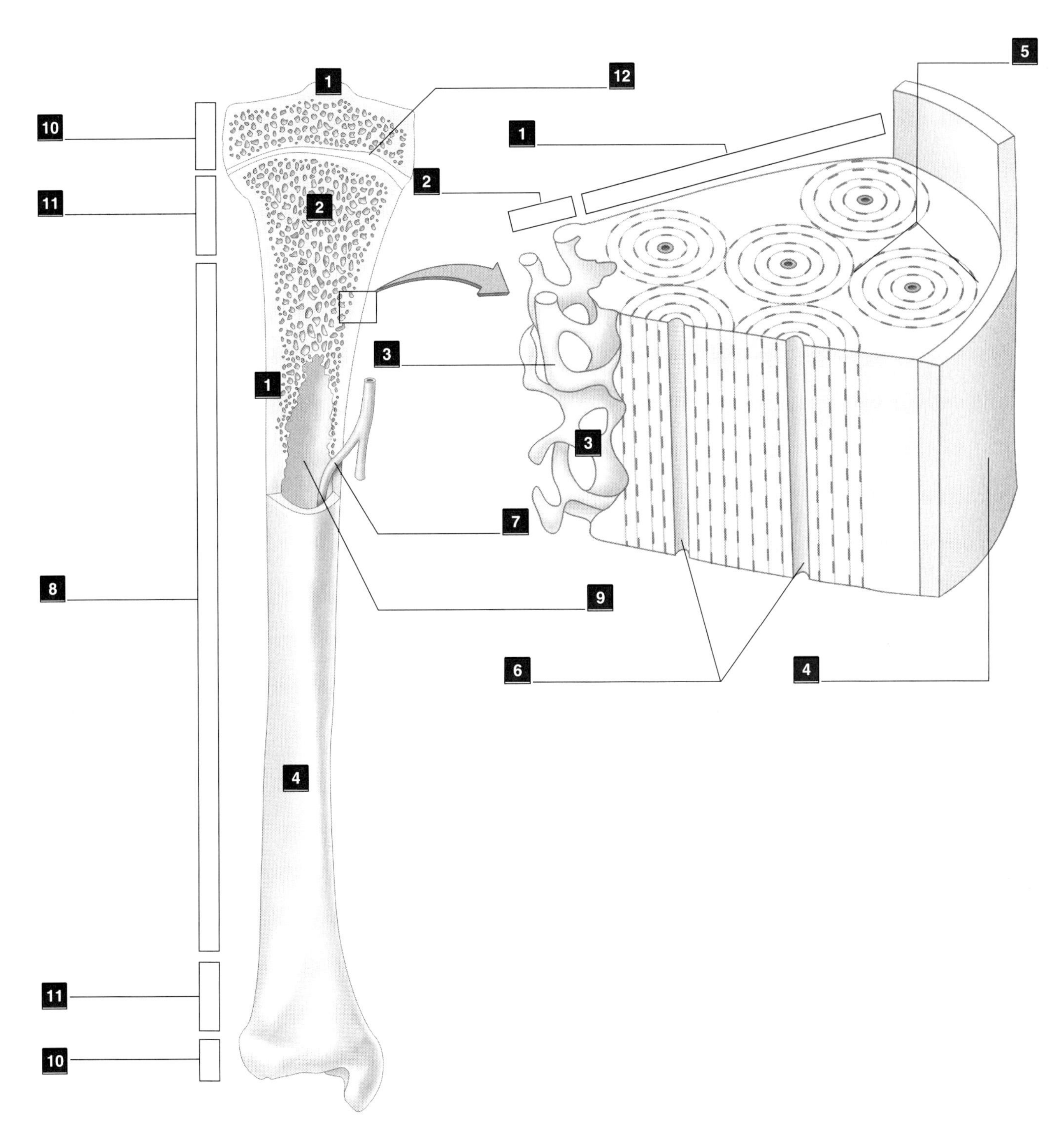

Figure 1.5 Common Osteological Features

Overview of the Skeleton

Figure 1.6 depicts the major bones or groups of bones that constitute the human skeleton. Identify each of these elements on a skeleton. What features permit you to distinguish each of the long bones of the arm, forearm, thigh, and leg?

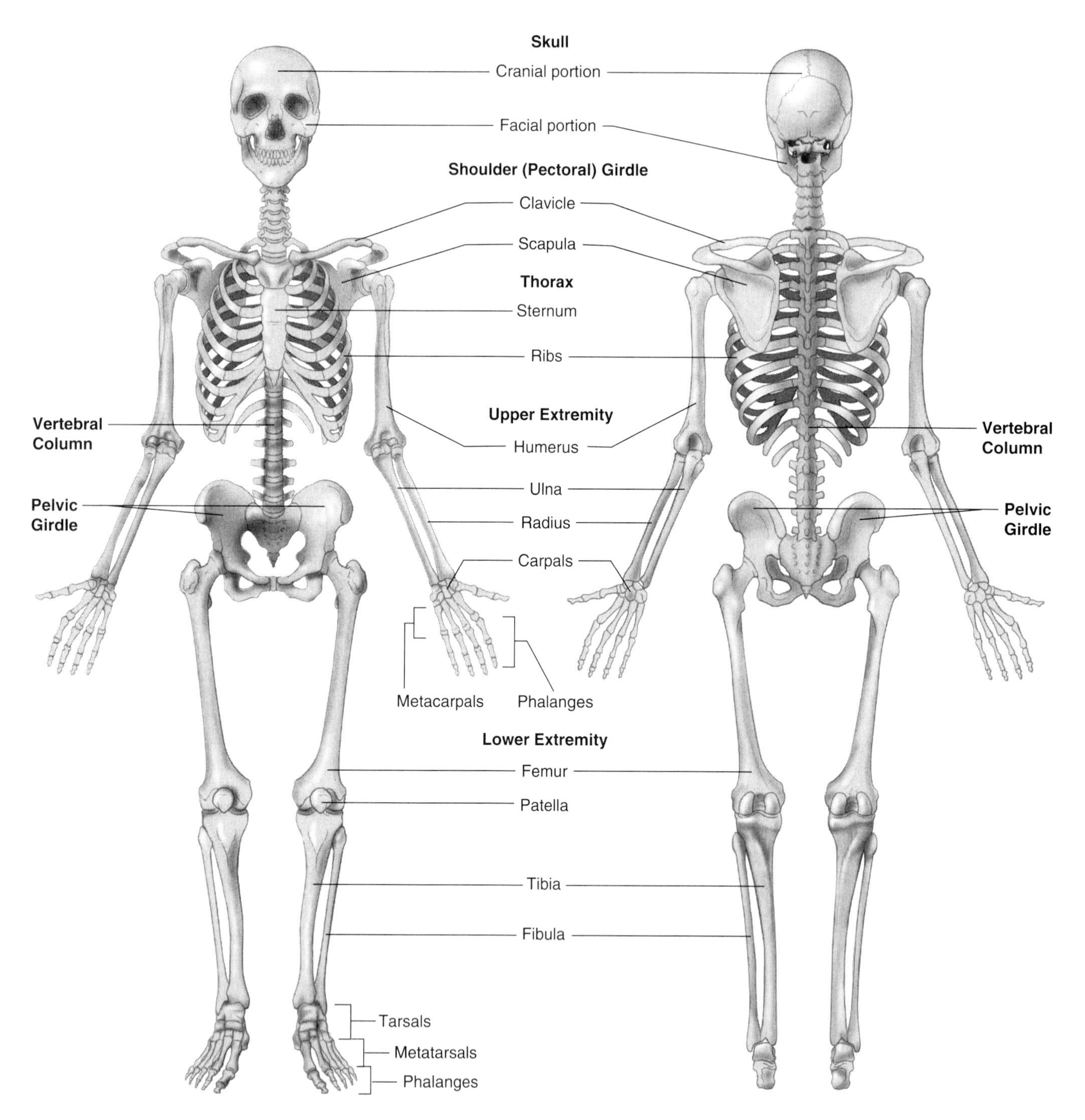

Figure 1.6 The Skeleton

Primary Division of the Skeleton

The skeleton is divided into two parts.

The **Axial Skeleton** comprises the skull, the vertebral column (including the sacrum and coccyx), the hyoid bone, and the thoracic cage, which includes the ribs, the costal cartilages, the manubrium, and the sternum.

The **Appendicular Skeleton** comprises the pectoral girdle (the scapula and clavicle) and upper limb bones, together with the pelvic girdle (the os coxae) and the lower limb bones.

In figure 1.7, color the axial elements red, and the appendicular elements blue.

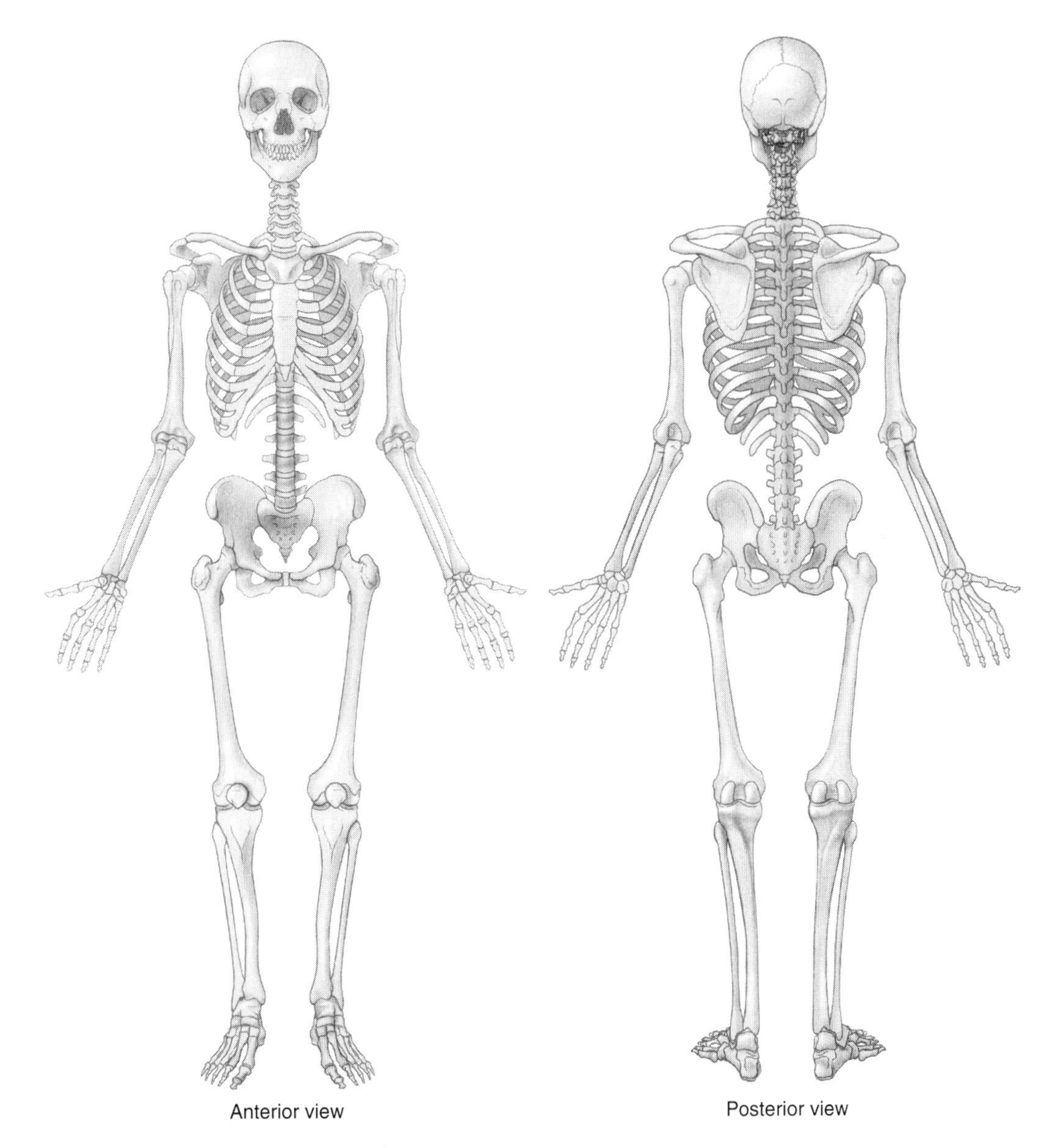

Figure 1.7 The Axial and Appendicular Skeletons

Common Osteological Terms

diaphysis	The tubular shaft of a long bone
epiphysis	The end portion of a long bone that has a separate ossification center
tuberosity	A large eminence
tubercle	A small eminence
trochanter	A large, very prominent eminence
process	A bony prominence
articular surface	A joint surface
facet	A small articular surface
condyle	A rounded articular surface
epicondyle	A nonarticular projection next to a condyle
head	A large, rounded end of a bone
neck	The part between the head and the diaphysis; there are usually separate surgical and anatomical necks
fossa	A shallow depression
foramen	A hole (a neurovascular opening)
sulcus	A long, broad groove

Epiphyses and Bone Growth

Long bones grow in length through proliferation of cartilage cells in a growth plate. The growth plate is also known as the ***Epiphyseal Plate.*** It is sandwiched between the metaphysis and the epiphysis. Most long bones have proximal and distal epiphyses, although only one is the major focus of growth.

Cartilage cells are replaced by bone in a process known as ***Ossification.*** This occurs on the metaphyseal side of the growth plate. Thus, the epiphyseal plate migrates away from the metaphysis during growth. The epiphysis and metaphysis remain separated by the growth plate until such time as cellular proliferation ceases and the cartilage becomes fully ossified. At this time, the epiphysis becomes fused with the diaphysis, and growth in the length of that bone ceases.

The epiphyses of various bones fuse at different times. Epiphyses of the foot bones (i.e., metatarsals and phalanges) may begin fusion as early as 12 years, but complete fusion may occur as late as 22 years. Similarly, the iliac crest may begin to fuse with the rest of the iliac blade as early as 16 years, but complete fusion may occur as late as 23 years. The last epiphysis of the skeleton to fuse is the medial end of the clavicle; it is usually united between 25 and 30 years of age.

This information is of immense value to a Forensic Anthropologist, who may use it to determine the age at death of an individual whose skeleton has been recovered by the police.

Figure 1.8 is a chart showing the times of epiphyseal union of various parts of the skeleton. The numbers represent years; the difference between each pair is the time span within which the particular epiphyses fuse.

Which epiphyses have already fused in your own skeleton?

Which epiphyses are likely undergoing fusion right now in your own skeleton?

In the lab, examine the long bone of an immature individual. Compare its epiphyses to those of an adult long bone. Examine radiographs in order to familiarize yourself with the appearance of an unfused epiphyseal growth plate.

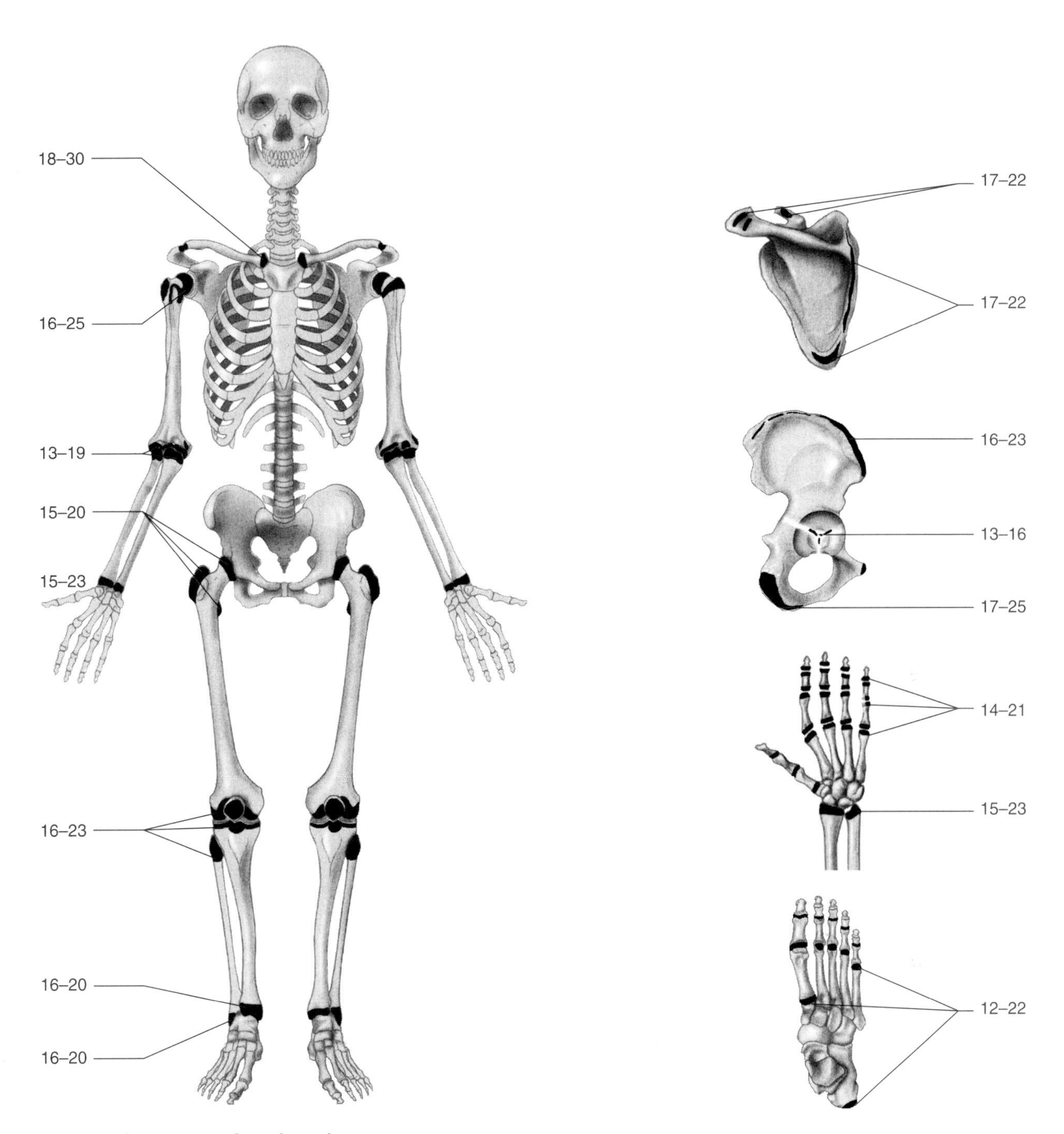

Figure 1.8 **Ages of Epiphyseal Fusion**
The numbers represent years of age; the difference between each pair is the time span within which the indicated epiphyses fuse.

1.3 GENERAL ARTHROLOGY

The junction of two bones is called a joint. The field of study that is concerned with joints is call *Arthrology*.

A joint's structure determines the degree of movement that is permitted between the bones. Not all joints are flexible. Joints are classified by the type of tissue that connects the bones. There are three principal kinds of joints: (1) Fibrous, (2) Cartilaginous, and (3) Synovial.

Fibrous Joints

The bones are joined by fibrous connective tissue. Little, if any movement takes place at this kind of joint, which is known also as a *Synarthrosis*. There are three kinds of fibrous joints: (1) Sutures, (2) Syndesmoses, and (3) Gomphoses.

Suture 1 The bones are connected by a dense layer of fibrous tissue that binds to the periosteum of each element. These are found in the skull.

Syndesmosis 2 The bones are connected by collagenous interosseous ligaments. Examples are the distal ends of the radius and ulna, and the distal ends of the tibia and fibula.

Gomphosis 3 This joint is between the teeth and the bones of the face (maxilla and mandible). The tooth root is connected to the walls of its socket by periodontal ligaments.

Cartilaginous Joints

The bones are joined either by hyaline cartilage or by fibrocartilage. There are two types of cartilaginous joints depending upon the type of cartilage that intervenes between the bones: (1) Synchondroses and (2) Symphyses.

Synchondrosis 4 The bones are connected by a plate of hyaline cartilage. This kind of joint is found between the diaphysis and epiphysis of a growing bone. The intervening hyaline cartilage forms the growth plate.

Symphysis 5 The bones are connected by a pad of fibrocartilage, which permits a limited amount of movement. Examples are the pubic symphysis of the pelvis, and the intervertebral discs.

Identify the different types of fibrous and cartilaginous joints in figure 1.9

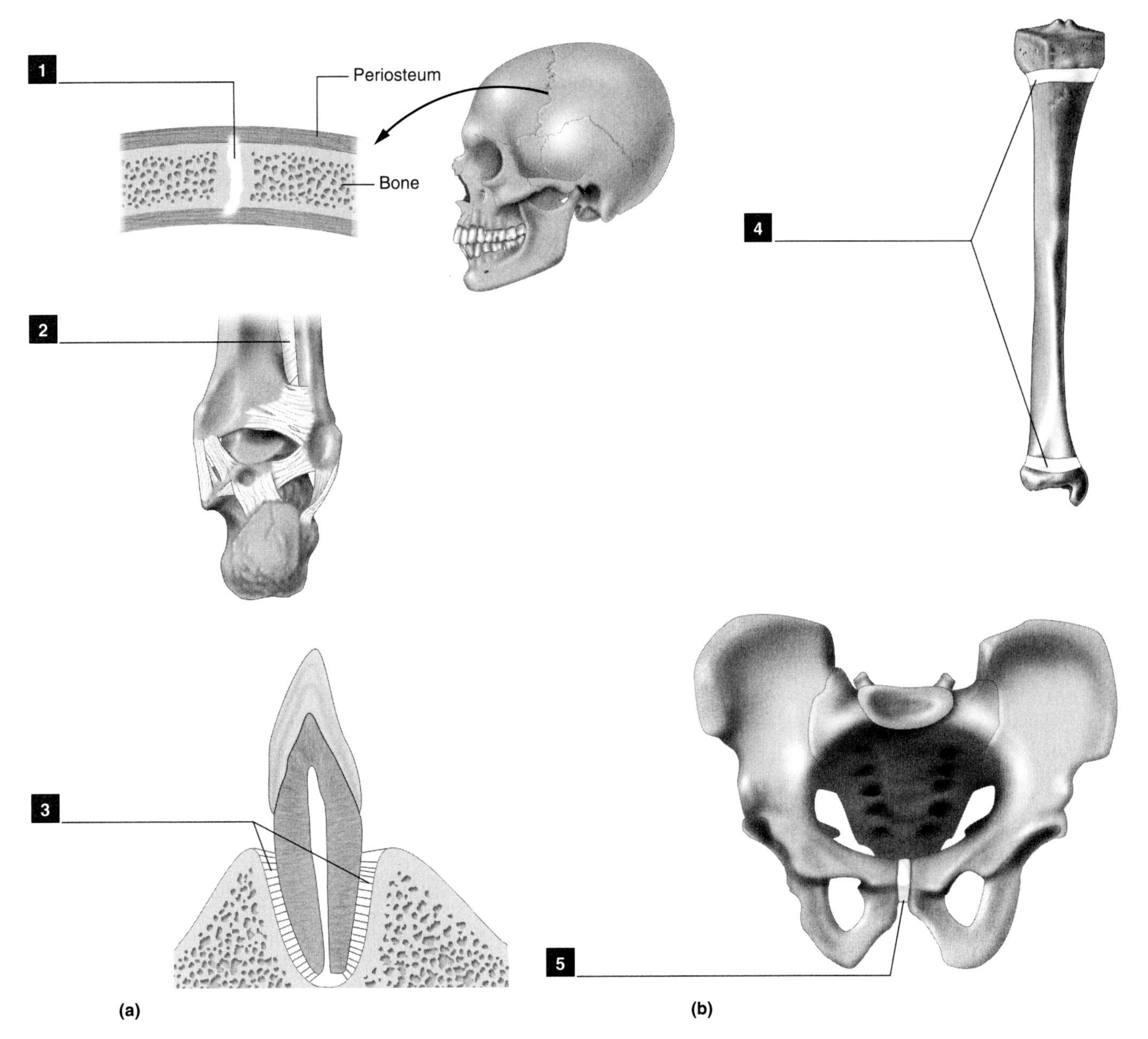

Figure 1.9 **Types of Joints**
(a) fibrous joints; (b) cartilaginous joints.

Synovial Joints

These are the most obvious joints. They are also the most complex. The adjoining bones are separated by a fluid-filled chamber that facilitates free movement between the articulating surfaces. The joint cavity is surrounded by a capsule of reinforcing ligamentous fibers. In some instances (such as in the sternoclavicular, knee, and temporomandibular joints), a disc of fibrocartilage may be interposed between the articular surfaces.

The range of motion and stability of a joint depend on three factors: (1) the shape of the articulating bony surfaces, (2) the strength of the ligaments that form and surround the joint capsule, and (3) the tone of the muscles around the joint.

Structure of a Typical Synovial Joint

The articular surfaces of the adjacent bones are covered with a layer of hyaline cartilage, known as the **Articular Cartilage 1.** The articular surfaces and the space between them are enclosed by a **Joint Capsule 2.** The external part of the capsule is composed of elastic fibers that stretch from one bone to the next. This is known as the **Capsular Ligament 3.** The internal part of the capsule is comprised by a **Synovial Membrane 4,** which secretes a lubricating *Synovial Fluid.* The synovial membrane does not cover the articular cartilage.

Ligaments provide support and important sensory information about joint movement and position. Ligaments that run between adjacent bones across the joint, but do not form part of the capsule, are known as *Extracapsular Ligaments.* The tibial and fibular collateral ligament of the knee are examples of such reinforcing bands. In some instances, *Intracapsular Ligaments* further strengthen the joint. These traverse the joint capsule between the adjacent bones. The cruciate ligaments of the knee are an example.

Bursae and tendon sheaths are fluid-filled fibrous sacs that are generally found close to joints where tendons cross the joint. A **Bursa** 5 is a lubricating device that cushions a tendon as it crosses over bony or ligamentous surfaces. An example is the subacromial bursa of the shoulder. A *Tendon Sheath* is a tubular bursa sac that surrounds a tendon over part of its length. The sac is invaginated by the tendon from one side.

Articular discs are plates of fibrocartilage that are sometimes interposed between the articular surfaces of the bones within a joint capsule. The **Articular Disc** 6 is attached firmly to the fibrous joint capsule. Its surfaces are bathed by synovial fluid. The medial and lateral menisci of the knee are examples of articular discs. The sternoclavicular and temporomandibular joints also have articular discs.

Identify the structures of a typical synovial joint in figure 1.10.

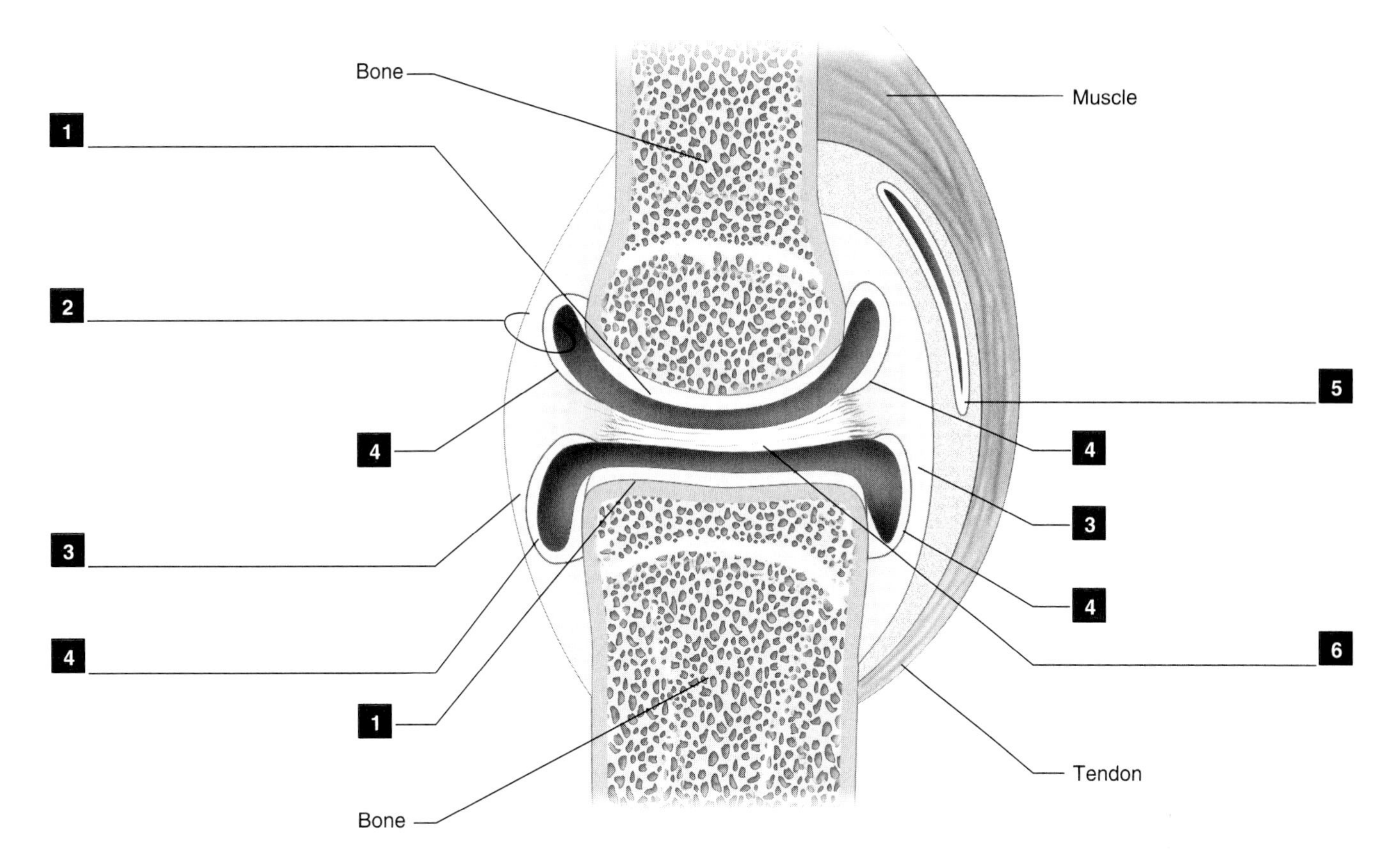

Figure 1.10 **Cross Section of a Typical Synovial Joint**

LABORATORY 2

The Back

2.1 THE INTEGUMENT

The integument covers an area of about 2 square meters; it is generally about 2 mm thick. It is divisible into two parts: (1) a cutaneous portion (the *Skin*), and a (2) subcutaneous layer that serves to attach the skin to underlying muscle or bone.

The cutaneous part consists of two layers: the **Epidermis 1**, and the deeper **Dermis 2**.

Underlying the dermis is the **Subcutaneous Layer 3**, which is also commonly known as the *Superficial Fascia*. (The subcutaneous layer is sometimes called the Hypodermis.) It contains abundant adipose tissue, which permits movement of the skin, and assists in thermal regulation.

The **Epidermis 1** has four separate layers, except on the palm of the hand and the sole of the foot, where there are five. The **Superficial Stratum 4** comprises the outer two (or three) layers of dead or dying cells. The outermost of these is the *Stratum Corneum*; deep to it is the *Stratum Granulosum*. Interposed between them on the palms of the hands and soles of the feet is the *Stratum Lucidum*. The **Stratum Germinativum 5** comprises the deepest two layers of the epidermis, where cell division and growth takes place. These layers are known as the *Stratum Spinosum* and *Stratum Basale*.

Melanocytes are usually located in the *Stratum Germinativum*. These cells are responsible for the production of a dark pigment, melanin, which serves to screen excessive UV rays.

The **Dermis 2** makes up the bulk of the thickness of the skin. It consists of a superficial **Papillary Layer 6**, the projections of which join it to the epidermis. Indistinctly separated from the papillary layer and deep to it is the **Reticular Layer 7**. The dermis contains the **Nerves 8**, the **Lymphatic Vessels 9**, the **Blood Vessels 10**, the **Hair Follicles 11**, the **Sweat (Sudoriferous) Glands 12**, and the **Sebaceous (Oil) Glands 13**.

There are two types of **Sweat (Sudoriferous) Glands 12**. *Eccrine Glands* produce sweat in response to heat. The larger *Apocrine Glands* produce sweat of a characteristic odor in response to stress. Apocrine glands in the female breast are specialized to secrete milk, and those within the canal of the outer ear produce the watery component of cerumen (ear wax). The **Sebaceous (Oil) Glands** are connected to hair follicles, and secrete an oily substance known as sebum.

The nerve endings responsible for touch reception are known as **Meissner's Corpuscles 14**; they are located in the papillae. **Nerve Plexuses 15** that surround the bulbs of hair follicles are also receptive to touch. **Pacinian Corpuscles 16** are responsible for pressure reception. **Free Nerve Endings 17** within the dermis are responsible for pain sensation.

Each hair follicle is surrounded by an epithelial root sheath. Attached to this sheath and the base of the epidermis is a smooth muscle known as the **Arrector Pili 18**. When it contracts, it pulls the follicle and its hair to an erect position, elevates the skin above, which produces a "goose bump", and forces sebum from the sebaceous gland. The smooth muscles of the blood vessels and the sweat glands of the dermis also receive autonomic innervation.

In figure 2.1, identify the aforementioned items. Color these separately. In anatomical illustrations, adipose tissue is usually rendered in light yellow, and muscle in light red. Remember that arteries are dark red, veins are blue, and nerves are bright yellow. Use different colors for the other structures.

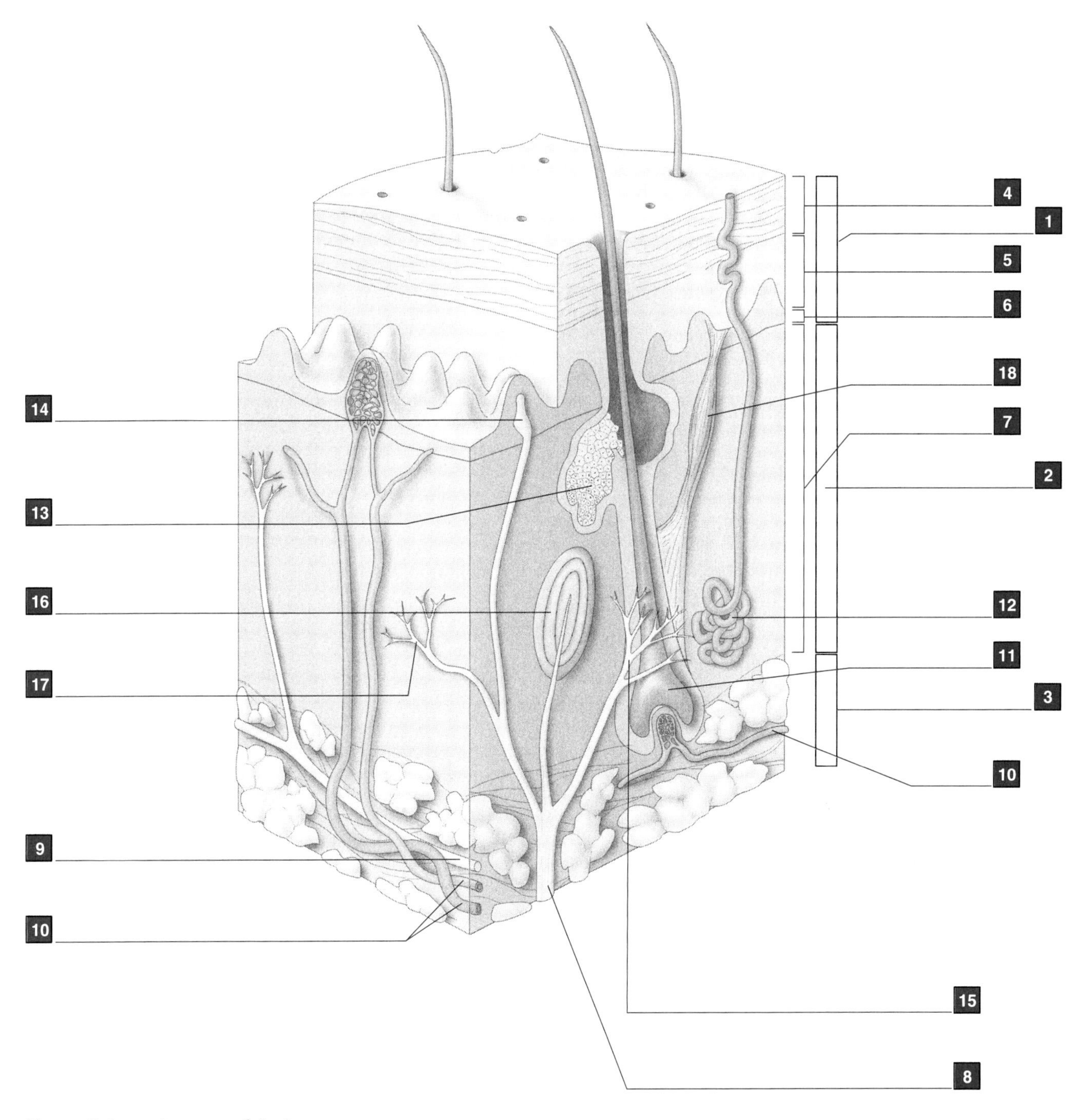

Figure 2.1 **Structure of the Integument**

2.2 THE VERTEBRAL COLUMN

The vertebral column generally consists of 33 vertebrae. These are arranged in five regions—cervical, thoracic, lumbar, sacral, and coccygeal—according to the morphology of the elements that comprise each region. Although the vertebrae of each region differ morphologically, they share a number of anatomical features.

Structure of a Typical Vertebra

Each vertebra except the first cervical (C1 = the atlas) has a weight-bearing **Body** 1. Projecting posteriorly from the body are two **Pedicles** 2. Each pedicle becomes a flattened area called the **Lamina** 3. Together, the laminae and pedicles form the **Neural Arch** 4. The neural arch and the posterior surface of the body form the borders around the **Vertebral Foramen** 5. The series of vertebral foramina is called the *Vertebral Canal*.

The **Transverse Process** 6 projects laterally from the pedicle. The **Spinous Process** 7 projects posteriorly from the midline junction of the laminae.

Projecting upward and downward from the neural arch are the **Superior Articular Processes** 8 and the **Inferior Articular Processes** 9. These form synovial joints between adjacent vertebrae.

Together, the vertebral bodies form a pressure-bearing rod that provides rigidity for the trunk. Between adjacent vertebral bodies is an **Intervertebral Disc** 10. This forms a cartilaginous joint between the bodies, and imparts a degree of mobility to the vertebral column. Each disc consists of a gelatinous core—the *Nucleus Pulposus*—that is surrounded by concentric layers of dense fibrous connective tissue known as the *Annulus Fibrosus*.

The pedicles are not as deep as the vertebral body. Thus there is a gap between the pedicles of adjacent vertebrae. This gap, which is bordered anteriorly by the vertebral bodies and intervertebral disc, is known as the **Intervertebral Foramen** 11.

Each intervertebral foramen transmits a *Spinal Nerve*. We will discuss spinal nerves later in this lab.

Identify and color the features of vertebral morphology named in figure 2.2. Leave the vertebral foramen and intervertebral foramina blank.

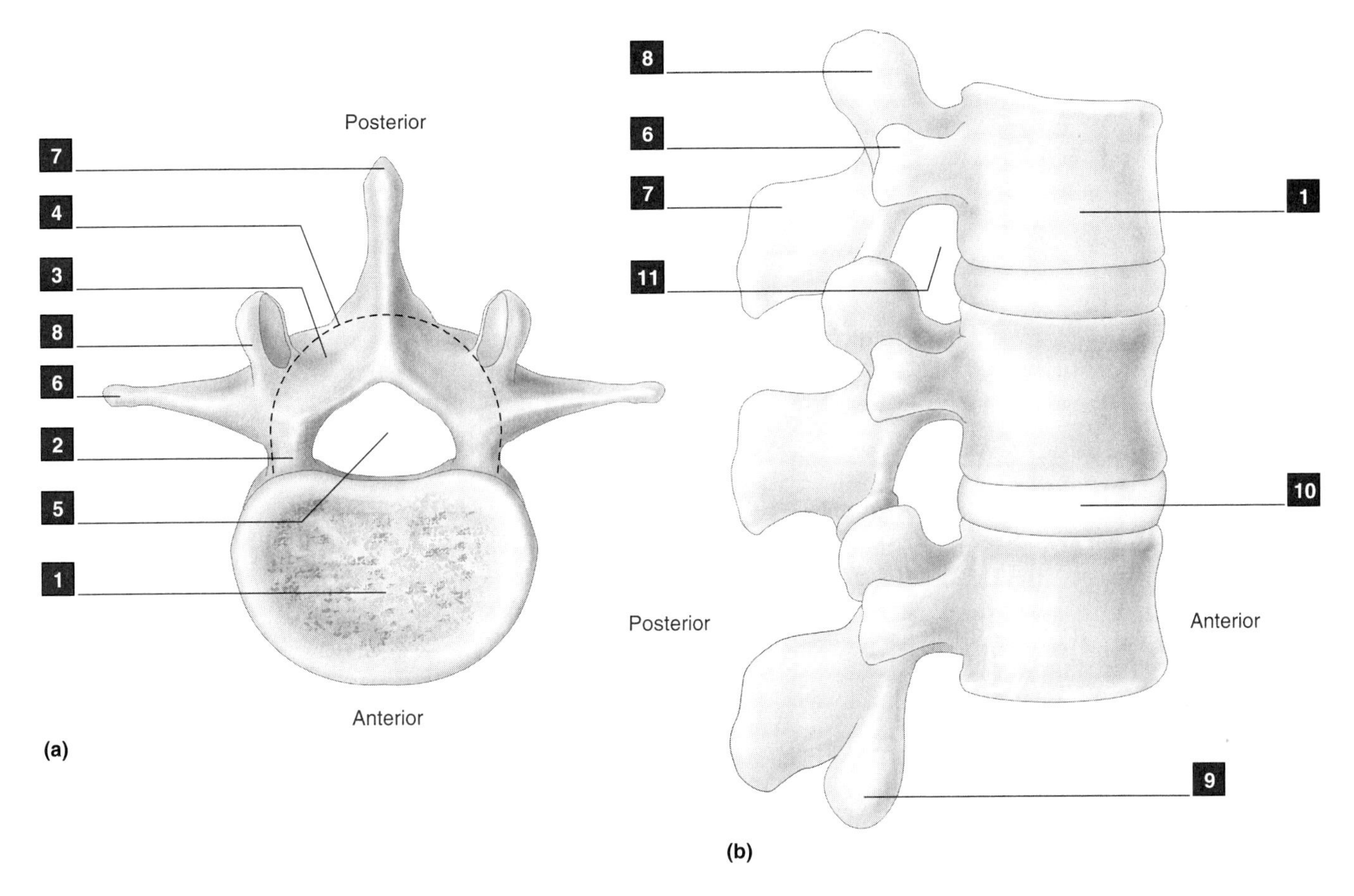

Figure 2.2 **Vertebral Structure**
(a) superior view; (b) lateral view.

Characteristics of Different Vertebrae

Each of the five types of vertebrae can be identified by certain characteristics.

Cervical Vertebrae

There are seven cervical vertebrae (C1–C7). The uppermost (C1) is known as the **Atlas 1,** so called because it supports the skull. It is readily identifiable because it lacks both a body and a spinous process. Its **Superior Articular Facet 2** is elongate and concave to accommodate the occipital condyles of the skull. You use this joint when nodding your head up and down. Like the other six cervical vertebrae, its **Transverse Process 3** is pierced by the **Foramen Transversarium 4,** which transmits the vertebral artery.

The second (C2) is known as the **Axis 5;** so called because the atlas rotates about it when turning the head. It is readily identified by the presence of the **Dens 6** (*Odontoid Process*), which projects superiorly to articulate with the back of the anterior arch of the atlas. Like the five cervical vertebrae below it, the **Spinous Process 7** of the axis is long and commonly bifid at the tip.

Like the atlas and axis, the other five ("typical") cervical vertebrae can be distinguished from all others by the presence of a foramen transversarium, and the commonly bifid spinous process. Also, in these five vertebrae, the body tends to be mediolaterally concave superiorly and convex inferiorly so that the body has a lateral **Superior Lip 8** and a lateral **Inferior Lip 9.**

Identify these features of a cervical vertebra in figure 2.3, and on isolated vertebrae in the lab.

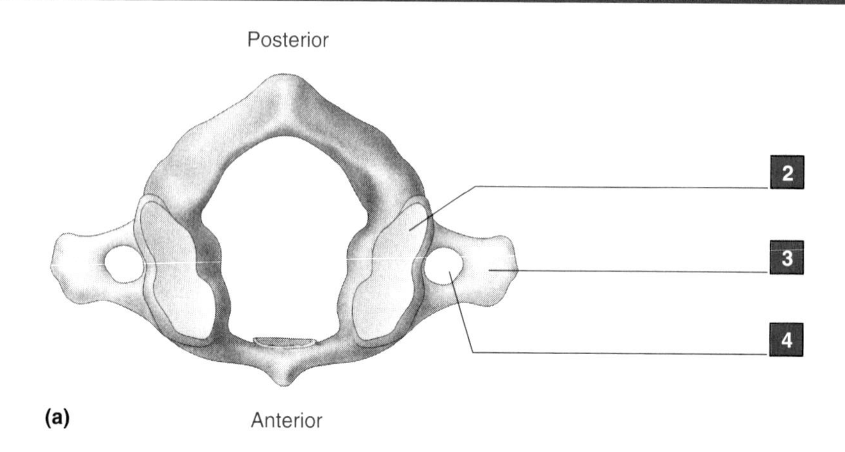

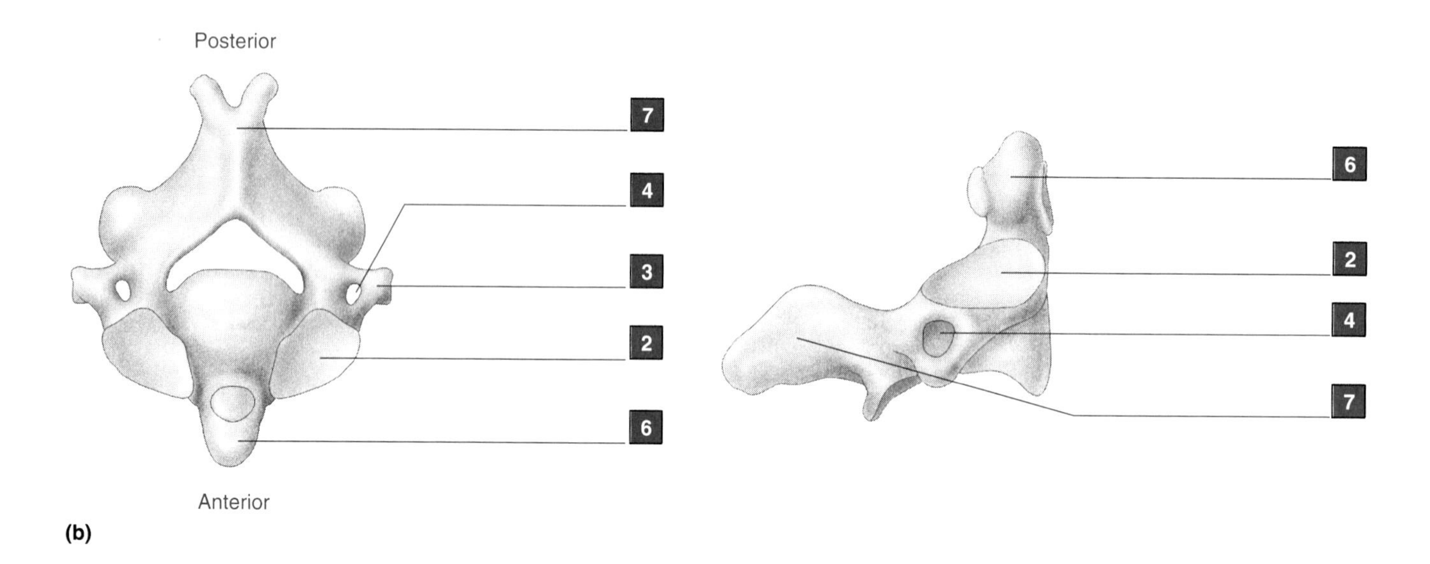

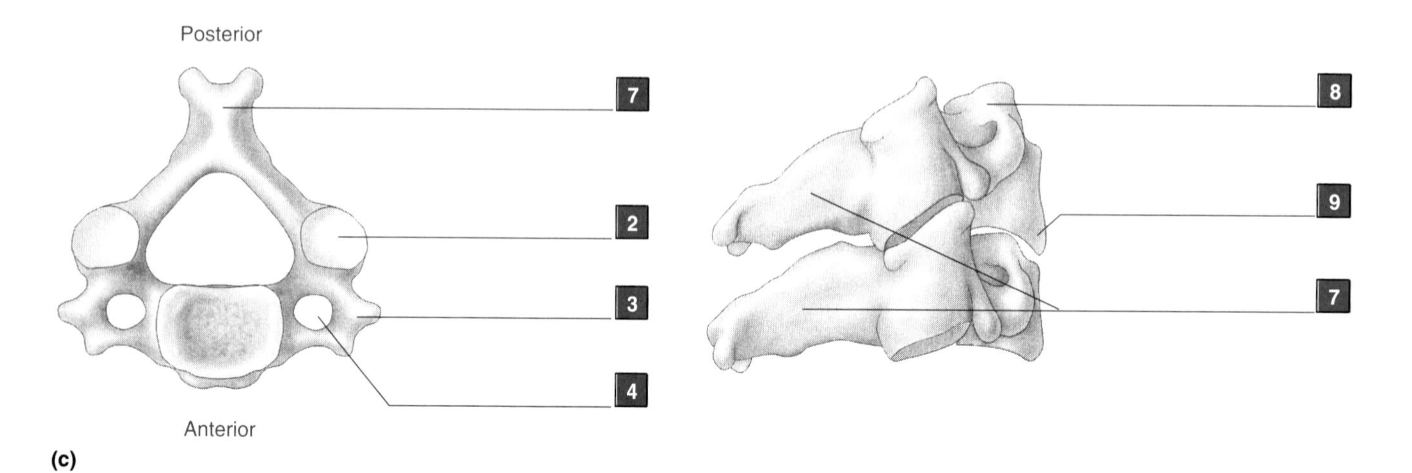

Figure 2.3 **Cervical Vertebrae**
(a) superior view of atlas, (b) superior and lateral views of axis, (c) superior and lateral views of typical cervical vertebra.

Thoracic Vertebrae

There are 12 thoracic vertebrae (T1–T12). They can be readily distinguished from all others by the presence of facets for the articulation of the ribs. **Costal Facets** 1 located on the transverse processes (except T11 and T12), articulate with the neck (tubercle) of the ribs. Costal facets located on the posterosuperior and posteroinferior corners of the lateral surface of the body (except T10–T12, which have only a single facet on the side of the body) articulate with the head of the ribs.

Thoracic vertebrae generally have a long, inferiorly projecting **Spinous Process** 2, and the **Superior Articular Facets** 3 tend to face posteriorly. The **Vertebral Body** 4 tends to have a heart-shaped body (when viewed superiorly), although the transverse diameter of the body increases toward the cranial and caudal ends of the series. The lowermost thoracic come to resemble their lumbar neighbors with a broad body and a stocky, horizontal spinous process.

Identify these features of a thoracic vertebra in figure 2.4, and on isolated vertebrae in the lab.

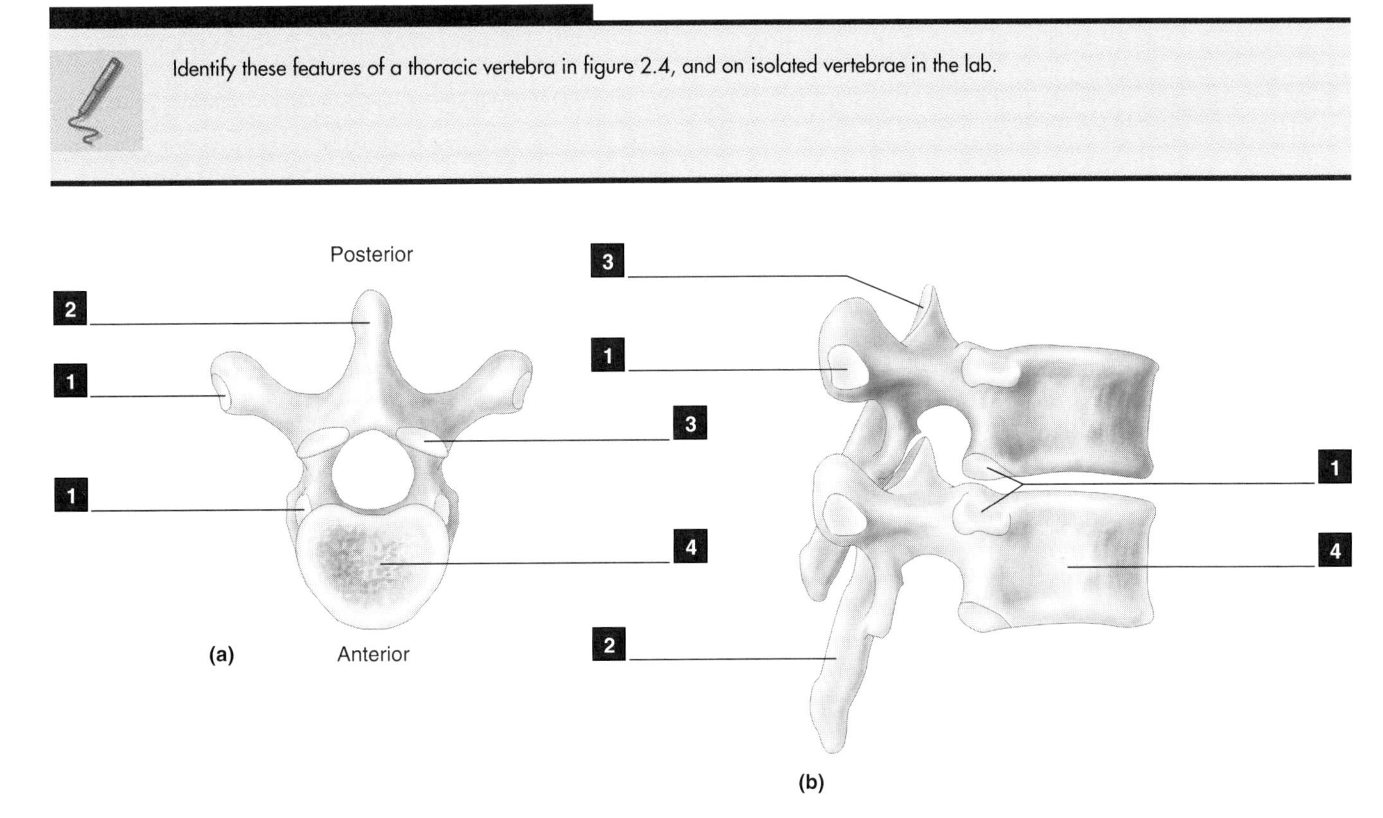

Figure 2.4 **Typical Thoracic Vertebra**
(a) superior view; (b) lateral view.

Lumbar Vertebrae

There are five lumbar vertebrae (L1–L5). They can be distinguished by the absence of both a foramen transversarium in the transverse process, and of costal facets on the body and/or transverse processes. The **Spinous Process** 2 is stocky and horizontal. The **Superior Articular Facets** 3 tend to face medially, and the **Vertebral Body** 4 is large.

Identify these features of a lumbar vertebra in figure 2.5, and on isolated vertebrae in the lab.

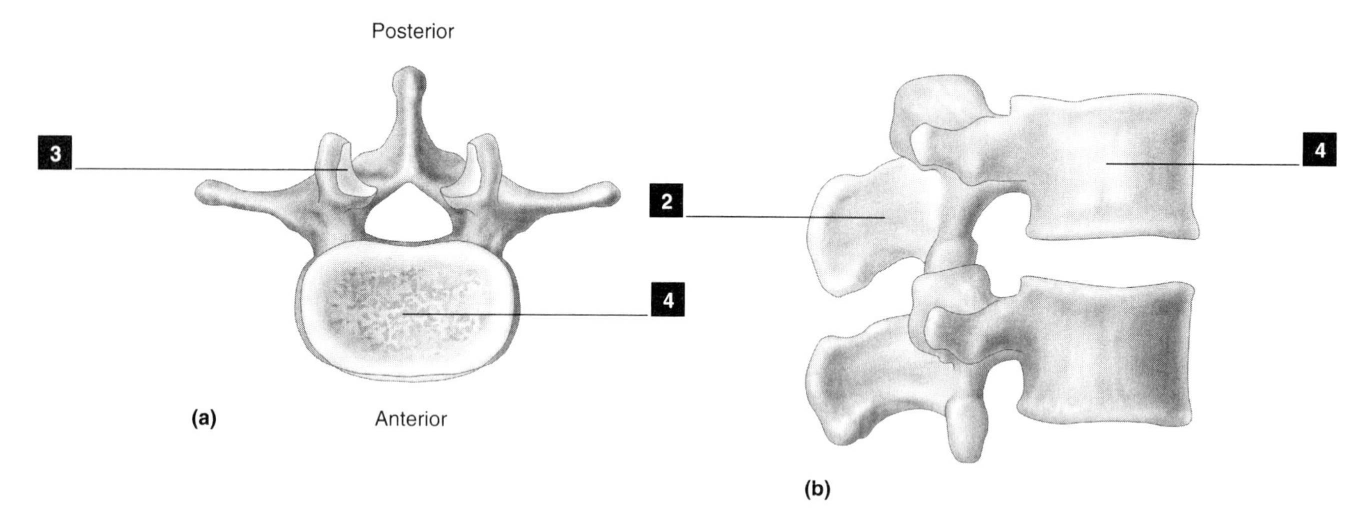

Figure 2.5 **Typical Lumbar Vertebra**
(a) superior view; (b) lateral view.

Sacral Vertebrae

There are five sacral vertebrae (S1–S5) that are fused into a single element by about age 23 or 24, although the first two may complete their union as late as 32 years.

Although the vertebrae are fused, each one is identifiable. The junction of adjacent **Bodies** 1 is marked by a raised transverse line. The intervertebral foramina are represented by four anterior and four posterior **Sacral Foramina** 2 by which the spinal nerves exit from the **Sacral Canal** 3, which is a continuation of the vertebral canal. The **Lateral Mass** 4 represents a fusion of the transverse processes. Remnants of the stubby spinous processes may also be distinguished.

The sacrum articulates laterally with the iliac portion of the os coxae (hip bone) at the *Sacroiliac Joint* by a large **Auricular Surface** 5.

Coccygeal Vertebrae

There are four coccygeal vertebrae (Co1–Co4). The **Coccyx** 6 of an adult usually consists of two pieces: the first vertebrae (Co1), and the fused inferior segments (Co2–Co4). In old individuals, these two may fuse into a single element, which may eventually unite with the sacrum.

Coccygeal vertebrae lack any component of the neural arch. Rather, they are simply rudimentary bodies, although Co1 has stubby transverse processes and vestigial superior articular processes, known as the coccygeal cornua.

Identify the aforementioned features of the sacrum and coccyx in figure 2.6

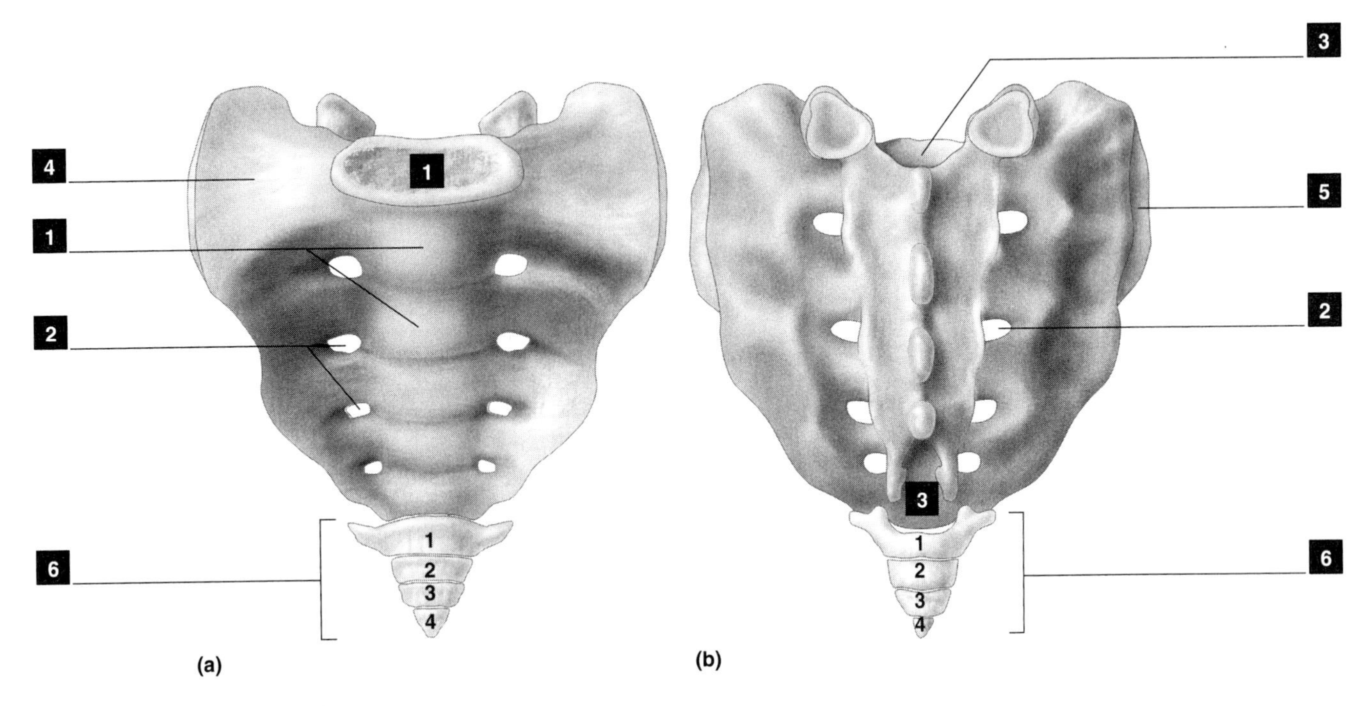

Figure 2.6 **Sacrum and Coccyx**
(a) anterior view; (b) posterior view.

Curvatures of the Vertebral Column

At birth, the entire vertebral column has a gentle, ventrally concave curve. With the achievement of more upright postures, it soon attains a sigmoid profile with two anteriorly concave curves and two posteriorly concave curves. An anterior concavity is known as a Kyphotic curvature; a posterior concavity is a Lordodic curvature.

As a child begins to raise its head, the intervertebral discs in the neck thicken anteriorly; thus, this region develops a posterior concavity. This is the **Cervical Lordosis 1.**

The thoracic portion of the column retains its original curve because the bodies are higher posteriorly. This is the **Thoracic Kyphosis 2.**

As a child begins to walk, the lumbar vertebral bodies and intervertebral discs thicken anteriorly. This induces a **Lumbar Lordosis 3.**

The original curve of the sacral and coccygeal vertebrae persists into adulthood because the sacrum is fixed to the os coxae. Thus, the adult skeleton has a **Sacral Kyphosis 4,** and a **Coccygeal Kyphosis 5.**

Stabilizing Ligaments of the Vertebral Column

Extension of the vertebral column eliminates the thoracic kyphosis. Flexion eliminates the cervical and lumbar lordoses. It is desirable to have some degree of movement of the spinal column, but excessive movement must be prevented. Postural stability is maintained in part by five Intervertebral Ligaments.

The **Supraspinous Ligament** 6 connects the tips of the spinous processes. In the neck it forms the powerful *Ligamentum Nuchae*, which attaches to the external occipital protuberance of the skull. The **Interspinous Ligaments** 7 run between the inferior and superior edges of adjacent spinous processes. The **Ligamentum Flavum** 8 runs between the inferior margin of one lamina to the superior edge of the next lower lamina. The **Posterior Longitudinal Ligament** 9 runs from the skull to the sacrum; it attaches to the posterior surface of each vertebral body and each intervertebral disk. The **Anterior Longitudinal Ligament** 10 attaches to the anterior and lateral aspects of the vertebral bodies and intervertebral disks.

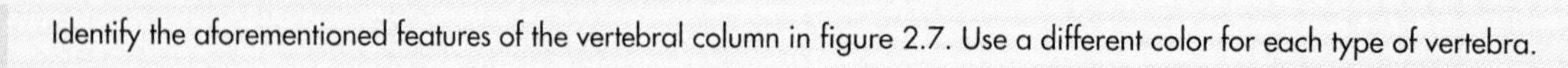

Identify the aforementioned features of the vertebral column in figure 2.7. Use a different color for each type of vertebra.

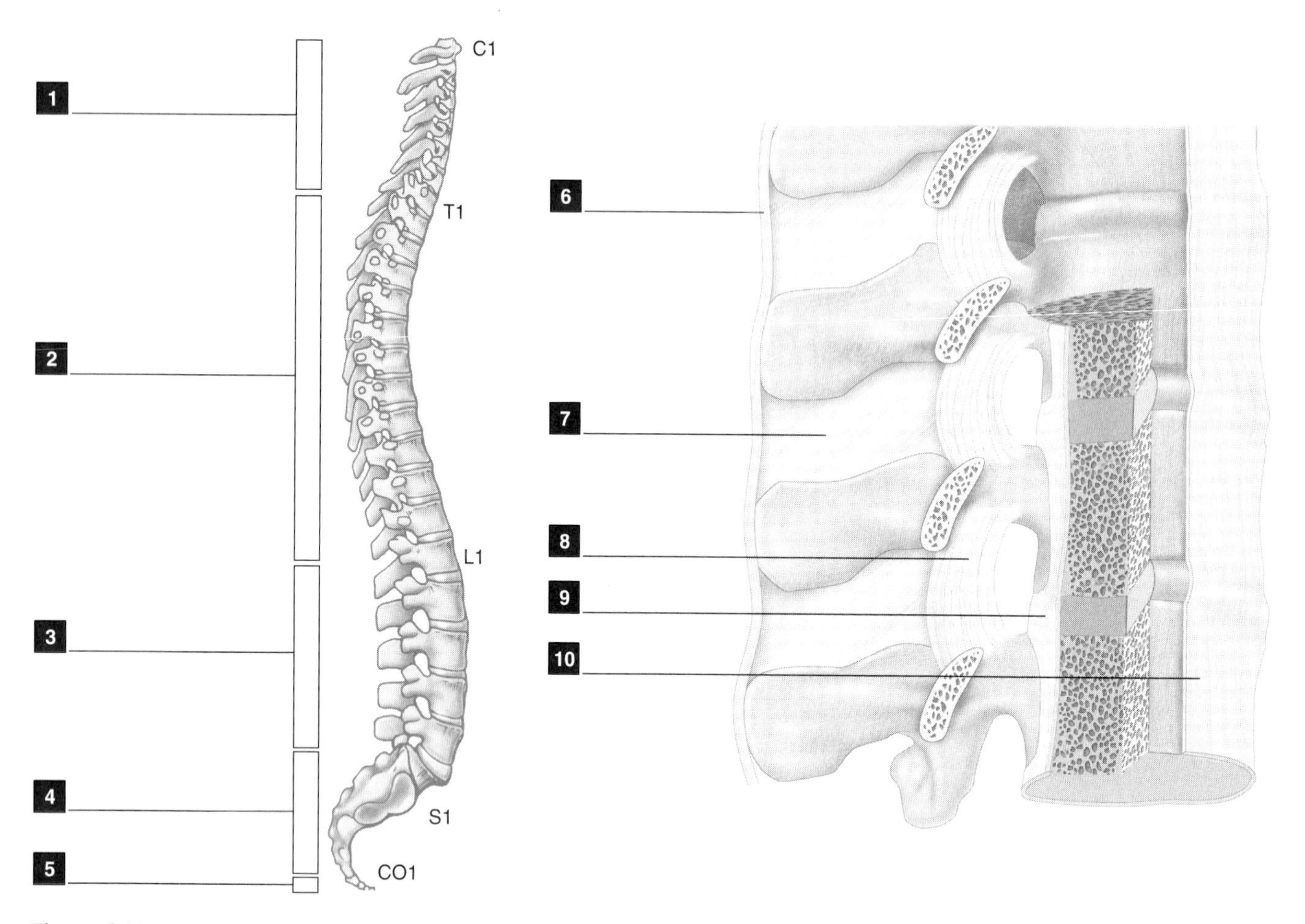

Figure 2.7 **Vertebral Curvature and Stabilizing Elements**

2.3 SPINAL CORD AND SPINAL NERVES

The nervous system is an integrated network that can be thought of as comprising central and peripheral components. The spinal cord, together with the brain, comprises the *Central Nervous System*. The spinal nerves, together with the cranial nerves, make up the *Peripheral Nervous System*.

Spinal Cord and Meninges

The **Spinal Cord** 1 runs through the vertebral canal. It is continuous with the *Medulla Oblongata* of the brain stem. It conveys (1) sensory information from most of the body to the brain, and (2) motor impulses to the muscles of most of the body. In adults, it runs to the level of the intervertebral disc between L1–L2, where it tapers to an end known as the **Conus Medullaris** 2.

The spinal cord is surrounded by three (3) protective membranes, called *Meninges*. The outermost layer, the **Dura Mater** 3, is tough and fibrous. The middle layer, the **Arachnoid Mater** 4, is delicate and cobweblike. The innermost **Pia Mater** 5 is intimately attached to the cord.

Between the dura mater and the vertebral periosteum is the **Epidural Space** 6. It contains fat and a plexus of veins. Between the layers of arachnoid and pia mater is the **Subarachnoid Space** 7. It contains *Cerebrospinal Fluid* (CSF).

In the embryo, the spinal cord and meninges are the same length as the vertebral column. The cord and, to a lesser degree, the sheaths of dura mater and arachnoid mater lag behind growth in vertebral column length. Thus, the spinal cord ends at the level of L1–L2 in an adult, but the dura/arachnoid sheath extends to the level of S2. The subarachnoid space between L2 and S2 is the *Lumbar Cistern*.

The pia mater extends from the end of the conus medullaris as a thin band called the **Filum Terminale** 8. It is attached to the coccyx.

The lower lumbar, the sacral, and the coccygeal spinal nerve fibers run through the vertebral canal caudal to the conus medullaris to form nerves that will eventually exit through lumbar intervertebral foramina or sacral foramina. The bundle of descending fibers is known as the **Cauda Equina** 9.

The spinal cord is made up of myelinated and unmyelinated neurons. Neuronal cell bodies, unmyelinated axons occupy a region of H-shaped **Gray Matter** 10 in the center of the cord. This is surrounded by myelinated axons that comprise the surrounding **White Matter** 11 of the cord.

Identify these features of the spinal cord and its meninges in figure 2.8.

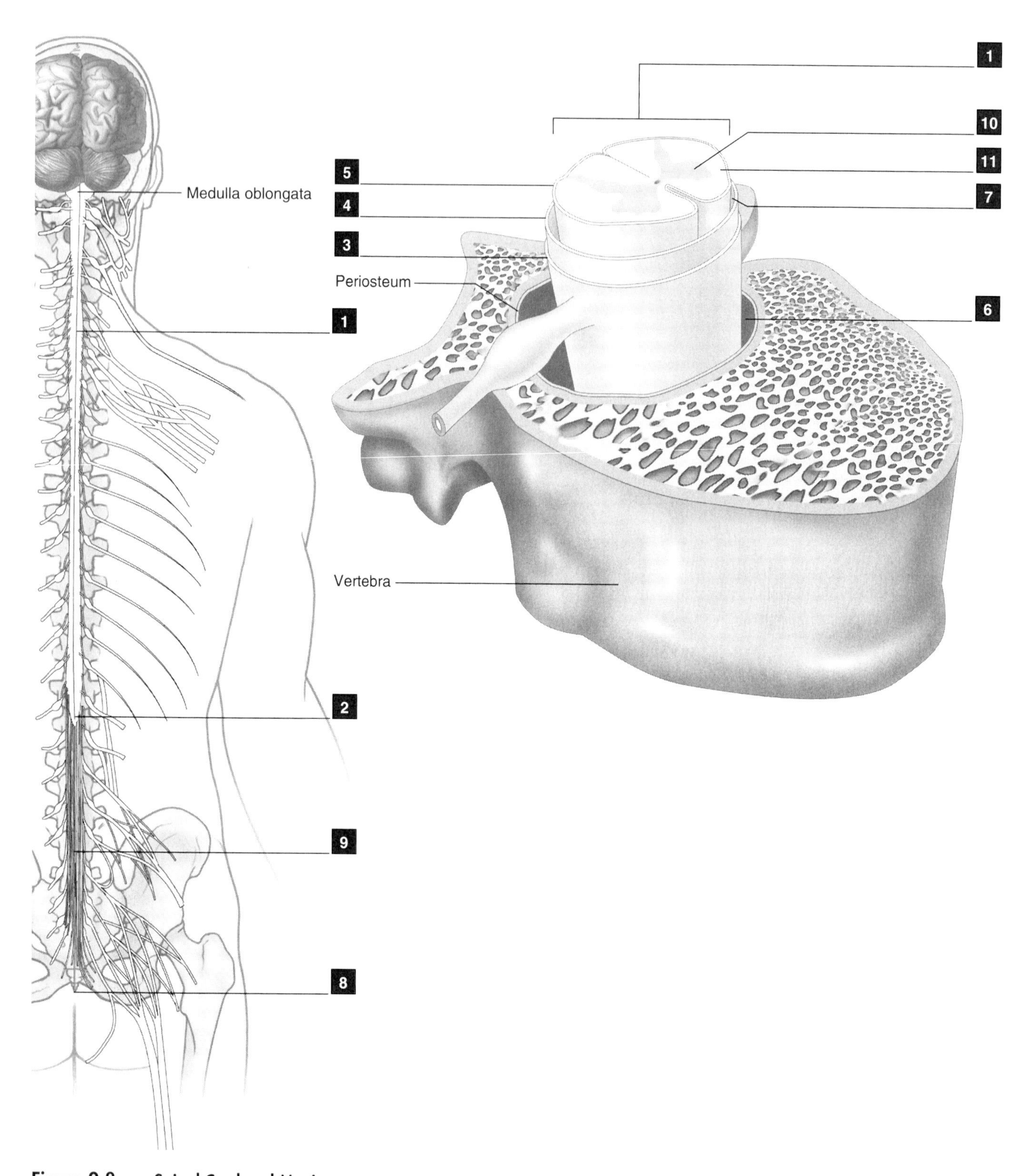

Figure 2.8 **Spinal Cord and Meninges**

Spinal Nerves

There are 31 pairs of spinal nerves that emerge from the spinal cord. They are named according to the region of the vertebral column with which they are associated.

Even though there are only seven cervical vertebrae, there are eight **Cervical Nerves 1**. This is because the first nerve exits between the skull and atlas, and the last nerve of the neck exits between the seventh cervical vertebra and the first rib-bearing (thoracic) vertebra. There are **12 Thoracic Nerves 2**, **5 Lumbar Nerves 3**, **5 Sacral Nerves 4**, and **1 Coccygeal Nerve 5**.

Motor neuron cell bodies are located in the ventral horn of gray matter of the spinal cord. Motor neurons thus send out axons through ventral rootlets. These unite to form a single **Ventral Root 6**. Each ventral root then divides into a **Dorsal Ramus 7**, which provides innervation to the intrinsic muscles of the back (*Epaxial Muscles*), and a **Ventral Ramus 8**, which provides innervation to all other skeletal muscles (*Hypaxial Muscles*).

Sensory neurons synapse in the dorsal horn of gray matter of the spinal cord. Impulses from epaxial structures travel along axons in the dorsal ramus; those from hypaxial structures travel along the ventral ramus. The sensory and motor fibers that run in the *dorsal and ventral rami* join together to form a short true **Spinal Nerve 9**. The sensory axons from the spinal nerve then diverge toward the dorsal horn of the spinal cord through the **Dorsal Root 10**. A portion of the dorsal root, the **Dorsal Root Ganglion 11**, is swollen by the cell bodies of the sensory neurons. The sensory neurons continue toward the spinal cord, where they divide into smaller bundles called dorsal rootlets.

Thus, the dorsal ramus, the ventral ramus, and the spinal nerve contain both sensory and motor neurons. The dorsal root contains only sensory fibers, and the ventral root contains only motor fibers.

In several instances the ventral rami of spinal nerves are braided to form a network known as a *Plexus*. Here, different spinal nerves are mixed so that a particular peripheral nerve will contain fibers from several. There are four plexuses:

Cervical Plexus 12 = nerves C1–C4
Brachial Plexus 13 = nerves C5–T1
Lumbar Plexus 14 = nerves T12–L4
Sacral Plexus 15 = nerves L5–S3

The lumbar and sacral plexuses are joined by small branches from L4 to L5 known as the *Lumbosacral Trunk*. This unit is known as the *Lumbosacral Plexus*.

Identify the spinal nerves and plexuses in figure 2.9, and the components of a spinal nerve in figure 2.10.

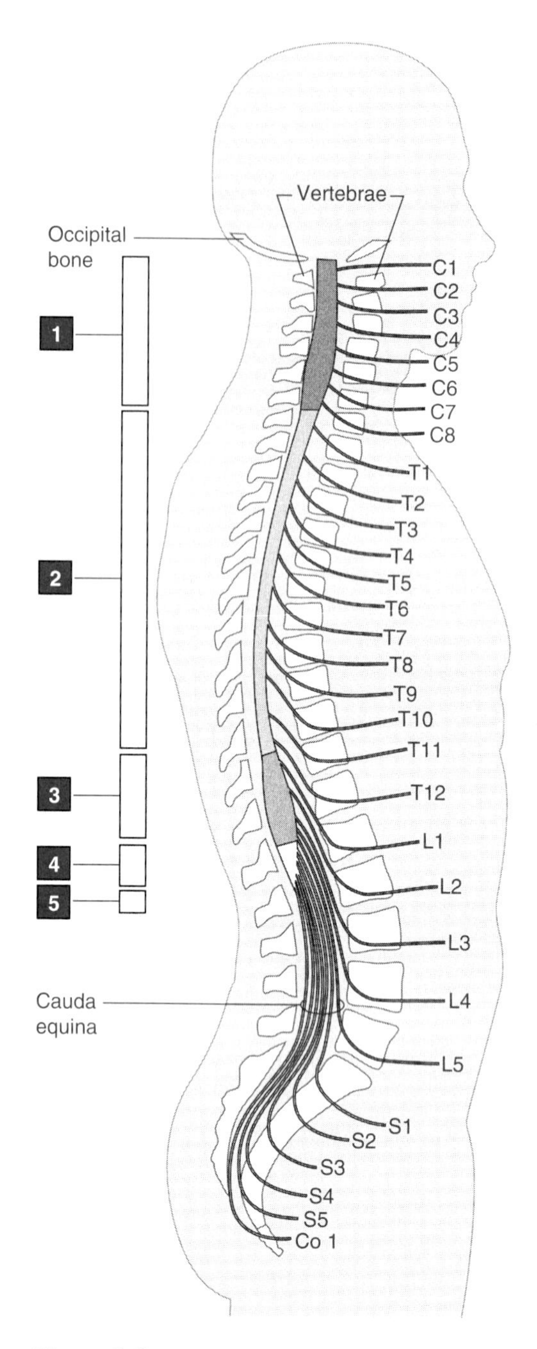

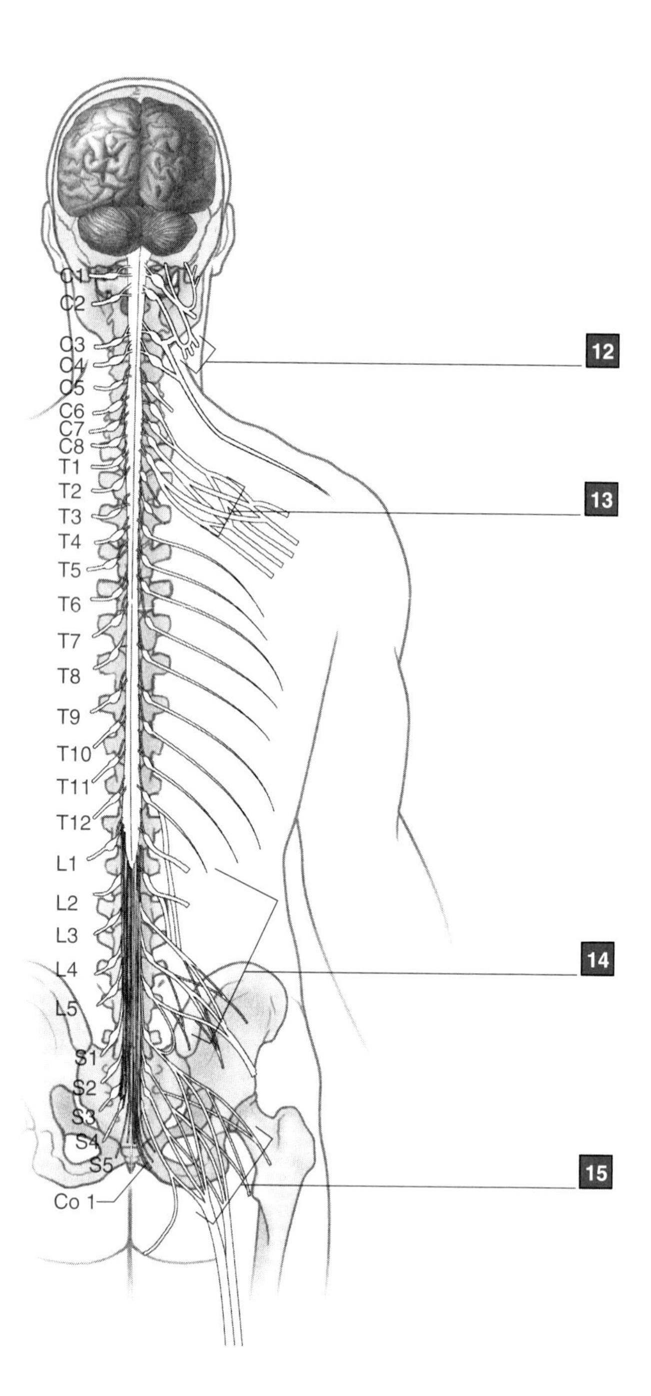

Figure 2.9 Spinal Nerves and Plexuses

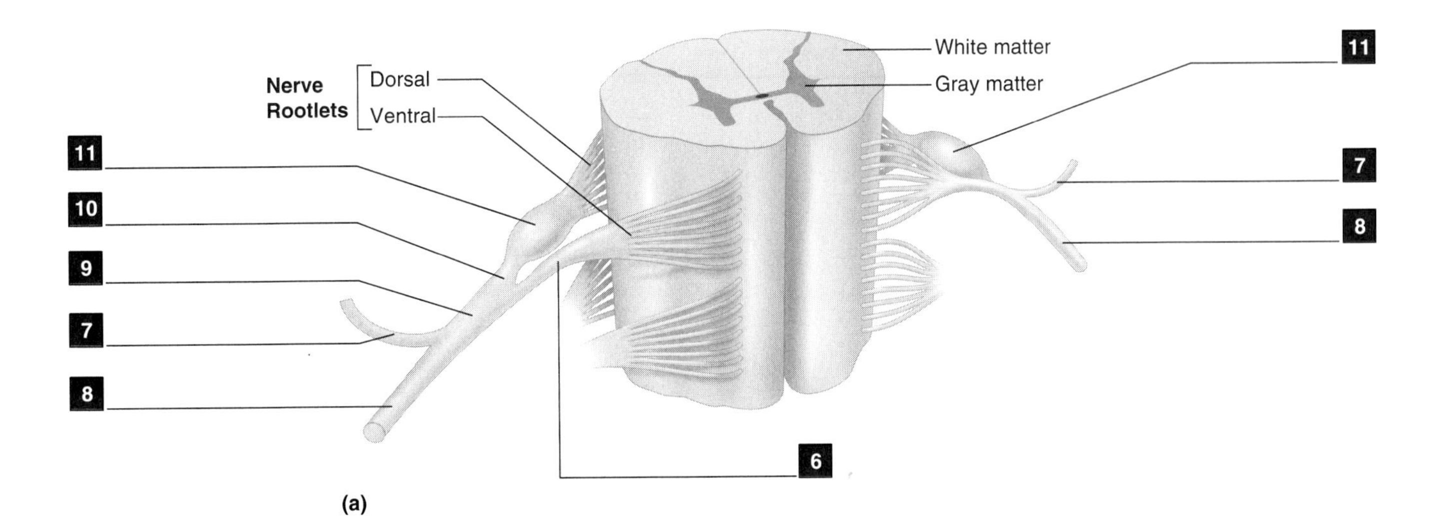

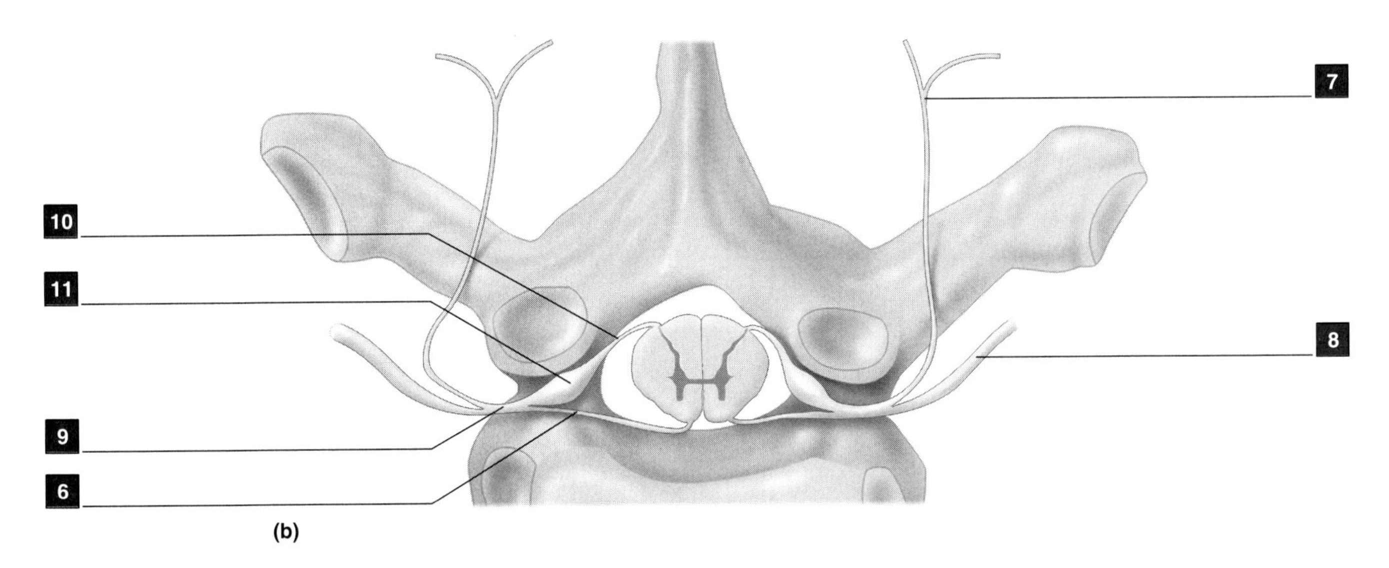

Figure 2.10 **Motor and Sensory Neurons**
(a) anterolateral view; (b) superior view.

2.4 MUSCLES OF THE BACK

Removal of the integument and superficial fascia from the dorsum of the trunk exposes the muscles of the back. These are attached to the vertebral column and/or the ribs. Some move the head, some move the scapula, and some move the vertebral column itself. One moves the humerus.

The muscles of the back may be divided into two groups according to their developmental origin and innervation.

Superficial (Hypaxial) Muscles innervated by *Ventral Rami* of spinal nerves.

Intrinsic (Epaxial) Muscles innervated by the *Dorsal Rami* of spinal nerves.

Superficial Muscles of the Back

Trapezius 1 attaches to the occipital bone of the skull, the ligmentum nuchae, and the spines of the thoracic vertebrae; it inserts onto the spine and acromion process of the scapula and onto part of the clavicle. It fixes, elevates, and rotates the scapula; it also extends the head.

Innervated by the *Accessory (XI) Cranial Nerve*.

Ligamentum nuchae is an expansion of the supraspinous ligament. Its anterior edge attaches to the spines of the cervical vertebra, its superior edge attaches to the superior nuchal line of the occipital bone, and its posterior margin extends from the external occipital protuberance to the spine of C7 (the so-called *vertebra prominens*).

Latissimus dorsi 2 attaches to the spines of vertebrae from T7 to the sacrum, and to the iliac crest and two ribs; it inserts into the intertubercular sulcus of the humerus. It extends, adducts, and medially rotates the arm.

Innervated by the *thoracodorsal nerve* (C6–C8).

Levator Scapulae 3 arises from the transverse processes of C1–C4, and inserts onto the medial border of the scapula above the root of the spine. It fixes and elevates the scapula; it also flexes the neck laterally.

Innervated by the *dorsal scapular nerve* (C5).

Rhomboideus comprises two muscles. **R. minor** 4 attaches to the spine of T1. **R. major** 5 attaches to the spines of T2–T5. Both insert onto the medial border of the scapula, and fix, retract, and rotate it.

Innervated by the *dorsal scapular nerve* (C5).

Quadratus lumborum 6 attaches to the iliolumbar ligament and iliac crest; it inserts onto the transverse processes of L1–L4 and the T12 and L1 rib. It laterally flexes the spine. It forms part of the posterior abdominal wall.

Innervated by *spinal nerves*.

Identify the superficial muscles of the back in figure 2.11. Examine an articulated, painted skeleton to satisfy yourself of their bony attachments.

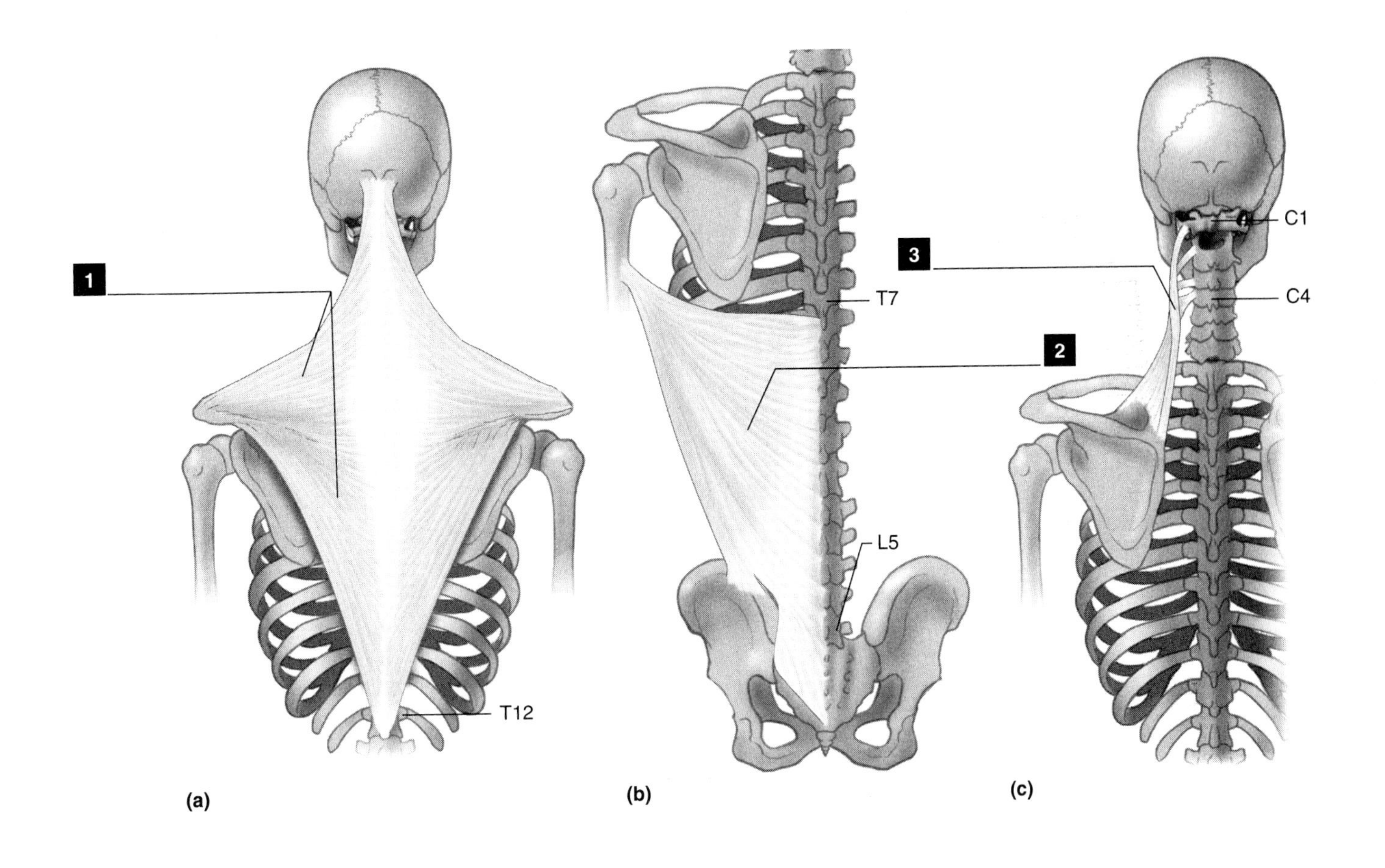

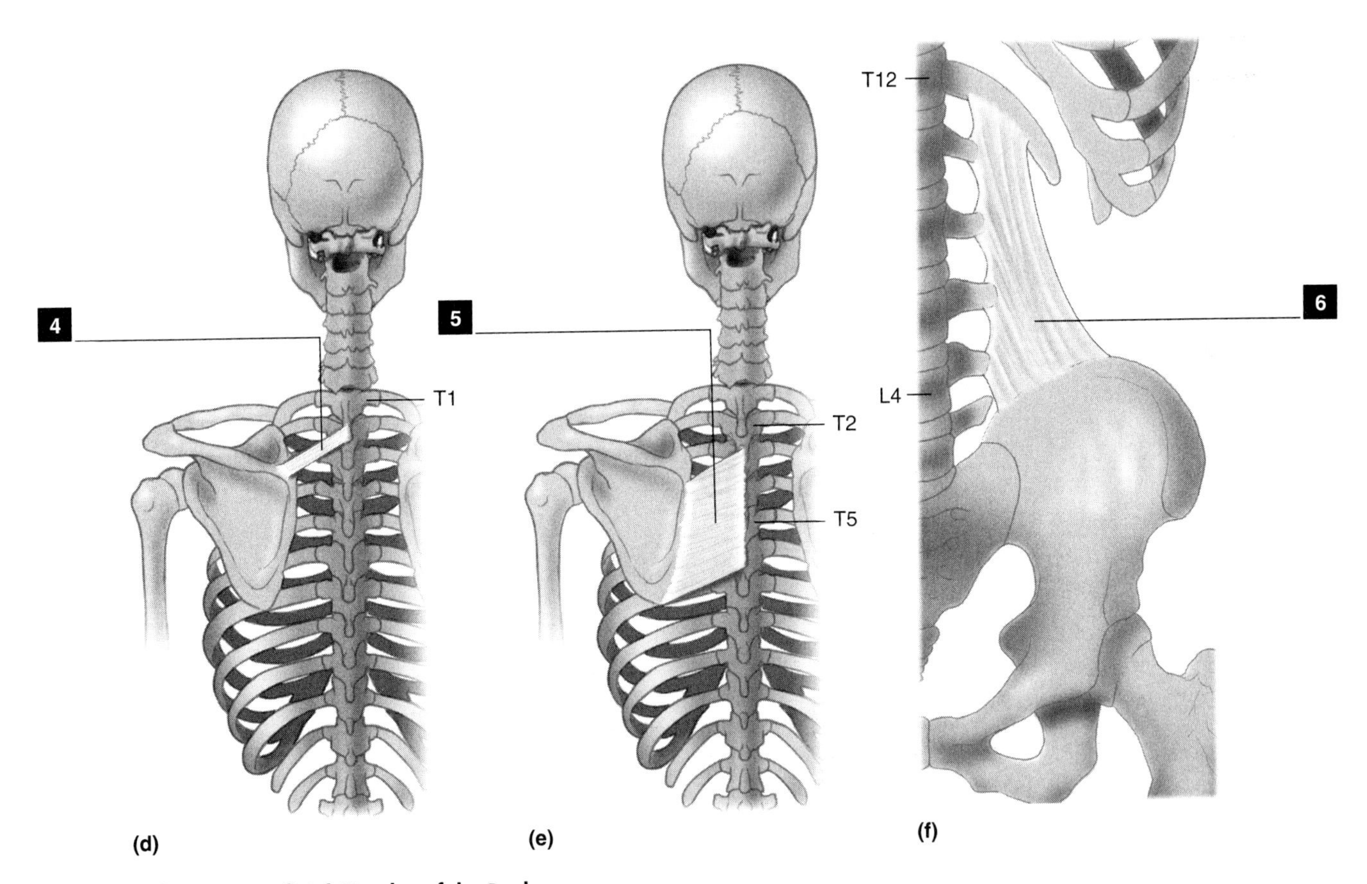

Figure 2.11 **Superficial Muscles of the Back**
(a–e) posterior view; (f) anterior view.

Intrinsic Muscles of the Back

The intrinsic muscles of the back may be divided into two groups according to their anatomical relationships. The groups are superficial intrinsic muscles and deep intrinsic muscles.

Superficial Intrinsic Muscles

Splenius comprises two groups of muscle fibers that attach to the spines of T1–T6 and the lower part of the *ligamentum nuchae*. One group of fibers, **S. cervicis** 1 runs to the transverse processes of C1–C4. The other group, **S. capitis** 2, runs to the lateral aspect of the mastoid process and lateral part of the superior nuchal line of the skull. It laterally flexes and rotates the head and neck, or extends the head and neck. Innervated by *middle and lower cervical nerves*.

Erector spinae comprises three vertical columns of overlapping muscle fascicles that mount up the vertebral column and ribs from a common tendinous origin to the sacrum, the posterior part of the iliac crest of the pelvis, and the spines of the lowermost lumbar vertebrae. The individual columns are descriptively recognized by the regions that they traverse. Innervated by *spinal nerves* that emerate from levels crossed by the muscle fibers.

Iliocostalis 3 forms the *Lateral Column*. It runs from the common origin to the lower six ribs, from these to the upper six ribs, and from these to the transverse processes of the lower cervical vertebrae. It extends and laterally flexes the spine.

Longissimus 4 forms the *Intermediate Column*. It runs from the common origin to the transverse processes of the thoracic vertebrae, from these to the transverse processes of the cervical vertebrae, and from these to the mastoid process of the skull. Its lower portions extend and laterally flex the spine; its upper part extends or rotates the head.

Spinalis 5 forms the *Medial Column*. It runs from the spines of lumbar vertebrae to those of upper thoracic vertebrae between L2 and T2. Occasionally, its fibers may run to cervical vertebrae. It extends the vertebral column.

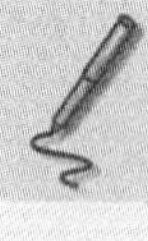

Identify these intrinsic muscles of the back in figure 2.12. Examine an articulated, painted skeleton to satisfy yourself of their bony attachments.

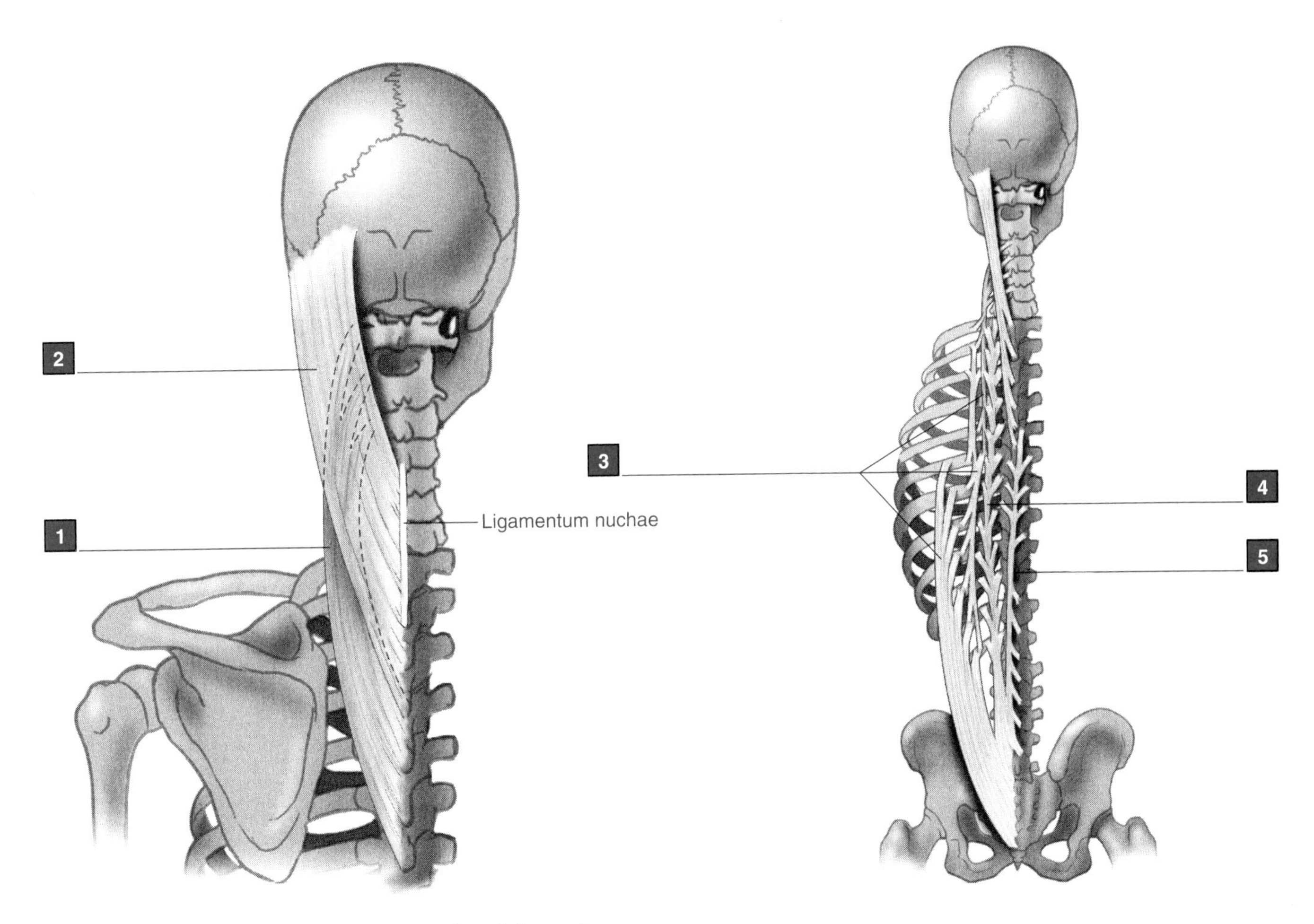

Figure 2.12 **Superficial Intrinsic Muscles of the Back**

Deep Intrinsic Muscles

There are a number of deep intrinsic back muscles, but most are small segmental muscles that bridge successive vertebrae, or small suboccipital muscles that connect the atlas and axis to the bottom of the skull. There are two that you should be familiar with: *Semispinalis* and *Multifidus*.

Semispinalis is the most superficial of the deep intrinsic muscles. It has two components. Innervated by *spinal nerves* that enervate from levels crossed by the muscle fibers.

Semispinalis 1**,** forms the lower portion. Its fibers run upwards and medially between T12 and C2 from the transverse processes to the spines of superior vertebrae. Each fiber bundle spans from four to six vertebrae. Because it spans the thoracic and cervical regions, it is commonly divided into two parts—*S. thoracis* and *S. cervicis*. This component rotates and extends the vertebral column of the trunk.

Semispinalis capitis 2**,** forms the upper portion. Its fibers run from the transverse processes of T1–T6 to insert onto the medial part of the occipital bone between the superior and inferior nuchal lines. This component is a powerful rotator and extensor of the head and neck.

Multifidus 3 is the most powerful of the deep intrinsic muscles. It runs from the vertebral arches to the spines along the entire vertebral column from S4 to C2. Each fiber bundle spans from one to three vertebrae. It laterally flexes, rotates, and extends the vertebral column. Innervated by *spinal nerves* that enervate from levels crossed by the muscle fibers.

Small Segmental Muscles. There are four groups of rather trivial deep back muscles that bridge the gaps between successive vertebrae. These muscles are used principally in postural "steadying" because they have little leverage. The four are: *Rotatores, Intertransversarii, Interspinales, and Levatores costarum.*

Suboccipital Muscles. There are four muscles that either connect the axis to the atlas, or connect one of these vertebrae to the occipital bone of the skull. These muscles are mainly postural. The intervertebral muscle is *Obliquus capitis inferior*. The three that attach to the skull are *Obliquus capitis superior*, *Rectus capitis posterior major*, and *Rectus capitis posterior minor*.

Identify the important deep intrinsic muscles of the back in figure 2.13. Study an articulated skeleton to trace their bony attachments.

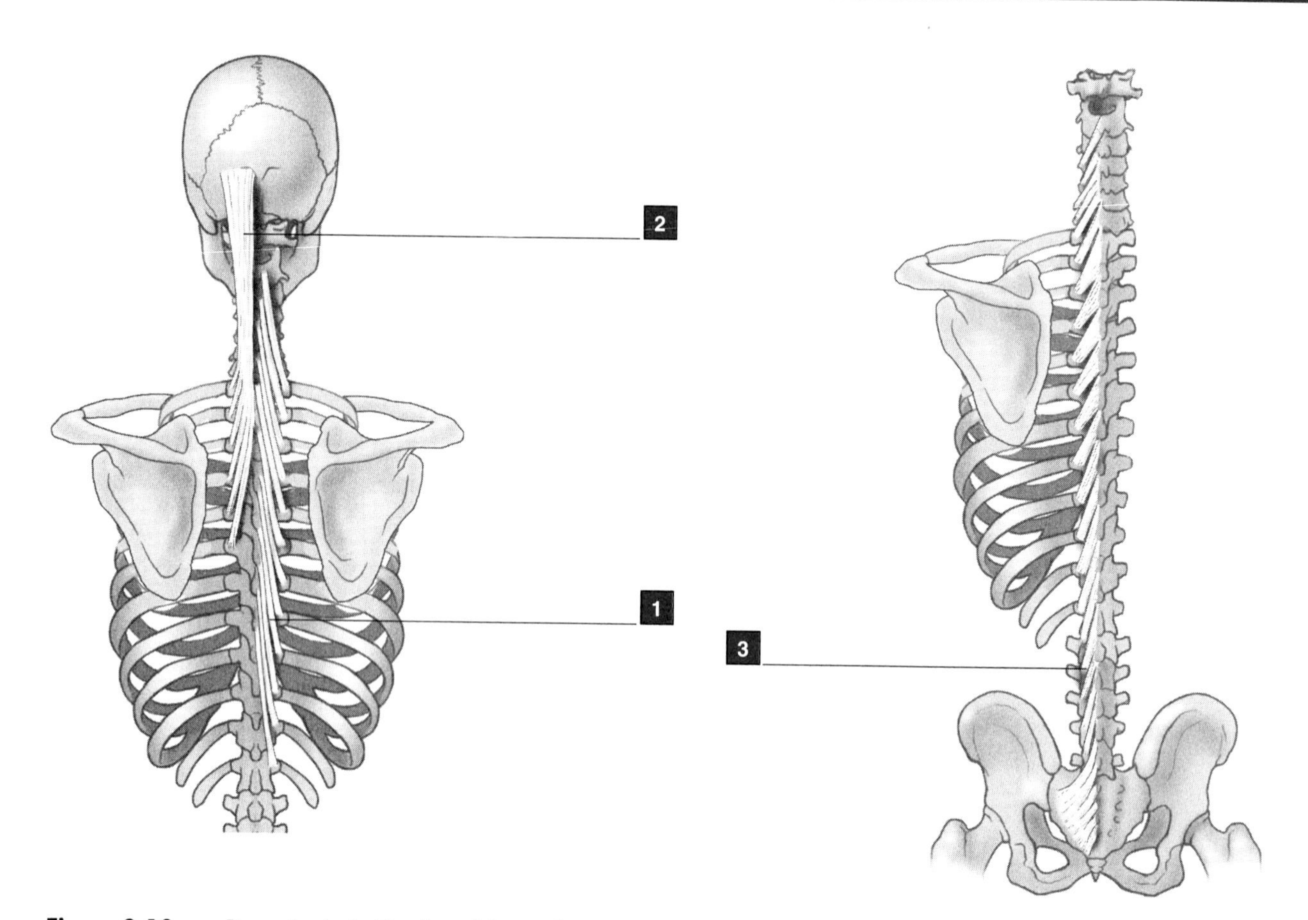

Figure 2.13 Deep Intrinsic Muscles of the Back

LABORATORY 3

The Upper Limb

3.1 BONES OF THE UPPER LIMB

The bones of the upper limb comprise two groups: (1) those that form the free part of the limb, and (2) those that connect it to the body wall. The latter form the pectoral girdle.

Pectoral Girdle

The pectoral girdle is made up by the clavicle and scapula. The clavicle is the only bony link between the upper limb and the axial skeleton.

Clavicle 1. This is an "S-shaped" bone; its medial third is convex anteriorly, and its lateral third is concave anteriorly. It can be palpated along its entire length. Its expanded medial end articulates with the **Manubrium** 2 at the mobile **Sternoclavicular Joint** 3. (We will discuss the structure of joints next in this chapter.) The flattened lateral end of the clavicle articulates with the acromion of the scapula at the synovial **Acromioclavicular Joint** 4.

Scapula 5. This triangular bone covers the second to seventh ribs on the dorsum of the thorax. Its medial border parallels the vertebral column, and the long axillary border is thickened. The **Glenoid Cavity** 6 forms the joint surface for the humerus. The **Spine** 7 rises from the dorsal surface and projects laterally as the **Acromion** 8 which bends anteriorly to overhang the glenoid cavity. The fingerlike **Coracoid Process** 9 projects anterolaterally from the superior border.

Ligaments. The *Sternoclavicular Joint* is reinforced by two ligaments: (1) the **Interclavicular Ligament** 10 runs between the medial ends of both clavicles and the top of the manubrium; (2) the **Costoclavicular Ligament** 11 runs from the underside of the clavicle to the top of the **First Rib** 12. The *Acromioclavicular Joint* is reinforced by two sets of ligaments: (1) the **Coracoclavicular Ligaments** 13 extend from the coracoid process to the lateral third of the clavicle, (2) the **Coracoacromial Ligament** 14 runs from the coracoid process to the acromioclavicular joint.

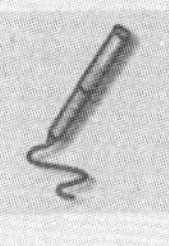

Identify the aformentioned structures of the pectoral girdle in figure 3.1.

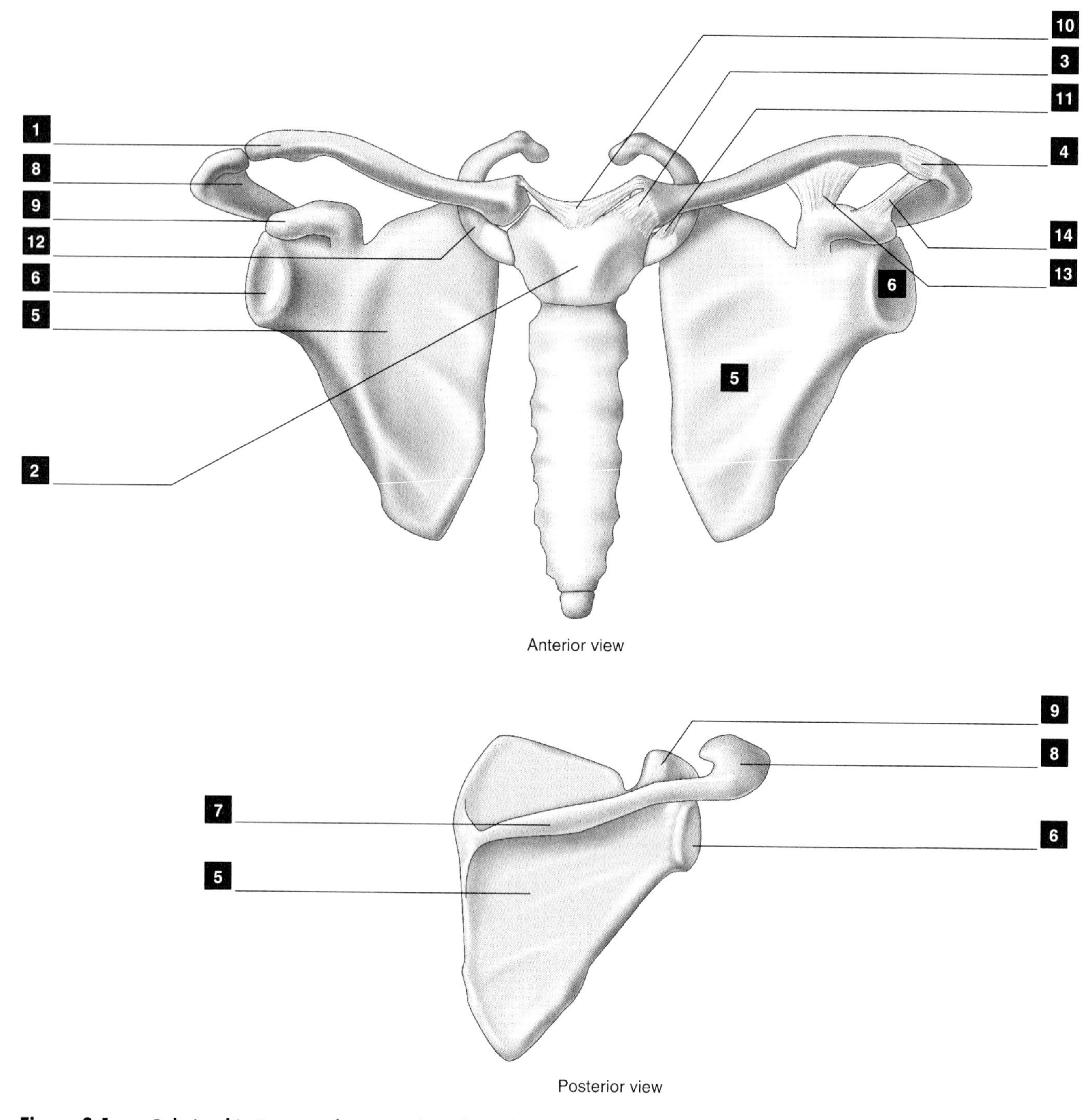

Figure 3.1 Relationship Between the Pectoral Girdle and Axial Skeleton

Arm

The arm, or brachium, has a single bone: the humerus.

Humerus. At the proximal end is the hemispherical **Head** 1 that articulates with the *glenoid fossa* of the scapula. Next to the head are a large lateral **Greater Tubercle** 2 and a smaller, anterior **Lesser Tubercle** 3. At about midshaft below the greater tubercle is the prominent, roughened **Deltoid Tuberosity** 4. The distal end of the bone is expanded medially to form the prominent **Medial Epicondyle** 5, which can be easily palpated. The **Lateral Epicondyle** 6 is smaller. The distal end has two articular surfaces. Medially, the spool-shaped **Trochlea** 7 articulates with the ulna. Laterally, the bulbous **Capitulum** 8 articulates with the radius. On the posterior surface there is a large triangular depression just proximal to the trochlea known as the **Olecranon Fossa** 9, which receives the olecranon of the ulna when the forearm is fully extended.

Forearm

The forearm, or antebrachium, has two bones: the radius (lateral) and ulna (medial).

Radius. At the proximal end, there is a disclike **Head** 10 that articulates with both the ulna (medially) and the capitulum of the humerus (proximally). Distal to the head on the anteromedial side of the shaft is the very prominent **Radial (Bicipital) Tuberosity** 11. The distal end of the radius is enlarged. Laterally the **Styloid Process** 12 projects distally; it is easily palpated. On the medial side, there is a shallow notch for articulation with the head of the ulna. The concave distal end of the radius articulates with two of the wrist (carpal) bones—the *lunate* (medially) and the *scaphoid* (laterally).

Ulna. The proximal end of the ulna is greatly expanded. It articulates with both the radius and humerus. The posterior aspect of this end is the **Olecranon** 13, which is easily palpated. It fits into the olecranon fossa of the humerus when the forearm is fully extended. The anterior aspect of the proximal end has a large **Trochlear Notch** 14 that articulates with the trochlea of the humerus. Distal and lateral to the trochlear notch is the shallow **Radial Notch** 15, which articulates with the head of the radius. The distal end of the ulna is known as the **Head** 16. It articulates medially and anteriorly with the radius, and distally with one of the carpal bones—the triquetrum. Posterior to the head, the short, fingerlike **Styloid Process** 17 projects distally.

Identify these features of the arm and forearm in figures 3.2 and 3.3.

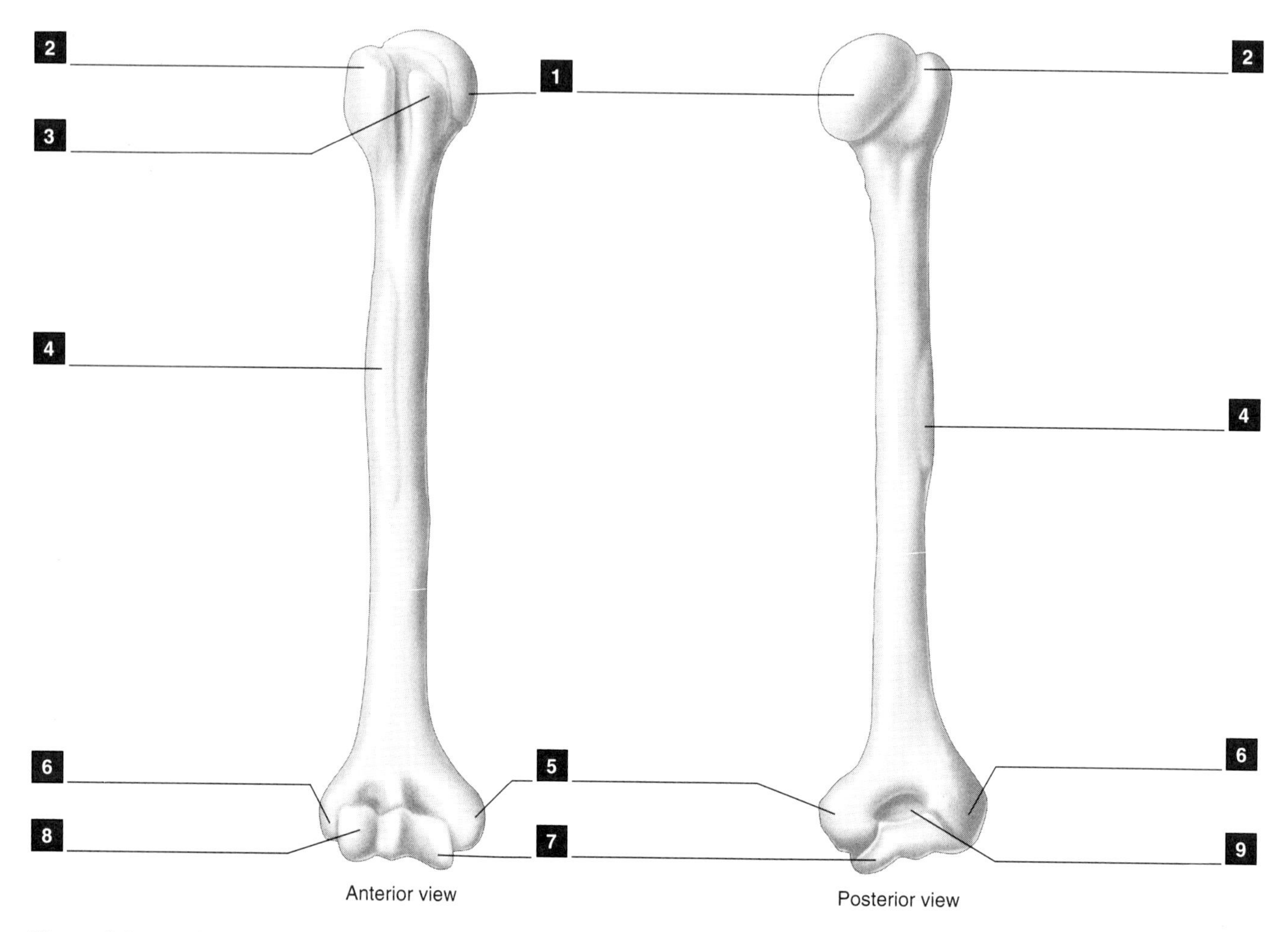

Figure 3.2 **The Arm Bone**
anterior and posterior views of a right humerus

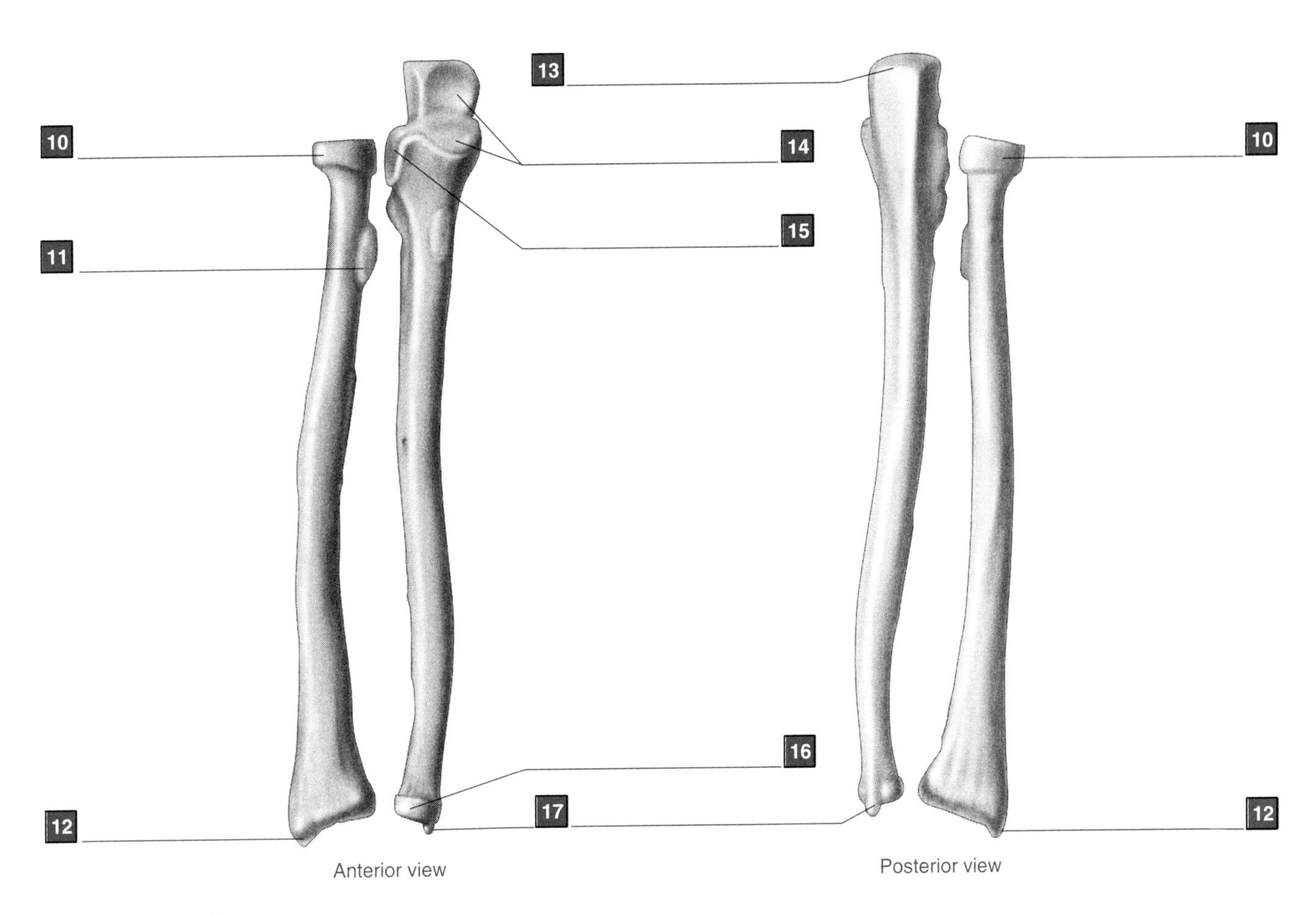

Figure 3.3 **The Forearm Bones**
anterior and posterior views of a right radius ulna

Hand

The hand contains a total of 27 individual bones. These are divided into three groups: *Carpals* (8 bones), *Metacarpals* (5 bones), and *Phalanges* (14 bones).

Carpals **1**. Eight small carpal bones form the *wrist*. They are aligned in two transverse rows of four bones each.

The four bones that make up the *proximal* row are (from lateral to medial): *Scaphoid*, *Lunate*, *Triquetrum*, and *Pisiform*. Three of these—the scaphoid, lunate, and triquetrum—participate in movement of the wrist through their articulation with the radius and ulna. The pisiform is set anterior to the others. It sits on the triquetrum.

The four bones that constitute the *distal* row are (from lateral to medial): *Trapezium*, *Trapezoid*, *Capitate*, and *Hamate*. All four participate in movement of the wrist through their articulation with the metacarpal bones of the palm.

Metacarpals **2**. Five metacarpals form the palm. The first element of each ray (numbered I–V from the thumb to the little finger) is its metacarpal. The proximal end is expanded to form a *base* that articulates with one or more of the carpal bones. The distal end has a rounded *head* that articulates with the proximal phalanx. Each metacarpal is slightly concave on its palmar side and gently convex on its dorsal side.

Phalanges **3**. Fourteen phalanges constitute the *digits*. Each of the four *Fingers* has a **Proximal Phalanx** **4**, a **Middle Phalanx** **5** and a **Distal Phalanx** **6**. The *Thumb* (*Pollex*) has only a **Proximal Phalanx** **7** and a **Distal Phalanx** **8**. The distal end of each distal phalanx is expanded to form an *Apical Tuft* (*Ungual Tuberosity*), which is smooth on its dorsal (nail) side and roughened on its volar (pad) side. It provides the support that enables the broad, fleshy fingertips to grip in opposition with the thumb.

Identify the aforementioned bones of the hand in figure 3.4. Use separate colors for the proximal, middle, and distal phalanges. Examine an articulated skeleton of the hand to satisfy yourself of the articular relationships of its bones.

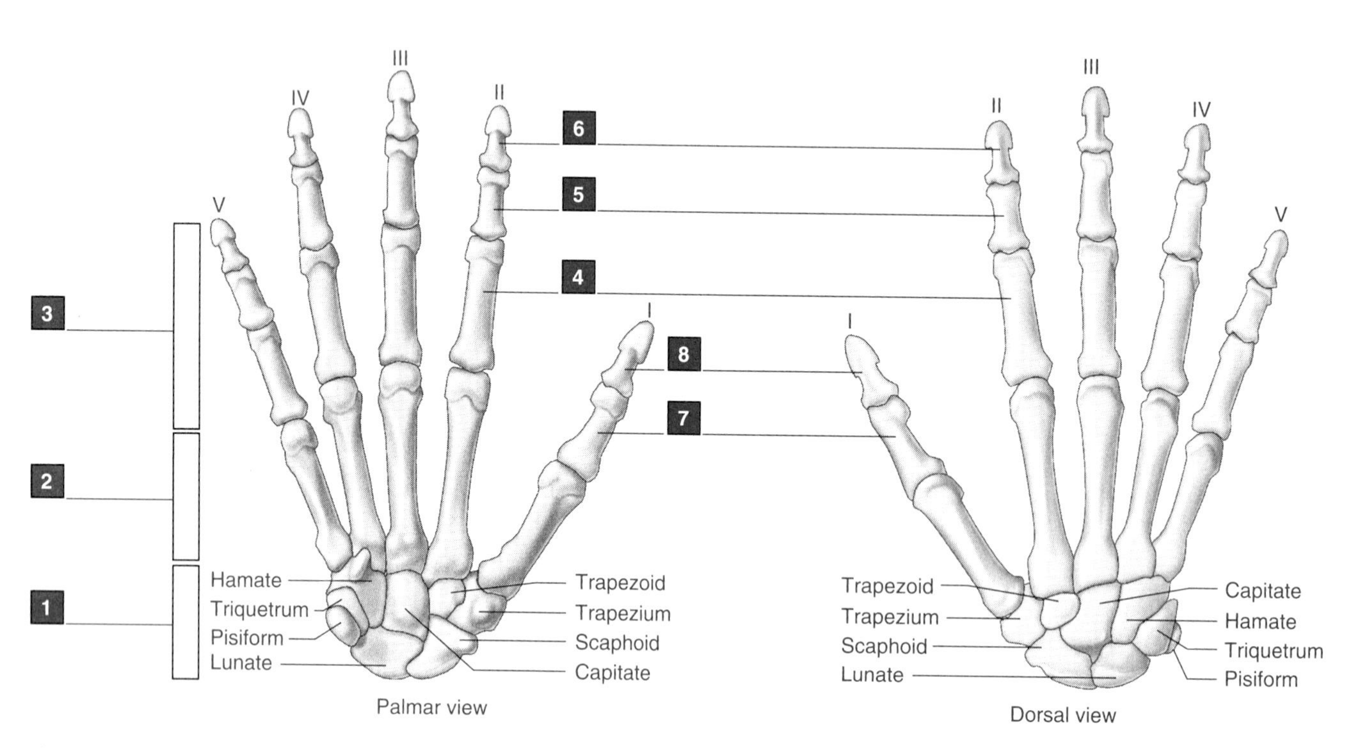

Figure 3.4 **The Hand Bones**
palmar and dorsal views of a right hand

3.2 JOINTS OF THE UPPER LIMB

In this section we will study the anatomy of three upper limb joints: the shoulder, the elbow, and the wrist.

Shoulder Joint

The very shallow glenoid cavity of the scapula and the large humeral head endow this joint with the greatest degree of mobility of any in the body.

The head of the humerus is covered with **Articular Cartilage** 1. The **Glenoid Articular Cartilage** 2 is expanded around the rim to form a **Labrum** 3. The **Joint Capsule** 4 attaches proximally to the rim of the glenoid and distally to the anatomical neck of the humerus. The capsule is lined with a **Synovial Membrane** 5. Its fibrous outer part is formed by the **Glenohumeral Ligaments** 6.

The joint capsule is pierced by the **Tendon of the Long Head of Biceps Brachii** 7, which attaches to the top of the glenoid rim. The tendon runs through the capsule over the head of the humerus; it exits the capsule into the *Intertubercular Groove* on the anterior aspect of the humerus. As the tendon passes through the joint capsule, the synovial membrane becomes wrapped around it as a **Tendon Sheath** 8. This permits the tendon free movement under the *Transverse Humeral Ligament* between the greater and lesser tubercles of the humerus.

The glenohumeral ligaments are relatively thin, and weak. The joint capsule is reinforced by the tendons of four muscles that cross the joint. One muscle—**Supraspinatus** 9—passes the joint superiorly. One—**Subscapularis** 10—passes the joint anteriorly. Two—**Infraspinatus** 11 and **Teres Minor** 12—pass the joint posteriorly. These four form an incomplete cuff around the shoulder joint. They are known as the *Rotator Cuff Muscles*. We will discuss them in greater detail a bit later. They help to prevent shoulder joint dislocation.

Shoulder Dislocation and Shoulder Separation

Dislocation of a joint is the displacement of the bones from their natural articular position. This usually involves tearing of ligaments, tendons, and the joint capsule. Shoulder dislocation relates to the *glenohumeral joint*. It is the one most commonly dislocated, especially when the arm is abducted and laterally rotated.

Shoulder Separation relates to the *acromioclavicular joint* rather than the glenohumeral joint. This involves the lateral end of the clavicle riding up and over the acromion of the scapula. It is caused by forceful trauma to the lateral side of the shoulder, such as the shoulder striking the ground in the course of a fall.

There are two prominent bursae around the shoulder joint. One is superior, the other is anterior to the joint. Superiorly, the **Subacromial Bursa** 13 lies beneath the acromion of the scapula and above the tendon of supraspinatus. Anteriorly, the **Subscapular Bursa** 14 is located between the tendon of the subscapularis muscle and the joint capsule. This bursa is really an extension of the synovial membrane of the joint capsule, which has herniated through two openings around one of the glenohumeral ligaments.

Color and label the structures of the shoulder joint in figure 3.5.

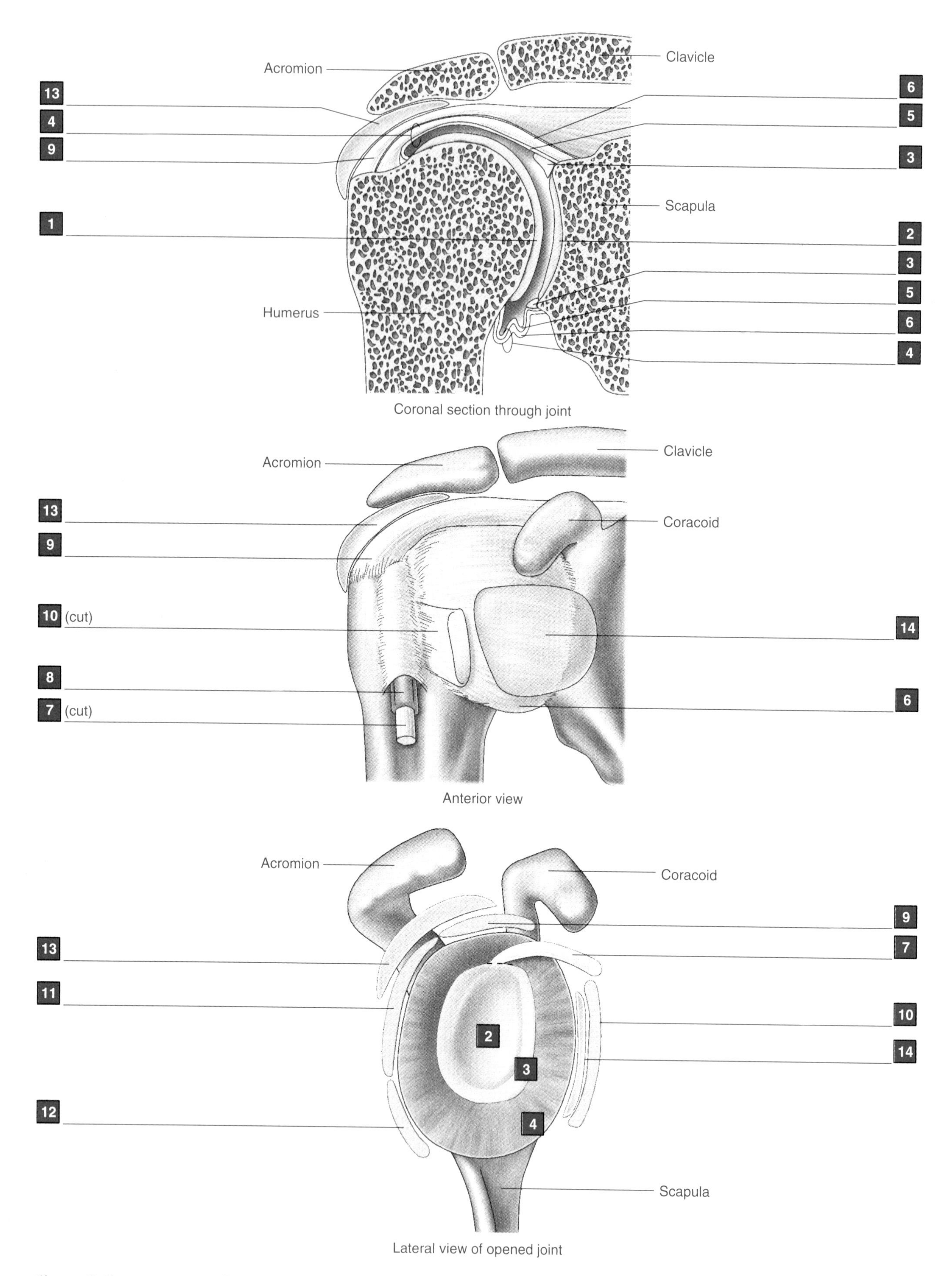

Figure 3.5 **Structure of the Shoulder Joint**
coronal section, anterior view, and lateral view of opened joint of right shoulder.

Elbow Joint

The elbow joint involves three articulations:

1. A hinge joint between the humerus and ulna
2. A ball-and-socket joint between the humerus and radius
3. A pivot joint between the radius and ulna

All three are contained within the same joint capsule and synovial cavity.

Articulations 1 and 2 permit the ulna and radius to flex and extend; articulations 2 and 3 permit medial and lateral rotation (pronation and supination) of the radius.

Articular cartilage covers the **Trochlea of the Humerus** 1 and the opposing **Trochlear Notch of the Ulna** 2. It also covers the **Capitulum of the Humerus** 3 and the opposing **Head of the Radius** 4. It also covers the **Radial Notch of the Ulna** 5 and the opposing head of the radius.

The fibrous **Joint Capsule** 6 of the elbow is lined with a **Synovial Membrane** 7.

Three extracapsular ligaments reinforce the joint. The **Annular Ligament** 8 holds the head of the radius tightly against the ulna. It sweeps around the circumference of the radial head from one margin of the ulnar radial notch to the other. The radius and ulna are held against the humerus by ligaments on the medial and lateral sides of the joint. The **Ulnar Collateral Ligament** 9 arises from the medial eipcondyle of the humerus and fans out to attach to the medial side of the ulna. The **Radial Collateral Ligament** 10 arises from the lateral epicondyle of the humerus and fans out to insert into the annular ligament and the lateral aspect of the ulna.

There are two prominent bursae around the elbow joint. Both are situated posterior to the joint. The **Subtendinous Olecranon Bursa** 11 lies between the joint and the tendon of the *Triceps brachii* muscle, which inserts onto the apex of the olecranon of the ulna. The subcutaneous **Olecranon Bursa** 12 lies immediately below the skin at the back of the ulnar olecranon.

Identify these features of the elbow joint in figure 3.6.

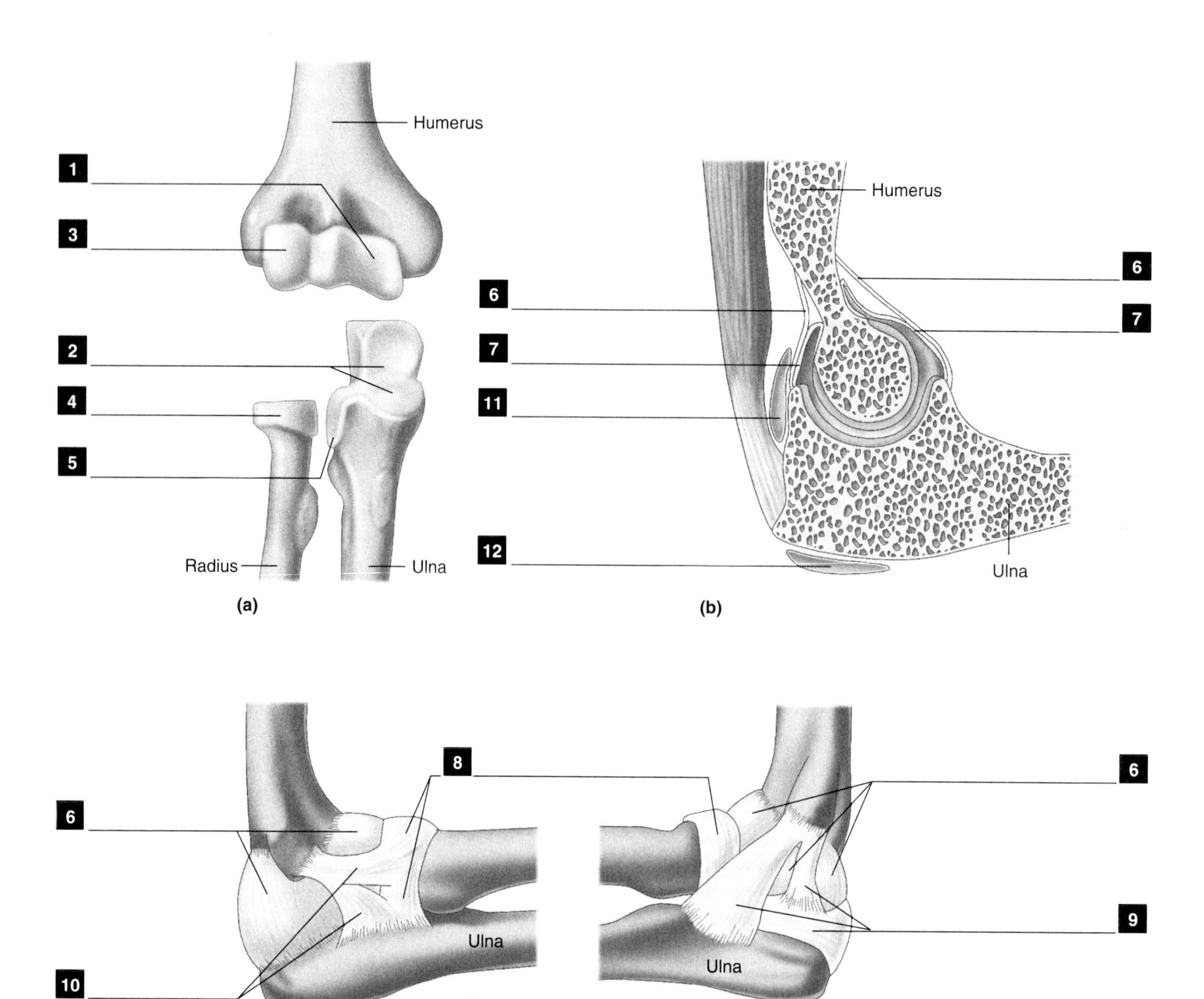

Figure 3.6 **Structure of the Elbow Joint**
(a) anterior view of bones; (b) median longitudinal section through joint; (c) lateral view of joint; (d) medial view of right elbow joint.

Wrist Joint

The wrist joint involves two articulations:

1. Between the distal ends of the radius and ulna
2. Between the forearm bones and the proximal row of carpals

Articulation 1 is known as the ***Distal RadioUlnar Joint.*** It permits *pronation and supination* of the forearm and hand. In these motions, the concave ulnar notch of the radius sweeps around the head of the ulna.

Stability of this joint is provided by a **Triangular Disc** 1 of fibrocartilage that covers the distal end of the ulna. It extends from the edge of the ulnar notch of the radius to the styloid process of the ulna, and is interposed between the ulna and the carpal bones (lunate and triquetral). The disc glides across the ulnar head as the radius moves around it.

Articulation 2 is known as the ***RadioCarpal Joint.*** It permits *flexion* and *extension* as well as *abduction* and *adduction* of the hand. It is formed by the distal end of the radius and the triangular disc, which articulate with the **Scaphoid** 2**, Lunate** 3 and **Triquetral** 4 distally.

More specifically, the radius articulates with the scaphoid and lunate, and the triangular disc articulates with the lunate in anatomical position, abduction, and throughout pure flexion and extension. When the hand is adducted the triangular disc articulates with the triquetral.

Stability of this joint is provided by four sets of ligaments: (1) **Radiocarpal Ligaments** 5 dorsally and ventrally, (2) **Ulnocarpal Ligaments** 6 dorsally and ventrally, (3) a **Radial Collateral Ligament** 7 between the styloid process of the radius and the scaphoid, and (4) an **Ulnar Collateral Ligament** 8 between the styloid process of the ulna and the triquetral.

The Flexor Retinaculum

The tendons of the muscles that flex the fingers (*Flexor digitorum superficialis* and *Flexor digitorum profundus*) and thumb (*Flexor pollicis longus*) pass into the hand superficial to the palmar radiocarpal and palmar ulnocarpal ligaments, and beneath a strong transverse ligament that extends from the pisiform and hamate medially to the scaphoid and trapezium laterally. This ligament is known as the **Flexor Retinaculum** 9**.** It creates a **"Carpal Tunnel"** that prevents the flexor tendons from "bowstringing" when the muscles contract.

Carpal Tunnel Syndrome

Even though the nine flexor tendons run through the carpal tunnel surrounded by a common synovial tendon sheath, if they become inflamed they will impinge upon the Median Nerve, which also runs beneath the flexor retinaculum. This results in a painful condition known as *"Carpal Tunnel Syndrome."*

Color and label the structures of the wrist joint in figure 3.7.

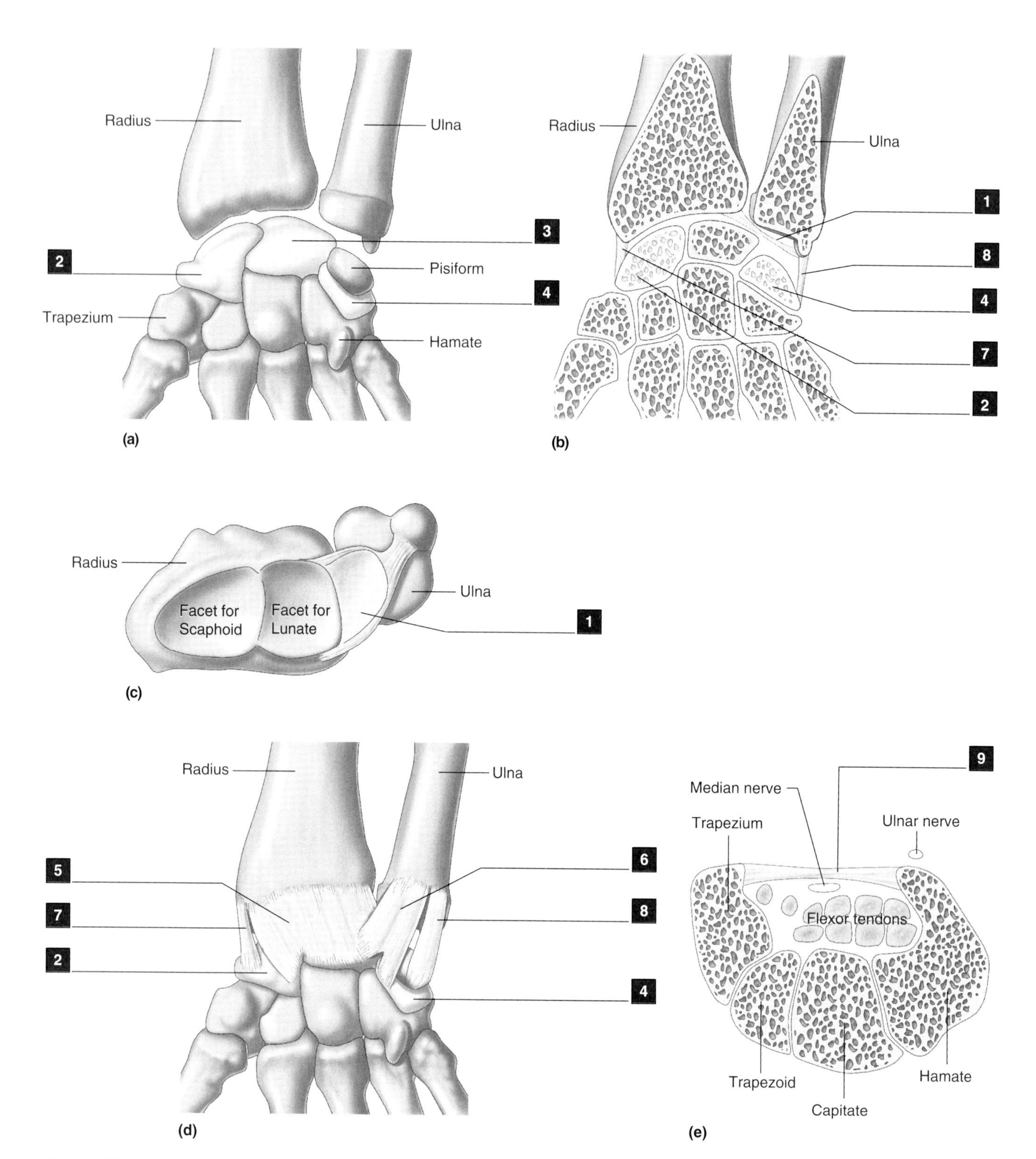

Figure 3.7 **Bones of the Wrist Joint**
(a) palmar aspect of right wrist; (b) coronal section through right wrist; (c) distal view of right forearm bones; (d) palmar aspect of right wrist; (e) section through distal row of carpal bones.

3.3 INNERVATION OF THE UPPER LIMB

In almost all textbooks, the muscles of the upper limb are listed alongside the source(s) of innervation. This results in a list of seemingly meaningless associations that requires memorization. However, the muscles and the nerves that supply them are organized according to a developmental pattern. An appreciation of this pattern is the key to remembering the associations.

The muscles of the upper limb occupy one of two compartments. One compartment is **Dorsal 1**. The other compartment is **Ventral 2**.

In figure 3.8, color the *ventral* compartments *blue*, and the *dorsal* compartments *red*.

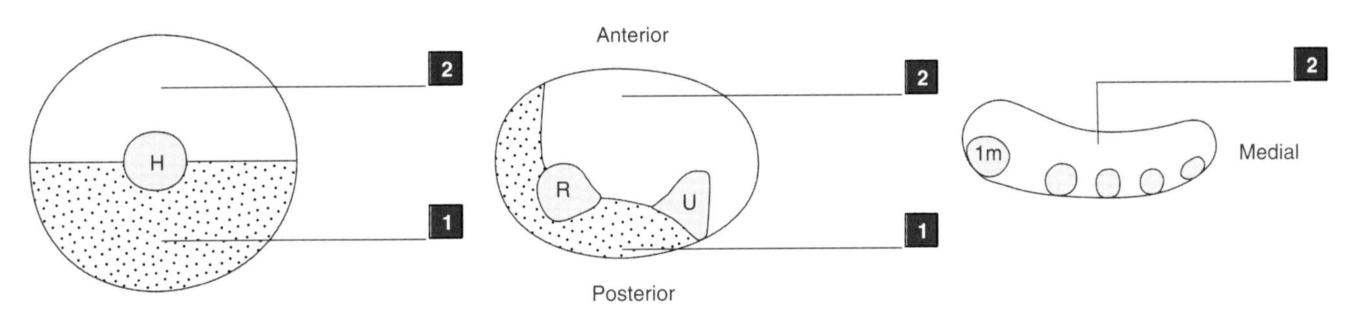

Figure 3.8 **Muscle Compartments of the Upper Limb**
Schematic transverse sections through a left arm, forearm, and hand.
(H = humerus, R = radius, U = ulna, 1m = first metacarpal)

Dorsal Compartment Muscles generally stabilize the shoulder joint, adduct and *extend* the arm, *extend* and supinate the forearm, *extend* the wrist, and *extend* the fingers and thumb. Muscles that originate on the vertebral column, the back of the rib cage, and the scapula (except those that arise from the coracoid process) are members of the dorsal group.

Ventral Compartment Muscles generally *flex* the arm, *flex* and pronate the forearm, *flex* the fingers, and *flex* and oppose the thumb. Muscles that originate on the front of the rib cage, and the coracoid process of the scapula are members of the Ventral group. Feel your own hand. There are no fleshy muscle fibers on the dorsal side of the bones. The intrinsic hand muscles that move the thumb and fingers are all ventral.

Recall that all upper limb muscles are *Hypaxial*. As such, they are supplied by ***Ventral Rami of Spinal Nerves***. Each ventral ramus carries nerve fibers that will innervate a *Ventral Compartment* muscle, and those that will innervate a *Dorsal Compartment* muscle. In order to provide the muscles with nerve fibers from more than one spinal nerve, the ventral rami are braided together in the *Brachial Plexus*.

The Brachial Plexus

The muscles of the upper limb are served by the *Ventral Rami* of five *Spinal Nerves C5–T1* via the brachial plexus. Recall that a peripheral nerve that emerges from a plexus has axons from more than one ventral ramus.

Roots

The five ventral rami comprise the *five roots* of the plexus.

Trunks

The roots combine to form ***three Trunks*****—Superior, Middle,** and **Inferior.** Each trunk has axons that will go to dorsal and ventral compartment muscles.

Cords

From each trunk, the axons that will supply a dorsal compartment split apart from those that will supply a ventral compartment. The *split* forms ***three Cords*****—Lateral, Medial,** and **Posterior.** The axons of each cord travel only to dorsal compartment or only to ventral compartment muscles.

The **Lateral** and **Medial Cords** carry *Ventral Compartment* axons.

The **Posterior Cord** carries *Dorsal Compartment* axons.

Peripheral Nerves

The axons of each cord separate to form the principal peripheral nerves that innervate the muscles. Some of the fibers from the lateral and medial cords combine to form the median nerve. The peripheral nerves are derived as follows:

Lateral Cord yields the *Lateral Pectoral Nerve*, the *Musculocutaneous Nerve* and part of the *Median Nerve*.

Medial Cord yields the *Medial Pectoral Nerve*, the *Ulnar Nerve* and the other part of *Median Nerve*.

Posterior Cord yields the *Subscapular Nerves*, the *Radial Nerve* and the *Axillary Nerve*.

Superior Trunk yields the *Suprascapular Nerve*. This is the only nerve to emerge directly from a trunk. It innervates dorsal compartment muscles.

Now that we have established the basis by which the muscles of the upper limb are innervated, we can proceed to examine the distribution of peripheral nerves from the brachial plexus.

Color figure 3.9 according to whether the part carries dorsal or ventral compartment axons. Again, use *blue for ventral* and *red for dorsal.*

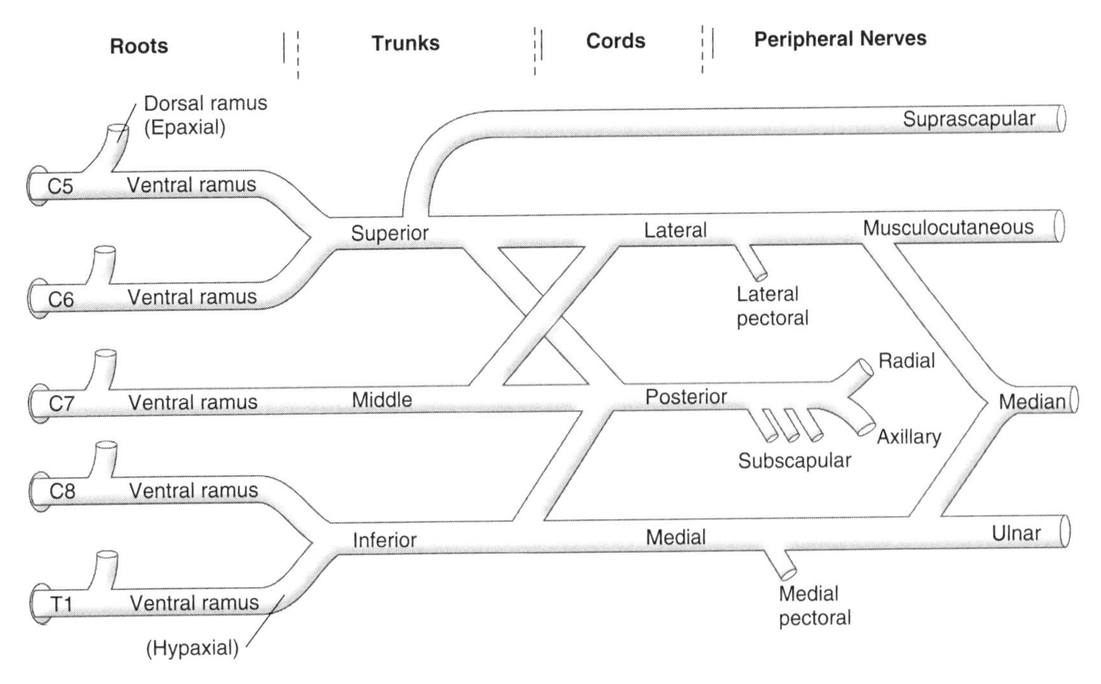

Figure 3.9 **The Brachial Plexus**

Nerves to Muscles That Move the Arm

These muscles arise from the vertebral column and ribs, and insert on the humerus. With only two exceptions, they are *Dorsal Compartment Muscles*. They are supplied by the

Suprascapular Nerve

Axillary Nerve

Subscapular Nerves

One of the two exceptions is the **Pectoralis major** muscle. It arises from the ventral surface of the thorax and is, therefore, supplied by *Ventral Compartment* nerves. They are the

Lateral and Medial Pectoral Nerves

The other exception is the **Coracobrachialis** muscle. It arises from the coracoid process of the scapula and runs through the ventral compartment of the arm to the humeral shaft. It is, therefore, supplied by a *Ventral Compartment* nerve. This is the

Musculocutaneous Nerve

Nerves to Muscles That Move the Forearm, Wrist, and Digits

The innervation of these muscles can be worked out very easily by determining whether they lie in the ventral compartment or the dorsal compartment of the arm and forearm.

A muscle whose fleshy belly lies in the *Ventral Compartment* of the *Arm* will be supplied by the **Musculocutaneous Nerve** 1.

A muscle whose fleshy belly lies in the *Ventral Compartment* of the *Forearm or Hand* will be supplied by either the **Median Nerve** 2 or the **Ulnar Nerve** 3.

A muscle whose fleshy belly lies in the *Dorsal Compartment* of the *Arm or Forearm* will be supplied by the **Radial Nerve** 4.

Label the areas of the upper limb that are supplied by the aforementioned nerves in figure 3.10.

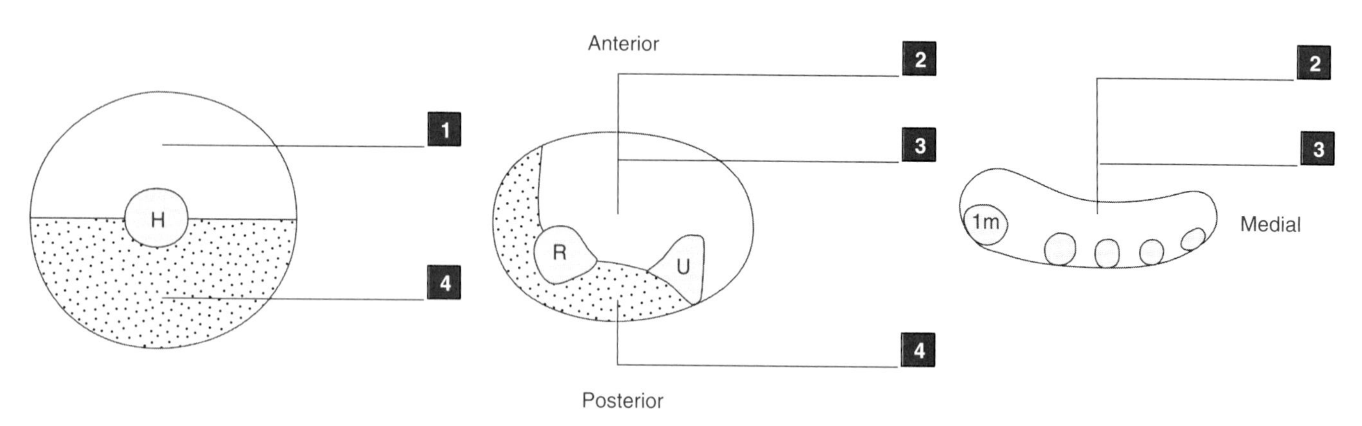

Figure 3.10 **Nerves to Muscles That Move the Forearm, Wrist, and Digits**
Schematic transverse sections through a left arm, forearm, and hand (H = humerus, R = radius, U = ulna, 1m = first metacarpal)

Courses of the Peripheral Nerves

You should be familiar with the paths followed by the six principal nerves that supply the muscles of the upper limb. Their paths can be followed by the landmarks that they pass.

Ventral Compartment Nerves

Musculocutaneous 1. It arises about an inch below the coracoid process of the scapula, pierces the Coracobrachialis muscle and then crosses the anterior compartment of the arm obliquely, supplying three muscles of that compartment and the elbow joint.

Median 2. It descends with the brachial artery along the medial aspect of the arm next to the Biceps brachii muscle to cross the front of the elbow. It follows a "median" course along the forearm, crossing the wrist through the *Carpal Tunnel.*

Ulnar 3. It arises about an inch below the coracoid process of the scapula and runs with the median nerve and brachial artery to about the middle of the arm, where it turns posteriorly to pass behind the medial epicondyle of the humerus. Here it runs in contact with the capsule of the elbow joint just to the medial side of the ulna. This is the "*funny bone*" nerve. It continues distally by turning anteriorly to run along the ulna. It crosses the wrist anterior to the carpal tunnel.

Dorsal Compartment Nerves

Suprascapular 4. It curves over onto the dorsal surface of the scapula through the notch in the upper border of that bone adjacent to the coracoid process.

Axillary 5. It runs across the anterior surface of the Subscapularis muscle and exits the axilla between it and the Teres major muscle. It runs behind the humerus.

Radial 6. It is the largest branch of the brachial plexus. It descends along the back of the humerus and then curves around to emerge at the elbow anterior to the lateral epicondyle of the humerus. Here it divides into two branches, one of which continues distally along the lateral side of the radius to supply the back of the hand.

Identify these peripheral nerves in figure 3.11. Use shades of red and blue as appropriate for dorsal and ventral compartment nerves.

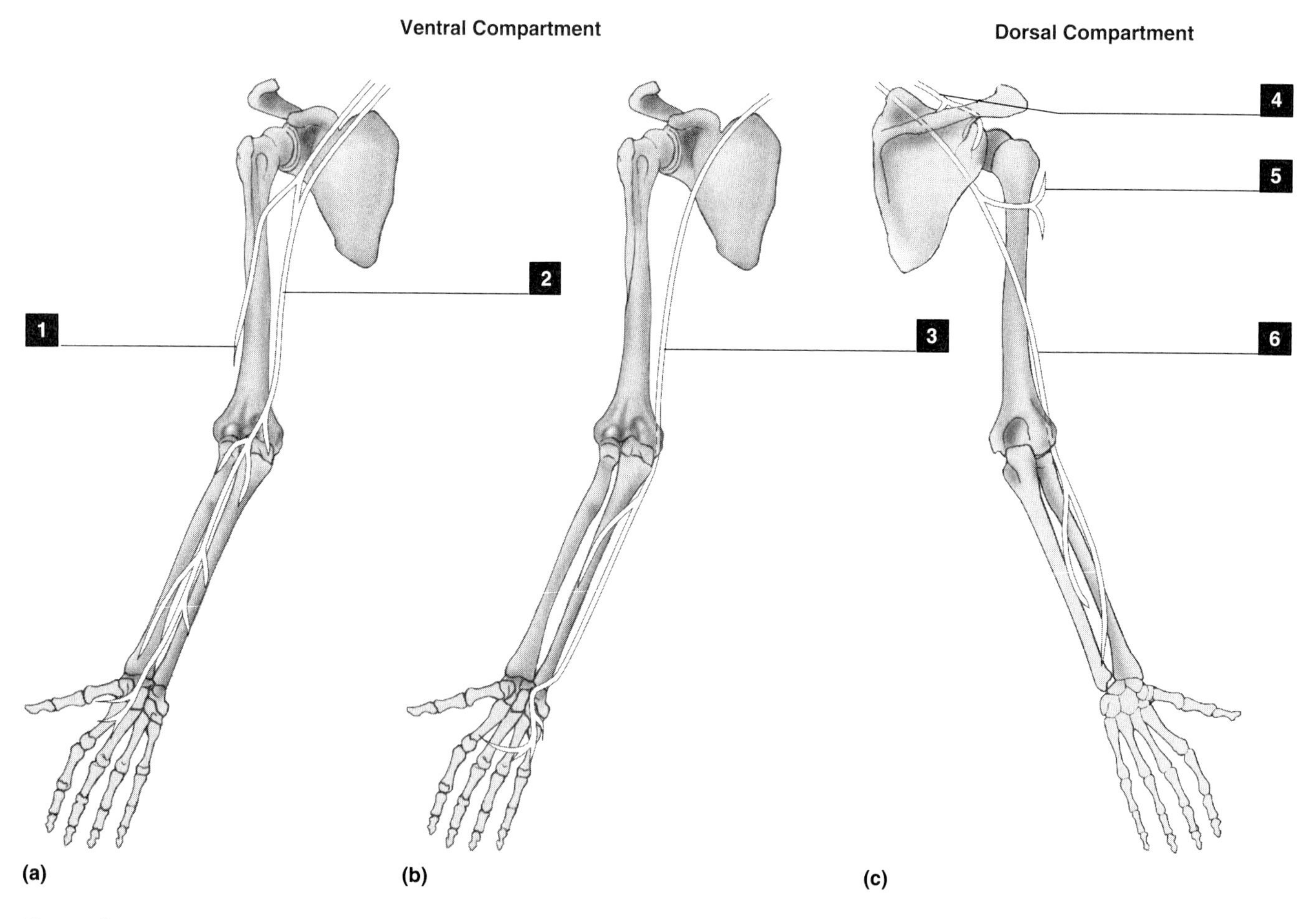

Figure 3.11 **Peripheral Nerves of the Upper Limb**
(a) and (b) anterior view; (c) posterior view.

3.4 MUSCLES OF THE UPPER LIMB

Recall that the muscles of the upper limb may be divided into two groups according to their location and innervation:

1. Ventral compartment muscles that are innervated by ventral division nerves: lateral pectoral, medial pectoral, musculocutaneous, median, ulnar
2. Dorsal compartment muscles that are innervated by dorsal division nerves: suprascapular, subscapular, axillary, radial

We will examine these muscles according to the anatomical region upon which they act and by the action they produce.

Muscles That Move the Arm

Muscles that insert onto the humerus may arise from the thoracic axial skeleton, or from the scapula and clavicle.

Thoracohumeral Muscles

There are two muscles that run from the thoracic axial skeleton to the humerus. We studied one of these—**Latissimus dorsi**— in Laboratory 1 (p. 13).

Pectoralis major 1 is a large muscle on the front of the chest. It arises from the medial half of the clavicle, the sternum and part of the 7th rib. It inserts onto the greater tubercle of the humerus. It adducts and medially rotates the humerus. The clavicular part also flexes the shoulder joint whereas the sternal part extends an already flexed arm.

Ventral compartment = *Medial* and *Lateral Pectoral nerves*.

Scapulohumeral Muscles

Seven muscles run from the scapula to the humerus. Four of these constitute the rotator cuff.

Coracobrachialis 2 runs from the tip of the coracoid process of the scapula to the medial side of the middle of the humeral shaft. It adducts and flexes the arm.

Ventral compartment = *Musculocutaneous nerve*

Teres major 3 runs from the dorsal surface of the scapular blade to the lesser tubercle of the humerus. It adducts, medially rotates and extends the arm.

Dorsal compartment = *lower Subscapular nerve*

Deltoid 4 attaches to the spine and acromion of the scapula, and the lateral third of the clavicle. It inserts onto the deltoid tuberosity of the humerus. It flexes, abducts, and extends the arm.

Dorsal compartment = *Axillary nerve*

Rotator Cuff Muscles

Recall that the tendons of four muscles adhere to the outer surface of the shoulder joint capsule as they pass it anteriorly, superiorly, and posteriorly. They form a cuff around the joint that reinforces it (see p. 49 of this lab).

Supraspinatus 5 runs from the supraspinous fossa of the scapula across the top of the shoulder joint to insert onto the greater tubercle of the humerus. It elevates and abducts the arm.

Dorsal compartment = *Suprascapular nerve*

Infraspinatus 6 runs from the infraspinous fossa of the scapula across the back of the shoulder joint to the greater tubercle of the humerus. It laterally rotates the arm.

Dorsal compartment = *Suprascapular nerve*

Teres minor 7 arises from the dorsal surface of the scapula along its axillary border just superior to the origin of Teres major. It runs across the back of the shoulder joint to the greater tubercle of the humerus. It laterally rotates the arm, but only when it is adducted.

Dorsal compartment = *Axillary nerve*

Subscapularis 8 arises from the anterior surface of the blade of the scapula (the subscapular fossa), runs across the front of the shoulder joint, and inserts onto the lesser tubercle of the humerus. It medially rotates the arm.

Dorsal compartment = *upper* and *lower Subscapular nerves*

The illustration on page 62 shows the relationship of the tendons of the four rotator cuff muscles to the shoulder joint when the arm is *adducted* (left) and when it is *abducted* to 90 degrees and laterally rotated (right). Anterior displacement is the most common type of shoulder dislocation, but it usually occurs only when the arm is abducted and laterally rotated. Why would this be so? Identify and color the tendons of the four rotator cuff muscles in this illustration.

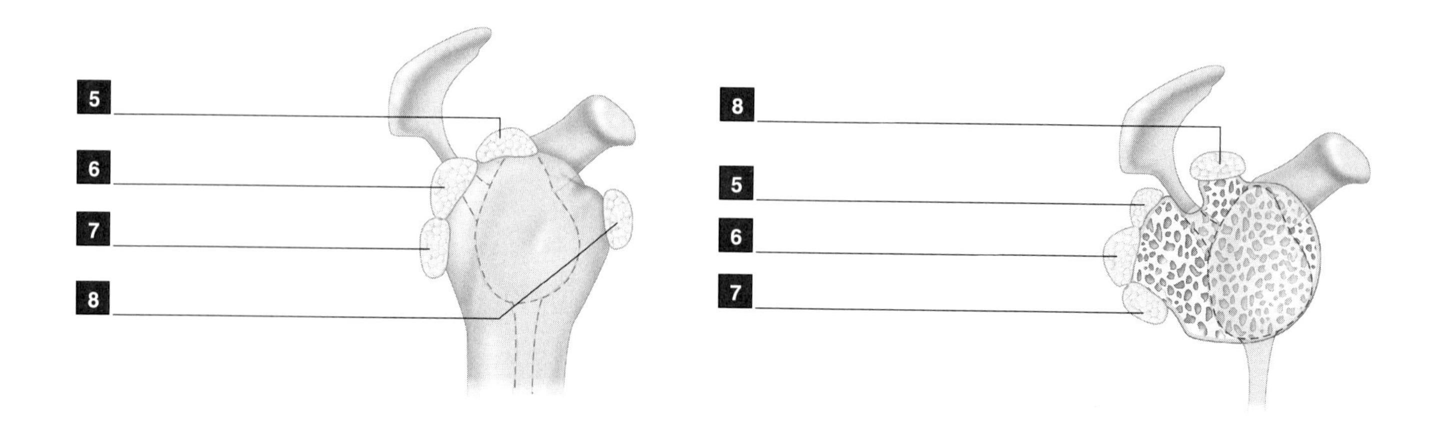

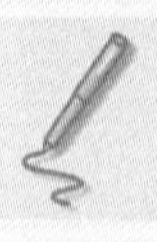

Identify the muscles that move the arm in figures 3.12 and 3.13.

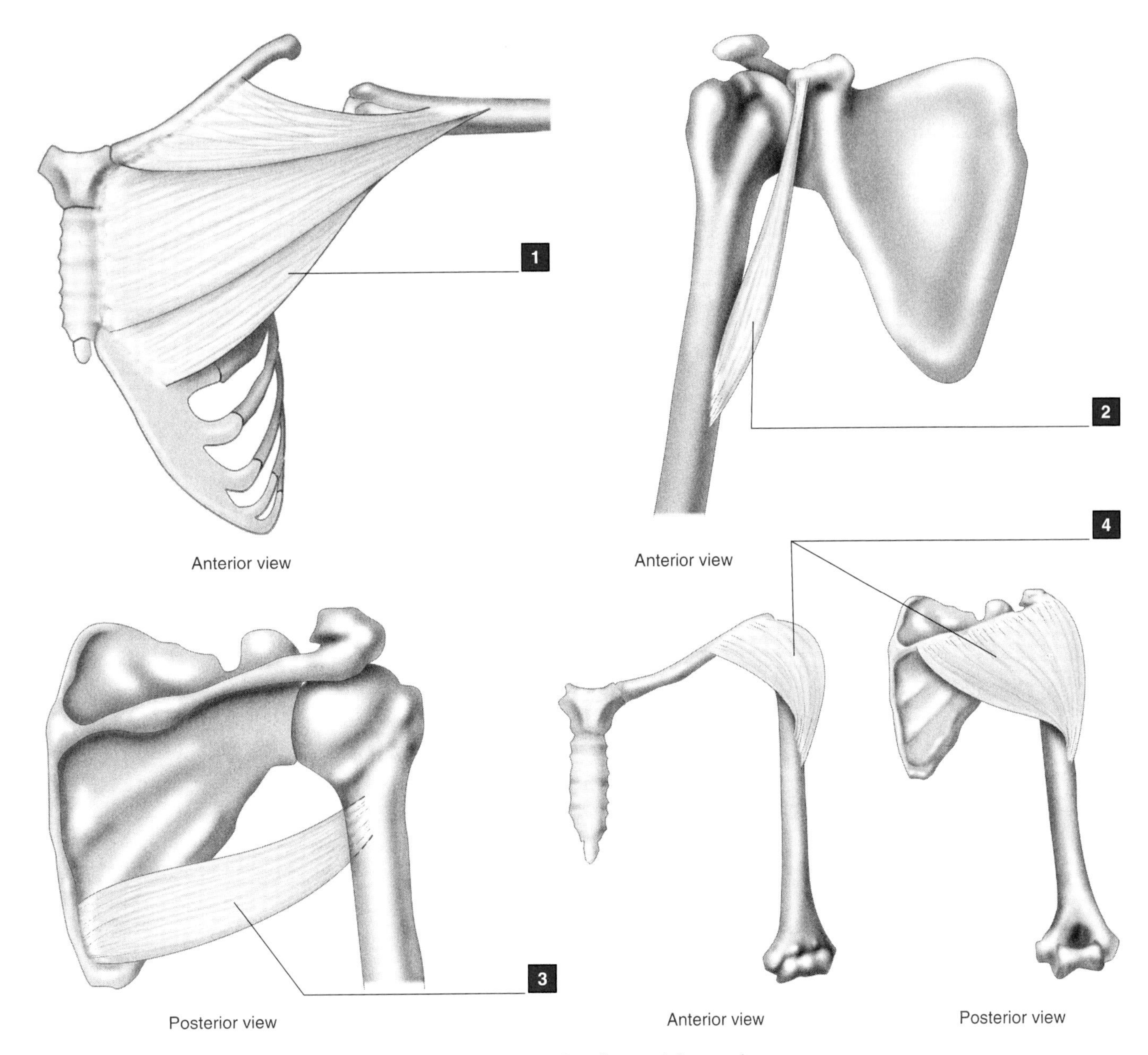

Figure 3.12 Muscles That Move the Arm: Thoracohumeral and Scapulohumeral

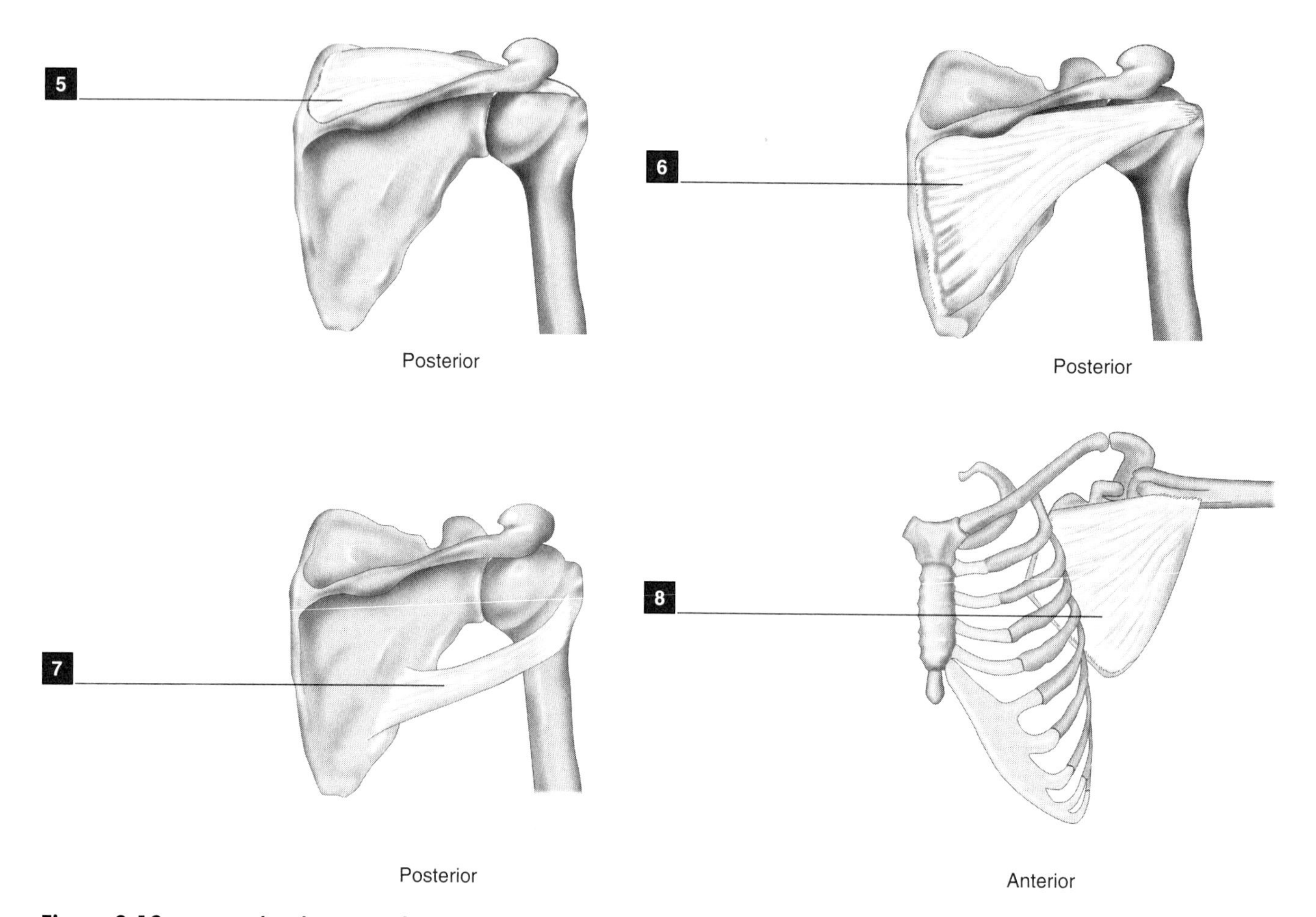

Figure 3.13 **Muscles That Move the Arm: Rotator Cuff**

Muscles That Move the Forearm

These muscles insert onto the radius or ulna. They arise from the humerus and scapula. They may be grouped into four categories depending upon their action, although some produce more than one movement.

Elbow Flexors

Biceps brachii **1** arises by two separate scapular attachments that merge to form a single belly. ***Long Head*** arises from a tubercle above the glenoid fossa; its tendon passes through the shoulder joint capsule and runs along the intertubercular groove of the humerus. ***Short Head*** arises from the coracoid process. The unified belly inserts onto the tuberosity of the radius. Some tendinous fibers separate from it above the elbow and sweep medially to form the ***Bicipital Aponeurosis*** that merges with the deep fascia. In addition to flexing the elbow, it is an important forearm supinator.

Ventral compartment = *Musculocutaneous nerve*

Brachialis **2** arises from the front of the humeral shaft, and inserts onto the ulna at the base of the coronoid process. It flexes the elbow when the forearm is pronated.

Ventral compartment = *Musculocutaneous nerve*

Brachioradialis **3** originates on the lateral edge of the humerus above the lateral epicondyle; it inserts onto the radius close to the styloid process. It flexes the elbow.

Dorsal compartment = *Radial nerve*

Elbow Extensors

Triceps brachii **4** has three bellies that merge to insert onto the olecranon of the ulna. ***Long Head*** arises from the inferior rim of the glenoid fossa of the scapula. ***Lateral Head*** arises from the humerus between the greater tubercle and deltoid tuberosity. ***Medial Head*** arises from the posterior surface of the humerus. In addition to extending the forearm, the long head adducts and extends the humerus.

Dorsal compartment = *Radial nerve*

Anconeus **5** runs from the back of the lateral epicondyle of the humerus to the lateral side of the olecranon of the ulna. It extends the forearm.

Dorsal compartment = *Radial nerve*

Identify the elbow flexor and extensor muscles in figure 3.14.

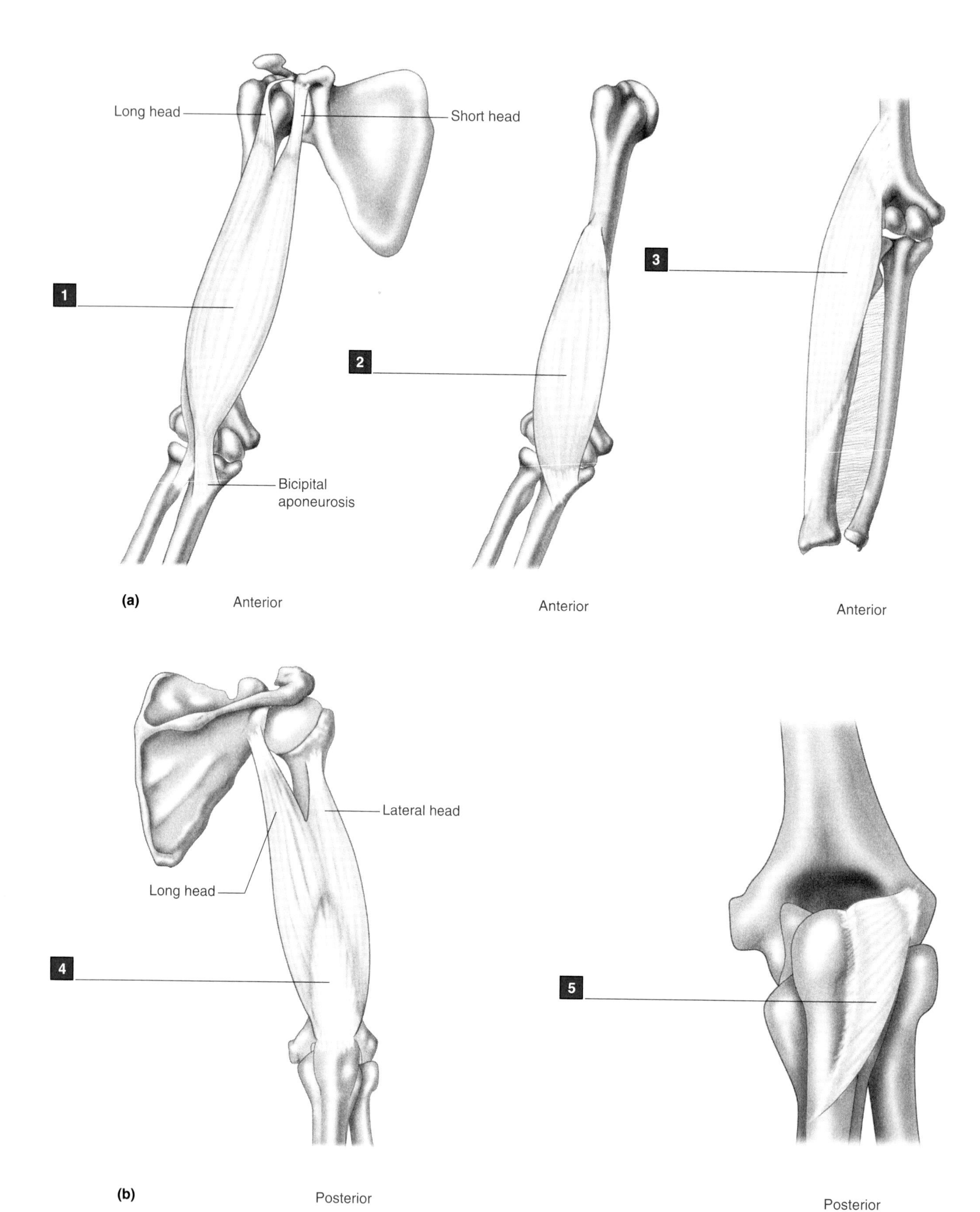

Figure 3.14 **Muscles That Move the Forearm**
(a) elbow flexors; (b) elbow entensors.

Pronators of the Forearm

Pronator teres 1 arises from the medial epicondyle of the humerus and crosses the elbow to insert onto the lateral side of the radius shaft. In addition to being a pronator, it is a powerful flexor of the elbow.

Ventral compartment = *Median nerve*

Pronator quadratus 2 is located just above the wrist. It runs from the anteromedial edge of the ulna to the anterior surface of the radius.

Ventral compartment = *Median nerve*

Supinators of the Forearm

Supinator 3 is arranged in two layers. The superficial arises from the lateral epicondyle of the humerus; it runs behind the radial head and turns forward to insert onto the front of the radial shaft. The deep layer arises from the lateral side of the upper end of the ulna; it runs behind the radius and turns forward to insert on the front of the upper part of the radial shaft.

Dorsal compartment = *Radial nerve*

Biceps brachii was discussed on page 65. Recall that in addition to being a powerful flexor of the elbow, it is an important supinator of the forearm.

Ventral compartment = *Musculocutaneous nerve*

Identify the forearm pronator and supinator muscles in figure 3.15.

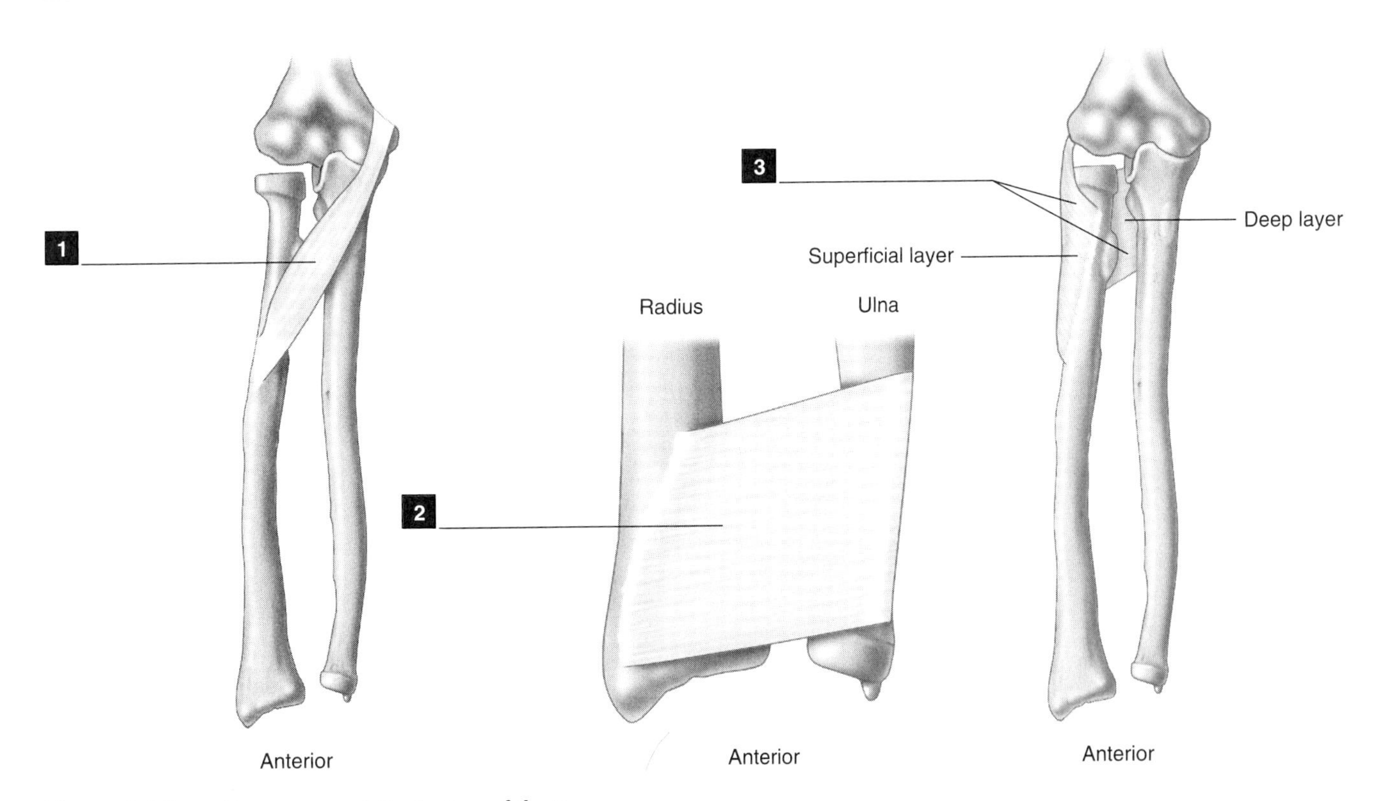

Figure 3.15 Pronators and Supinators of the Forearm

Muscles That Move the Wrist

There are six muscles whose sole action it is to produce movement at the wrist joint. Other muscles that cross the radiocarpal joint may move the wrist, but their principal function relates to the digits. We will consider them later.

Flexors of the Wrist

These arise from the ***medial epicondyle*** of the humerus by a common tendon. All are ventral compartment muscles.

Flexor carpi radialis 1 crosses to the radial side of the forearm to attach to the bases of the second and third metacarpals. It also abducts the wrist.

Median nerve

Flexor carpi ulnaris 2 runs along the ulnar side of the forearm to attach to the pisiform and then onto the base of the fifth metacarpal. It also adducts the wrist.

Ulnar nerve

Palmaris longus 3 runs between the other two flexors to attach to the front of the *Flexor Retinaculum*. From here it fans out to form four separate bands that join the flexor sheaths of the fingers. The bands are connected by fibrous tissue that forms a sheet known as the ***Palmar Aponeurosis***.

Median nerve

Flex your wrist to 45 degrees with your fingers extended and apply pressure to your palm. The **Palmaris longus** tendon makes a sharp ridge under the skin in the middle of your wrist. What tendon makes the less prominent ridge lateral to it? If you have only one ridge, don't worry. Not everyone has a Palmaris longus!

Extensors of the Wrist

These arise from the ***lateral epicondyle*** of the humerus. All are dorsal compartment muscles.

Extensor carpi radialis longus 4 runs along the lateral side of the forearm to insert onto the base of the second metacarpal. It also abducts the wrist.

Radial nerve

Extensor carpi radialis brevis 5 runs along the lateral side of the forearm to insert onto the base of the third metacarpal. It also abducts the wrist.

Radial nerve

Extensor carpi ulnaris 6 also arises from the back of the ulnar shaft. The conjoined fibers run along the medial side of the forearm to insert onto the base of the fifth metacarpal. It also adducts the wrist.

Radial nerve

Identify the aforementioned muscles that move the wrist in figure 3.16.

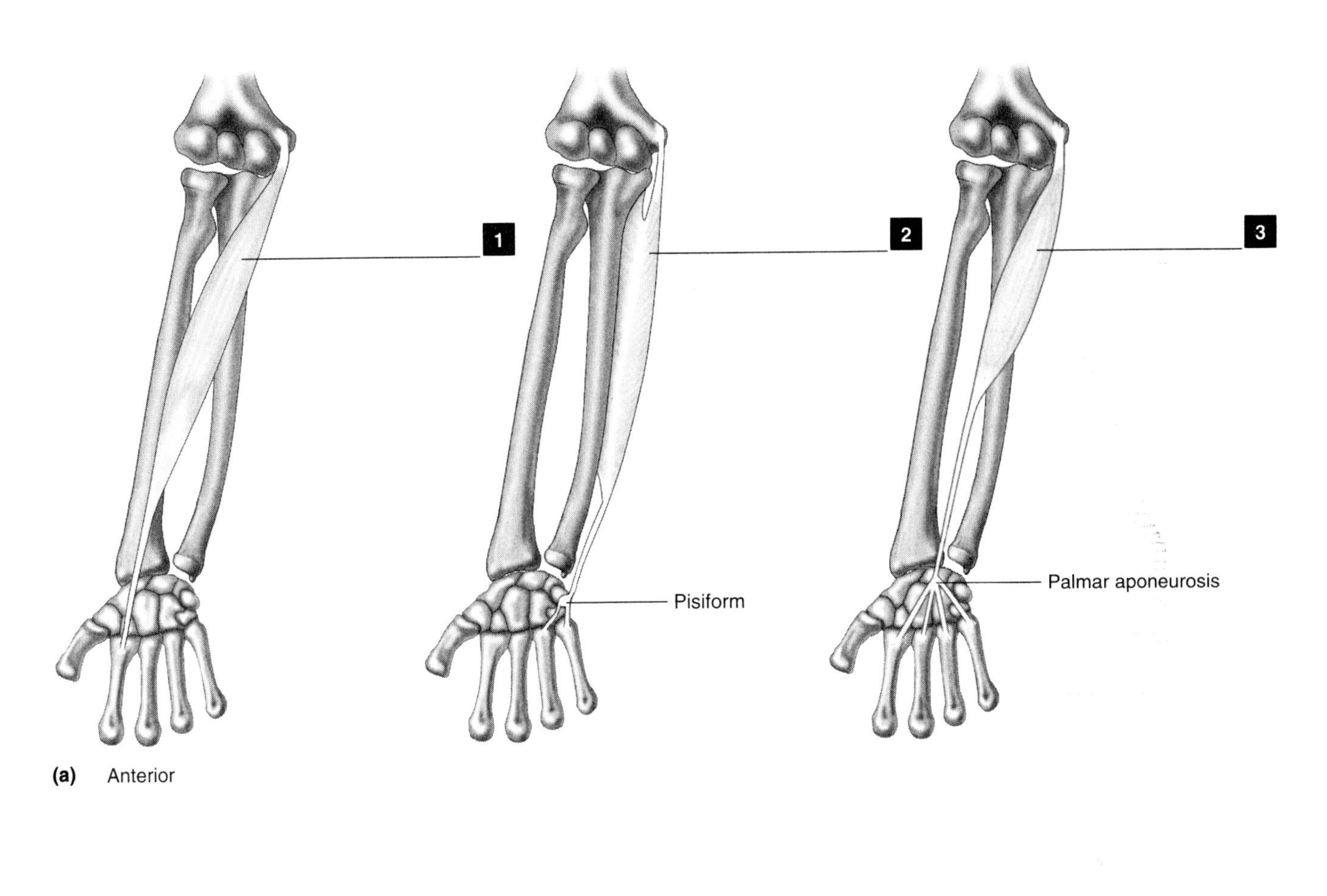

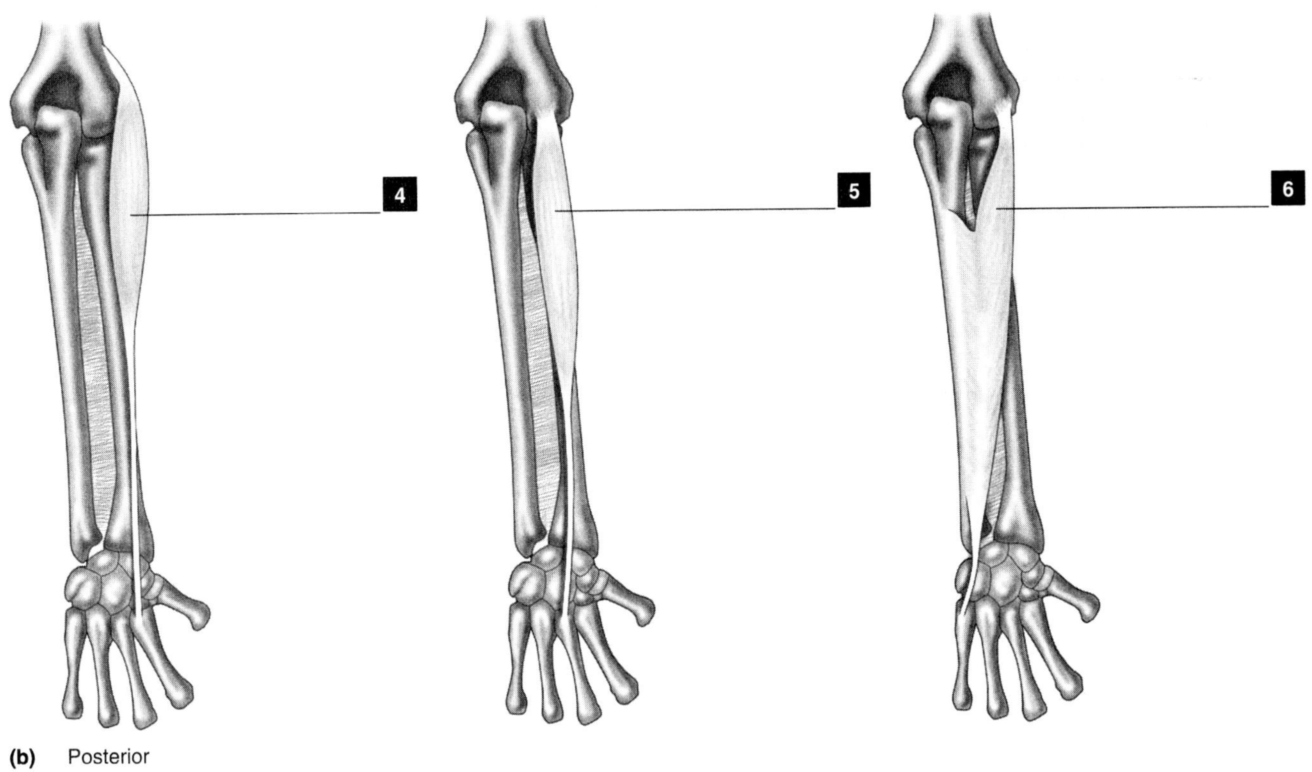

Figure 3.16 **Muscles That Move the Wrist**
(a) wrist flexors; (b) wrist extensors.

Muscles That Move the Fingers

There are 10 muscles, or groups of muscles that move the fingers. These can be divided into two groups of five muscles each. (1) Extrinsic muscles arise from the arm or forearm. (2) Intrinsic muscles originate from within the hand itself.

Extrinsic Flexors of the Fingers

The tendons of these two muscles run through the carpal tunnel. Both are ventral compartment muscles.

Flexor digitorum superficialis 1 arises from the medial epicondyle of the humerus and the medial side of the coronoid process of the ulna; it also arises from the shaft of the radius. The conjoined fibers run down the forearm deep to the wrist flexors. Just above the wrist, it gives rise to four tendons, each of which inserts onto a middle phalanx after separating around the *Flexor digitorum profundus* tendon over the proximal phalanx.

Median nerve

Flexor digitorum profundus 2 arises from the ulnar shaft; it also gives rise to four tendons, each of which inserts onto a *distal phalanx.*

Median and *Ulnar nerves*

Extrinsic Extensors of the Fingers

These three are dorsal compartment muscles.

Extensor digitorum communis 3 arises from the lateral epicondyle of the humerus; it gives rise to four tendons, each of which inserts onto a distal phalanx.

Radial nerve

Extensor digiti minimi 4 arises from the lateral epicondyle of the humerus and runs to the fifth digit.

Radial nerve

Extensor indicis 5 is a deep muscle that originates along the distal part of the ulnar shaft and runs to the second digit.

Radial nerve

Identify the aforementioned extrinsic muscles that move the fingers in figure 3.17.

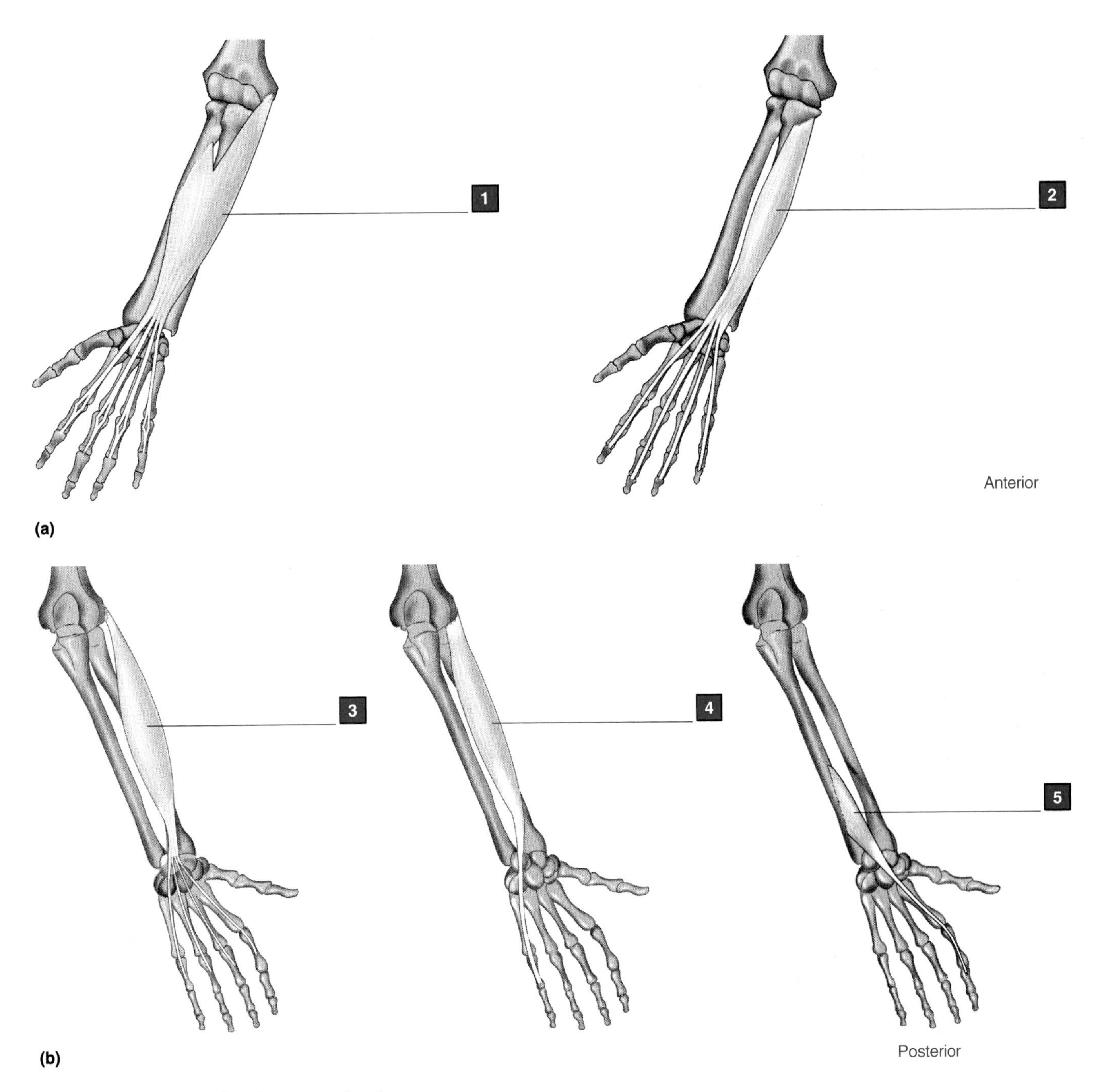

Figure 3.17 **Muscles That Move the Fingers**
(a) extrinsic flexors; (b) extrinsic extensors.

Intrinsic Muscles of the Fingers

Three of the five intrinsic muscles run from the flexor retinaculum to the little finger; they form the ***Hypothenar Eminence*** on the medial side of the palm. The other two comprise groups of muscles. All five are ventral compartment muscles.

Flexor digiti minimi 1 arises from the flexor retinaculum and inserts onto the base of the proximal phalanx of the fifth digit. It flexes the fifth digit.

Ulnar nerve

Abductor digiti minimi 2 arises from the flexor retinaculum and also inserts onto the base of the proximal phalanx of the fifth digit. It adducts and flexes the fifth digit.

Ulnar nerve

Opponens digiti minimi 3 arises from the flexor retinaculum and inserts onto the fifth metacarpal. It flexes and laterally rotates the metacarpal to a slight degree. This causes the ventral surface of the little finger to face the thumb when the tips of these digits are opposed.

Ulnar nerve

Lumbricals 4. There are four lumbrical muscles—one for each finger. Each arises from the tendon of *Flexor digitorum profundus*. It runs along the lateral side of that tendon to insert onto the base of the proximal phalanx. It then turns dorsally to join the common extensor tendon. It flexes the metacarpophalangeal (MP) joint and extends the proximal interphalangeal (IP) joint.

Median and *Ulnar nerves*

Interossei 5. There are seven interosseous muscles—so-called palmar and dorsal. The *three Palmar Interossei adduct* the fingers toward the midline. Since the third digit cannot be adducted, it has no palmar interosseus muscle. Each runs from a metacarpal to the base of a proximal phalanx. (There is occasionally a fourth interosseus muscle of the thumb.) The *four Dorsal Interossei abduct* the fingers away from the midline (the third digit requires two muscles). Each arises from the opposing surfaces of two adjacent metacarpals to insert onto the base of the proximal phalanx.

Ulnar nerve

Identify the aforementioned intrinsic muscles that move the fingers in figure 3.18.

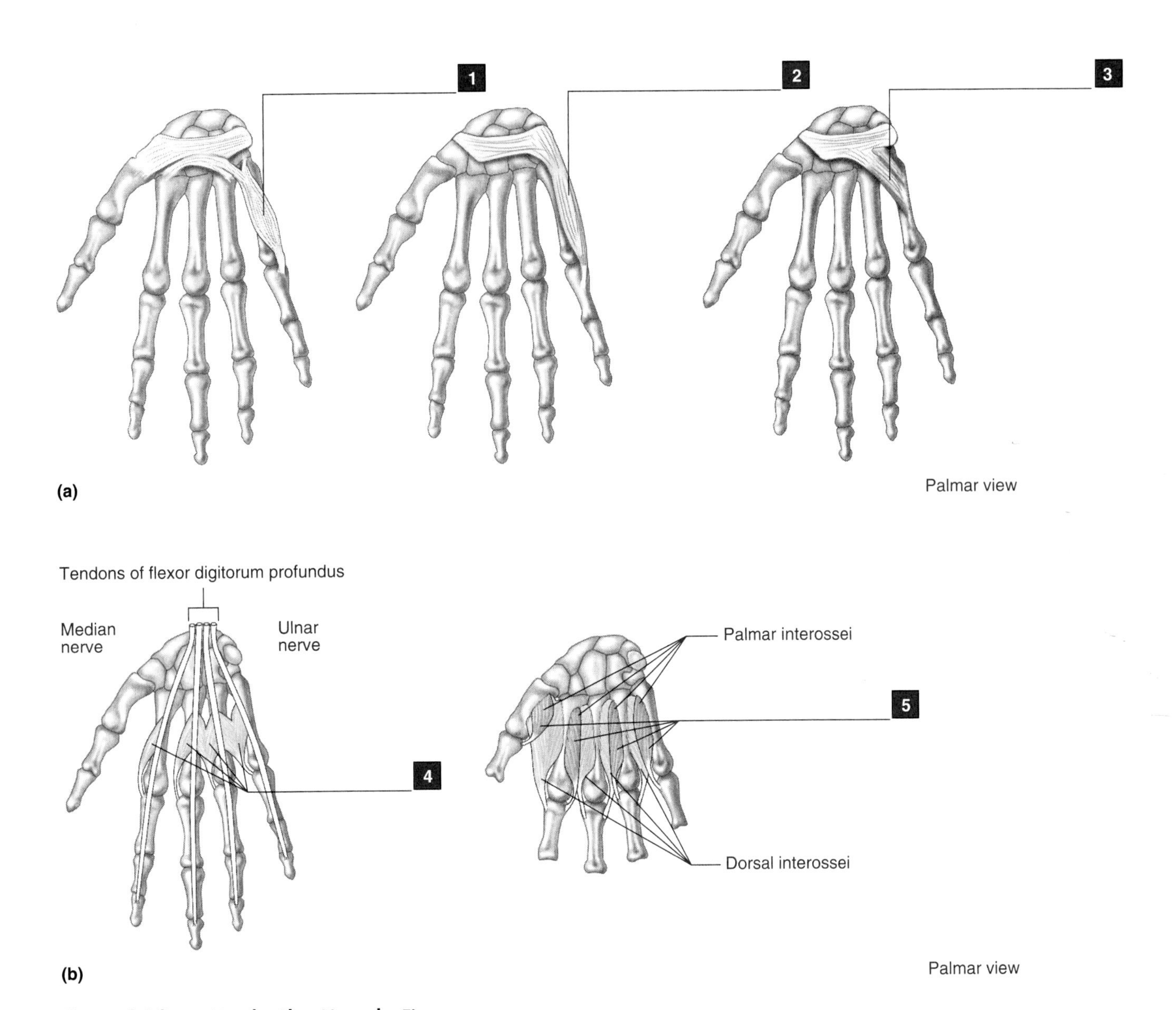

Figure 3.18 **Muscles That Move the Fingers**
(a) intrinsic hypothenar muscles; (b) other intrinsic muscles.

Muscles That Move the Thumb

Eight muscles produce movement of the thumb. These may be divided into two groups of four each. (1) Extrinsic Muscles originate from the arm or forearm. (2) Intrinsic Muscles originate from within the hand itself. All of the muscles that move the thumb have the term *Pollicis* in their name.

Extrinsic Muscles of the Thumb

Flexor pollicis longus 1 arises from the anterior surface of the radial shaft, the interosseous membrane, and the medial side of the coronoid process of the ulna. Its long tendon runs through the carpal tunnel to inserts onto the ventral surface of the base of the distal phalanx. It is the only extrinsic thumb muscle in the anterior compartment of the forearm.

Ventral compartment = *Median nerve*

Abductor pollicis longus 2 arises from the posterior surfaces of the radius, ulna, and interosseous membrane. Its tendon runs past the styloid process of the radius to insert onto the lateral side of the base of the first metacarpal. Its tendon forms part of the anterior margin of the "anatomical snuff-box." It also laterally rotates the thumb.

Dorsal compartment = *Radial nerve*

Extensor pollicis longus 3 arises from the back of the ulnar shaft and inserts onto the dorsal surface of the distal phalanx. Its tendon forms the posterior margin of the "anatomical snuff box."

Dorsal compartment = *Radial nerve*

Extensor pollicis brevis 4 arises from the posterior surface of the radius and interosseous membrane, and inserts on the dorsal surface of the base of the proximal phalanx. Its tendon forms part of the anterior margin of the "anatomical snuff box."

Dorsal compartment = *Radial nerve*

The "Anatomical Snuff Box"

If you extend your thumb and look at the lateral surface of your wrist, you will see a triangular hollow known as the "anatomical snuff box." The base of the hollow lies along the distal edge of the radius; its apex points toward the thumb. The anterior margin of the "snuff box" is formed by the tendons of *Abductor pollicis longus* and *Extensor pollicis brevis.* Its posterior margin is formed by the tendon of *Extensor pollicis longus.*

The "snuff box" is of importance because the ***Radial Artery*** courses through it.

Identify the extrinsic muscles of the thumb in figure 3.19.

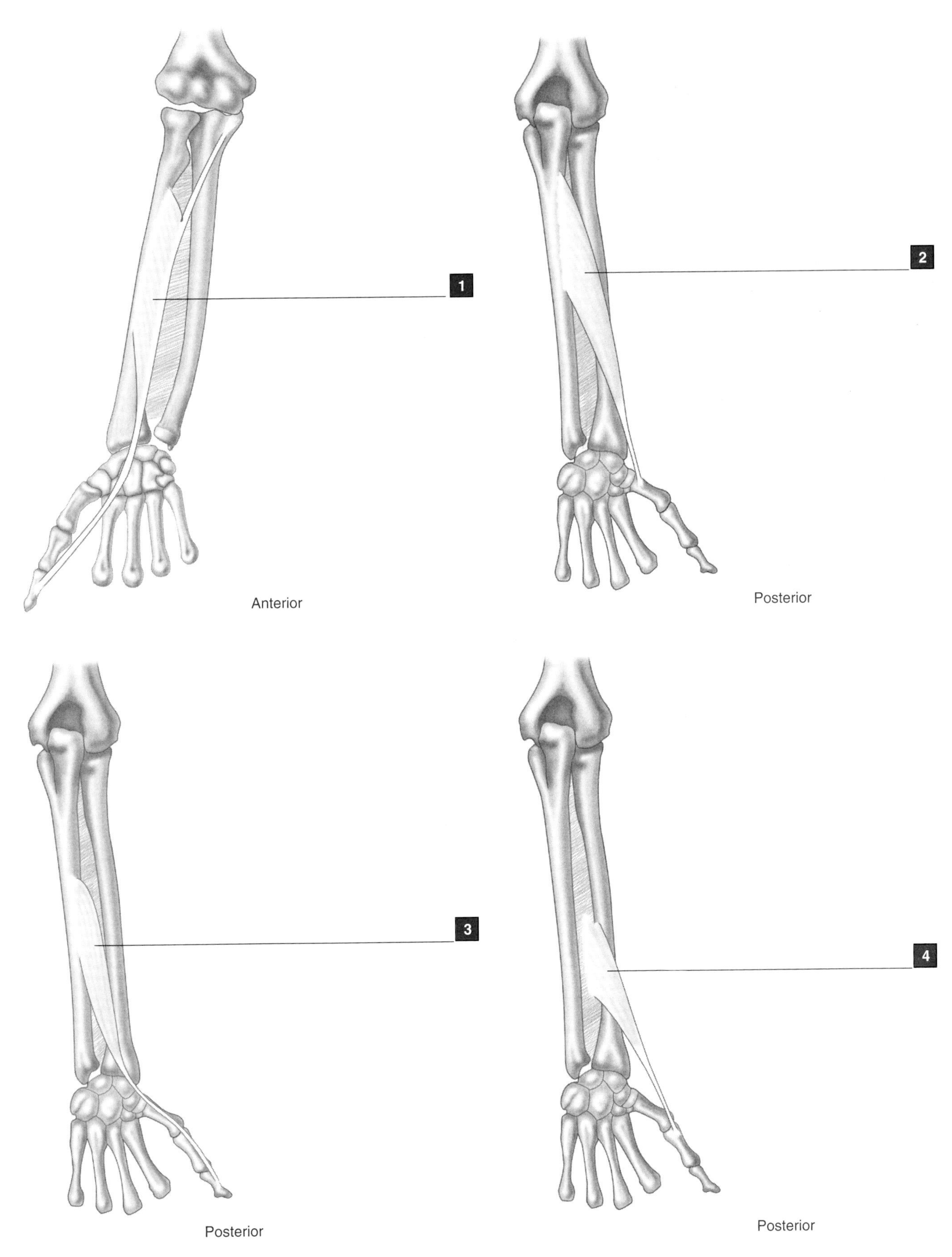

Figure 3.19 Extrinsic Muscles That Move the Thumb

Intrinsic Muscles of the Thumb

There are four intrinsic muscles of the thumb. They are all ventral compartment muscles.

Three arise from the flexor retinaculum and form the ***Thenar Eminence*** on the lateral side of the palm. They act together to produce opposition of the thumb. The other muscle—*Adductor pollicis*—lies deep in the palm.

Flexor pollicis brevis 1 arises from the flexor retinaculum and the trapezium; it inserts onto the lateral aspect of the base of the proximal phalanx. It opposes and flexes the thumb.

Median and *Ulnar nerves*

Abductor pollicis brevis 2 arises from the flexor retinaculum as well as the scaphoid and trapezium; it inserts with the tendon of flexor pollicis brevis onto the lateral side of the base of the proximal phalanx. It opposes and abducts the thumb.

Median nerve

Opponens pollicis 3 arises from the flexor retinaculum and trapezium; it inserts onto the lateral surface of the first metacarpal. It produces opposition.

Median nerve

Adductor pollicis 4 arises from the capitate, and from the shaft of the third metacarpal. The fibers converge to insert onto the medial side of the base of the proximal phalanx. It flexes the thumb when it is in an adducted position.

Ulnar nerve

Identify the intrinsic muscles of the thumb in figure 3.20.

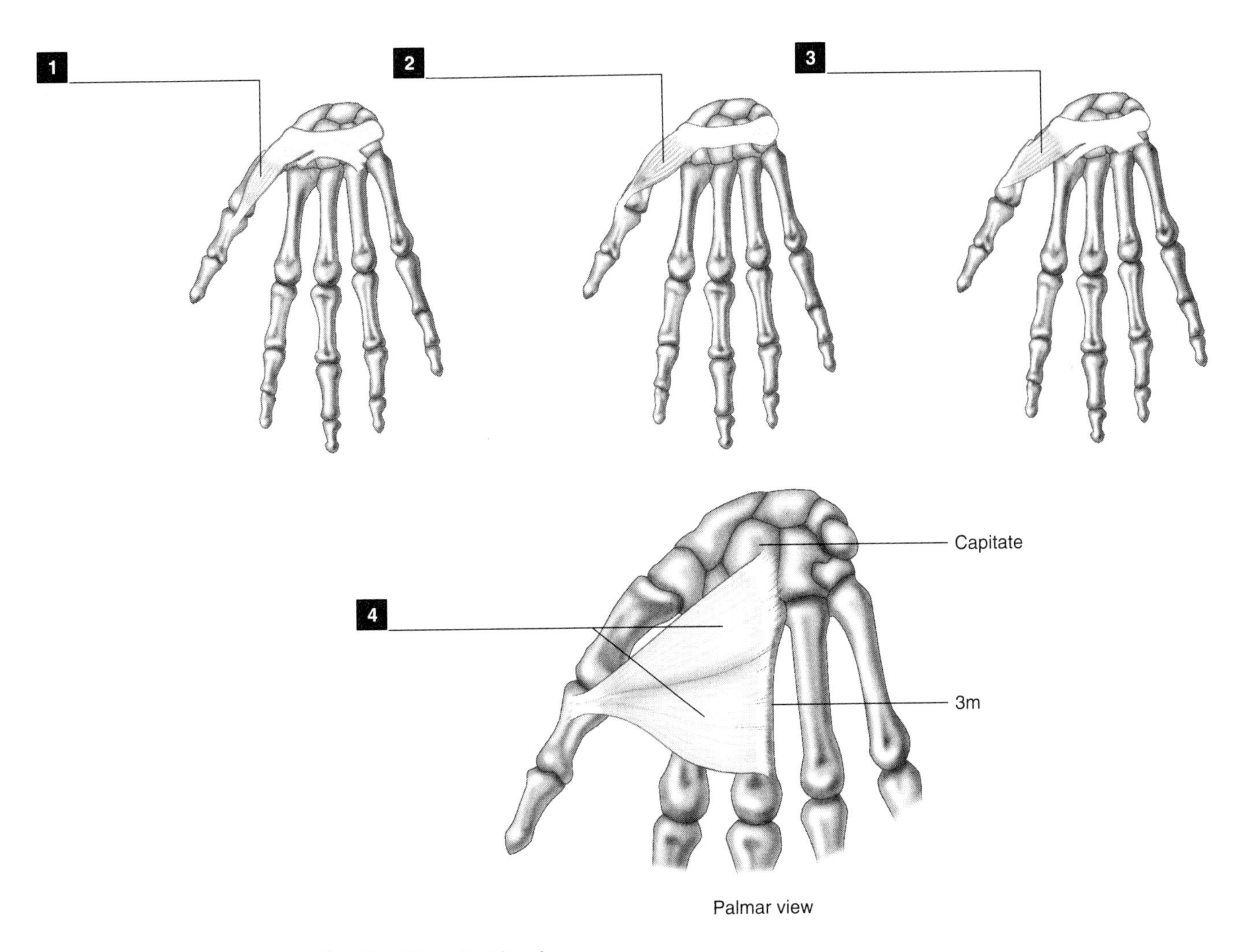

Figure 3.20 **Intrinsic Muscles That Move the Thumb**

3.5 BLOOD VESSELS OF THE UPPER LIMB

Arteries

The arterial supply of the upper limb on both sides of the body derives from the **Aortic Arch** 1 just above the heart.

To get to the ***right*** upper limb, the blood travels from the aortic arch into the **Brachiocephalic Trunk** 2**.** This trunk divides into the common carotid artery and the right **Subclavian Artery** 3**.**

To get to the ***left*** upper limb, the blood travels from the aortic arch directly into its third principal branch, the left **Subclavian Artery** 3**.** This arises from the aortic arch immediately after the origin of the common carotid artery.

On ***Both Sides*** of the body, the **Subclavian Artery** runs laterally between the clavicle and the first rib. Past the edge of the rib it takes on another name—the **Axillary Artery** 4**.** This runs through the axilla. Past the lower edge of the *Teres major* muscle it takes on yet another name—the **Brachial Artery** 5**.** This runs along the medial side of the arm. These name changes (subclavian—axillary—brachial) simply identify different portions of the same vessel.

Just above the elbow, the brachial artery emerges from underneath the *Biceps brachii* muscle onto the front of the *Brachialis* muscle. It is here that the pulse is easily felt. The artery continues distally into the *Cubital Fossa,* where it divides into two branches just below the elbow.

One branch of the brachial artery is the **Radial Artery** 6**.** It runs along the tendon of the *Brachioradialis* muscle to cross the distal end of the radius. The pulse is most easily felt here, as the artery lies superficially on the lateral edge of the front of the radius. It crosses the wrist through the "anatomical snuff box" and then divides into two main branches. One turns posteriorly to supply the back of the hand. The other, the **Deep Palmar Arch** 7**,** supplies the deep hand muscles.

The other branch of the brachial artery is the **Ulnar Artery** 8**.** It runs medially and deeply, underneath the *Pronator teres* muscle and between the *Flexor carpi ulnaris* and *Flexor digitorum profundus* muscles. It runs with the *ulnar nerve* across the *flexor retinaculum* of the wrist as the **Superficial Palmar Arch** 9 to supply the superficial hand muscles.

There are well-developed anastomoses around the elbow between branches of the brachial artery (profunda brachii and ulnar collaterals) and the radial and ulnar arteries. There are also important anastomoses in the hand between the radial and ulnar arteries.

Identify the aforementioned arteries in figure 3.21.

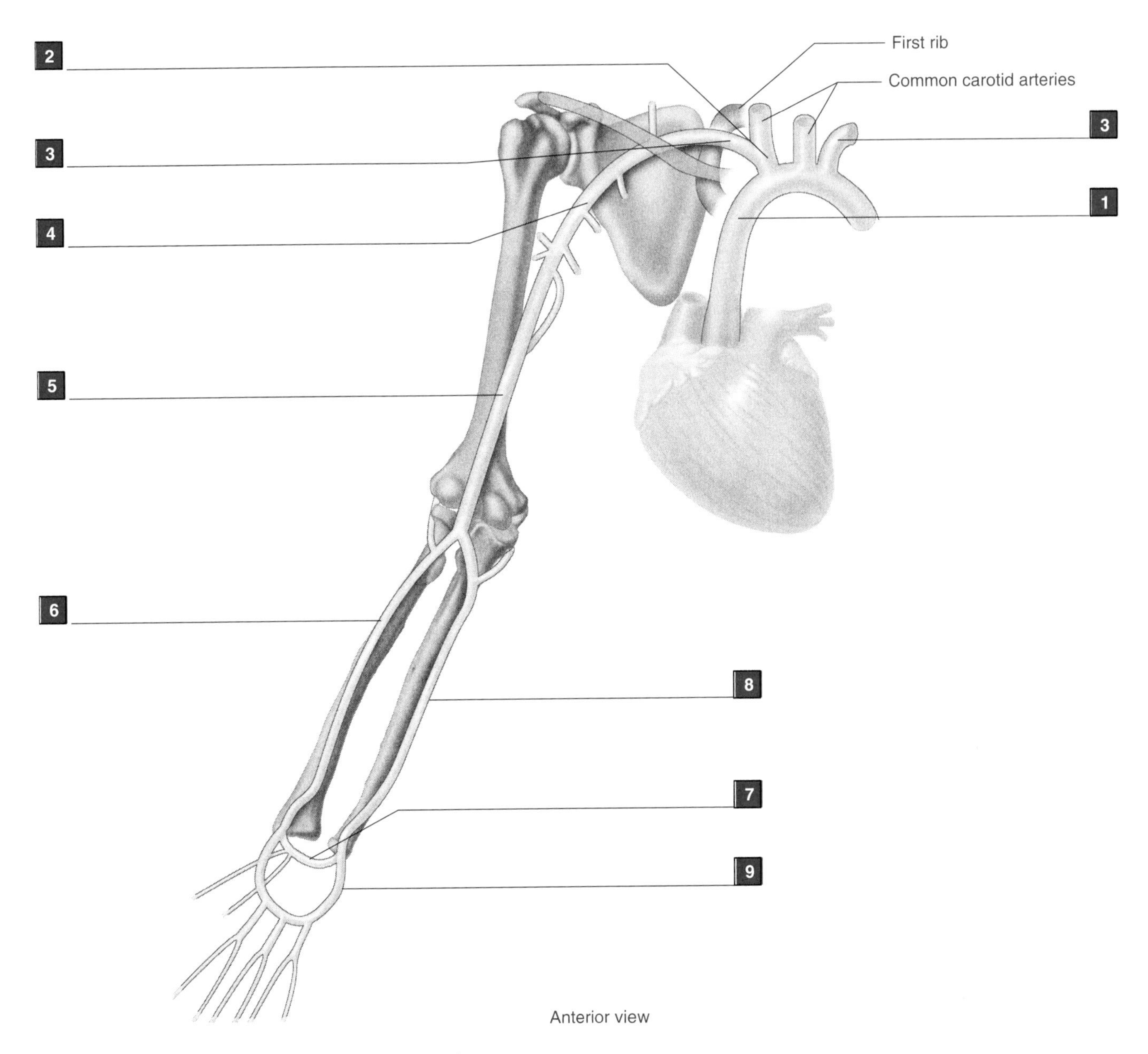

Figure 3.21 **Major Arteries of the Upper Limb**

Veins

Deep Veins

The deep veins of the upper limb run alongside the arteries. These veins are known as ***Venae Comitantes.*** They are named according to the artery that they accompany. Thus, for example, the **Brachial Vein** 1 is the venae comitantes of the brachial artery. The brachial vein is similarly continuous with the **Axillary Vein** 2**.** The axillary vein is likewise continuous with the **Subclavian Vein** 3**.**

Unlike the subclavian arteries, however, the ***Subclavian Veins*** on ***both sides*** of the body are symmetrical in terms of their structure. Thus each subclavian vein joins the roots of the external and internal jugular veins to form the **Brachiocephalic Trunk** 4**.** The left and right brachiocephalic trunks join to form the **Superior Vena Cava** 5**,** which drains directly into the heart.

Superficial Veins

These veins run in the *Superficial Fascia.*

The **Cephalic Vein** 6 drains the back of the hand. It arises on the posterolateral side of the wrist, and then turns to run up the anterior and lateral aspect of the forearm. It crosses the elbow in front of the lateral epicondyle of the humerus and runs upward along the lateral edge of the *Biceps brachii*. It then turns onto the front of the shoulder between the *Deltoid* and *Pectoralis major* muscles. It empties into the axillary vein.

The **Basilic Vein** 7 arises on the medial side of the wrist and runs upward along the *Flexor carpi ulnaris* muscle. It crosses the elbow in front of the medial epicondyle of the humerus, and continues proximally between the *Biceps brachii* and *Brachialis* muscles. It joins with the brachial vein to form the axillary vein.

The **Median Cubital Vein** 8 is a large anastomotic channel between the cephalic and basilic veins. It leaves the cephalic vein just distal to the elbow, crosses the *cubital fossa*, and joins the basilic vein just above the medial epicondyle of the humerus. *It is a favorite vein used to draw blood.*

The **Median Antebrachial Vein** 9 drains the palm of the hand. It arises on the ventral surface of the wrist and runs straight up the forearm to empty into the median cubital vein

Identify the veins of the upper limb in figure 3.22.

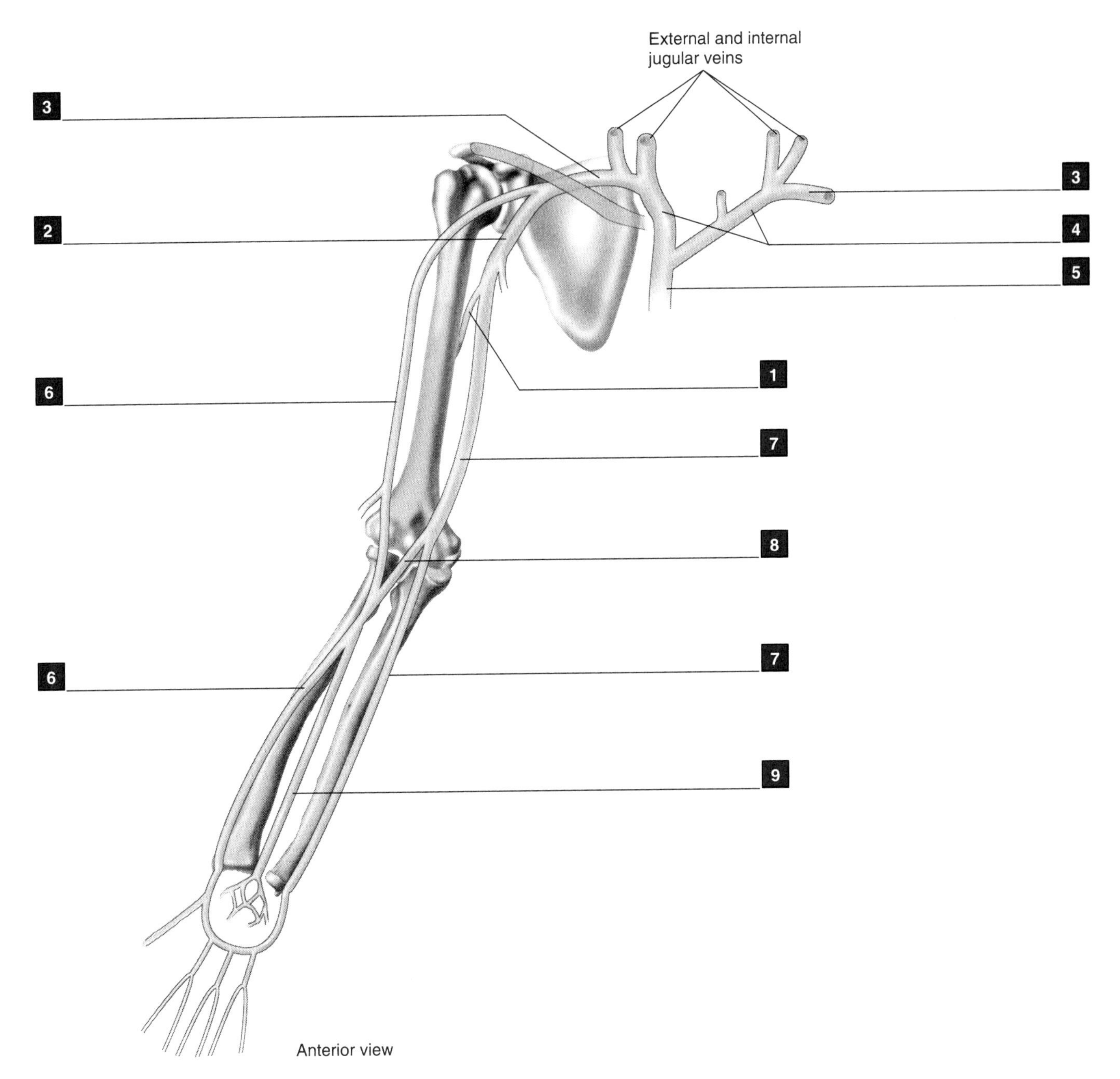

Figure 3.22 **Major Veins of the Upper Limb**

LABORATORY

The Lower Limb

4.1 BONES OF THE LOWER LIMB

The bones of the lower limb comprise two groups: (1) those that form the free part of the limb, and (2) those that attach it to the trunk. The latter form the pelvic girdle.

Pelvic Girdle

The bony pelvis is made up of the **Sacrum and Coccyx** 1 and by the left and right **Os Coxae** 2. The os coxae links the free part of the lower limb and the axial skeleton. We have already studied the sacrum (laboratory 2, p. 28).

Os coxae. This is a large, irregularly shaped bone. It is sometimes (erroneously) referred to as the innominate. Each os coxae consists of three bones that fuse in adulthood. The superior element is the **Ilium** 3. The posterior element is the **Ischium** 4. The anterior element is the **Pubis** 5. Prior to their fusion, these bones are separated by the *Triradiate Cartilage* in the **Acetabulum** 6, or hip socket. The pubis and ischium are separated by cartilage below the large **Obturator Foramen** 7. In life, the obturator foramen is covered by a tendinous membrane.

Posteriorly, each os coxae articulates with the sacrum by the large, ear-shaped **Auricular Surface** 8. This articulation is bounded anteriorly and posteriorly by very strong *Sacroiliac Ligaments*. Anteriorly, the os coxae articulate with each other in the midline at the cartilaginous **Pubic Symphysis** 9.

The superior margin of the os coxae is known as the **Iliac Crest** 10. It projects ventrally as the **Anterior Superior Iliac Spine** 11.

Below the auricular surface, the ischium projects as a prominent **Ischial Spine** 12. Below this, the ischium broadens laterally as the **Ischial Tuberosity** 13.

Between the auricular surface and the ischial spine is the deep **Greater Sciatic Notch** 14. Between the ischial spine and ischial tuberosity is the shallow **Lesser Sciatic Notch** 15. In life, ligaments stretch across the posterior margins of both notches, transforming them into foramina.

The superior ramus of the pubis forms a sharp crest (the *Pecten Pubis*) that runs posteriorly from near the symphysis to the ilium. The line of the crest continues on the inner surface of the ilium as a ridge, known as the *Arcuate Line*, to the sacroiliac joint. The pecten pubis and arcuate line separate the pelvis into two parts: the so-called *False Pelvis* above, and the *True Pelvis*, or *Obstetrical Pelvis* below.

Identify the aforementioned features of the pelvic girdle in figure 4.1.

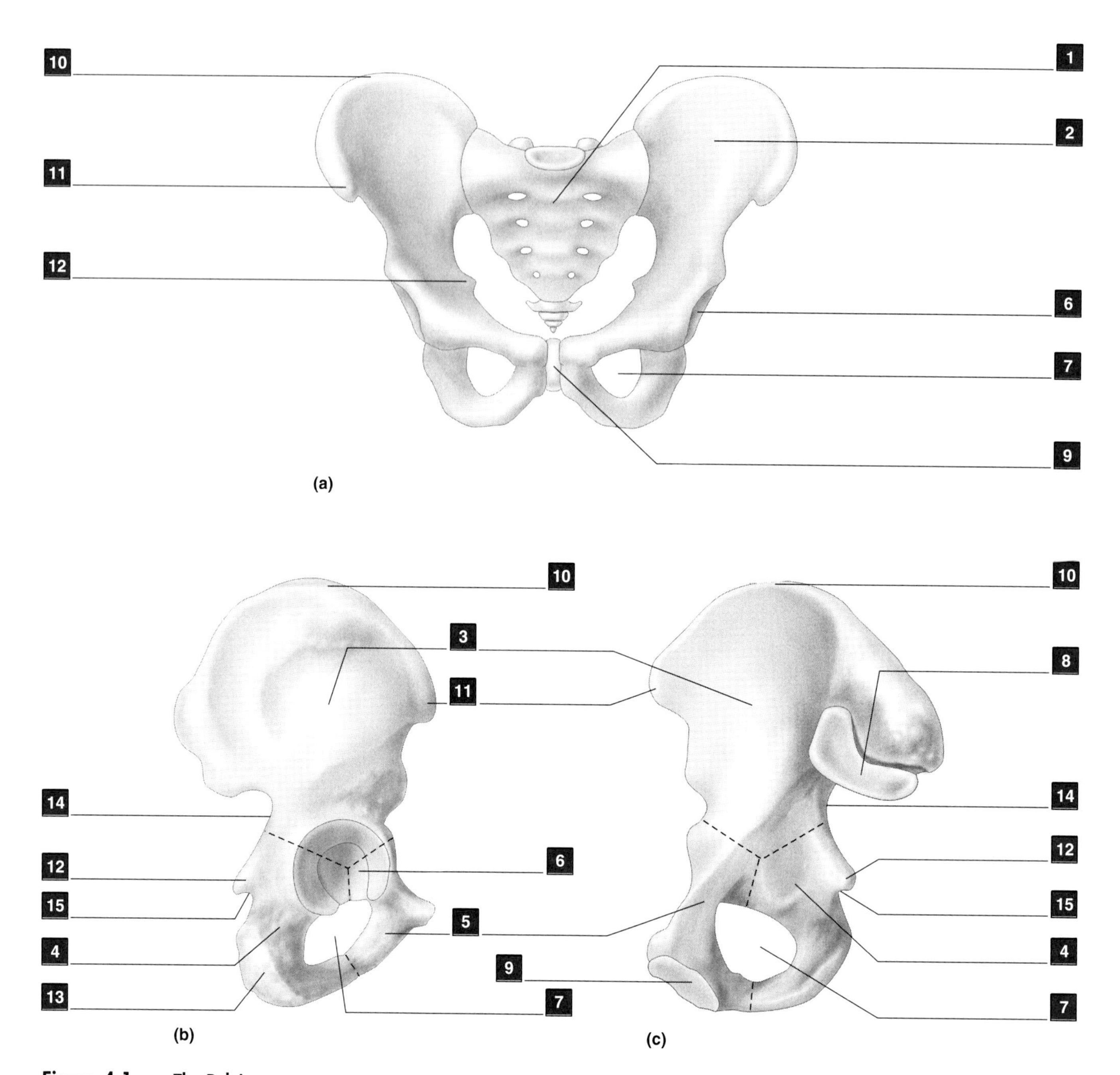

Figure 4.1 **The Pelvis**
(a) anterior view of pelvis; (b) lateral view of right os coxae; (c) medial view of right os coxae.

Thigh

The thigh has a single bone: the femur. Suspended in a tendon between the front of the thigh and leg is a small bone, the patella.

Femur. At the proximal end is a hemispherical **Head** 1 that articulates in the acetabulum of the os coxae. The head is connected to the shaft by a long **Neck** 2. At the top of the junction of the neck and shaft is the **Greater Trochanter** 3; it has a deep fossa on its medial side. The **Lesser Trochanter** 4 projects medially from the back of the shaft where it meets the neck. The trochanters are connected on the back of the shaft by a ridge, the **Intertrochanteric Crest** 5. Lateral to the lesser trochanter is a roughened area, the *Gluteal Tuberosity*, which continues down the back of the shaft as a prominent ridge, the **Linea Aspera** 6. The medial and lateral margins of the linea aspera diverge distally around the slightly concave *Popliteal Surface*. The distal end of the femur has two articular surfaces—the **Lateral Condyle** 7 and **Medial Condyle** 8—that articulate with the proximal end of the tibia. The condyles are separated posteriorly; anteriorly they merge to form a spool-shaped **Patellar Surface** 9 over which the patella glides. The distal epiphysis is expanded above the articular surface to form **Lateral** 10 and **Medial Epicondyles** 11.

Patella. This triangular sesamoid bone forms in the tendon of the *Quadriceps femoris* muscle. Most of its posterior surface articulates with the patellar surface of the femur.

Identify the aforementioned features of the femur in figure 4.2.

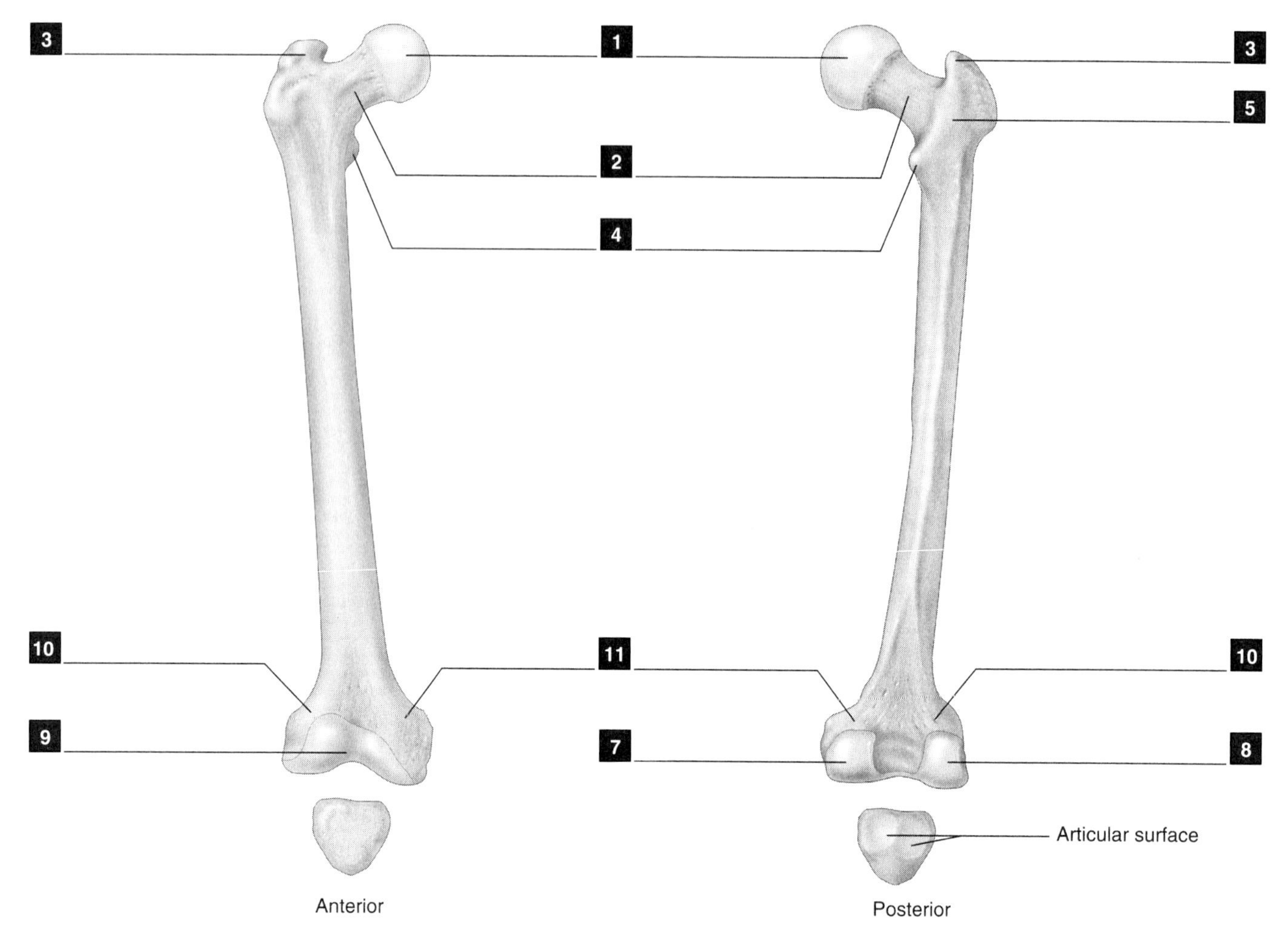

Figure 4.2 **Femur and Patella**
anterior and posterior views of right femur and patella

Leg

The leg comprises two bones: the tibia (medial) and the fibula (lateral).

Tibia. The proximal end is expanded to form the *Tibial Plateau,* which articulates with femur by separate Medial 1 and Lateral Condyles 2. They are separated in the center of the plateau by the Intercondylar Eminence 3. The shaft is triangular in cross section; its anterior edge (the "shin") lies just below the skin. There is a roughened *tuberosity* on the front of the shaft just below the plateau where the tendon of the *Quadriceps femoris* muscle attaches. The medial side of the distal end of the bone projects as the Medial Malleolus 4. The talus articulates with the distal surface of the tibia, and with the lateral surface of the medial malleolus.

Fibula. It has a slightly expanded proximal end, the **Head 5,** which articulates with a facet on the tibia just below its lateral condyle. The shaft is very slender, because the bone bears practically no weight. The slightly enlarged distal end is held tightly against a notch in the tibia by ligaments; it forms the **Lateral Malleolus 6.** Its medial surface articulates with the talus.

Identify the aforementioned structures of the tibia and fibula in figure 4.3.

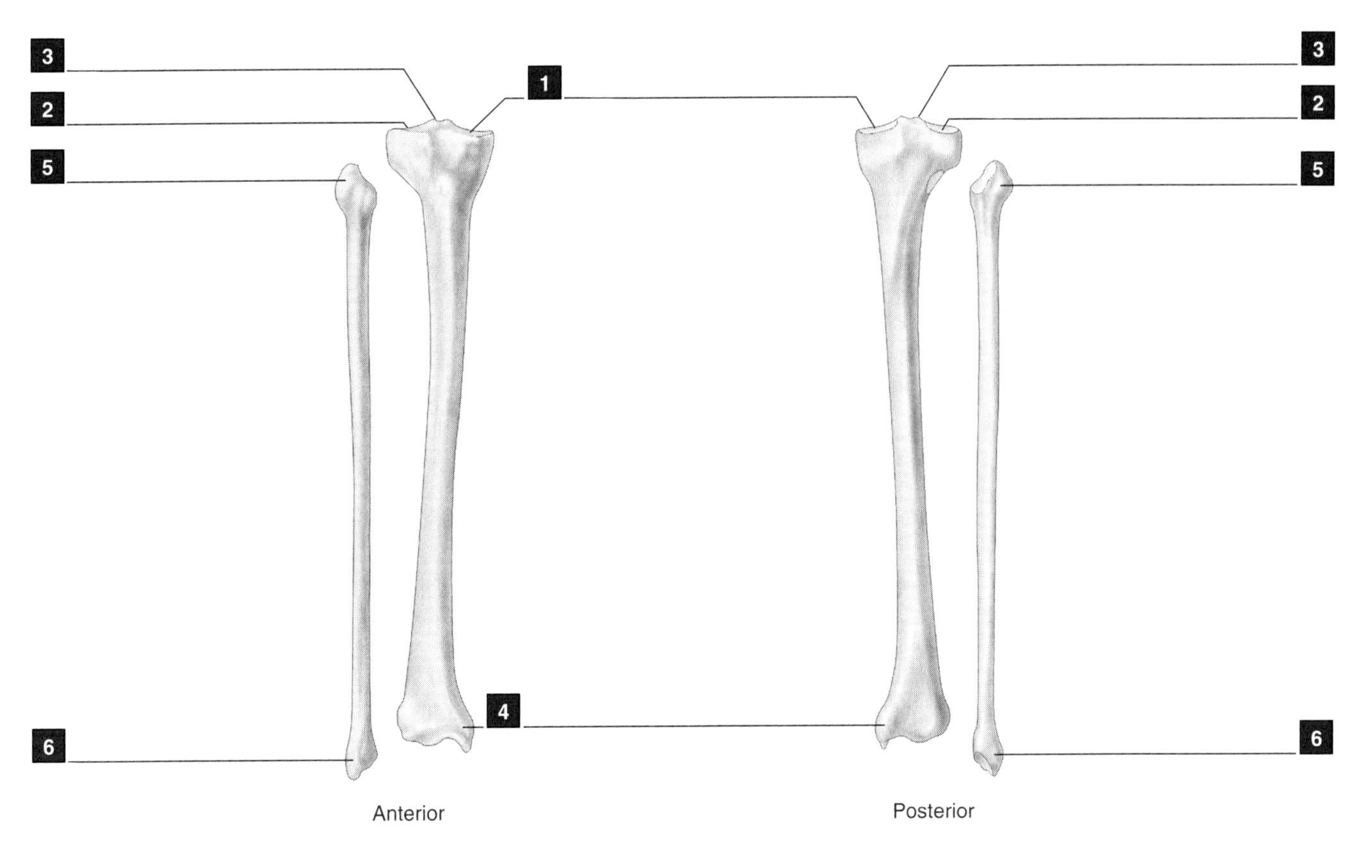

Figure 4.3 **Tibia and Fibula**
anterior and posterior views of right tibia and fibula

Foot

The foot contains a total of 26 individual bones. These are divided into three groups: *Tarsals* (7 bones), *Metatarsals* (5 bones), and *Phalanges* (14 bones).

Seven irregularly shaped **Tarsal Bones** 1 form the ankle, heel and the back of the sole of the foot. They are arranged into a proximal group and a distal group.

The three *proximal* bones are: talus, calcaneus, and navicular. The talus and calcaneus are the largest and most irregularly shaped of the tarsal bones.

The **Talus** 2 has a spool-shaped superior surface that articulates with the distal end of the tibia and the medial and lateral malleoli of the tibia and fibula. This (*Ankle*) joint permits the foot to be plantarflexed and dorsiflexed. The talus sits atop the calcaneus, and articulates anteriorly with the navicular.

The **Calcaneus** 3 articulates superiorly with the talus. This (*Subtalar*) joint allows the foot to be inverted and everted. It articulates anteriorly with the cuboid bone. The posterior half of the calcaneus is expanded into a tuberosity (the heel) that makes initial contact with the ground (heel strike) during walking and running. It also serves as the attachment of the large calf muscles by the *Achilles tendon*.

The **Navicular** 4 is an elliptical bone. It has a concave proximal surface that articulates with the talus, and a convex distal surface that articulates with each of the three cuneiform bones.

The four *distal* bones are the **Cuboid** 5 and three (*medial, intermediate* and *lateral*) **Cuneiforms** 6**.** They form a row that articulates with the metatarsals.

Five **Metatarsals** 7 form most of the sole. The first element of each ray (numbered I–V from the big toe to the little toe) is its metatarsal. The proximal end is expanded to form a *Base* that articulates with one or more of the tarsal bones. The distal end forms a rounded *Head* that articulates with a proximal phalanx. The first (big toe) metatarsal is very robust; the other four are long and slender.

Fourteen **Phalanges** 8 constitute the *Toes*. Each of the four lateral toes has a **Proximal Phalanx** 9**, Middle Phalanx** 10**,** and **Distal Phalanx** 11**.** The *Big Toe* (= *Hallux*) has only a **Proximal Phalanx** and a **Distal Phalanx.** The hallucial phalanges are notably the most robust, because the big toe is a very important weight-bearing element during walking and running.

Identify the bones of the foot in figure 4.4.

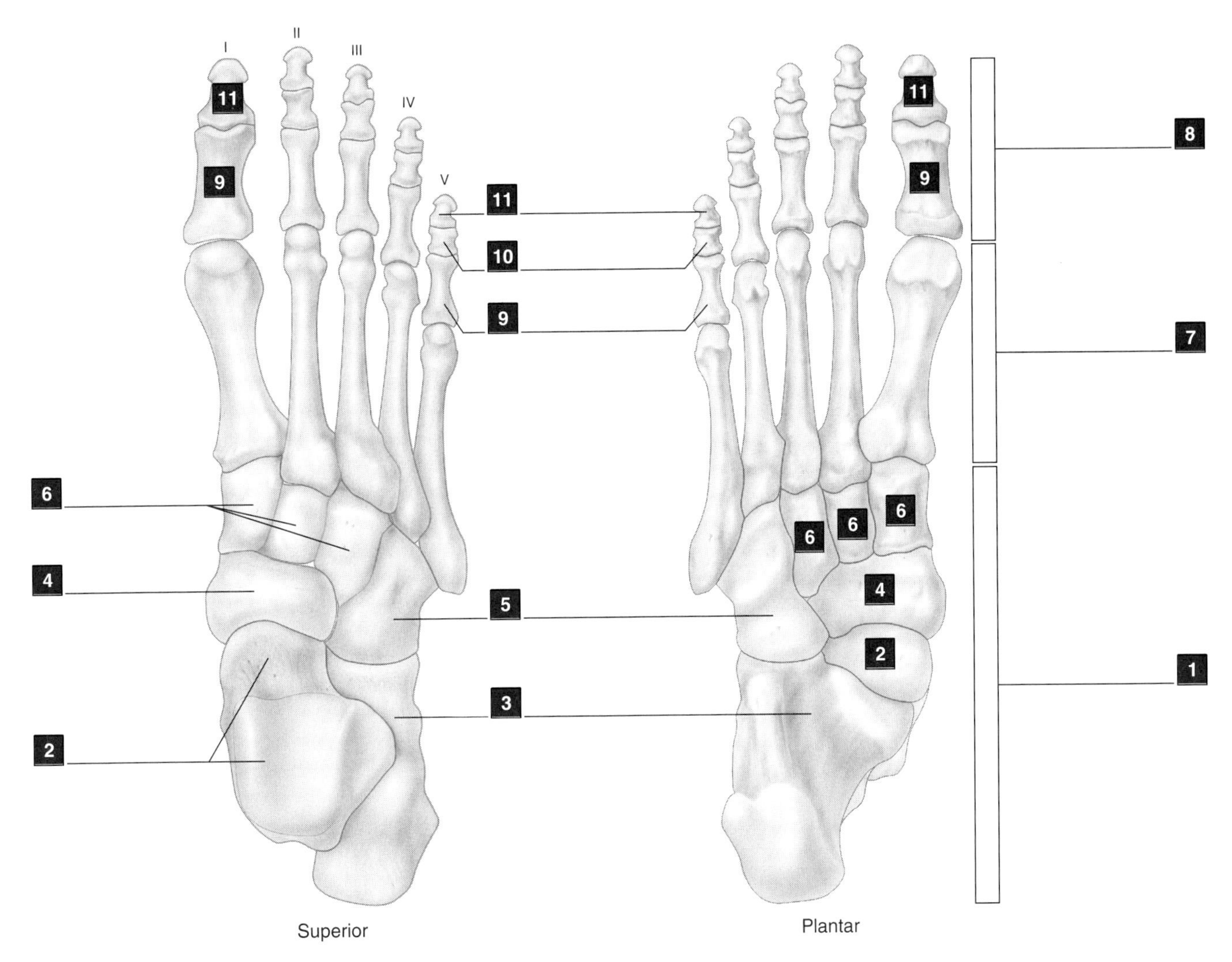

Figure 4.4 **Foot Bones**
superior and plantar (inferior) views of right foot

The Arch of the Foot

A feature that is peculiar to humans is a foot with a plantar arch. An arched foot was present in our ancestral lineage at least 3.7 million years ago. We know this because humanlike footprints are preserved in a volcanic ash layer in Tanzania.

The arch resists the pressures of deformation that are generated during walking and running. In the latter part of the support phase of locomotion, the body's weight is borne by the front part of the foot. This would cause the foot to dorsiflex were it not for the arch.

A prominent ***Longitudinal Arch*** runs along the medial side of the foot between the tuberosity of the calcaneus (the heel) and the metatarsal-phalangeal joint (the ball). A footprint emphasizes this observation. There is also a transverse arch, but this is largely a by-product of the fact that the longitudinal arch is higher medially than laterally.

The arch is maintained by (1) the shapes and fit of the bones, (2) the ligaments that bind the tarsal bones, and (3) the plantar aponeurosis. Some of the intrinsic muscles of the foot also provide some support for the arch.

1. *Bones*. When the foot is examined from its medial side, the bones are seen to form an arch. The **Talus** 1 is its keystone. It is wedged between the **Calcaneus** 2 and the **Navicular** 3. The back of the arch is formed by the calcaneus. The front of the arch is formed by the navicular, one of the three cuneiforms (principally the medial), and the more medial metatarsals (principally the first).

2. *Ligaments*. The body's weight is transmitted directly to the talus. This would cause the arch to collapse were it not for the fact that the tarsals are bound together by a number of strong ligaments. These provide a relatively rigid structure that is also capable of some degree of movement. The most important of these is the **Spring Ligament 4.** It arises from the edge of a medially projecting shelf of bone from the calcaneus (the *Sustentaculum tali*), runs underneath the talus, and inserts onto the navicular.

3. *Plantar Aponeurosis*. Just deep to the superficial fascia of the sole of the foot is the thick, fibrous **Plantar Aponeurosis 5.** It arises from the tuberosity of the calcaneus, and fans out to insert onto the base of the proximal phalanx of each toe. The aponeurosis is tightened as the toes are dorsiflexed, and this causes the arch to be maintained.

The *Abductor hallucis* muscle runs from the tuberosity of the calcaneus to the base of the proximal phalanx of the big toe. It also helps sustain the medial longitudinal arch during locomotion. We will study it a bit later in this lab (p. 125).

Identify the features related to the arch of the foot in figure 4.5.

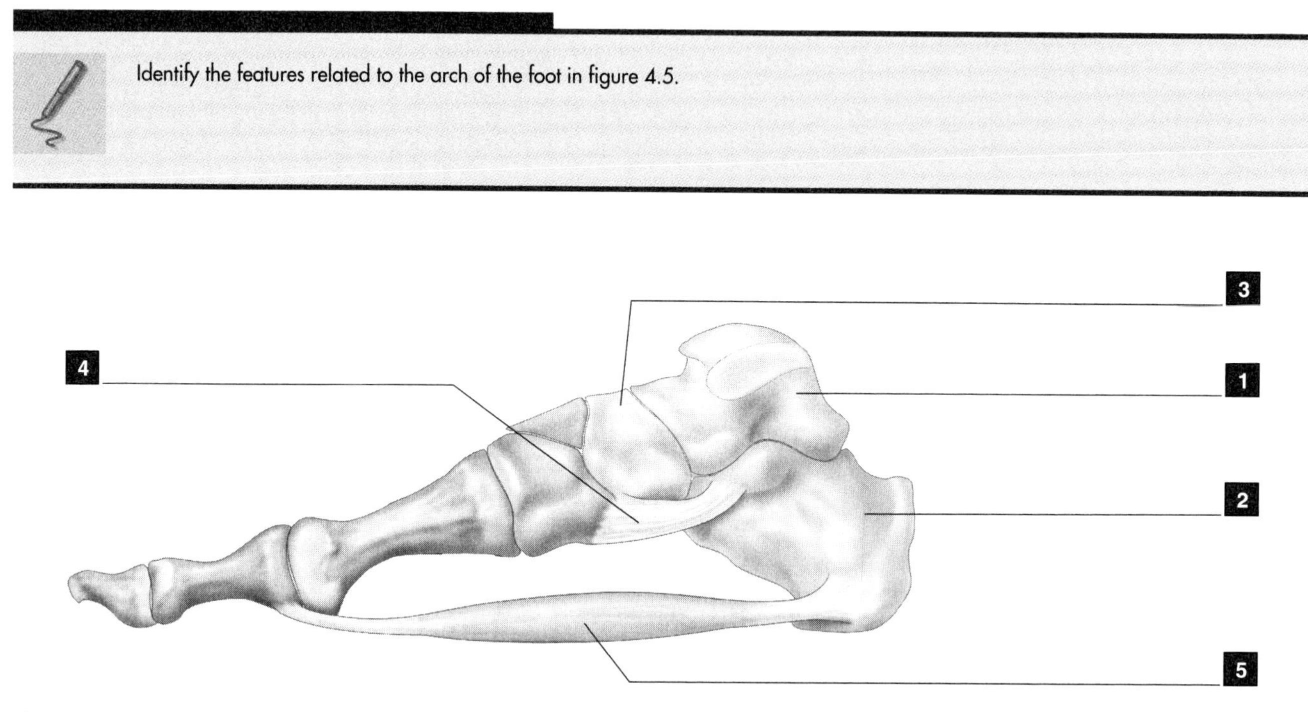

Figure 4.5 Medial View of Pedal Skeleton

4.2 JOINTS OF THE LOWER LIMB

In this section we will study the anatomy of the three lower limb joints: the hip, the knee, and the ankle.

Hip Joint

The deeply cupped *Acetabulum* of the os coxae and the hemispherical *Head of the Femur* form a snugly stable ball-and-socket joint.

The articular surface of the acetabulum is a horseshoe-shaped surface known as the **Lunate Surface** **1**. It is open inferiorly (because it bears no weight here), but its two horns are connected by the **Transverse Acetabular Ligament** **2**. This completes the socket, which is deepened further by a fibrocartilaginous **Acetabular Labrum** **3**.

Almost the entire head of the femur is covered by an articular surface except for a small pit known as the **Fovea Capitis** **4** through which blood vessels enter the head. A short ligament fans out from this pit to attach to the horns of the lunate surface and the transverse acetabular ligament. This is known as the **Ligament of the Head of the Femur (= Ligamentum Teres)** **5**. It helps secure the stability of the joint. The ligament of the head of the femur is covered by **Synovial Membrane** **6**, and runs through the synovial cavity of the hip joint.

The **Joint Capsule** **7** is attached medially to the acetabular labrum, and laterally to the base of the femoral neck. The fibrous part of the capsule is very thick over most of its coverage. Because its fibers have different orientations, *three separate ligaments* can be identified.

The first is the **Iliofemoral Ligament** **8**. It arises from the anterior edge of the ilium between the acetabulum and anterior inferior iliac spine, and fans out to insert along a line on the front of the femur between the greater and lesser trochanters. This very strong ligament prevents overextension of the lower limb during standing.

The second is the **Ischiofemoral Ligament** **9**. It arises from the back of the acetabulum adjacent to the ischial tuberosity, and runs laterally to attach onto the back of the femoral neck. This ligament also limits extension of the lower limb.

The third is the **Pubofemoral Ligament** **10**. It arises from the superior pubic ramus, and fans out laterally to insert along the lower part of the intertrochanteric line on the front of the femoral neck. This weak ligament probably limits abduction of the lower limb.

Identify the aforementioned features of the hip joint in figure 4.6.

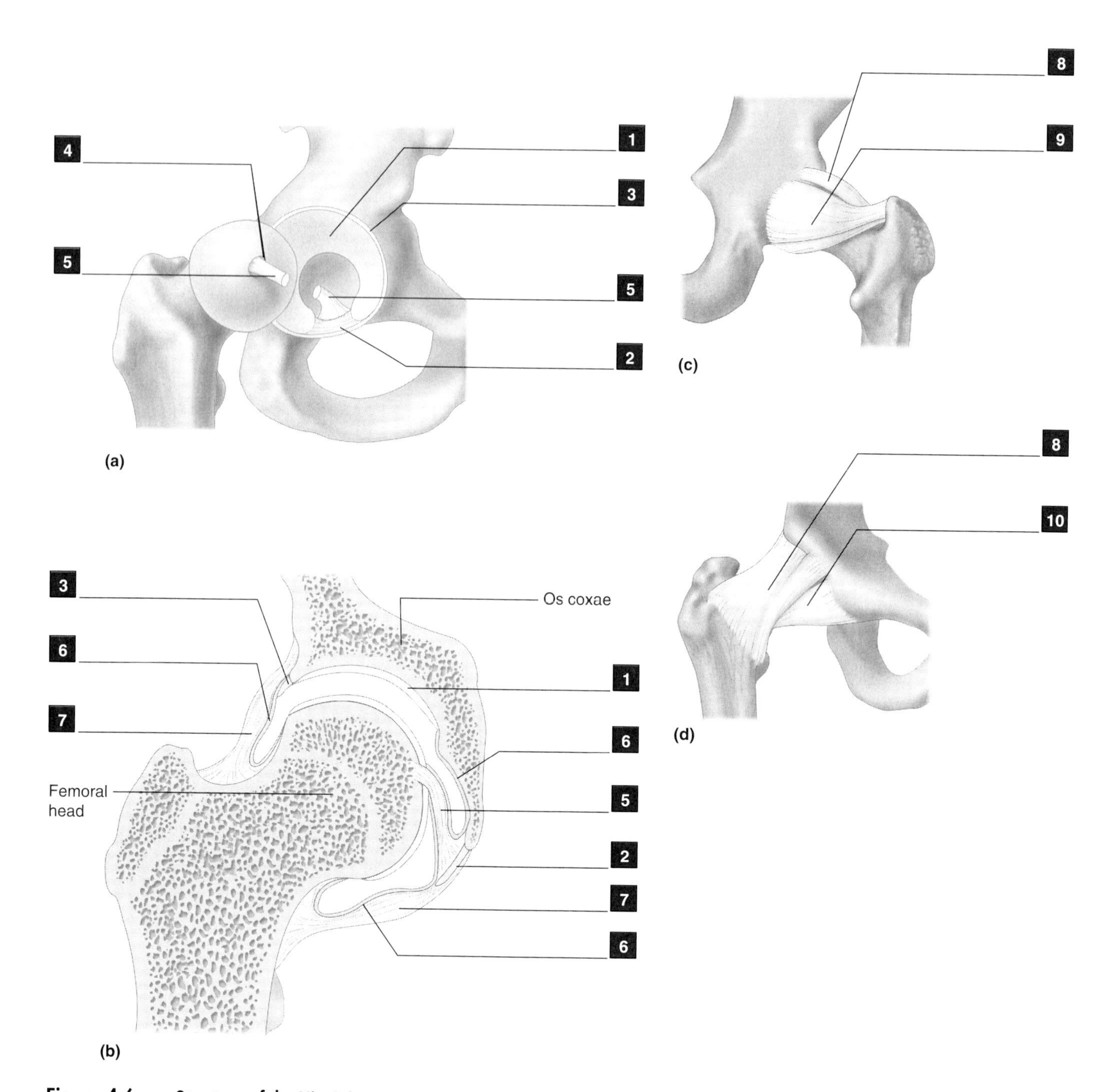

Figure 4.6 **Structure of the Hip Joint**
(a) lateral view with femur rotated out of joint; (b) coronal section; (c) posterior view; (d) anterior view.

Knee Joint

The knee joint is the most complex in the body. It involves two articulations:

1. A uniaxial hinge between the femur and tibia
2. A gliding plane between the femur and patella

Both are contained within a single synovial cavity. Within the joint capsule there are fibrocartilage disks and ligaments between the femur and tibia.

Articular cartilage covers the **Femoral Condyles** 1, the **Tibial Condyles** 2, and the posterior surface of the **Patella** 3.

The flat articular surfaces of the tibia do not match the rounded condyles of the femur. In order to overcome this mismatch, a C-shaped fibrocartilage disc—called a ***Meniscus***—is placed atop each tibial condyle. Thus, there is a **Medial Meniscus** 4 and a **Lateral Meniscus** 5. They are located within the joint capsule.

The **Fibrous Joint Capsule** 6 attaches to the sides and back of the articular surfaces of the femur and tibia. The **Synovial Membrane** 7 lines the fibrous capsule on the sides and back of the joint between the edges of the menisci and the opposing articular surfaces. The fibrous joint capsule is absent from the front of the joint; this permits the synovial membrane to bulge upward beneath the tendon of the *Quadriceps femoris* muscle. This synovial pouch is the **Suprapatellar Bursa** 8.

Because the bony articular surfaces cannot limit movement of the joint, it has strong ligaments both inside and outside the joint capsule that stabilize the knee.

Intracapsular (Cruciate) Ligaments. The **Posterior Cruciate Ligament** 9 runs forward from the back of the intercondylar eminence of the tibia to insert into the inner surface of the medial femoral condyle. The **Anterior Cruciate Ligament** 10 runs backward from the front of the intercondylar eminence of the tibia to insert into the inner surface of the lateral femoral condyle. They get the name cruciate because they cross each other in the center of the joint.

Extracapsular (Collateral) Ligaments. The **Patellar Ligament** 11 runs from the bottom of the patella to the tibial tuberosity; it is a continuation of the Tendon of the **Quadriceps Femoris Muscle** 12. It is separated from the synovial membrane of the joint by a pad of fat. The **Medial (Tibial) Collateral Ligament** 13 runs along the capsule from the medial epicondyle of the femur to the tibia. It is attached also to the edge of the medial meniscus. The **Lateral (Fibular) Collateral Ligament** 14 runs across the capsule from the lateral epicondyle of the femur to the head of the fibula.

Identify these features of the knee joint in figure 4.7.

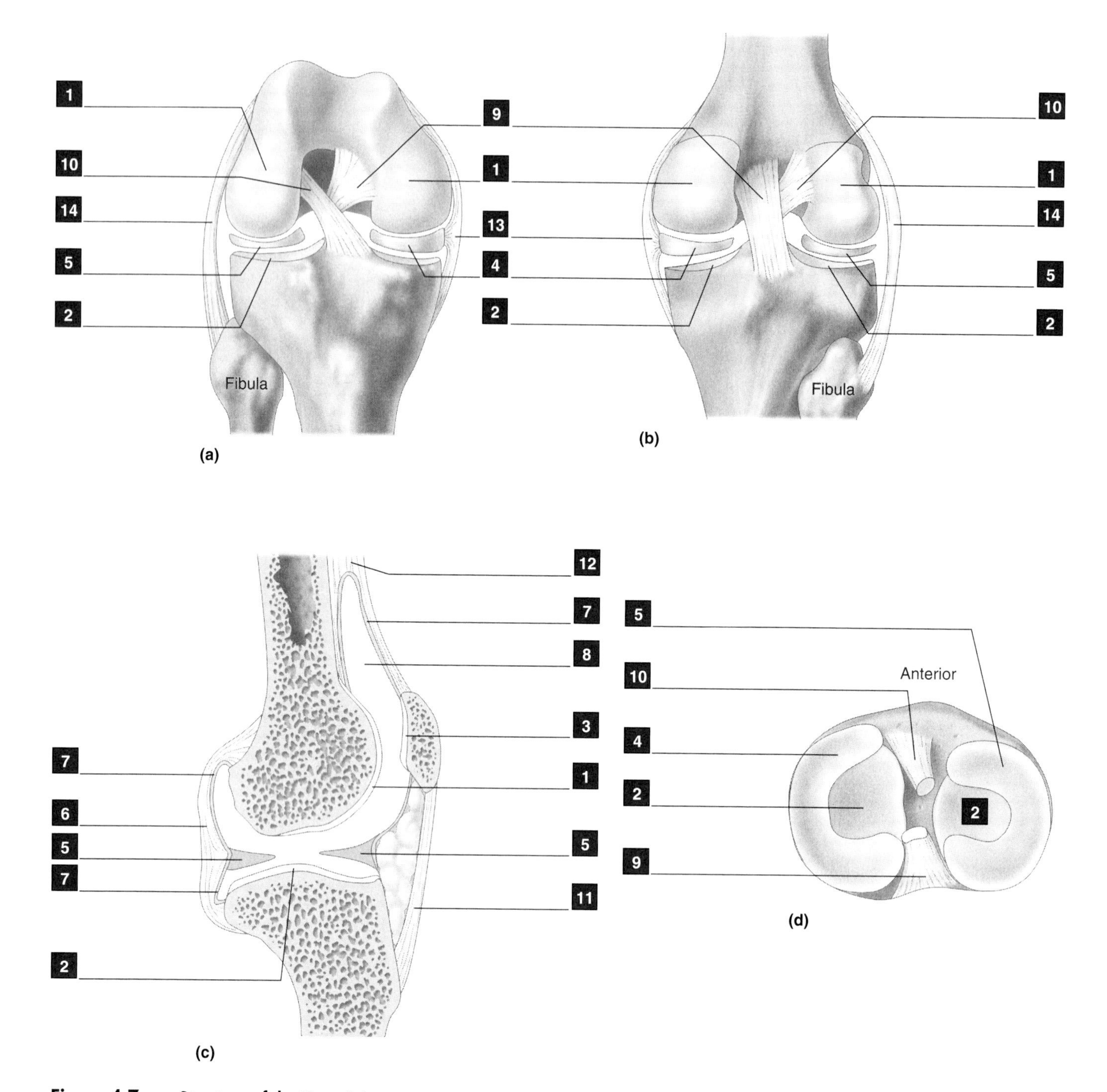

Figure 4.7 **Structure of the Knee Joint**
(a) anterior view of right knee in flexion; (b) posterior view of right knee in extension; (c) sagittal section (lateral to midline); (d) superior view of right tibial plateau.

Ankle Joint

Strictly speaking, the ankle joint involves only the articulation between the talus and the tibia and fibula. It permits dorsiflexion and plantarflexion of the foot. Other movements of the foot—eversion, inversion and rotation—are provided by joints between the tarsal bones. Because these intertarsal joints are involved together with the ankle in the principal movements of the foot, they will be considered together under the rubric of the "ankle." Both sets of joints are centered around the talus as the keystone of movement.

The **Talocrural (= Ankle) Joint** 1 involves the articulation of the body of the **Talus** 2 with the tibia and fibula. In dorsiflexion and plantarflexion, the superior surface of the talus glides over the distal surface of the tibia. The medial and lateral sides of the body of the talus articulate with the **Medial malleolus** 3 and the **Lateral malleolus** 4, which prevent side-to-side movement and rotation of the talus.

The talocrural joint has a single joint capsule. It is stabilized by ligaments that prevent the talus from moving anteriorly and posteriorly, and from being abducted or adducted. Three ligaments fan out from the medial malleolus to insert onto the talus and **Calcaneus** 5. This fan is known as the **Deltoid ligament** 6. Three ligaments also radiate from the lateral malleolus to insert onto the talus and calcaneus. These three—*Anterior talofibular*, *Calcaneofibular*, and *Posterior talofibular*—form what is referred to as the **Lateral collateral ligament** 7.

The ankle is the most frequently injured joint of the body. The lateral collateral ligament is commonly damaged because it is much weaker than the deltoid ligament. A sprained ankle results from twisting the talus within the tibiofibular mortise, and it nearly always involves inversion, with consequent damage to the weaker lateral collateral ligament.

The **Intertarsal Joints** involve articulations between the talus and calcaneus, and between them and the **Navicular** 8 and **Cuboid** 9. Two articulations (anterior and posterior) between the talus and calcaneus comprise the **Subtalar Joint** 10. Articulations between the talus and navicular and between the calcaneus and cuboid constitute the **Transverse Tarsal Joint** 11. The anterior subtalar joint relates the talus to both the calcaneus and navicular in a single capsule; thus, it is part of both the subtalar and transverse tarsal joints.

The subtalar and transverse tarsal joints are reinforced by a series of ligaments. The two most significant of these run on the plantar aspect between the calcaneus and cuboid (laterally) and between the calcaneus and navicular (medially). The latter is known as the ***Spring Ligament***—its significance was noted earlier in the discussion of the arch of the foot (p. 92).

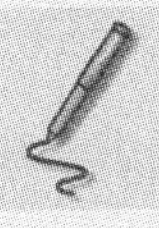

Identify the aforementioned structures of the ankle joint in figure 4.8.

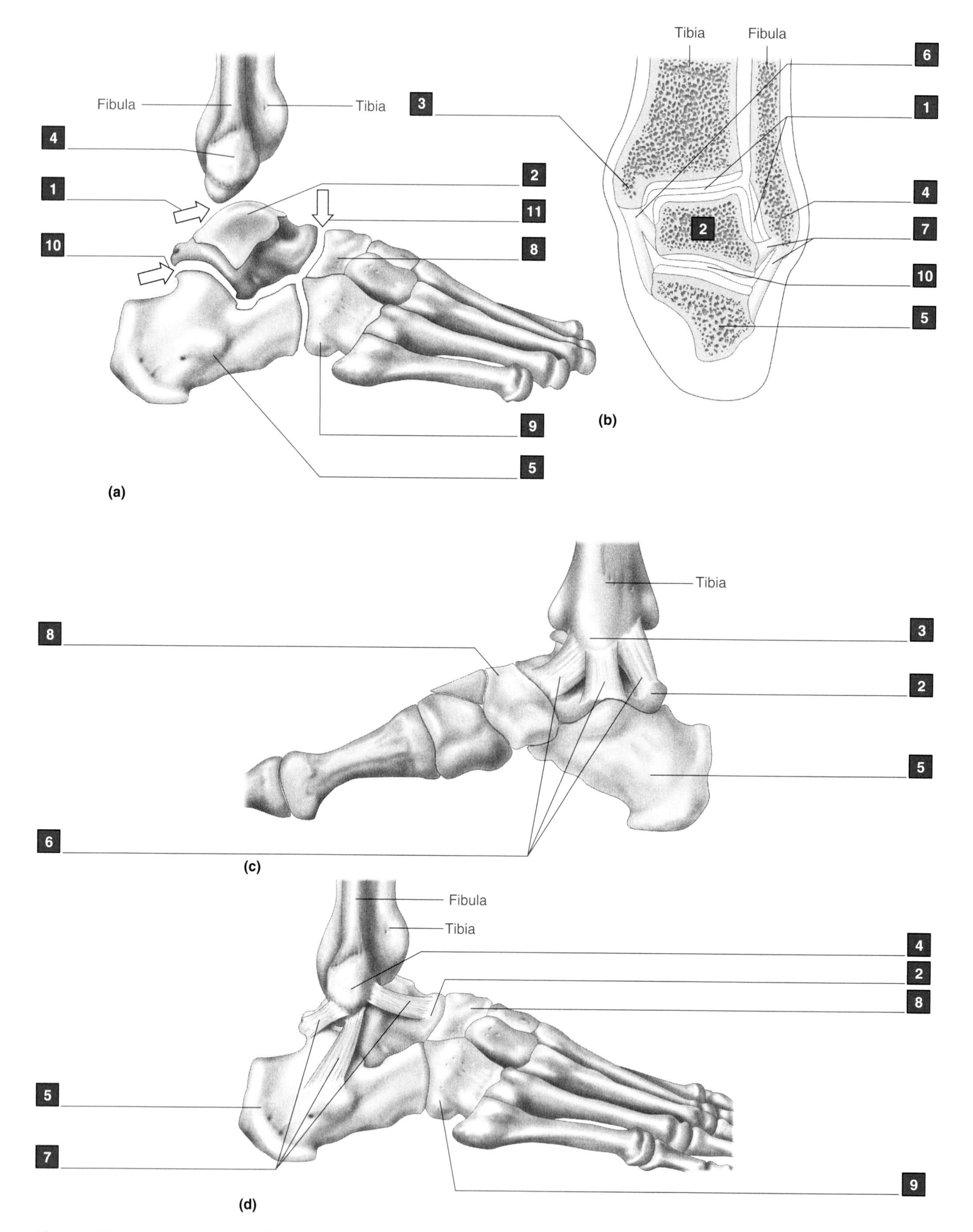

Figure 4.8 **Principal Joints of the "Ankle"**
(a) lateral view of right foot with the three joints opened; (b) coronal section through right foot; (c) medial view of right foot; (d) lateral view of right foot.

4.3 INNERVATION OF THE LOWER LIMB

There is a pattern to the innervation of the lower limb muscles. The key to remembering the nerves that supply the muscles is to appreciate this developmental pattern.

The muscles in the thigh, leg, and foot occupy one of two compartments. One is dorsal; the other is ventral. From your experience with the upper limb, and your knowledge of anatomical terminology, you might expect that the dorsal compartment would occupy the posterior aspect, and that the ventral compartment would be on the anterior side of the lower limb. This is not the case! Rather, the situation is exactly the reverse of what you might expect.

In the lower limb, the ***Dorsal Division Muscles*** occupy the anterior and lateral aspects of the thigh and leg; the ***Ventral Division Muscles*** occupy the posterior and medial aspects of the thigh and leg.

These compartments are illustrated in figure 4.9. Color the *ventral* compartments *blue*, and the *dorsal* compartments *red*.

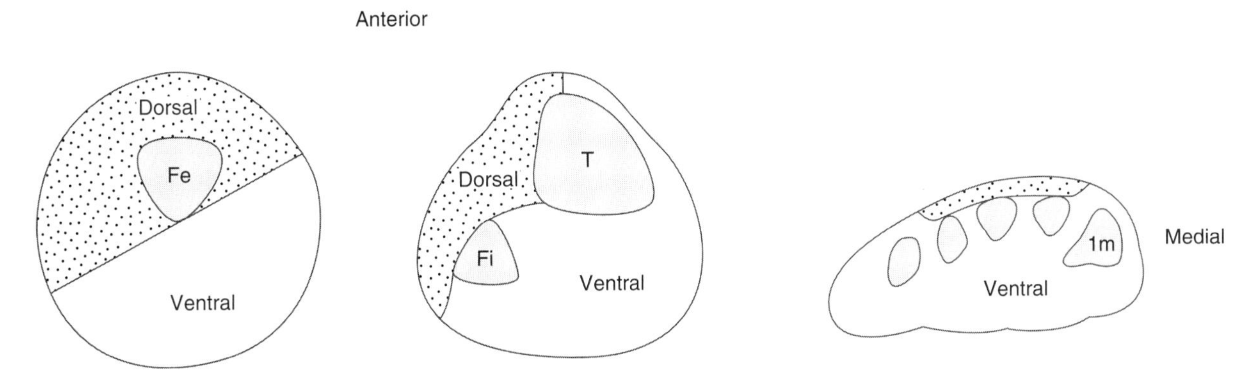

Figure 4.9 **Muscle Compartments of the Lower Limb**
Schematic transverse sections through a left thigh, leg, and foot.
(Fe = femur, Fi = fibula, T = tibia, 1m = first metatarsal)

How does this seemingly bizarre situation come to be? The explanation lies in the fact that the lower limb undergoes rotation and repositioning during embryonic development. This mimics the changes that took place during the course of evolution.

In the exercise that follows, we will relive our embryonic development (and recapitulate our evolutionary heritage) to better understand the pattern of lower limb innervation. The rotation and repositioning of the lower limb is essentially a four-stage process. Follow this by performing each exercise.

Rotation and Repositioning of the Lower Limb

Stage One

Hold your trunk horizontal and stick your lower limb out laterally. Your thigh should be horizontal, your leg vertical, and your foot horizontal.

The muscles that run across the top of your thigh, the laterally facing surface of your leg, and the top of your foot are **Dorsal Division Muscles** (*color the stippled cross section red*).

The muscles that run across the bottom of your thigh, the medially facing surface of your leg, and the bottom of your foot are **Ventral Division Muscles** (*color the open cross section blue*).

Now, rotate your lower limb 90 degrees so that your leg lies parallel to your backbone. This is accomplished by simply raising your trunk to a vertical position while holding your lower limb in its place. This brings you to stage 2.

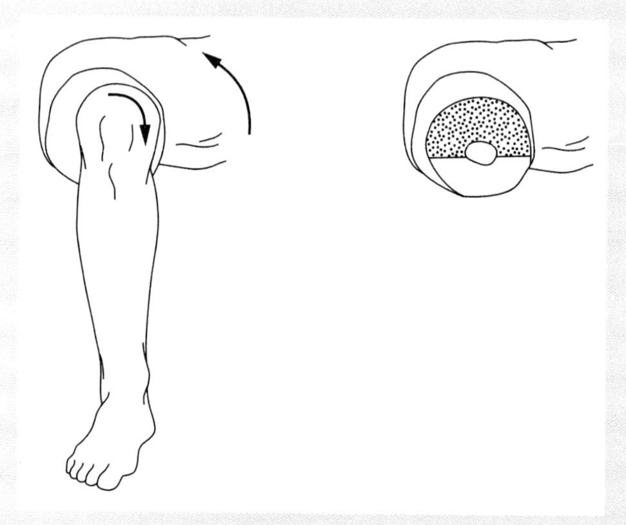

Stage Two

Your trunk is vertical. As before, your lower limb sticks out laterally. Your thigh is horizontal, your leg vertical, and your foot horizontal. (*Again, color the stippled cross section red, and the open cross section blue.*)

Now, bring your knee around toward the median sagittal plane, holding the thigh and leg in their flexed positions. This brings you to stage 3.

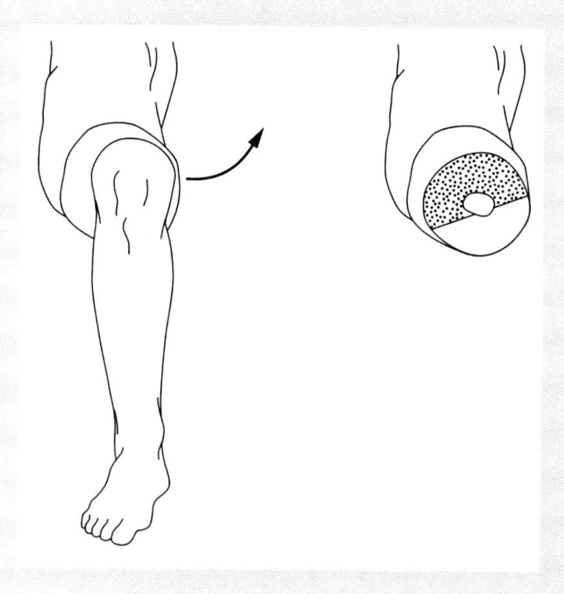

Stage Three

Your trunk is still vertical, but now your lower limb is flexed, and your knee and toes point forward. (*Again, color the stippled cross section red, and the open cross section blue.*)

Now, simply extend your thigh and leg. This brings you to stage 4.

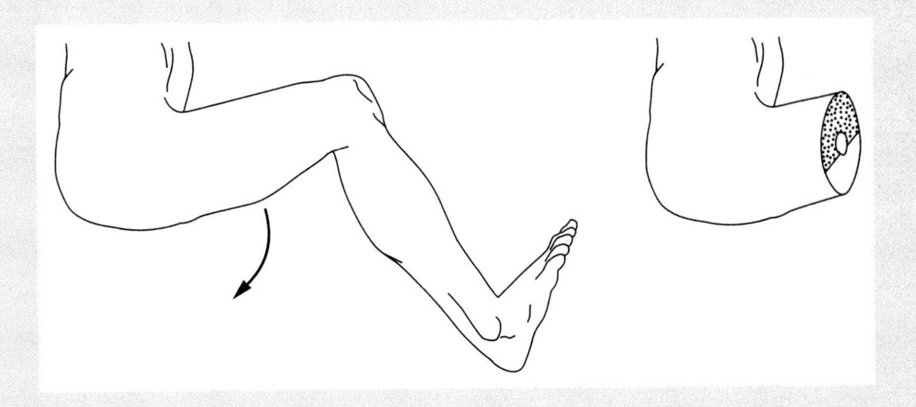

Stage Four

This is the Anatomical Position. (*Again, color the stippled cross sections red, and the open cross sections blue.*)

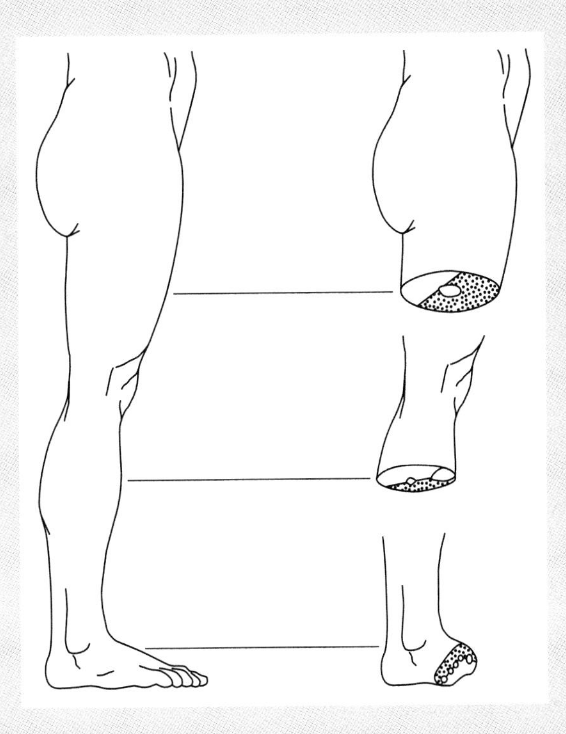

Note how the **Dorsal Division Muscles** that arise from the ilium and run onto what was originally the dorsal surface of your thigh have come to occupy its **anterior surface.** Note how the muscles that run from the dorsal surface of the thigh onto the lateral aspect of your leg now occupy its **anterior surface.** Note that the muscles that run from the lateral surface of the leg onto the top of the foot still occupy its **dorsal surface.**

Observe the changes to the **Ventral Division Muscles.**

The Lumbosacral Plexus

The muscles of the lower limb are served by the ventral rami of seven spinal nerves (*L2–S3*) via the lumbosacral plexus. The lumbosacral plexus actually has nine roots, but spinal nerves T12 and L1 do not provide fibers to lower limb muscles. Recall that a peripheral nerve that emerges from a plexus has axons from more than one ventral ramus.

Each ventral ramus carries fibers that will serve a ventral compartment muscle, and fibers that will serve a dorsal compartment muscle.

Two nearly separate entities, known as the ***Lumbar Plexus*** (T12–L4), and the ***Sacral Plexus*** (L5–S3), are connected by a single twig between L4 and L5, the ***Lumbosacral Trunk.***

The lumbosacral plexus differs from the brachial plexus in that there are *no trunks or cords*. Instead, the ventral and dorsal compartment fibers of each root split off from one another and then combine directly to form the peripheral nerves.

Peripheral Nerves

Ventral compartment fibers from L2–L4 form:

Obturator nerve

Dorsal compartment fibers from L2–L4 form:

Femoral nerve

These two nerves serve only hip and thigh muscles.

Ventral compartment fibers from L4–S3 form:

Tibial nerve
Nerve to Obturator Internus
Nerve to Quadratus Femoris

Dorsal compartment fibers from L4–S3 form:

Common Fibular nerve*
Superior Gluteal nerve
Inferior Gluteal nerve

These six nerves serve hip and thigh muscles, and all muscles below the knee.

The ***Tibial nerve*** and ***Common Fibular nerve*** run together in a single sheath through much of the thigh as the **Sciatic Nerve.** Thus, the sciatic nerve carries both ventral and dorsal compartment fibers.

* Note that older anatomy texts refer to the *Peroneal* nerve (i.e., Common Peroneal and its branches, the Deep Peroneal and Superficial Peroneal nerves). The term Peroneal has undergone a formal change to *Fibular.*

Color figure 4.10 using blue for ventral, and red for dorsal.

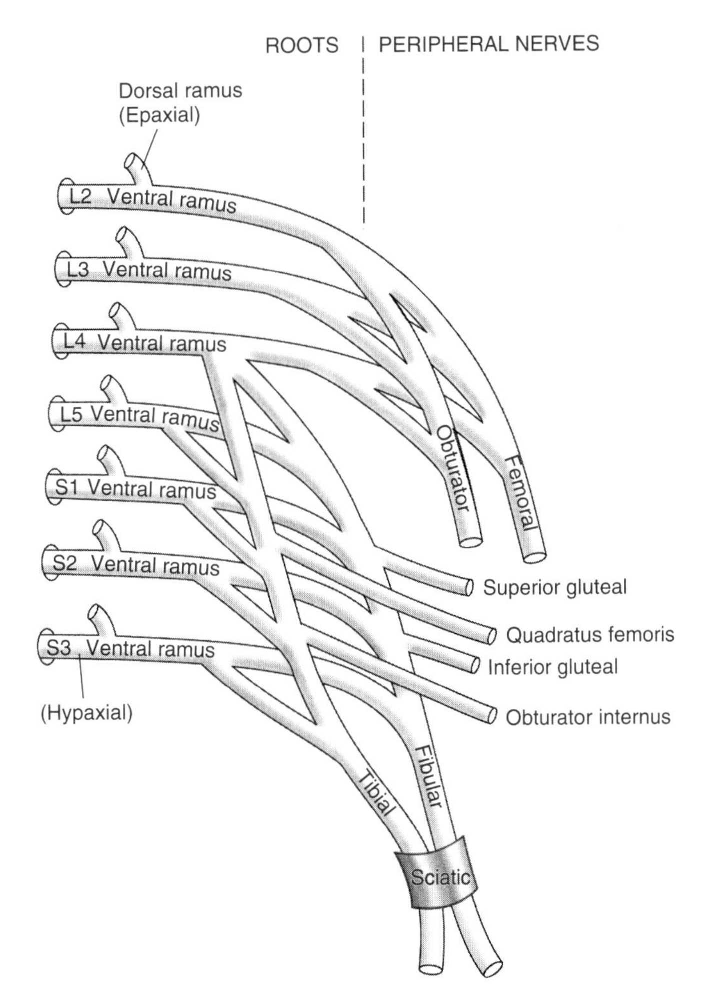

Figure 4.10 The Lumbosacral Plexus

Nerves to Muscles That Move the Thigh

The muscles that move the thigh arise from the pelvis and insert on the femur.

Dorsal Compartment Muscles run along the front and lateral sides of the hip and thigh. They are supplied by the

Femoral nerve 1
Superior Gluteal nerve 2
Inferior Gluteal nerve 3
Common Fibular nerve 4

Ventral Compartment Muscles run along the medial and back sides of the hip and thigh. They are supplied by the

Obturator nerve 5
Tibial nerve 6

Nerves to Muscles That Move the Leg, Ankle, and Digits

Muscles below the knee are supplied either by the tibial nerve or one of the two branches of the common fibular nerve.

Dorsal Compartment Muscles run along the front or lateral side of the leg. They are served by the

Deep Fibular nerve 7
Superficial Fibular nerve 8

Ventral Compartment Muscles run along the back of the leg. They are served by the

Tibial Nerve 6

Dorsal Compartment Muscles occupy the dorsum of the foot. They are served by the

Deep Fibular Nerve 7

Ventral Compartment Muscles run through the sole of the foot. They are served by the

Tibial Nerve 6

Identify the aforementioned nerves in figure 4.11.

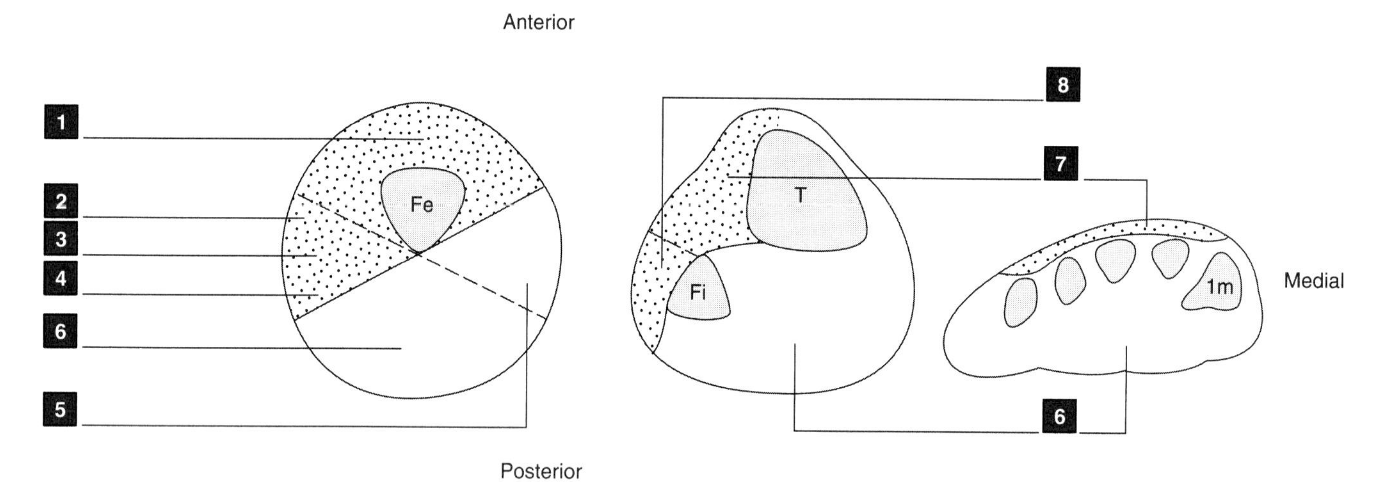

Figure 4.11 **Nerves to Muscles That Move the Leg, Ankle, and Digits**
Schematic transverse sections through the left thigh, leg, and foot.
(Fe = femur, Fi = fibula, T = tibia, 1m = first metatarsal)

Courses of the Peripheral Nerves

You should be familiar with the paths of the principal nerves of the lower limb.

Dorsal Compartment Nerves

Femoral 1. It supplies thigh flexors, such as *Iliopsoas* and *Rectus femoris*, and leg extensors, such as *Quadriceps femoris*. It descends through the back of the abdomen into the pelvis. It exits the pelvis underneath the inguinal ligament into a hollow known as the ***Femoral Triangle*** between the *Sartorius* and *Adductor longus* muscles. It then branches out.

Gluteals 2 3. The **Superior** 2 and **Inferior Gluteal** 3 nerves supply thigh extensors, abductors and lateral rotators. The nerves descend into the pelvis and exit it around the *Pyriformis muscle* through the ***Greater Sciatic Foramen.*** They then branch out to supply the gluteal muscles and the *tensor fascia latae*.

Common Fibular 4. It supplies only one leg flexor muscle, the *Short head of Biceps femoris*. It exits the pelvis as part of the **Sciatic Nerve** 5 through the ***Greater Sciatic Foramen.*** It descends the thigh deep to the *Gluteus maximus* and *Biceps femoris* muscles to emerge behind the knee. It crosses the lateral aspect of the joint and runs to the top of the fibula, where it divides into its two principal branches: deep fibular and superficial fibular.

Deep Fibular 6. It supplies foot dorsiflexors, and the extrinsic and intrinsic toe extensors. It descends the leg across the front of the interosseous membrane and the tibia, and branches onto the dorsum of the foot.

Superficial Fibular 7. It supplies foot plantarflexors, and much of the skin of the leg and foot.

Ventral Compartment Nerves

Obturator 8. It supplies the thigh adductors. It descends through the back of the abdomen into the pelvis, and exits through the ***Obturator foramen.*** It then branches out.

Tibial 9. It supplies one thigh adductor (hamstring part of *Adductor magnus*), most leg flexors and foot plantarflexors, and both extrinsic toe flexors. It innervates all but one of the intrinsic toe muscles through its terminal Plantar branches. It exits the pelvis with the common fibular nerve, as part of the ***Sciatic Nerve,*** through the ***Greater Sciatic Foramen.*** It descends the thigh as part of the sciatic nerve to emerge into the popliteal fossa behind the knee. It crosses the middle of the popliteal fossa and descends along the back of the leg deep to the calf muscles; it crosses the medial side of the ankle and then divides into the ***Medial Plantar*** and ***Lateral Plantar nerves.***

Identify these peripheral nerves in figure 4.12. Use shades of red and blue for dorsal and ventral compartment nerves respectively.

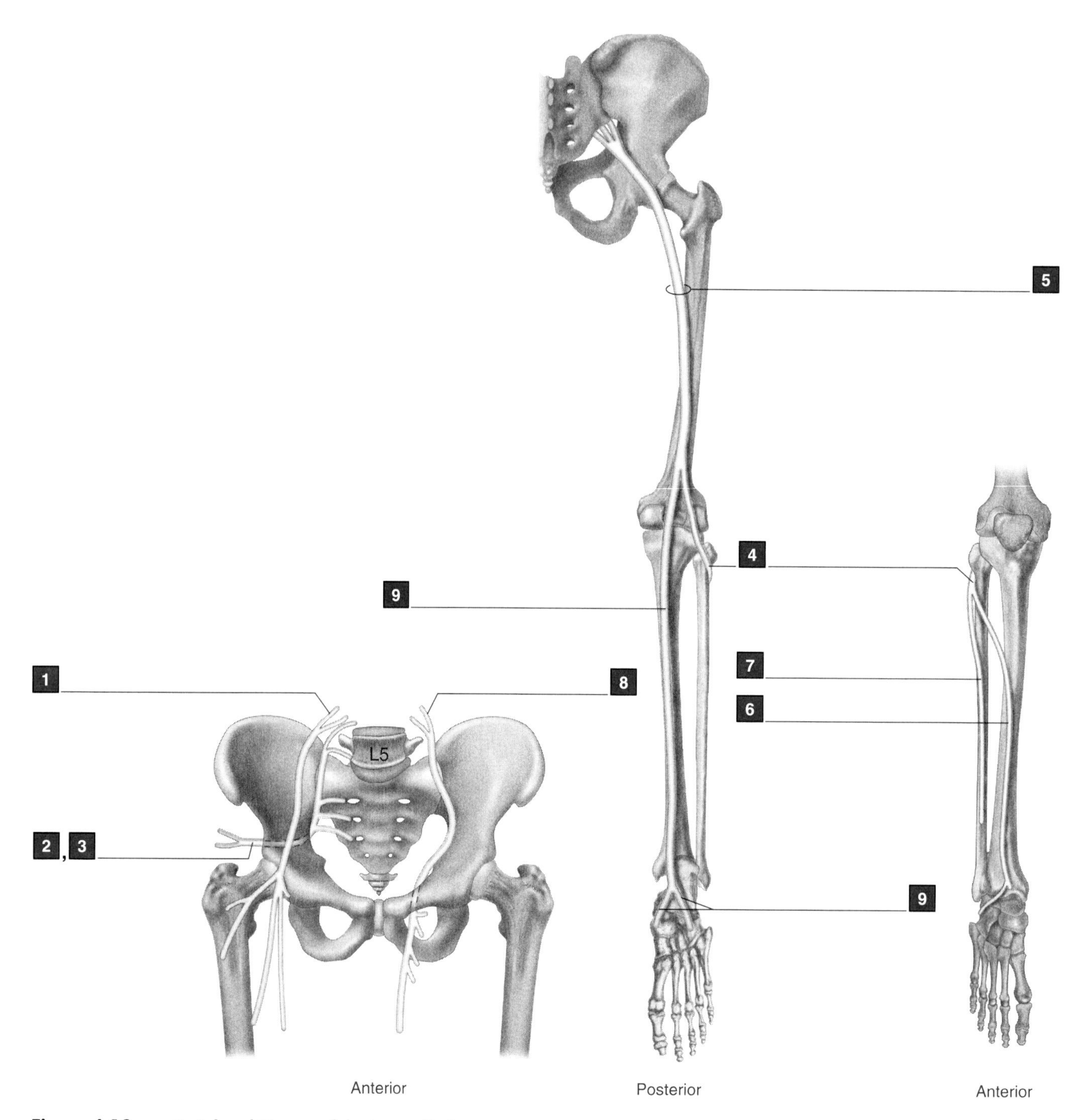

Figure 4.12 Peripheral Nerves of the Lower Limb

4.4 MUSCLES OF THE LOWER LIMB

Recall that the muscles of the lower limb may be divided into two groups according to their innervation:

1. Ventral Division innervated by obturator nerve, nerve to quadratus femoris, nerve to obturator internus, tibial nerve, medial plantar nerve, lateral plantar nerve
2. Dorsal Division innervated by femoral nerve, superior gluteal nerve, inferior gluteal nerve, common fibular nerve, deep fibular nerve, superficial fibular nerve

We will examine these muscles according to the anatomical region upon which they act and by the action they produce.

Muscles That Move the Thigh

Muscles that insert onto the femur arise from the pelvis or the lumbar vertebrae. Muscles that originate on the pelvis and attach to the leg bones may also move the thigh. All but one of these will be considered later because they are significant in producing movement of the leg.

Four groups of muscles move the thigh.

Iliopsoas Group Muscles

Iliopsoas 1 comprises two muscles whose fibers join to form a common tendon.

Psoas major 2 arises from the vertebral bodies and intervertebral discs from T12 to L5. Its fibers run into the pelvis where they merge with the fibers of **Iliacus** 3**,** which arises from the iliac fossa of the pelvis. The conjoined tendon runs across the superior pubic ramus and below the inguinal ligament, forming the back wall of the femoral triangle. It wraps around the femoral neck to insert onto the lesser trochanter. It flexes the hip joint.

Femoral nerve

Identify the iliopsoas group muscles in figure 4.13.

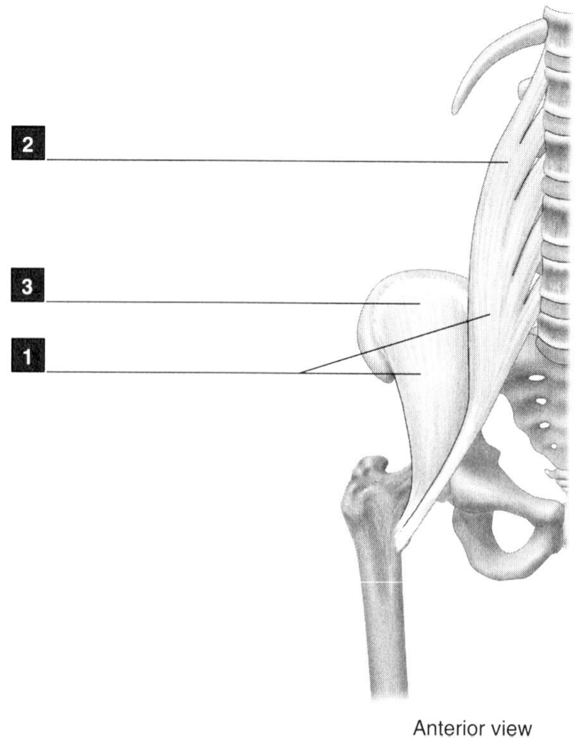

Figure 4.13 **Muscles That Move the Thigh: Iliopsoas group**

Gluteal Group Muscles

There are four gluteal group muscles. Three abduct and medially rotate the thigh; one extends and laterally rotates the thigh. All are *Dorsal Compartment* muscles, being innervated by one of the gluteal nerves.

Tensor fascia latae 1 arises from the outer lip of the anterior part of the iliac crest between the anterior superior iliac spine and the iliac tubercle. Its fibers become tendinous and form part of the ***Iliotibial Tract,*** which inserts onto the lateral condyle of the tibia. It flexes, medially rotates, and abducts the thigh.

Superior Gluteal nerve

The **Iliotibial Tract** is a very strong tendinous band that runs from the tubercle on the iliac crest to the anterolateral surface of the lateral condyle of the tibia. You can feel this band just above your knee when you stand on one leg. It tightens when the contralateral leg is lifted during walking and running to help prevent the pelvis from tilting over to the side when one leg is lifted from the ground. Some fibers of the iliotibial tract insert on the lateral edge of the patella. When these tighten, it may cause the patella to glide laterally out of the femoral trochlea, resulting in a condition known as *chondromalacia patella.*

Gluteus maximus 2 arises from the posterior part of the iliac crest and the lateral edge of the sacrum and coccyx. It inserts onto the gluteal tuberosity on the back of the femur just below and lateral to the lesser trochanter. Some of its fibers also insert into the iliotibial tract. The fibers of the superior half of the muscle abduct and laterally rotate the thigh; those in its lower half extend and laterally rotate the thigh. It is recruited as an extensor only when powerful force is required, such as in running and climbing stairs.

Inferior Gluteal nerve

Gluteus medius [3] arises from the posterior part of the iliac blade and inserts by a tendon onto the lateral aspect of the greater trochanter of the femur. It abducts and medially rotates the thigh.

Superior Gluteal nerve

Gluteus minimus [4] arises from the posterior part of the iliac blade and also inserts by a tendon onto the lateral aspect of the greater trochanter of the femur. It lies underneath the gluteus medius. It abducts and medially rotates the thigh.

Superior Gluteal nerve

Identify the gluteal muscles in figure 4.14.

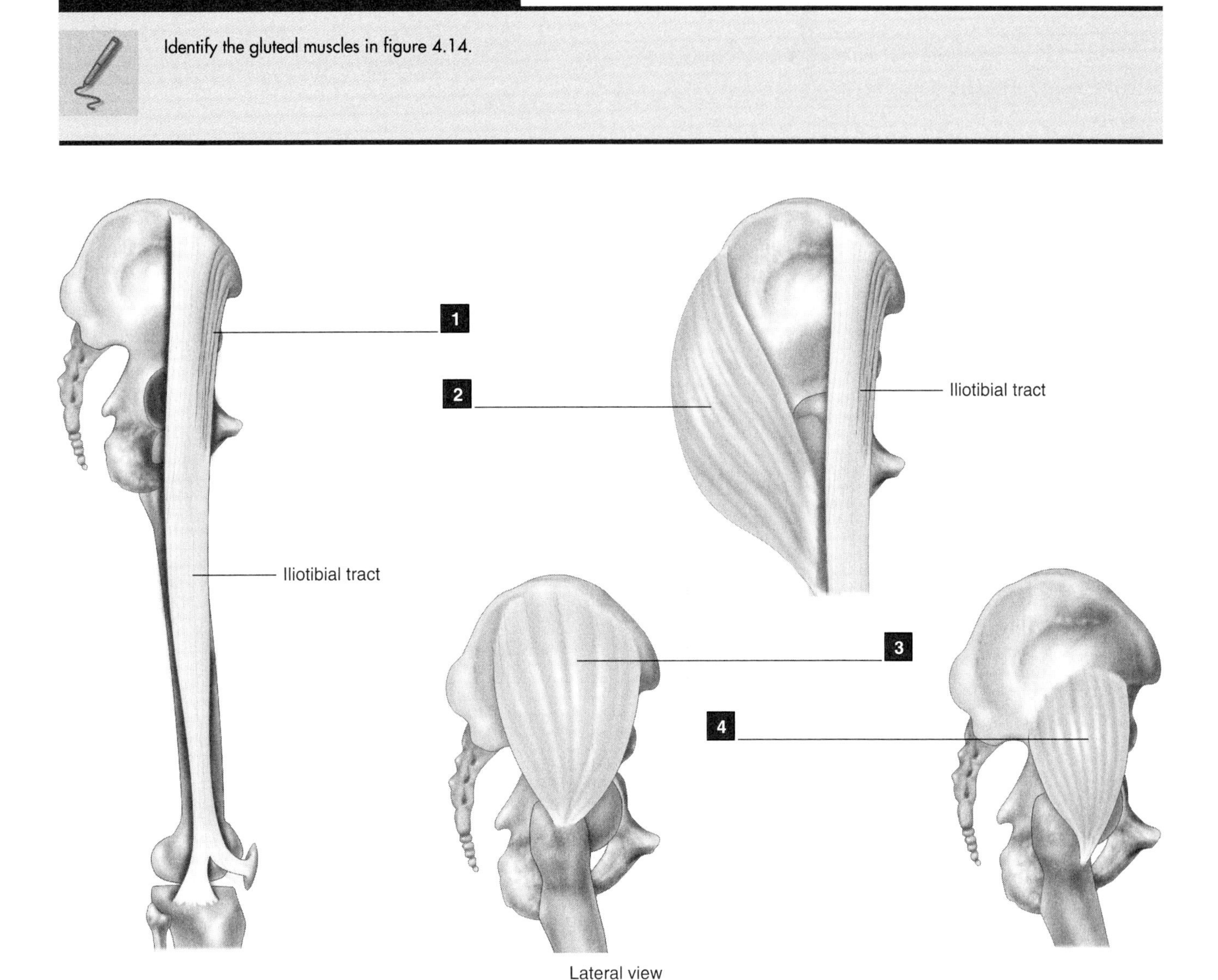

Figure 4.14 Muscles That Move the Thigh: Gluteal Group

Lateral Rotator Group Muscles

There are six lateral rotator muscles. In addition to acting as lateral rotators of the thigh, they may help maintain the integrity of the hip joint by steadying the femoral head in the acetabulum. All but one (*Piriformis*) are *Ventral Compartment* muscles.

Obturator internus 1 arises from the internal margin of the obturator foramen. Its tendon runs through the ***Lesser sciatic*** notch and makes a right angle around the ischium to insert on the top of the greater trochanter of the femur.

Nerve to Obturator internus

Gemellus superior 2 is a small muscle that arises from the ischial spine. Its tendon merges with that of *Obturator internus* to insert onto the greater trochanter of the femur.

Nerve to Obturator internus

Gemellus inferior 3 is a small muscle that arises from the upper part of the ischial tuberosity. Its tendon also merges with that of *Obturator internus* to insert onto the greater trochanter of the femur.

Nerve to Quadratus femoris

Quadratus femoris 4 arises from the lateral edge of the ischial tuberosity and runs laterally to insert onto the intertrochanteric crest on the back of the femur.

Nerve to Quadratus femoris

Piriformis 5 arises from the lateral aspect of the anterior surface of the sacrum. It runs laterally through the ***Greater sciatic notch*** to attach to the top of the greater trochanter of the femur.

Dorsal compartment = *branches from S1 and S2*

Obturator externus 6 attaches to the outer margin of the front of the obturator foramen. It runs behind the hip joint to insert into the fossa on the medial side of the greater trochanter.

Obturator nerve

Identify the lateral rotator muscles in figure 4.15.

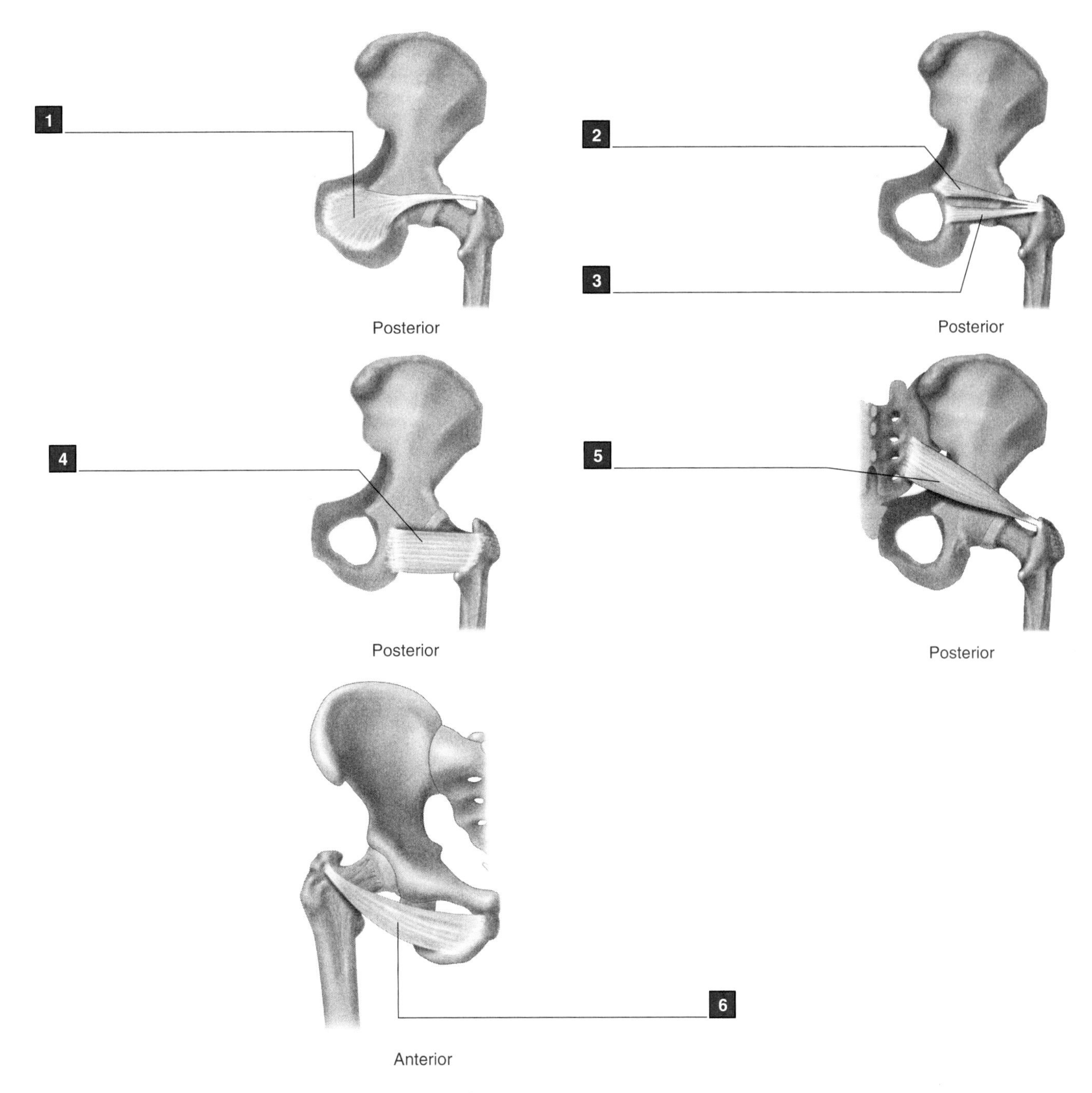

Figure 4.15 Muscles That Move the Thigh: Lateral Rotators

Adductor Group Muscles

There are five principal adductors of the thigh. Most of these muscles also medially rotate and flex the thigh. Four insert onto the femur, and one inserts onto the tibia. All but one (*Pectineus*) are *Ventral Compartment* muscles.

Adductor brevis 1 arises from the external surface of the pubis. It inserts onto the upper part of the linea aspera of the femur. It also flexes and medially rotates the thigh.

Obturator nerve

Adductor longus 2 arises from the front of the pubis. It inserts onto the distal two-thirds of the linea aspera of the femur. [Note: many painted lab skeletons mistakenly label the inferior attachment of adductor magnus as the insertion of adductor longus.] It also flexes and medially rotates the thigh.

Obturator nerve

Adductor magnus 3 is one of the largest muscles in the body. It has two parts that are separated inferiorly by a gap. Its two components are innervated separately.

The ***adductor (pubofemoral)*** component arises from the external surface of the ischiopubic ramus. It inserts onto the gluteal tuberosity, the linea aspera, and the medial supracondylar ridge of the femur. [Note: many painted lab skeletons mistakenly label the inferior attachment of adductor magnus as the insertion of adductor longus.] It also flexes the thigh.

Obturator nerve

The ***hamstring (ischicondylar)*** component arises from the ischial tuberosity and runs down the thigh to insert onto the adductor tubercle of the femur. It also extends the thigh.

Tibial (Sciatic) nerve

Pectineus 4 arises from the anterior surface of the superior pubic ramus. It runs laterally to insert on the back of the femur below the lesser trochanter. It also flexes and medially rotates the thigh.

Dorsal compartment = *Femoral nerve*

Gracilis 5 arises from the inferior part of the pubis. It runs straight down the medial surface of the thigh and inserts onto the upper part of the medial surface of the tibia. It medially rotates the lower limb, and also flexes the knee.

Obturator nerve

Identify the adductor muscles in figure 4.16.

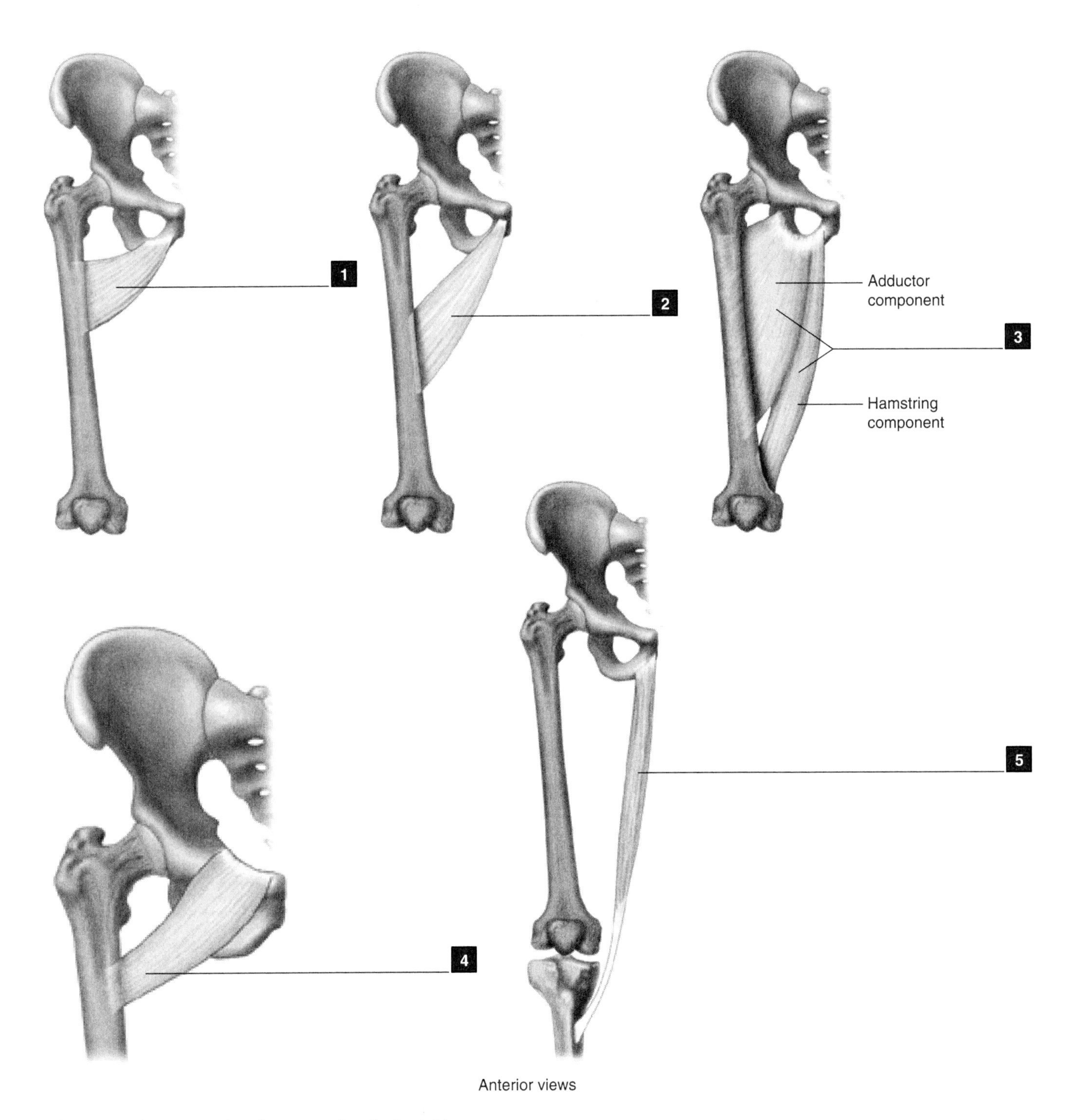

Figure 4.16 **Muscles That Move the Thigh: Adductors**

Muscles That Move the Leg

The 10 muscles that move the leg arise from the os coxae or femur, and insert onto the tibia or fibula. They may be grouped into two categories—flexors (6 muscles) and extensors (4 muscles).

Five of the leg flexors may also extend, or adduct the thigh. We have already examined one of these (*Gracilis*) in its capacity as a thigh adductor. One of the knee extensors may also produce flexion at the hip joint.

Leg Flexors

Six muscles flex the knee, and all are potential rotators of the leg. They are both dorsal and ventral compartment muscles. Five rotate the leg medially; one rotates it laterally. Some may also flex the hip joint, and others can extend the thigh. Three (*Biceps femoris*, *Semimembranosus*, and *Semitendinosus*) are known as the ***Hamstring Muscles.***

Biceps Femoris 1 has two heads. They differ in their origin and innervation. Both can laterally rotate the leg.

The **Long Head** 2 arises from the medial part of the ischial tuberosity. It runs around the back of the thigh, and its tendon curves posterolateral to the lateral femoral condyle to insert onto the head of the fibula. It also extends the lower limb.

Ventral compartment = *Tibial (Sciatic) nerve*

The **Short Head** 3 arises from the inferior two-thirds of the linea aspera and the lateral supracondylar ridge of the femur. It merges with the long head tendon just above the knee.

Dorsal compartment = *Common Fibular (Sciatic) nerve*

Semitendinosus 4 arises from the ischial tuberosity. It runs straight down the back of the thigh and over the back of the medial femoral condyle. Its tendon partially merges with that of the *Gracilis* muscle to insert onto the medial side of the proximal part of the tibia. It can also medially rotate the leg, and extend the thigh.

Ventral compartment = *Tibial (Sciatic) nerve*

Semimembranosus 5 arises from the ischial tuberosity. It runs deep to *Semitendinosus*, and crosses the back of the medial femoral condyle with it. Its tendon divides at the back of the knee to insert onto both the medial and posterior surfaces of the tibia. Other of its fibers turn upward and laterally into the back of the knee joint capsule. It can also medially rotate the leg, and extend the thigh.

Ventral compartment = *Tibial (Sciatic) nerve*

Identify the hamstring muscles in figure 4.17.

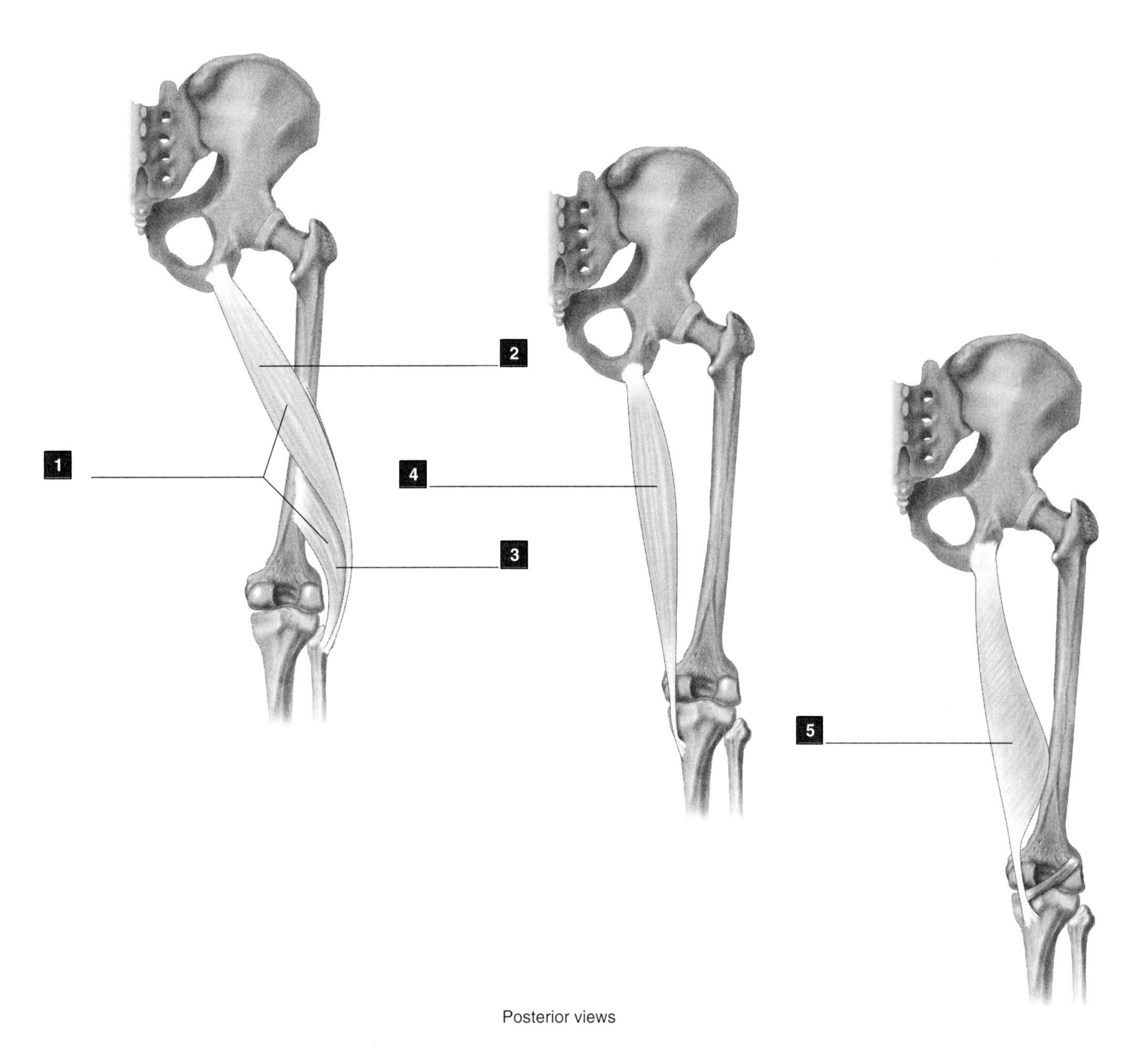

Figure 4.17 **Leg Flexors: the Hamstring Muscles**

Sartorius 1 arises from just below the anterior superior iliac spine of the os coxae. It spirals across the front and medial sides of the thigh, and inserts on the medial surface of the upper part of the tibia. It can also medially rotate the leg and flex the hip joint.

Dorsal compartment = *Femoral nerve*

The *Sartorius* muscle crosses over the *Adductor longus* muscle on the front of the thigh. Where they meet, they form the apex of an inferiorly pointing, triangular hollow. This hollow is known as the ***Femoral Triangle.*** You can see this on yourself on the medial aspect of the proximal part of your thigh when you flex your hip.

The Femoral Triangle

The triangle's lateral border is formed by *Sartorius;* its medial border by *Adductor longus.* Its base is formed by the *Inguinal Ligament,* which stretches between the anterior superior iliac spine and the pubic tubercle. The back wall of the triangular hollow is made up by the *Pectineus* and *Iliopsoas* muscles.

The **Femoral Nerve** 1**, Femoral Artery** 2 and **Femoral Vein** 3 run through the femoral triangle anterior to *Pectinius* and *Iliopsoas.* The hollow also contains the **Superficial Inguinal Lymph Nodes** 4**.** Identify these structures in the drawing below.

The hollow of the femoral triangle is covered only by skin and fascia. Thus its contents can be palpated easily. For example, the pulse of the femoral artery, and swelling of the inguinal lymph nodes can be detected by pressing on the femoral triangle.

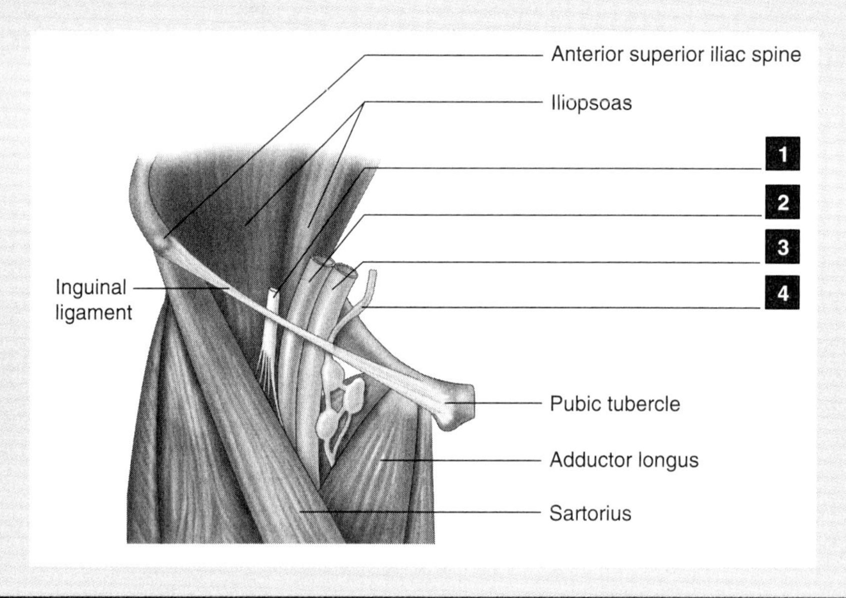

Popliteus 2 arises from the lateral surface of the lateral femoral condyle by a cordlike tendon. It inserts onto the medial aspect of the back of the tibia above the soleal line. It is principally a medial rotator of the leg.

Ventral compartment = *Tibial nerve*

Gracilis (see p. 112) was discussed earlier as an adductor of the thigh. It also flexes and medially rotates the leg.

Ventral compartment = *Obturator nerve*

Identify the Sartorius and Popliteus muscles in figure 4.18.

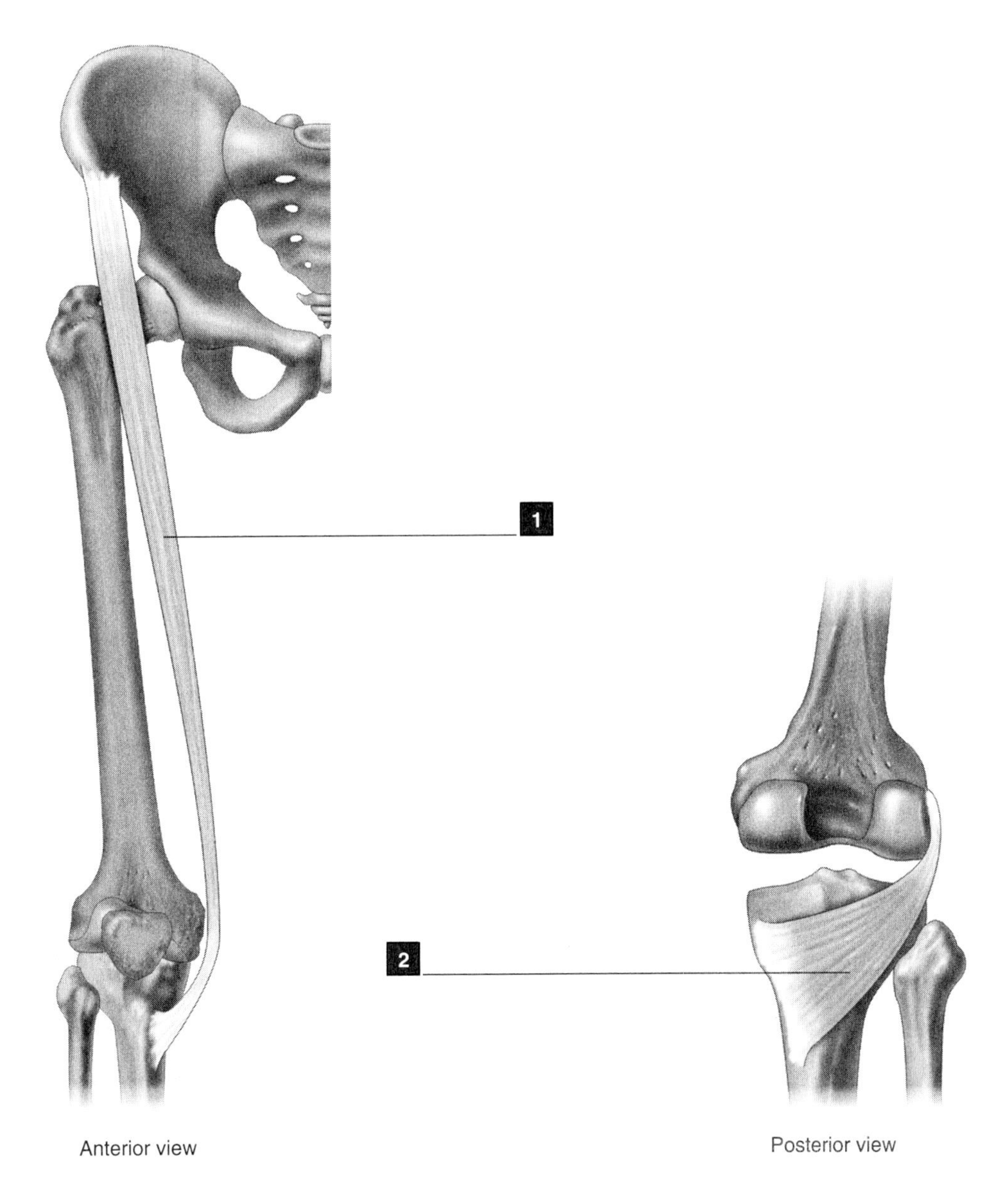

Figure 4.18 **Muscles That Move the Leg: Leg Flexors**

Leg Extensors

Four muscles extend the leg. They have no other action at the knee joint, although one also may produce flexion at the hip joint. All insert onto the patella and via the patellar tendon onto the tibial tuberosity. All are *Dorsal Compartment* muscles, innervated by the *Femoral nerve*.

Because of their similarities in action, insertion, and innervation, these four muscles are often regarded as a single muscle with four heads. This muscle is the ***Quadriceps femoris.*** Three of its components are superficial; one is deep.

Rectus femoris 1 is the middle of the three superficial components. It arises from the anterior inferior iliac spine, and from just above the acetabulum. Its fibers run down the middle of the thigh to insert onto the proximal margin of the patella. Because it crosses over the hip joint, it also flexes the thigh.

Vastus lateralis 2 is the lateral of the superficial components. It arises from the front and side of the greater trochanter, and from the femoral shaft along the lateral margin of the linea aspera. It inserts onto the patella; some fibers run onto the front of the lateral condyle of the tibia.

Vastus medialis 3 is the medial of the superficial parts. It arises from the intertrochanteric line, and the femoral shaft along the medial margin of the linea aspera. It inserts onto the patella; some fibers run onto the front of the medial condyle of the tibia.

Vastus intermedius 4 is the deep component. It arises from the upper three-fourths of the femoral shaft. It inserts onto the patella behind the tendons of the superficial vasti, although they may intermingle. A small bundle of fibers runs deep to the main tendon to insert onto the upper part of the articular capsule of the knee. These fibers constitute the ***Articularis genu*** muscle. It serves to draw the capsule proximally during extension of the knee. This prevents it from being caught between the patella and femur.

Identify the components of the Quadriceps femoris muscle in figure 4.19.

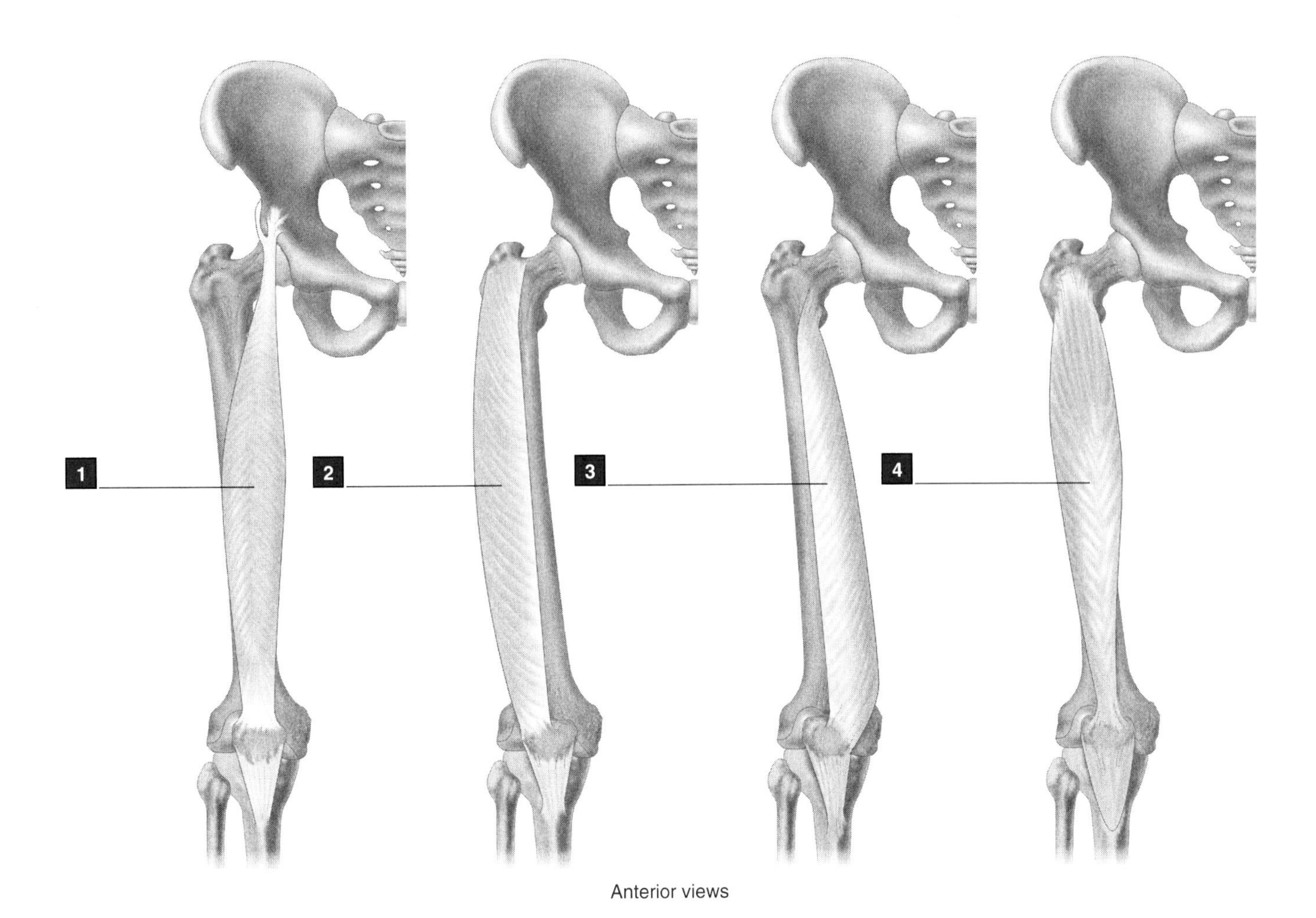

Figure 4.19 **Leg Extensors: the Quadriceps Femoris Muscle**

Muscles That Move the Foot

Twelve muscles cross the ankle joint to insert onto bones of the foot. Eight insert onto tarsal bones or metatarsals; thus, their actions are restricted to movement at the ankle joint. They plantarflex, dorsiflex, invert, and/or evert the foot.

Four muscles cross the ankle joint and insert onto the phalanges. These are the extrinsic muscles of the toes, and they act to extend or flex the toes. They can also move the foot. Toe movement is of little importance in walking and running, so the actions of these four muscles on the ankle joint are more important. Nevertheless, because of their insertions, the extrinsic toe muscles will be considered later (pp. 123–124).

Of these eight muscles, six are plantarflexors and only two are dorsiflexors. This reflects the observation that plantarflexion is very important in locomotion. All movement from initial heel-strike to toe-off involves plantarflexion.

Plantarflexors of the Foot

There are six plantarflexors. Three act only to plantarflex; the other three also either invert or evert the foot.

Gastrocnemius **1** is the most superficial of the calf muscles. It has two heads that arise from the medial and lateral epicondyles of the femur. Their fibers merge to form a common tendon that becomes prominent at about midcalf. This is the ***Achilles Tendon*** (= calcaneal tendon). It inserts onto the tuberosity at the back of the calcaneus. It acts to plantarflex the foot; it can also flex the leg.

Ventral compartment = *Tibial nerve*

Soleus **2** lies deep to gastrocnemius. It arises from the back of the head of the fibula, and the soleal line of the tibia. Its fibers merge with the Achilles tendon at about midcalf. Thus, it also inserts onto the calcaneal tuberosity. It acts only to plantarflex.

Ventral compartment = *Tibial nerve*

Plantaris **3** arises from the posterior surface of the femur above the lateral condyle. It runs medially across the back of the knee, and its long, thin tendon runs down the back of the tibia to insert onto the calcaneal tuberosity medial to the Achilles tendon. It acts only to plantarflex.

Ventral compartment = *Tibial nerve*

Tibialis posterior **4** arises from the lateral half of the upper part of the back of the tibia, and from the medial side of the upper part of the fibula. Its thin tendon curves around the back of the medial malleolus of the tibia and into the sole of the foot. It inserts principally onto the underside of the navicular bone, but its tendon fans out to insert also onto the other tarsal bones (except the talus) and onto the bases of the middle three metatarsals (2, 3, and 4). It also inverts the foot.

Ventral compartment = *Tibial nerve*

Peroneus longus **5** arises from the lateral surface of the proximal two-thirds of the fibula. Its thin tendon wraps around the back of the lateral malleolus of the fibula; it then turns underneath the calcaneus and crosses the sole of the foot to insert onto the base of the first metatarsal and the adjacent medial cuneiform bone. It also everts the foot.

Dorsal compartment = *Superficial Fibular nerve*

Peroneus brevis **6** arises from the distal two-thirds of the fibula. Its thin tendon wraps around the lateral malleolus anterior to the tendon of *Peroneus longus*. It then runs straight forward to insert onto the base of the fifth metatarsal. It also everts the foot.

Dorsal compartment = *Superficial Fibular nerve*

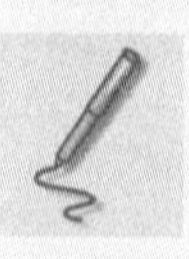

Identify the plantarflexors in figure 4.20.

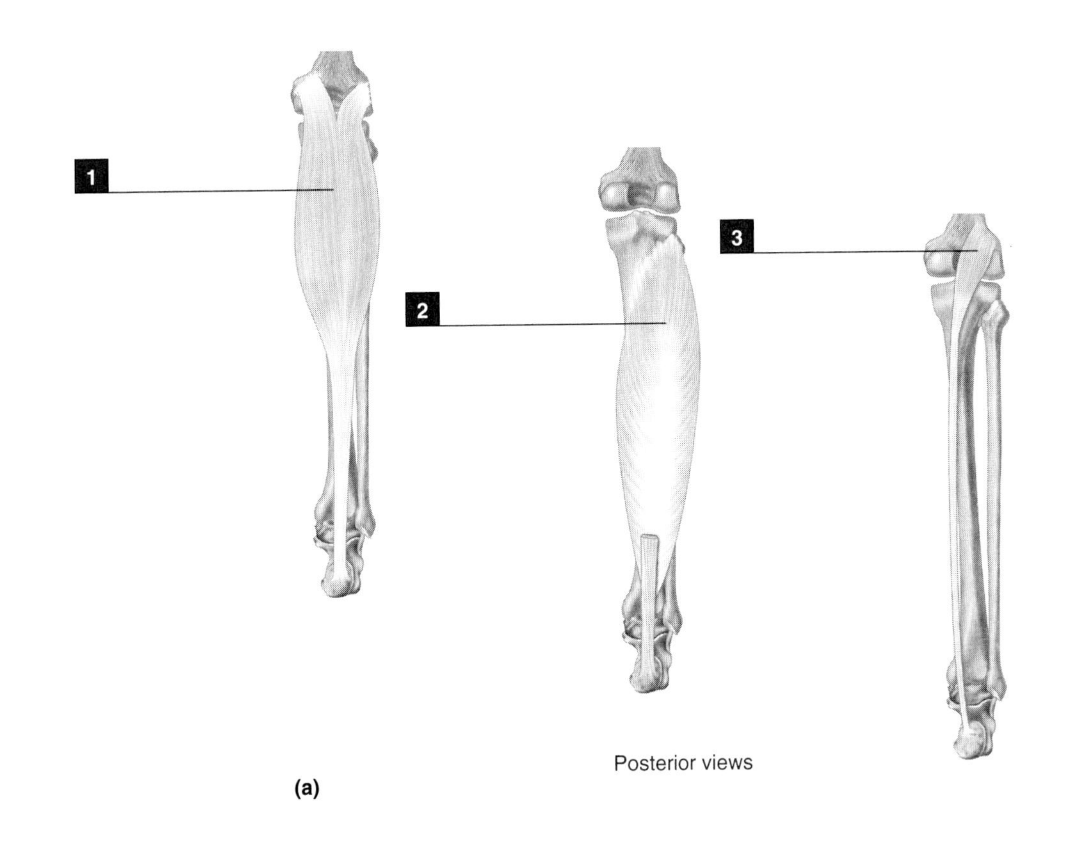

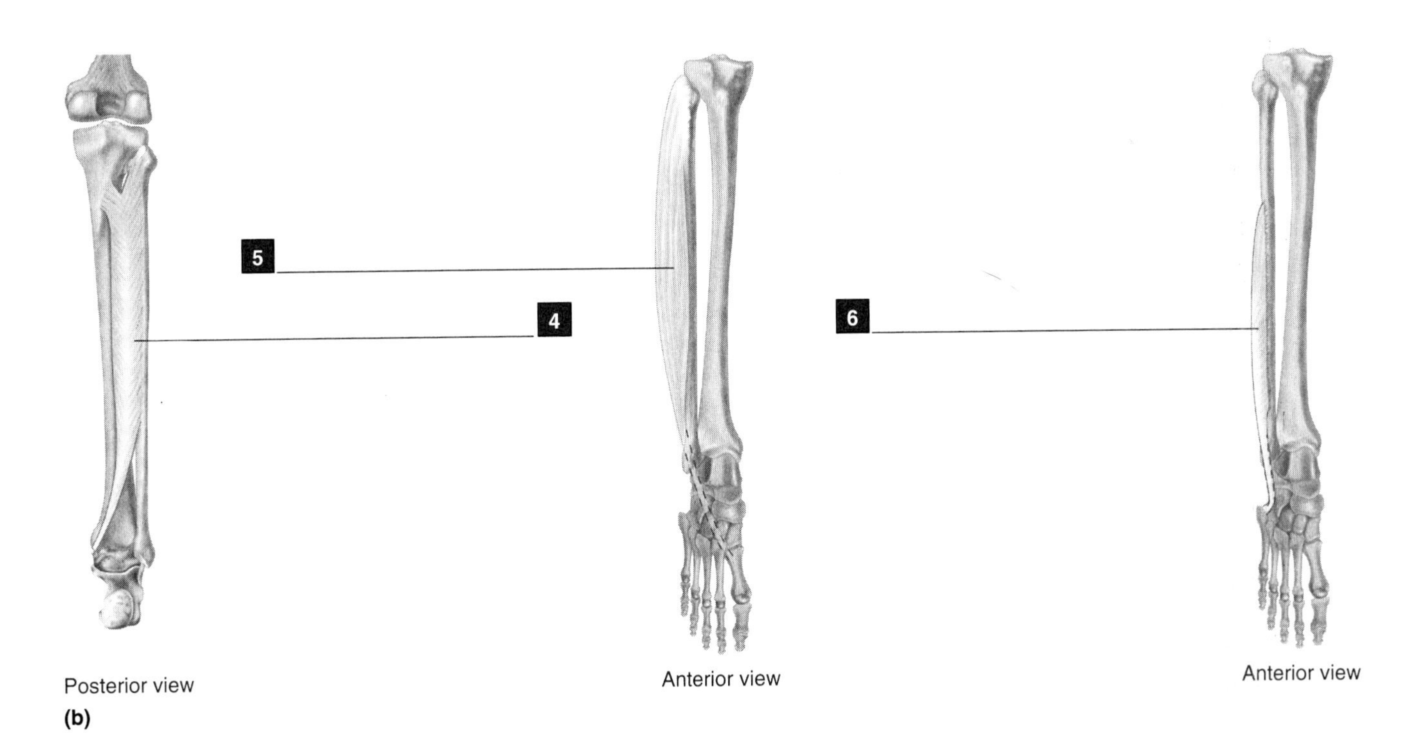

Figure 4.20 **Muscles That Move the Foot: Plantarflexors**
(a) pure plantarflexors; (b) plantarflexors that invert or evert.

Dorsiflexors of the Foot

There are two dorsiflexors. They also either invert or evert the foot. Both are *Dorsal* compartment muscles.

Tibialis anterior 1 arises from the lateral side of the proximal half of the tibial shaft. It angles medially across the front of the tibia and top of the foot to insert onto the base of the first metatarsal. It is also a powerful invertor of the foot.

Deep Fibular nerve

Peroneus tertius 2 arises from the anterior margin of the inferior quarter of the fibula, and inserts onto the base of the fifth metatarsal. Its tendon runs immediately lateral to that of *Extensor digitorum longus* [see p. 123], and it arises immediately below that muscle. Thus, Peroneus tertius is often regarded as simply a part of the larger Extensor digitorum longus. It also everts the foot.

Deep Fibular nerve

Identify the dorsiflexors in figure 4.21.

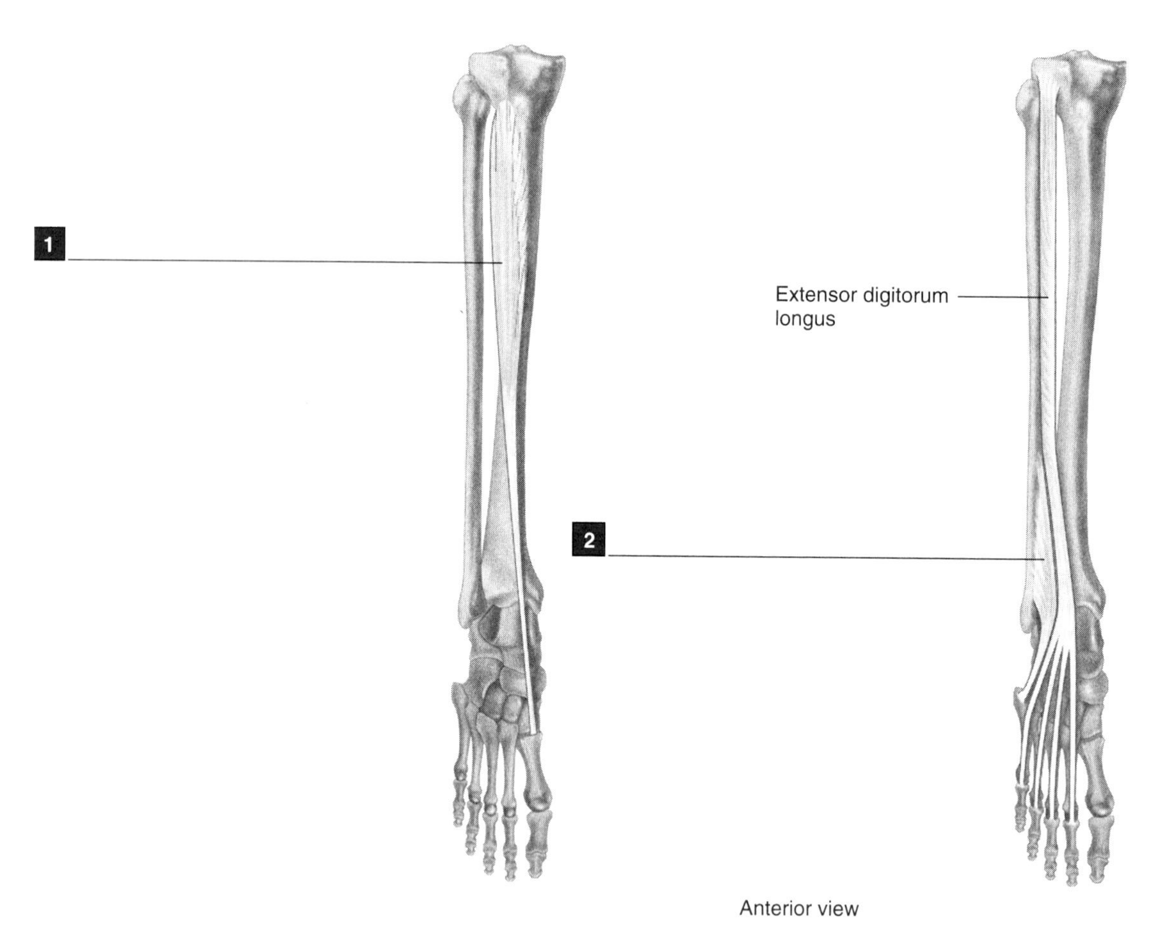

Figure 4.21 **Muscles That Move the Foot: Dorsiflexors**

Muscles That Move the Toes

Fifteen muscles or groups of muscles can move one or more of the toes. They can be divided into two groups. (1) Four *Extrinsic* muscles arise from bones of the leg. (2) Eleven *Intrinsic* muscles take origin from within the foot itself.

Muscles that insert onto the big toe (*Hallux*) have the term ***Hallucis*** in their names. Muscles that move the lateral toes (*Digits*) usually have ***Digitorum*** or ***Digiti*** in their names.

Extrinsic Flexors of the Toes

There are two extrinsic toe flexors. Both can also invert the foot at the ankle joint. Both are *Ventral Compartment* muscles, and are innervated by the *Tibial nerve*.

Flexor digitorum longus 1 arises from the medial part of the back of the tibial shaft and, by an aponeurosis, from the fibula. Its thin tendon turns under the medial malleolus, and then runs forward, dividing into four bands. Each band inserts onto the distal phalanx of one of the four lateral toes.

Flexor hallucis longus 2 arises from the lower two-thirds of the back of the fibula, and the fascia over the back of *Tibialis posterior*. Its thin tendon turns under the medial malleolus and talus. It crosses the inferior surface of the calcaneus, and inserts onto the distal phalanx of the big toe. It propels the body upward and forward at toe-off.

Extrinsic Extensors of the Toes

There are two extrinsic toe extensors. One can invert the foot, and the other can also evert the foot at the ankle joint. Both are *Dorsal Compartment* muscles, and are supplied by the *Deep Fibular nerve*.

Extensor digitorum longus 3 arises from the lateral condyle of the tibia and the proximal three-fourths of the fibula. Its tendon crosses the ankle anterior to the tibia, and then splits into three bands, which insert onto the middle and distal phalanges of the middle three toes. It also everts the foot.

Extensor hallucis longus 4 arises from the middle of the anterior edge of the fibula. Its tendon crosses the ankle anterior to the tibia, and inserts onto the distal phalanx of the big toe. It also inverts the foot.

Identify the extrinsic muscles of the toes in figure 4.22.

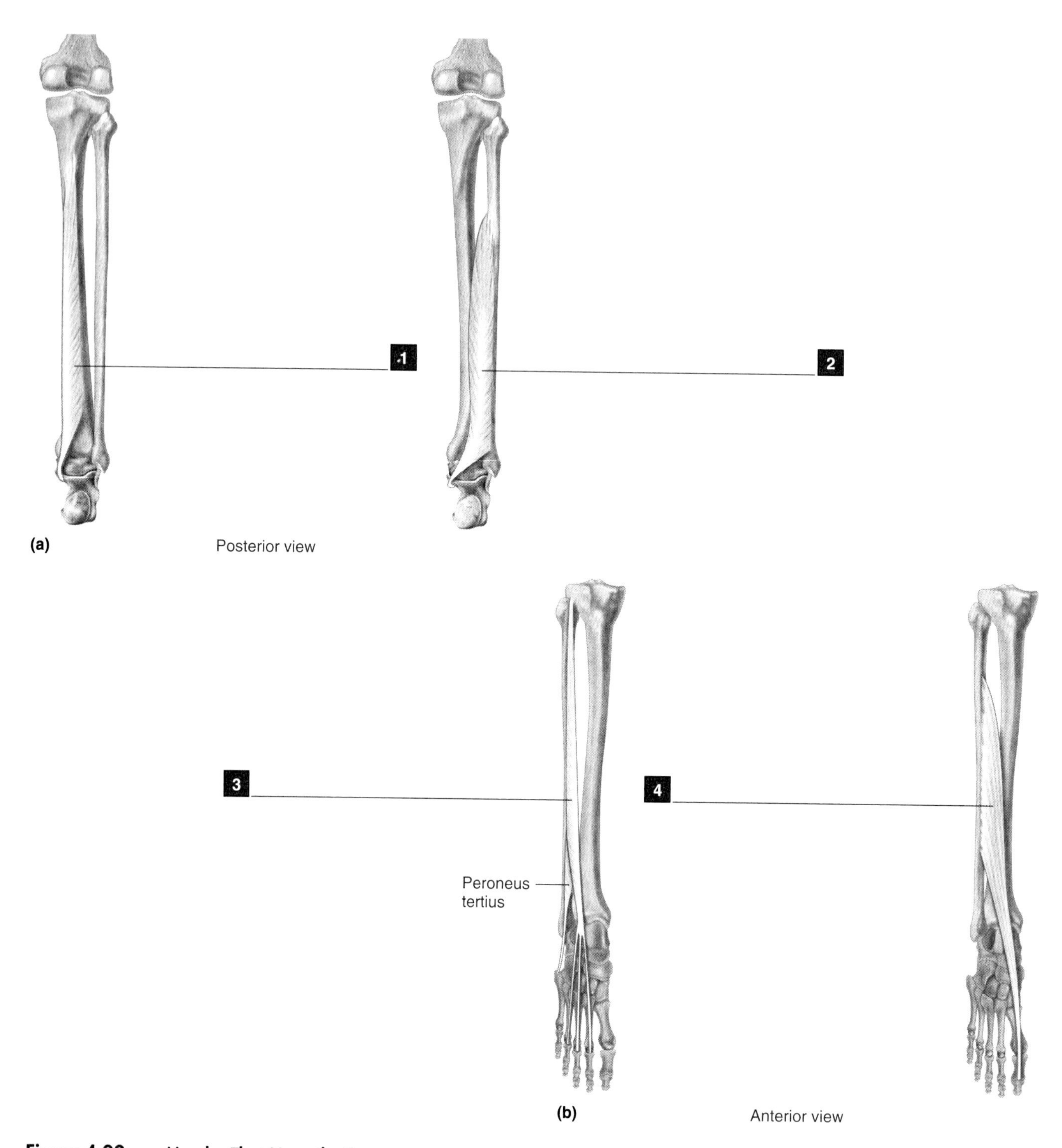

Figure 4.22 **Muscles That Move the Toes**
(a) extrinsic flexors; (b) extrinsic extensors.

Intrinsic Muscles of the Toes

There are 11 intrinsic muscles of the foot. All but one (or two) are located within the sole of the foot, and some help maintain its longitudinal arch during locomotion. This function is perhaps even more important than their ability to move the toes. Because movements of the toes are comparatively trivial (especially compared to movements of the fingers), we will concern ourselves only with the four most prominent intrinsic muscles of the foot.

Extensor digitorum brevis 1 is the only dorsal compartment muscle in the foot. It arises from the superior surface of the calcaneus, and separates into four bellies. The medial belly is the largest, and is commonly referred to as a separate muscle, the ***Extensor hallucis brevis.*** It inserts onto the proximal phalanx of the big toe. The other three bellies are commonly referred to collectively as the ***Extensor digitorum brevis.*** They insert onto the proximal phalanges of the second, third, and fourth toes. It extends the medial four toes.

Dorsal compartment = *Deep Fibular nerve*

Flexor digitorum brevis 2 is the most superficial intrinsic muscle in the sole of the foot. It arises from the calcaneal tuberosity and divides into four tendons that insert onto the middle phalanges of the four lateral toes. It flexes these toes.

Ventral compartment = *Medial Plantar nerve*

Abductor hallucis 3 is the most superficial muscle on the medial aspect of the sole of the foot. Most of the structures that enter the sole from the calf pass deep to it. It arises from the calcaneal tuberosity, and inserts onto the base of the proximal phalanx of the big toe. It can abduct and slightly flex the big toe, but its significance lies in its ability to help sustain the medial side of the longitudinal arch of the foot during locomotion.

Ventral compartment = *Medial Plantar nerve*

Flexor hallucis brevis 4 arises by two heads from the plantar surfaces of the distal tarsal bones. It inserts onto the base of the proximal phalanx of the hallux by way of the two sesamoid bones that lie on the ventral side of its metacarpophalangeal (MP) joint. It flexes the hallux, which is important in toe-off during walking and running.

Ventral compartment = *Medial Plantar nerve*

Identify the intrinsic muscles of the toes in figure 4.23.

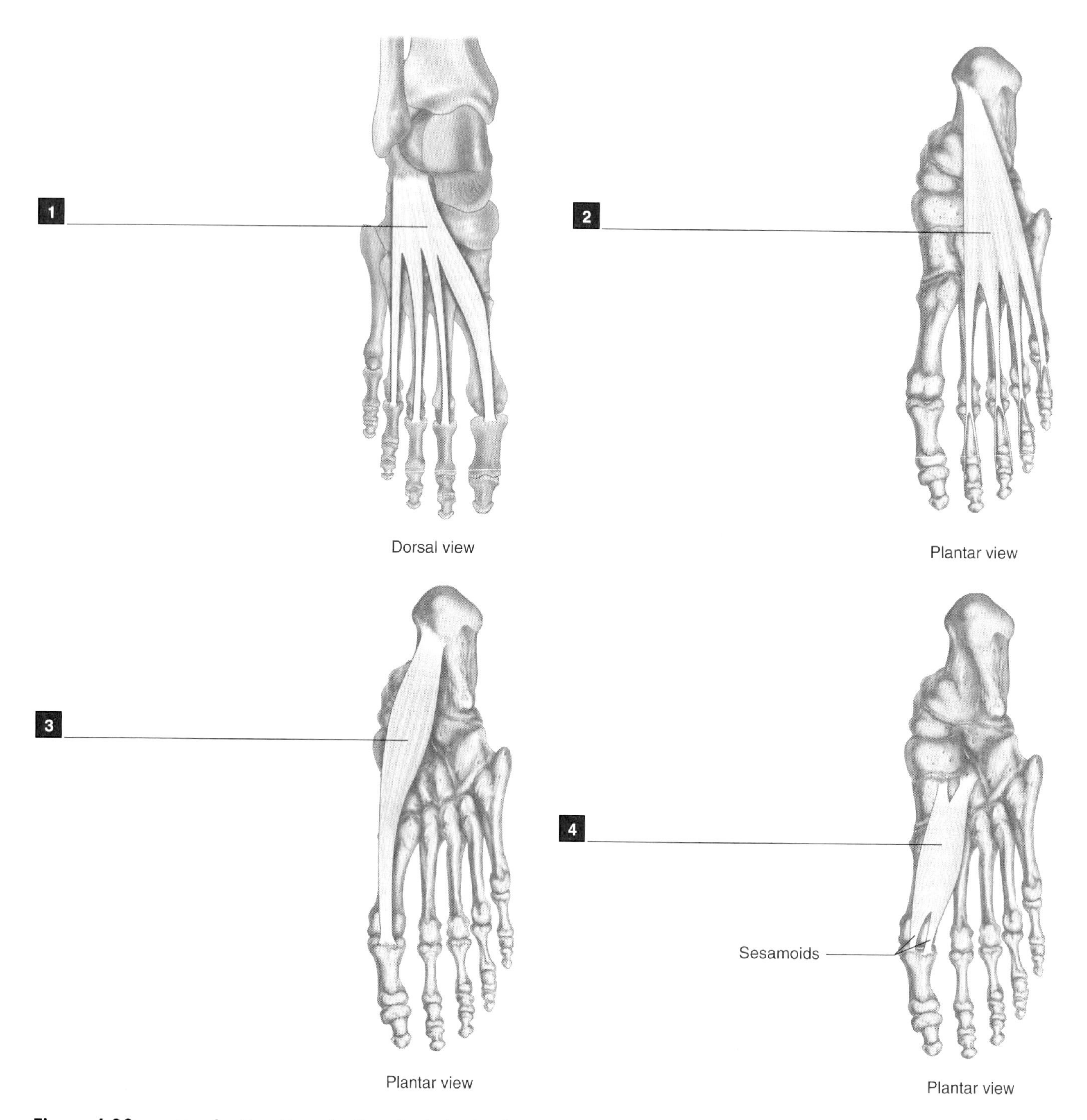

Figure 4.23 Muscles That Move the Toes: Intrinsic Muscles

4.5 BLOOD VESSELS OF THE LOWER LIMB

Arteries

The blood supply of the lower limb derives from the **Aorta** 1 at its terminal bifurcation in front of L4.

The aorta divides to form a **Common Iliac Artery** 2 on the left and right sides of the body. Within the pelvis, the common iliac divides into the **Internal Iliac Artery** 3 and **External Iliac Artery** 4.

The principal blood supply to the lower limb is provided by the external iliac artery. Some of the blood to the lower limb derives from the internal iliac artery via the obturator artery, and the superior and inferior gluteal arteries.

The external iliac artery exits the pelvis next to the *Iliopsoas* muscle between the inguinal ligament and the superior pubic ramus. Once the external iliac artery passes below the inguinal ligament into the *Femoral Triangle* it changes its name to the **Femoral Artery** 5.

An inch or so below the inguinal ligament, the femoral artery gives off its largest branch, the **Profunda (Deep) Femoris Artery** 6. This descends along the posteromedial aspect of the thigh, supplying the thigh adductor and hamstring muscles.

The femoral artery continues down the posteromedial side of the thigh in a canal through the adductor muscles. Toward the distal end of the back of the femur it pierces *Adductor magnus* and becomes known as the **Popliteal Artery** 7. This crosses the back of the knee, giving off several major ***Genicular branches*** that anastomose around the joint.

Just below the back of the knee (at the lower border of *Popliteus*), the popliteal artery divides into the **Anterior Tibial Artery** 8 and the **Posterior Tibial Artery** 9.

The anterior tibial artery passes anteriorly between the tibia and fibula and runs inferiorly along the interosseous membrane. When it crosses the ankle it becomes the **Dorsalis Pedis Artery** 10.

The posterior tibial artery gives off a branch—the **Fibular Artery** 11—that runs along the back of the fibula to the heel. The posterior tibial artery continues down the back of the leg, deep to *Soleus*. After crossing the ankle medial to the calcaneus, it divides into the medial and lateral **Plantar Arteries** 12.

Identify the principal arteries of the lower limb in figure 4.24.

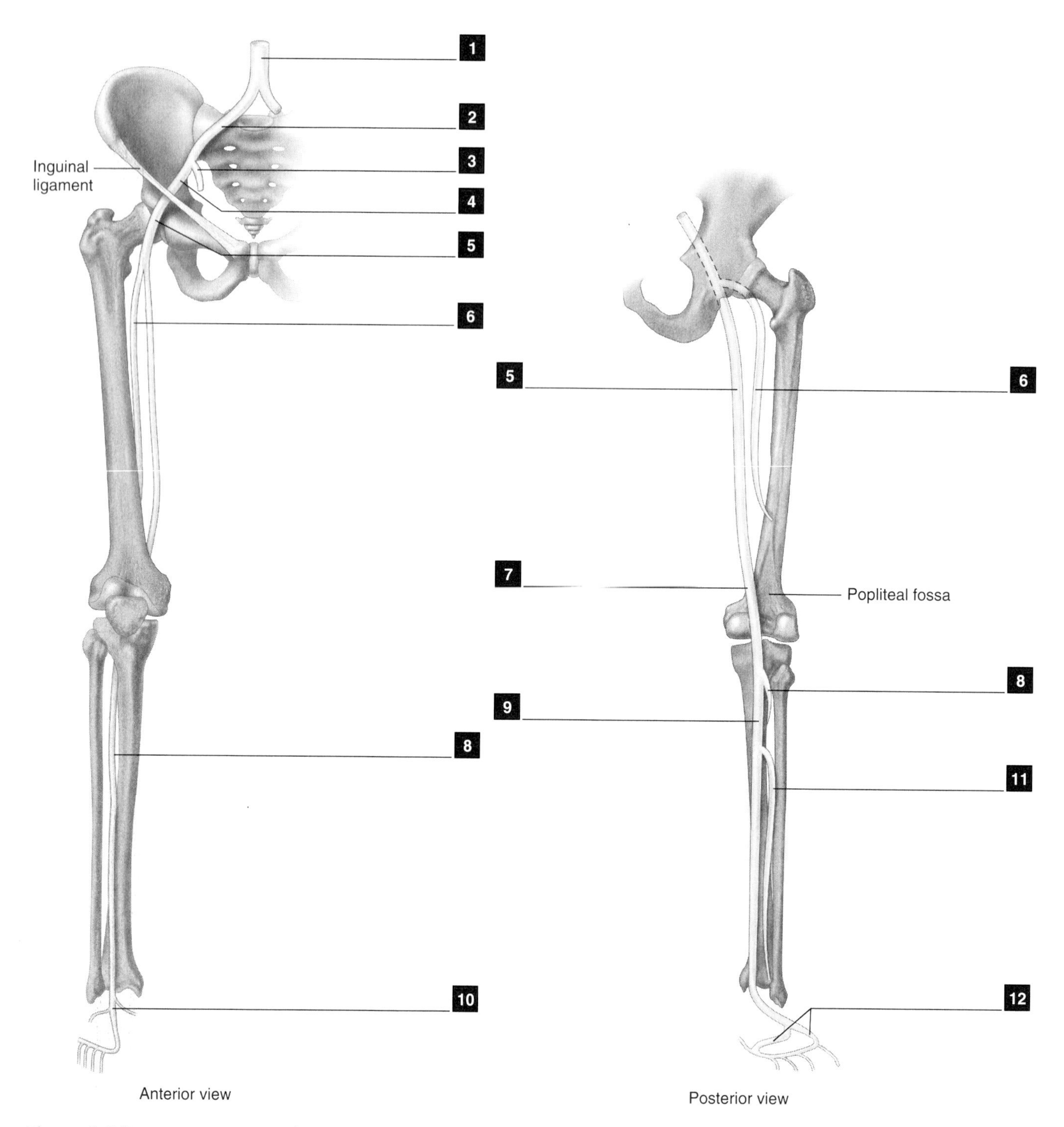

Figure 4.24 Major Arteries of the Lower Limb

Veins

Deep Veins

The deep veins of the lower limb run alongside the arteries. All the arteries mentioned have one or two veins that run alongside them. These are the ***Venae Comitantes.*** They are named according to the arteries they accompany.

Thus, the **Fibular Vein** 1 joins the **Posterior Tibial Vein** 2 to form the **Anterior Tibial Vein** 3**.** This becomes the **Popliteal Vein** 4**,** which runs upward as the **Femoral Vein** 5**.**

The femoral vein continues proximally below the inguinal ligament to become the **External Iliac Vein** 6**.** This is joined by the **Internal Iliac Vein** 7 to form the **Common Iliac Vein** 8**.** The left and right common iliac veins join to form the **Inferior Vena Cava** 9 at the level of the intervertebral disc between L5 and S1.

Superficial Veins

These veins form a network in the *Superficial Fascia.* They are not related to any arteries.

The **Great Saphenous Vein** 10 is the lower limb's version of the cephalic vein. It drains the dorsum of the foot, runs anterior to the medial malleolus of the tibia, and superiorly along the posteromedial surface of the calf. It crosses around behind the knee and continues proximally along the medial side of the thigh to join the ***Femoral Vein*** within the *Femoral Triangle*.

Because of its length and superficial location, the Great Saphenous Vein is the favorite choice for grafts during coronary bypass surgery.

The **Small Saphenous Vein** 11 begins behind the lateral malleolus of the fibula and runs upward in the middle of the calf to the *Popliteal fossa.* Here it dives deeply to join the popliteal vein.

The veins of the lower limb have more valves per unit length than those of the upper limb. Deep veins possess more valves than superficial veins. The valves help prevent the backflow of blood that occurs because of gravity. Venous return in deep veins of the lower limb may be assisted by muscular contractions during walking (e.g., the "soleal pump"). If valves of the superficial veins become incompetent, the vessels become *varicose.*

Identify the principal veins of the lower limb in figure 4.25.

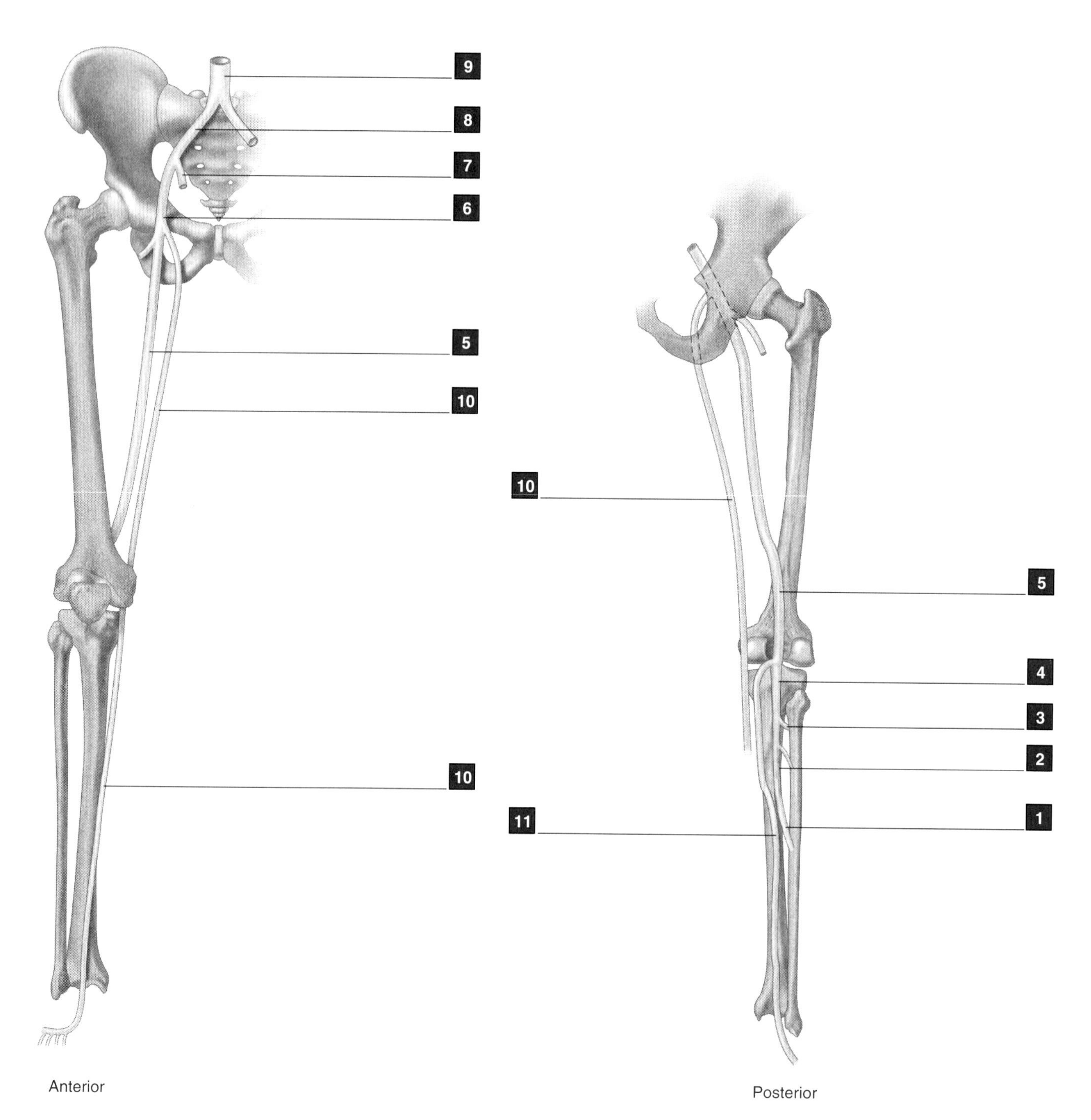

Figure 4.25 **Major Veins of the Lower Limb**

LABORATORY

The Neck

5.1 BONES AND CARTILAGES OF THE NECK

The cervical vertebrae and hyoid constitute the bones of the neck. Cartilages comprise the framework of the larynx, and the incomplete rings that support the trachea.

Bones

Cervical Vertebrae

There are seven **Cervical Vertebrae** **1**. Refer to laboratory 2 (p. 25) to review these bones.

Hyoid Bone

The **Hyoid** **2** is a small U-shaped bone, suspended by ligaments and muscles below the mandible and above the thyroid cartilage. It lies at the level of C3. Its midline *Body* is extended posterolaterally by long, slender *Greater Horns*. *Lesser Horns* project superiorly at the junction of the body and the greater horns.

It is suspended from the bottom of the skull (the styloid process of the temporal bone) by the *Stylohyoid Ligament*. The hyoid provides attachment for 10 muscles that connect it to the mandible, the pharynx, the larynx, and the sternum.

Cartilages

Tracheal Cartilages

The **Trachea** **3** runs down the front of the neck from the larynx into the thorax. The wall of this tube is comprised of 16 to 20 C-shaped cartilaginous rings that are joined to one another by fibroelastic connective tissue. The cartilage rings keep the tube open despite the pressure changes that occur during breathing. The rings are open posteriorly, where the trachea lies against the esophagus. Here, the tracheal wall is made up by a smooth muscle (*Trachealis*), which permits the esophagus to expand anteriorly as food passes through it, and which can decrease the diameter of the tracheal lumen.

Larynx

The **Larynx** **4**, also known as the "voice box," is an intricate arrangement of cartilages connected by ligaments and membranes. It lies at the level of C4 to C6. Part of it serves to keep the airway open by routing food and drink into the esophagus, and it plays an important role in voice production. Because of its intricacy and importance, the anatomy of the larynx deserves detailed examination. There are three unpaired, and three paired laryngeal cartilages. They are:

Unpaired	*Paired*
Thyroid	Arytenoids
Cricoid	Corniculates
Epiglottis	Cuneiforms

Thyroid Cartilage **5** This is the largest laryngeal cartilage. It lies at the level of C4 and C5. It is composed of two rectangular laminae that meet in the midline at an acute angle (the *Thyroid Angle*) to form the *Laryngeal Prominence*, or "Adam's apple." The posterior edge of each lamina is prolonged upward as a long *Superior Horn*, and downward as a shorter *Inferior Horn*.

Cricoid Cartilage **6** This is shaped like a signet ring. It lies at the level of C6, directly below and partly overlapped by the thyroid cartilage. The narrow part of the ring is anterior; the back of the ring is expanded upward. The paired *Arytenoid Cartilages* articulate with the posterolateral corners of the cricoid cartilage along its upper margin.

Epiglottis **7** This leaf-shaped cartilage is attached by its narrow stem by a strong ligament to the back of the angle of the thyroid cartilage. The epiglottis projects upward behind and beyond the hyoid bone. Its free upper edge sits behind the tongue.

Arytenoids 8 They are roughly pyramidal in shape. Each articulates inferiorly with the posterolateral corners of the cricoid cartilage. The arytenoids are capable of sliding across and rotating on the cricoid at their articulation. The base of the pyramid projects anteriorly as the *Vocal Process*, which serves as an attachment for the *Vocal Ligament*. The apex of the arytenoid cartilage supports the corniculate cartilage.

Corniculates 9 Each small cartilage sits atop the arytenoid. It serves as an attachment for the upper free margin (*Aryepiglottic Fold*) of the *Quadrangular Membrane* that extends anteriorly to the epiglottis.

Cuneiforms These tiny, trivial cartilages lie in the *Aryepiglottic Folds*. These cartilages need not concern us further.

Identify the cervical vertebrae, hyoid bone, trachea, and laryngeal cartilages in figures 5.1 and 5.2.

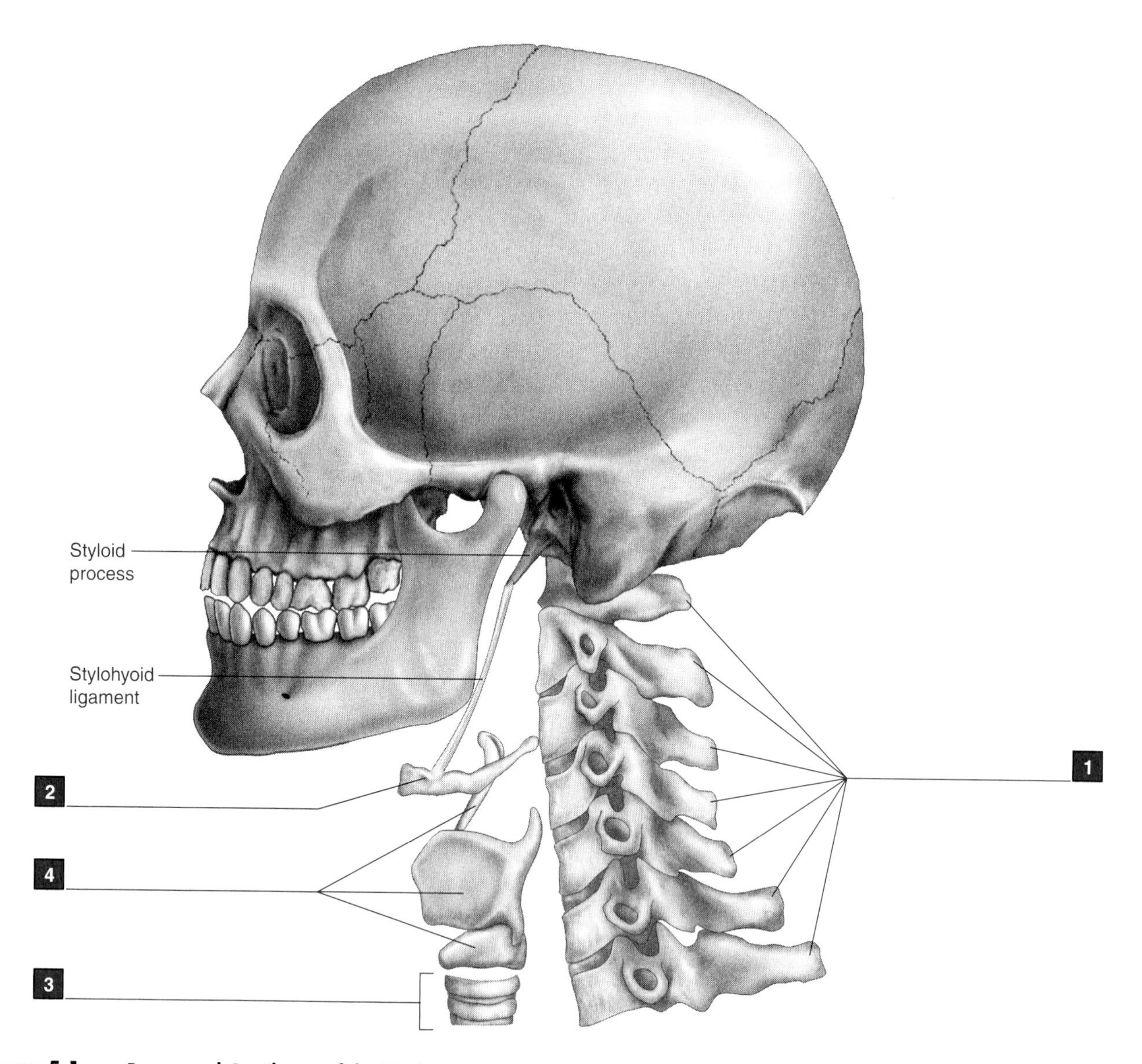

Figure 5.1 Bones and Cartilages of the Neck

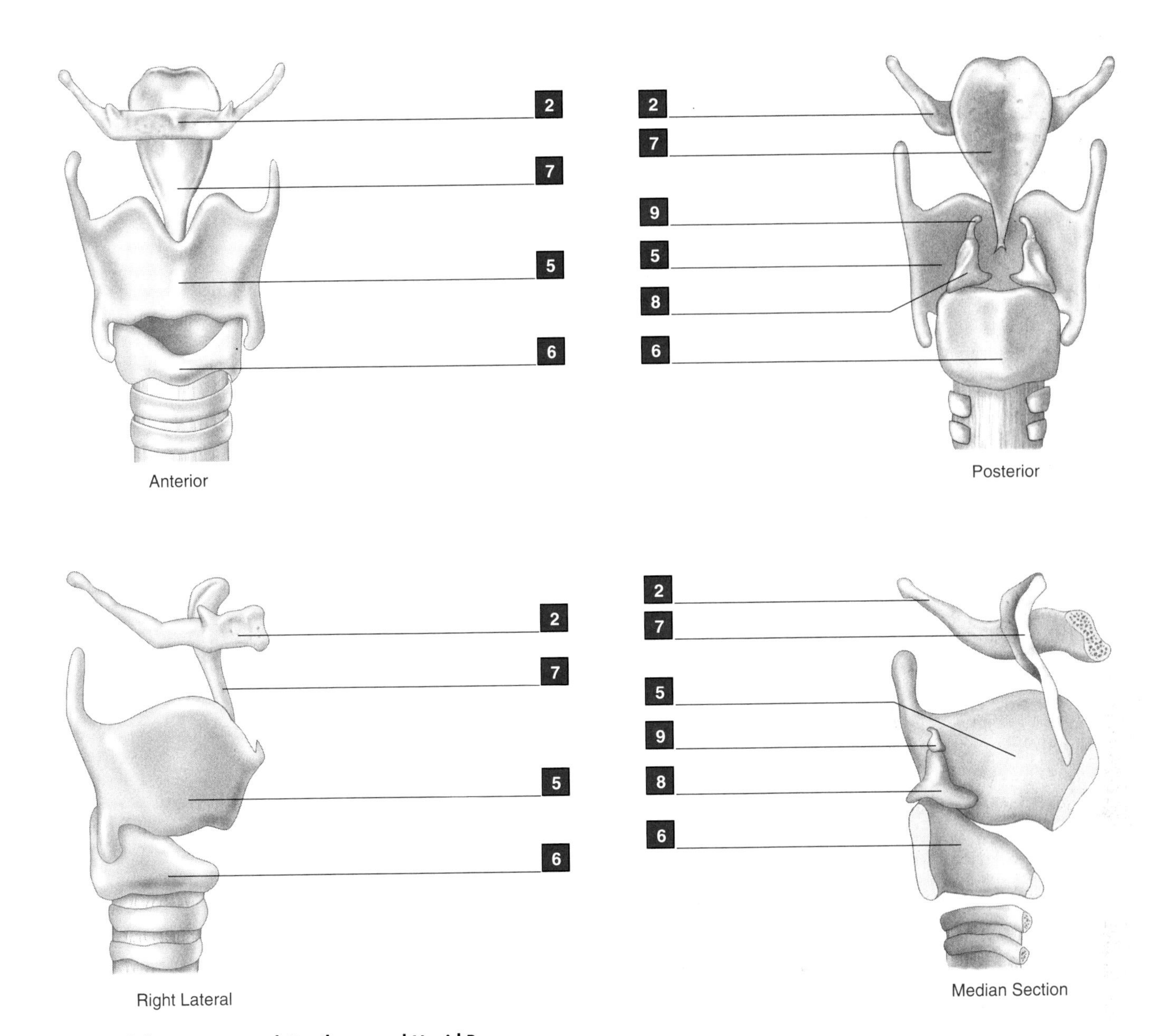

Figure 5.2 **Laryngeal Cartilages and Hyoid Bone**

Laryngeal Membranes

The laryngeal cartilages are connected to one another by elastic membranes that permit a certain degree of movement, while maintaining overall anatomical integrity. In some instances, the free margin of a membrane is slightly thickened; this is referred to as a *Ligament*. Some of the membranes are covered by squamous epithelium, which lines the interior of the larynx. This covering, together with the underlying ligament is referred to as either a *Fold* or a *Cord*.

The principal membranes, ligaments, and folds are:

The **Thyrohyoid Membrane** 1 runs from the superior border of the thyroid cartilage to the bottom of the hyoid bone.

The **Quadrangular Membrane** 2 runs posteriorly from the lateral sides of the epiglottis to the corniculate and arytenoid cartilages.

The superior border of the quadrangular membrane forms the **Aryepiglottic Fold** 3. The two folds form an oval opening to the larynx. As food passes from the tongue into the throat, the epiglottis is pushed posteriorly, partially blocking this opening. This prevents food and drink from entering the respiratory tract.

The inferior border of the quadrangular membrane is thickened to form the **Ventricular Fold** 4. These folds are also known as the *False Vocal Cords*.

The **Conus Elasticus** 5 is a triangular membrane that runs from the internal surface of the back of the thyroid cartilage angle to the superior border of the cricoid and arytenoid cartilages. It is certainly the most important of the laryngeal membranes because of its role in speech.

The superior free margin of the conus elasticus is thickened to form the **Vocal Cord** 6. The space between the vocal cords is known as the *Rima Glottidis*.

Identify the laryngeal membranes, folds and cords in figure 5.3.

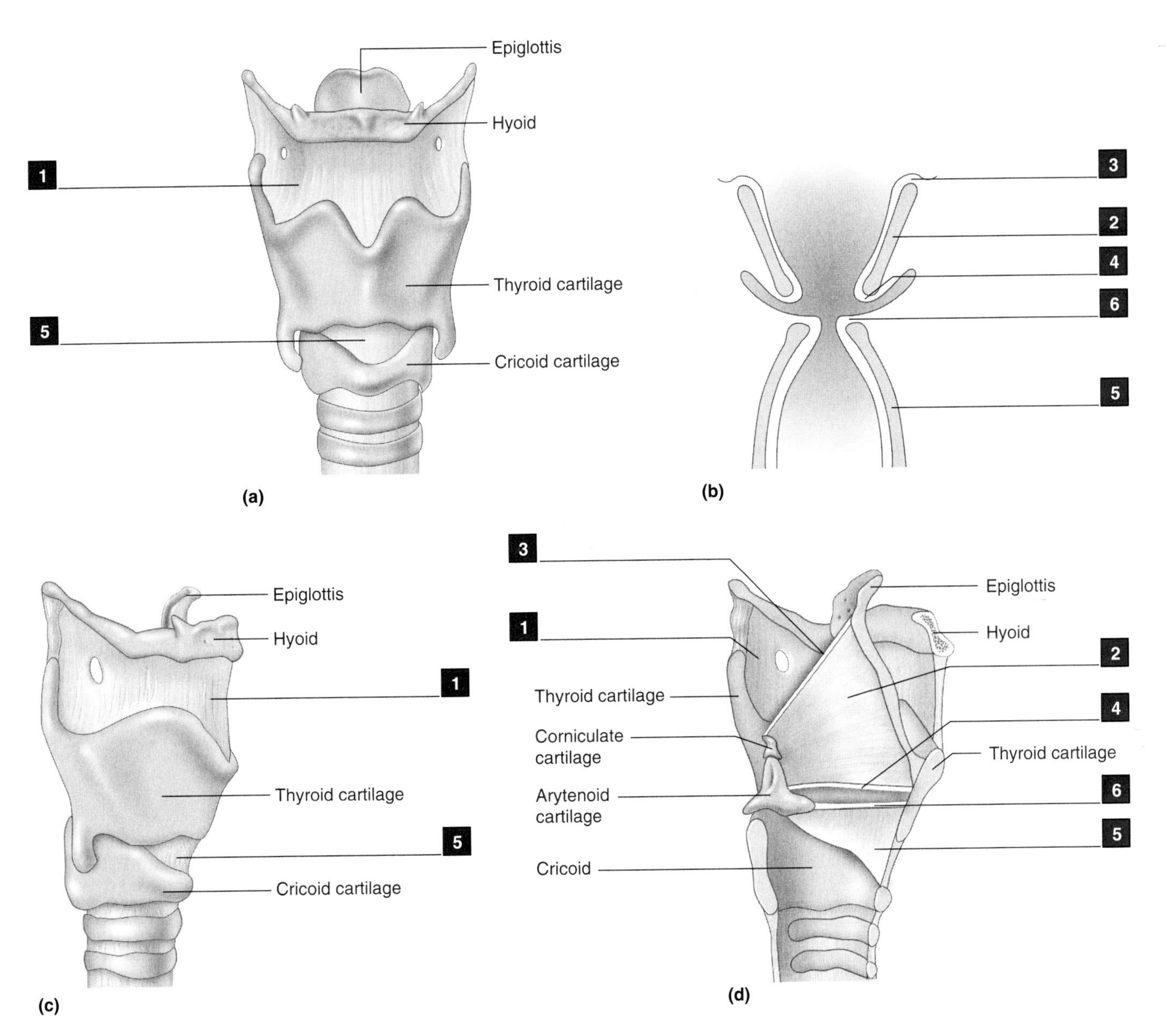

Figure 5.3 **Laryngeal Membranes**
(a) anterior view; (b) schematic coronal section; (c) right lateral view; (d) median section.

5.2 NERVES OF THE NECK

The muscles of the neck may be supplied by spinal nerves or by cranial nerves. Some structures in the neck, such as the thyroid and parathyroid glands, are supplied by autonomic fibers—*parasympathetic* from *cranial nerves* and *sympathetic* from the *cervical sympathetic ganglia*.

Spinal Nerves of the Neck

Spinal Nerves C1 to C4 supply muscles in the neck directly or via the cervical plexus. Cutaneous branches from the plexus innervate the skin over the neck, part of the head, the shoulder, and part of the chest.

Recall that the ventral rami of the lower cervical spinal nerves (C5–C8, together with T1) form the *Brachial Plexus*.

The Cervical Plexus

The *four roots* of the cervical plexus are the *Ventral Rami of C1–C4*. Each root gives off a branch that joins with one from its immediate neighbor, creating a loop between them. Thus, there are loops between C1–C2, C2–C3, and C3–C4. This plexus is deep to the internal jugular vein. Nerves from C1 and the C2–C3 loop join together around the internal jugular vein in a long loop known as the *Ansa Cervicalis*.

Cutaneous Nerves

Three nerves supply the neck and head from the loop between C2–C3. These are the **Lesser Occipital Nerve 1, Great Auricular Nerve 2,** and **Transverse Cervical Nerve 3.** One nerve supplies the shoulder and chest from the loop between C3–C4. It is the **Supraclavicular Nerve 4.** These four nerves emerge behind the posterior margin of the *Sternocleidomastoid* muscle.

Muscular Nerves

Numerous small, unnamed branches come directly from the ventral rami of C1–C4 to supply muscles nearby. Two motor nerves are named.

The **Ansa Cervicalis** 5 provides branches to the *infrahyoid* and *infrathyroid* muscles. Fibers from C1 run with the *Hypoglossal Nerve* (*CN XII*) for a short distance, and then descend between the internal jugular vein and internal carotid artery into the neck, where they join fibers from the C2–C3 loop that descend along the back of the internal jugular vein.

The **Phrenic Nerve** 6 derives from the ventral rami of C3–C5. It provides the *only motor nerve supply to the abdominal diaphragm*. It runs across the *Scalenus anterior* muscle to enter the thorax between the subclavian artery and brachiocephalic vein.

Identify the components of the cervical plexus in figure 5.4. Color cutaneous nerves yellow and muscular nerves purple.

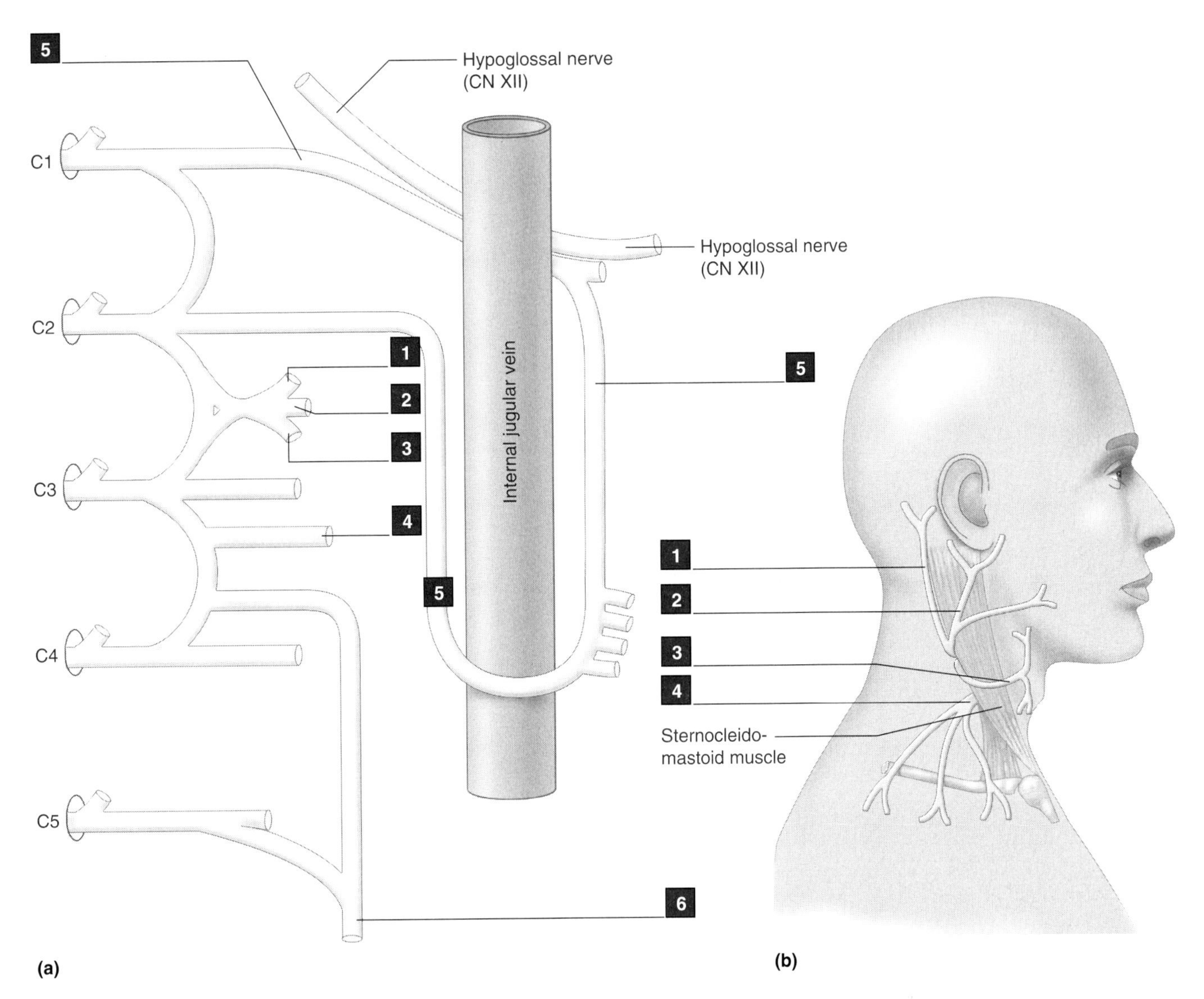

Figure 5.4 **The Cervical Plexus**
(a) schematic diagram; (b) distribution of cutaneous nerves.

Cranial Nerves in the Neck

The 12 cranial nerves are the "peripheral nerves of the brain." They are numbered by Roman numerals in the order that they emerge from the brain. Five of them supply structures in the neck. We will deal with the cranial nerves in more detail in laboratory 7. Here we will examine those that enter the neck and explore what they do there.

Trigeminal Nerve (V) 1 has three main divisions. One of these, the ***Mandibular Division,*** has two branches that enter the neck. One of these is the **Lingual Nerve** 2. It runs across the lateral surface of the *Hyoglossus* muscle to relay sensation from the anterior two-thirds of the tongue. The other branch is the **Mylohyoid Nerve** 3. It runs across the *Hyoglossus* and *Mylohyoid* muscles supplying the latter, and the anterior belly of *Digastric* muscle.

Facial Nerve (VII) 4 runs from beneath the ear to enter the parotid gland in the cheek, where it divides into several branches. One of these supplies the posterior belly of *Digastric* and the *Stylohyoid* muscle. Another runs along the lower border of the mandible to innervate the *Platysma* muscle.

Glossopharyngeal Nerve (IX) 5 runs from between the internal carotid artery and the internal jugular vein onto the back of the pharynx, to relay sensation from the *Pharyngeal Constrictor* muscles. A branch continues on to supply the posterior one-third of the tongue; it relays taste and sensation.

Vagus Nerve (X) 6 runs between the internal carotid artery and internal jugular vein throughout the length of the neck. It gives off a **Pharyngeal Branch** 7 to a plexus on the back of the pharynx that supplies the *Pharyngeal Constrictors*. It provides two branches to the larynx. The first is the **Superior Laryngeal Nerve** 8. It divides into the **Internal Laryngeal Nerve** 9, which relays sensation from the larynx above the vocal cords, and the **External Laryngeal Nerve** 10, which supplies motor to *Cricothyroid*. The second is the **Recurrent Laryngeal Nerve** 11. On the *left side* of the body, it comes off the vagus in the thorax, curving around under the arch of the aorta to run back up into the neck between the trachea and esophagus. On the *right side*, it comes off the vagus at the level of the subclavian artery, curving around underneath it to run back up along the trachea. The recurrent laryngeal nerves supply all but one of the *Intrinsic Laryngeal Muscles*.

Accessory Nerve (XI) 12 runs below the posterior belly of digastric to supply *Sternocleidomastoid*. It descends within this muscle for about half its length, where it turns to innervate *Trapezius*.

Hypoglossal Nerve (XII) 13 runs into the neck with the vagus for a short distance. It then runs anteriorly across the external carotid artery and medial to *Digastric* and onto *Hyoglossus* muscle, which it innervates. It also supplies all but one of the tongue muscles.

Identify the aforementioned branches of cranial nerves in figure 5.5.

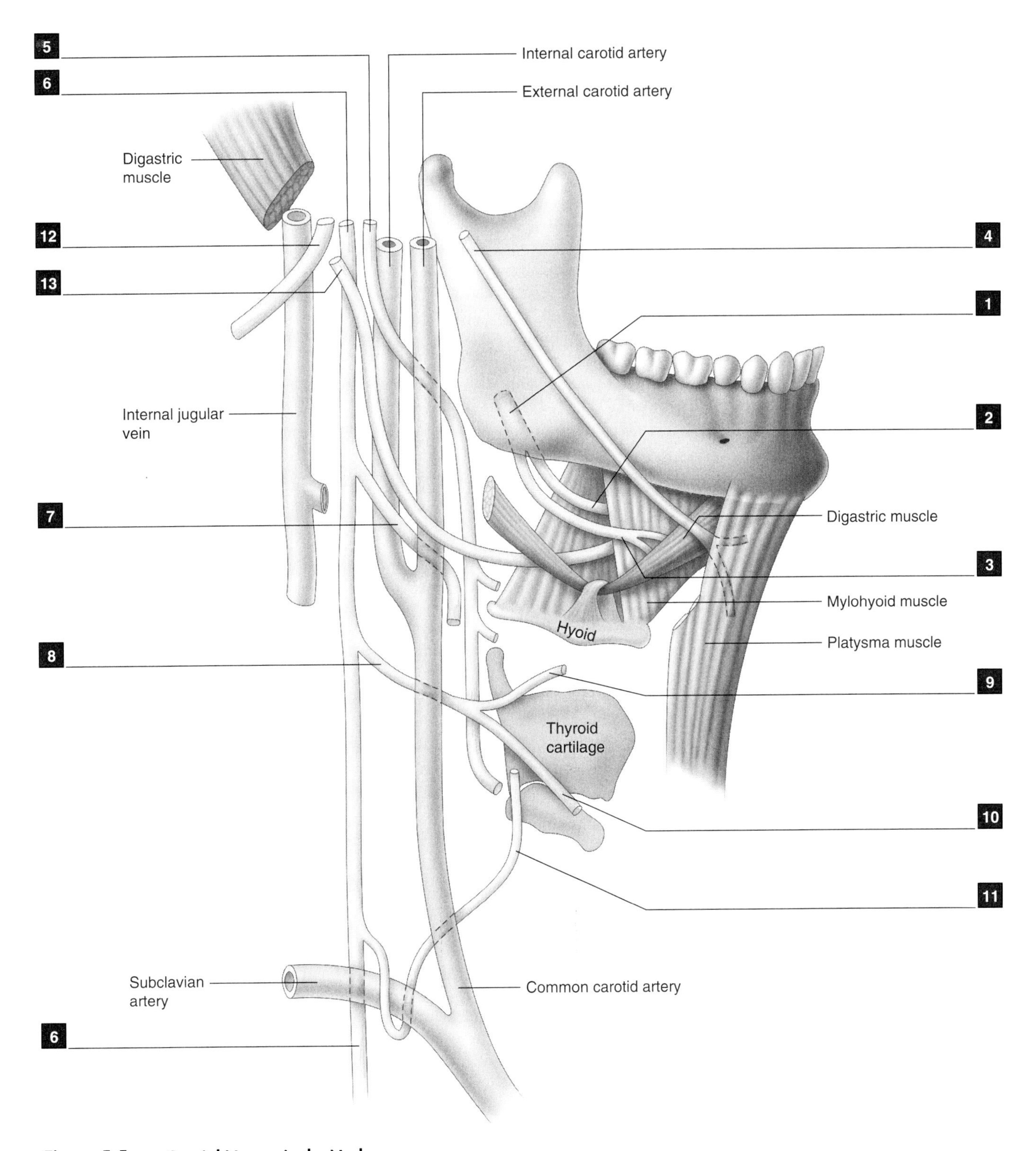

Figure 5.5 **Cranial Nerves in the Neck**
(Right side. Note that the course of nerve 11 will be different on the left side of the body.)

Autonomic Nerves in the Neck

The Autonomic Nervous System is concerned with the control and/or modification of involuntary activities such as smooth and cardiac muscle contraction, and glandular secretion. This system has two divisions: (1) *Parasympathetic* and (2) *Sympathetic*.

Both divisions involve two neurons in the transmission of motor impulses to a target organ. The first neuron (preganglionic) has its cell body in the *Central Nervous System*. It synapses with a second (postganglionic), whose cell body is located in an *Autonomic Ganglion* either close to the spinal cord (Sympathetic) or the target organ (Parasympathetic).

Sympathetic stimulation usually comes about via mass activation, whereas parasympathetic stimulation may involve individual nerve activation. The different physiological effects (cholinergic vs. adrenergic) of the two divisions are related to the neurotransmitters (acetylcholine vs. norepinephrine) released by their postganglionic neurons.

Parasympathetic Fibers

Parasympathetic innervation of structures in the neck derives from *Cranial Nerves*. Only one cranial nerve supplies glands that are in (or nearly in) the neck.

Facial Nerve (VII) (figure 5.5, item 4) provides parasympathetic fibers to the *Submandibular* and *Sublingual Salivary Glands* via a branch (the *Chorda Tympani*) that runs alongside the lingual nerve (figure 5.5, item 2). The Chorda Tympani synapses in the *Submandibular Ganglion*.

The Sympathetic Chain

The sympathetic nervous system is dealt with in greater detail in laboratory 9 (pp. 286–287), but because it innervates glands in the neck, and is readily "dissectable" there, we will pay it some attention now.

Mass discharge of the sympathetic division is conducted largely through a chain of fibers and ganglia located on either side of the vertebral column. Preganglionic neurons of the thoracic and lumbar region of the spinal cord send their axons out into the ventral rami of the spinal nerves. Some of these axons enter the chain of ***Sympathetic Ganglia*** via myelinated ***White Rami Communicantes***. Some axons synapse there, whereas others turn up or down the chain to synapse in higher or lower ganglia.

There is no sympathetic outflow from the cervical spinal nerves. But, the sympathetic chain extends into the neck. From the sympathetic ganglia in the neck, some postganglionic neurons run to a target organ in the neck or (usually) back in the thorax. Others jump back onto the ventral rami of the cervical spinal nerves for distribution to targets (e.g., *Arrector pili*) in the body wall and limbs. Postganglionic neurons travel to the spinal nerves via unmyelinated ***Gray Rami Communicantes***.

There are gray rami communicantes in the neck, but no white rami communicantes. Why is this so?

Let us examine the sympathetic chain and nerves in the neck.

The **Sympathetic Chain** 1 enters the neck deep to the subclavian artery, and runs upward through the length of the neck on the front of the *Longus colli* and *Longus capitis* muscles. It runs adjacent to the common carotid and internal carotid arteries, which it supplies. These postganglionic fibers follow the internal carotid artery into the skull.

The sympathetic trunk in the neck usually has three ganglia. The lowest is the ***Inferior Cervical Ganglion.*** It is located at or just below the level of C7, and is commonly fused with the uppermost thoracic ganglion to form the **Stellate Ganglion** 2. The **Middle Cervical Ganglion** 3 lies somewhere between the levels of C4 and C6. The long **Superior Cervical Ganglion** 4 stretches from C1 to C2.

At the level of the subclavian artery, the sympathetic chain sends out a branch that courses around the artery and upward to rejoin the chain. This loop is the **Ansa Subclavia** 5**.**

The chain communicates with the ventral rami of the cervical spinal nerves by thin **Gray Rami Communicantes** 6 that extend from the ganglia.

There are usually two (although there may be more) bundles of postganglionic sympathetic axons that travel back down the neck and into the thorax to innervate the heart. These are the **Cervical Sympathetic Cardiac Nerves** 7**.**

The chain provides innervation to the thyroid gland via unnamed branches. These **Unnamed Thyroid Nerves** 8 travel to the gland from the superior and middle cervical ganglia in company with the superior and inferior thyroid arteries.

Identify the components of the sympathetic nervous system in figure 5.6.

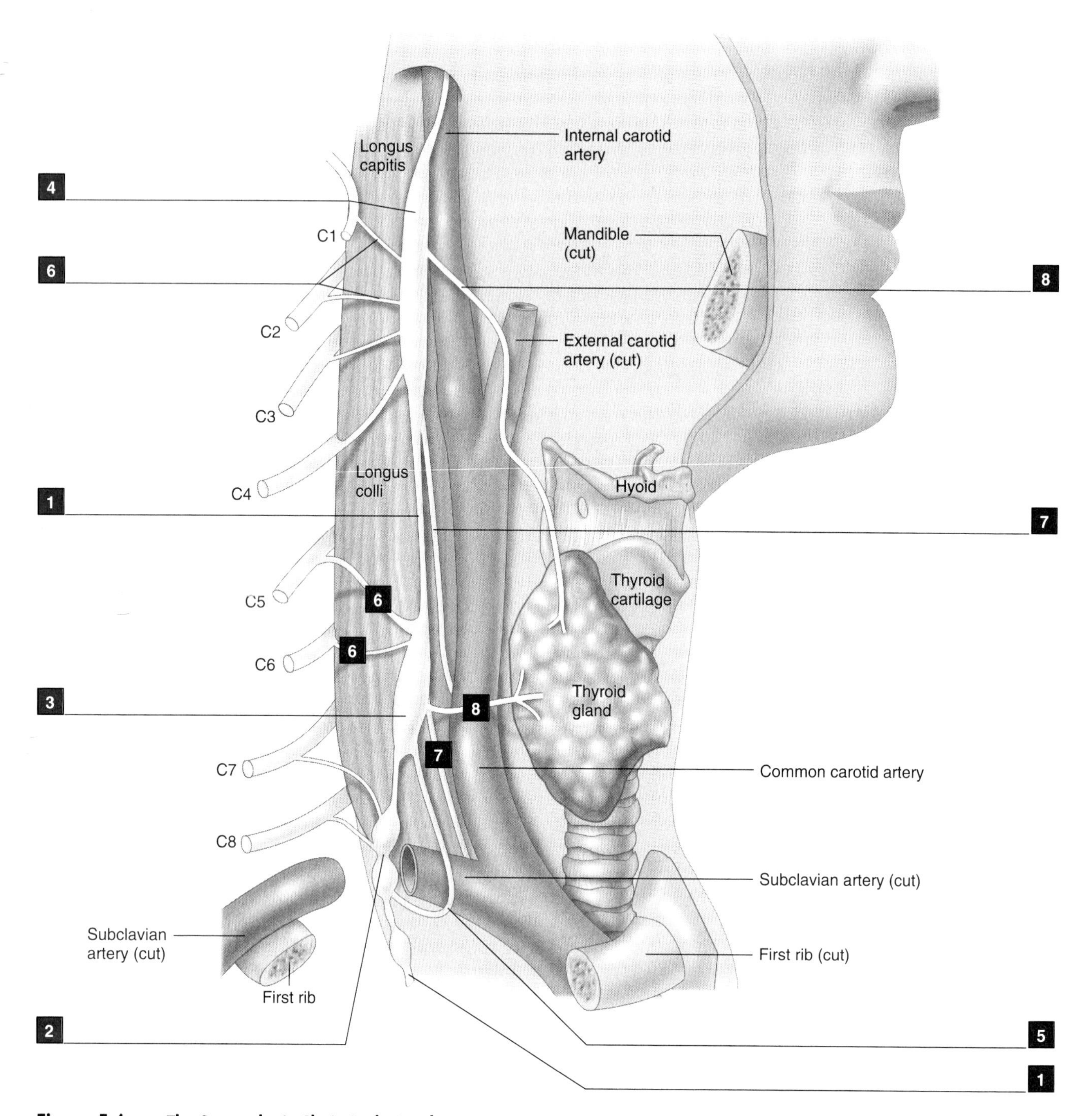

Figure 5.6 **The Sympathetic Chain in the Neck**

5.3 MUSCLES OF THE NECK

Muscles of the neck may attach to one or more of the bony or cartilaginous elements in it, or simply pass by on their way between the head and trunk. Some participate in movements of the shoulder (scapula). Others move the head, the spinal column, or the larynx and pharynx.

Muscles That Move the Scapula

Three muscles of the neck move the scapula. They are:

Trapezius
Rhomboideus
Levator Scapulae

We have examined these already in laboratory 2. Review them by studying figure 2.11 on p. 37.

Muscles That Move the Head

Seven muscles arise in, or pass through the neck to insert onto the skull. We have already examined three of them:

Splenius capitis
Erector spinae (Longissimus capitis part)
Semispinalis capitis

Review them by studying figures 2.12 and 2.13 on pp. 39–40.

Two of the other four muscles that move the head, **Rectus capitis anterior** and **Rectus capitis lateralis,** are minor, and need not concern us further. The other two are significant. They are *Longus capitis* and *Sternocleidomastoid.*

Longus capitis 1 arises from the transverse processes of C3–C6, and runs upward and medially to insert onto the base of the occipital bone anterior to the foramen magnum. It flexes the head, and rotates it to the same side.

Innervated by *Ventral Rami of C1–C6*

Sternocleidomastoid 2 arises from the medial third of the clavicle, and by a strong tendon from the front of the manubrium. It runs upward and backward on the side of the neck to insert onto the mastoid process and the lateral half of the superior nuchal line of the skull. It rotates the head to face toward the opposite side, and flexes the neck laterally. The two muscles acting together flex the neck.

Innervated by *Accessory Nerve* (*CN XI*), which communicates with *Spinal Nerves C2–C4*

Identify the longus capitis and sternocleidomastoid muscles in figure 5.7.

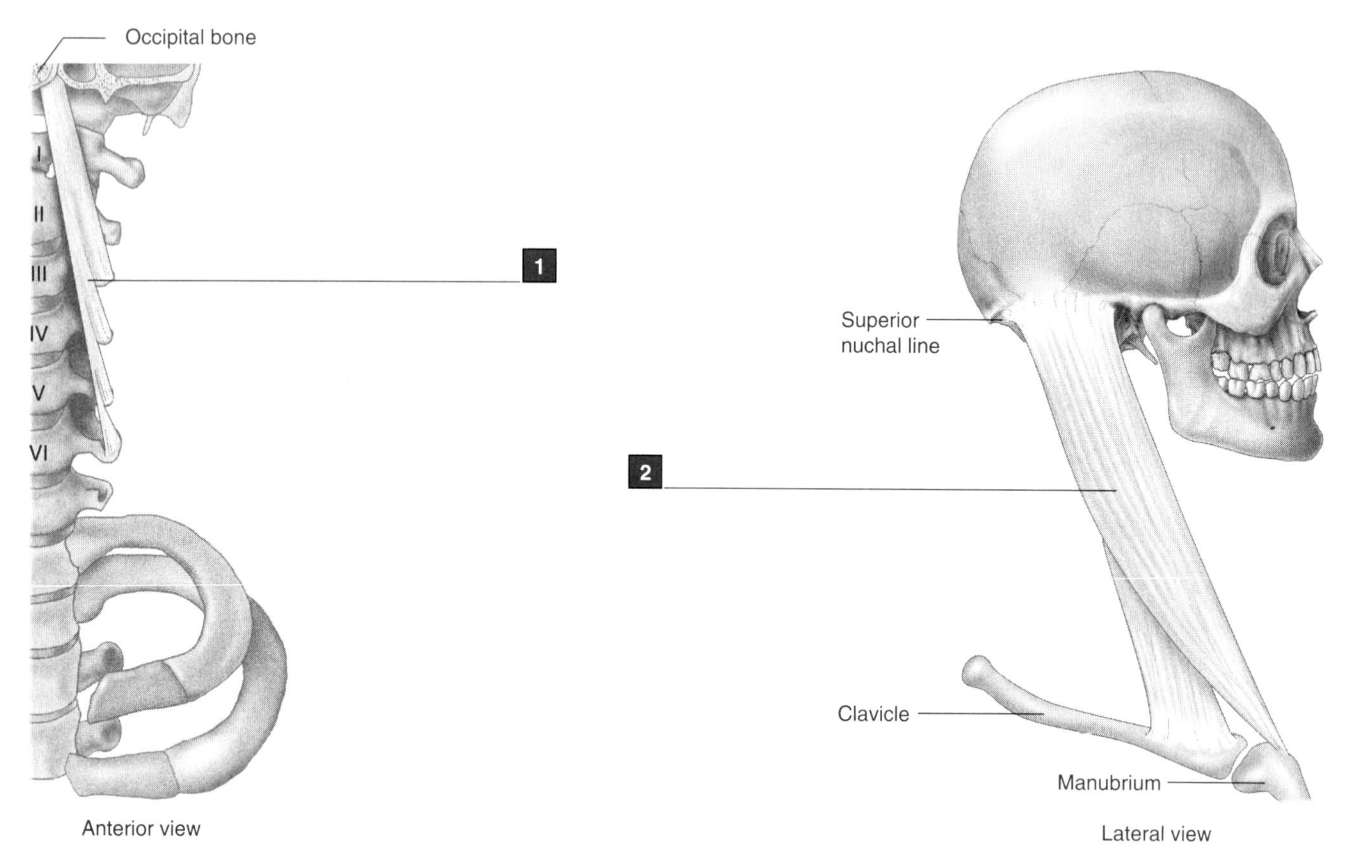

Figure 5.7 Muscles That Move the Head

Muscles That Move the Neck

Six muscles move the neck. We have already examined two of them:

Splenius (cervicis)

Semispinalis (cervicis)

Review them by studying figures 2.12 and 2.13 on pp. 39–40. The other four are: *Longus colli*, *Scalenus anterior*, *Scalenus medius*, and *Scalenus posterior*.

Longus colli 1 is a flat muscle that clothes the fronts of the cervical and first three thoracic vertebrae. It has three parts. The vertical part attaches to the 10 vertebral bodies. The superior oblique part arises from the transverse processes of C3–C5 and inserts onto the front of the atlas. The inferior oblique part arises from the bodies of T1–T3 and runs upward and laterally to insert onto the transverse processes of C5–C6. It flexes the neck forward or laterally.

Innervated by *Ventral Rami of C1–C7*

Scalenus anterior 2 arises from the transverse processes of C3–C6, and runs inferolaterally to insert on a bump called the scalene tubercle onto the superior surface of the first rib. It flexes the neck forward or laterally, and also assists in raising the first rib during forced inspiration.

Innervated by *Ventral Rami of C5–C6*

Scalenus medius 3 arises from the transverse processes of C2–C6, and runs inferolaterally to insert onto the superior surface of the first rib toward its posterior margin. It flexes the neck laterally.

Innervated by *Ventral Rami of C3–C8*

Scalenus posterior 4 is a little muscle that may blend with the fibers of *Scalenus medius*, or it may be absent. It arises from the transverse processes of C4–C6, and runs inferolaterally to insert onto the outer surface of the second rib. It flexes the neck laterally, and assists in raising the second rib during forced inspiration.

Innervated by *Ventral Rami of C4–C7*

Identify the aforementioned four neck muscles in figure 5.8.

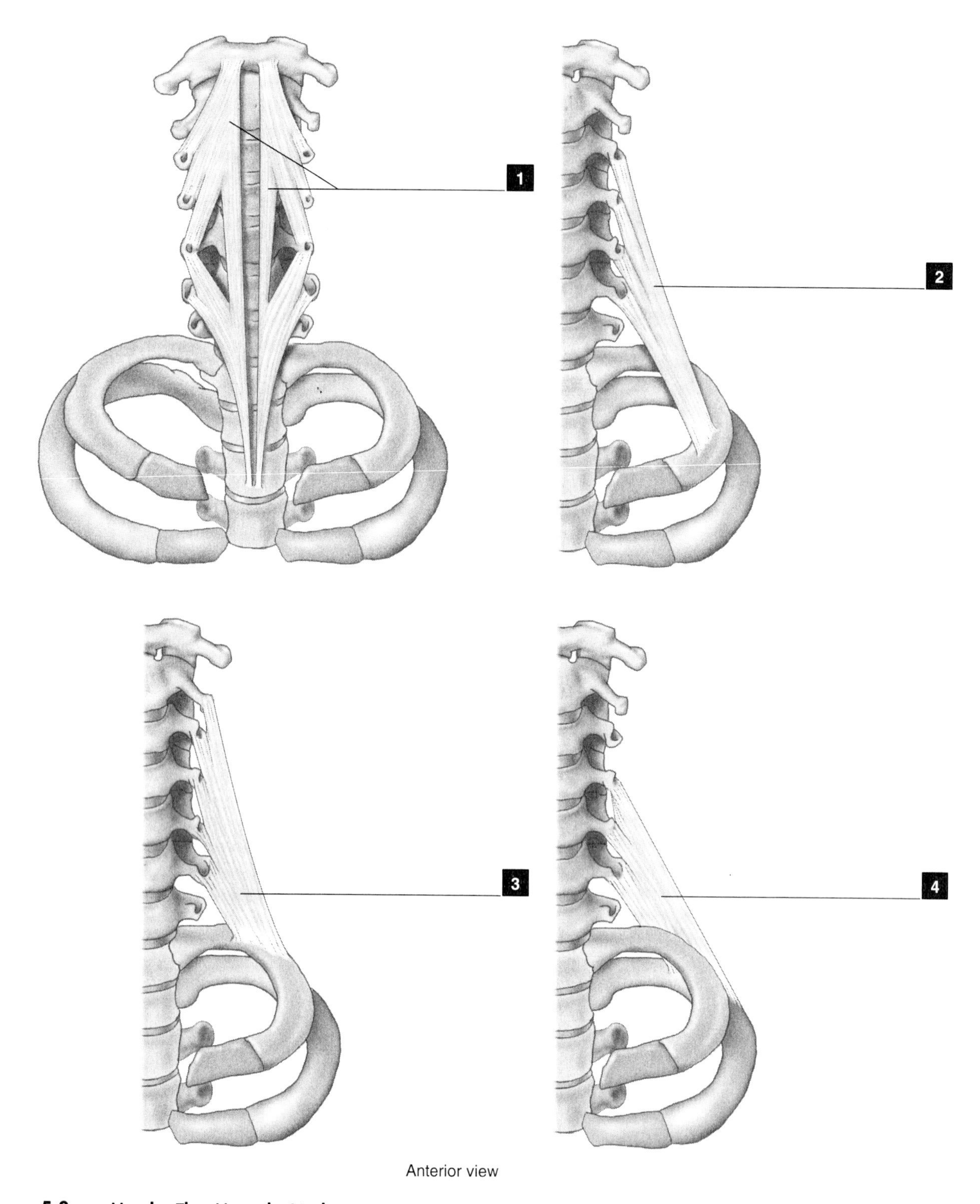

Figure 5.8 **Muscles That Move the Neck**

A Muscle of "Facial" Expression

One of the numerous muscles of "facial" expression is situated almost entirely in the neck. Like the others, it produces its effect by pulling on the skin. It has no effect on the skeleton. This muscle, like all others of facial expression, is supplied by the *Facial Nerve (CN VII)*.

Platysma **1** is a flat, thin muscular sheet that covers the anterolateral surface of the neck. It arises inferiorly from the fascia over the clavicle and runs upward across the lower border of the mandible to insert into the fascia below the mouth. Its uppermost medial fibers decussate (cross over) with those of the muscle from the opposite side. It draws the corners of the mouth downward (as in terror). When it contracts, its lateral margins define clear ridges in the skin on the side of the neck.

Innervated by *Cervical Branch of Facial Nerve (CN VII)*

Label and color the platysma muscle in figure 5.9.

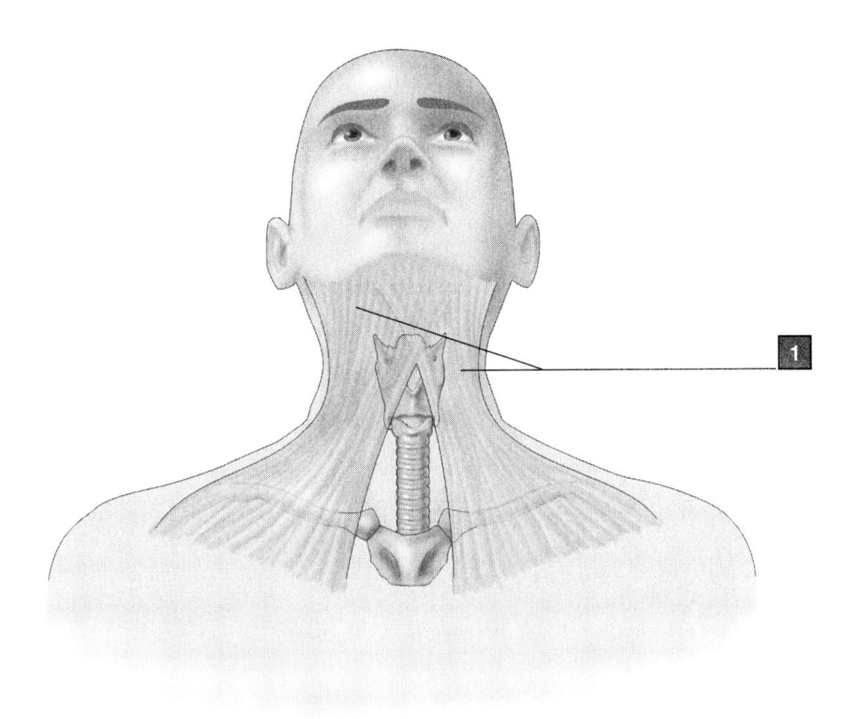

Figure 5.9 A Muscle of "Facial" Expression

Muscles That Move the Hyoid and Larynx

Two groups of muscles move the hyoid and larynx:

Hyolaryngeal Muscles (eight muscles). Six attach to the hyoid bone, and its movement translates into movement of the larynx because of their membranous/muscular connection. One runs between the hyoid bone and thyroid cartilage, and one attaches to the thyroid cartilage. They move the hyoid and larynx up and down during swallowing.

Intrinsic Laryngeal Muscles (five muscles). These run between the laryngeal cartilages to change vocal cord tension, which is an integral part of voice production.

Hyolaryngeal Muscles

Four run from the skull to the hyoid (*suprahyoid*), two run from the hyoid to the thoracic skeleton (*infrahyoid*), one runs between the thyroid cartilage and hyoid (*suprathyroid*), and one runs from the thyroid cartilage to the thoracic skeleton (*infrathyroid*). We will deal with them in that order.

Stylohyoid 1 arises from the styloid process of the skull, and runs alongside the *Stylohyoid Ligament* to insert onto the hyoid greater horn. It elevates the hyoid during swallowing.

Innervated by *Posterior Auricular Nerve* = branch of *Facial Nerve (CN VII)*

Digastric 2 has two bellies connected by an intermediate tendon that is attached to the hyoid by a loop of fascia. The posterior belly arises from the bottom of the skull medial to the mastoid process. The anterior belly inserts onto the lower border of the mandible near its midline.

The **posterior belly** elevates the hyoid during closing of the mouth.

Innervated by *Posterior Auricular Nerve* = branch of *Facial Nerve (CN VII)*

The **anterior belly** elevates the hyoid during swallowing and opening of the mouth. It also depresses the mandible.

Innervated by *Mylohyoid Nerve* = branch of *Trigeminal Nerve (CN V3)*

Mylohyoid 3 arises from a line on the inside of the mandible mandibular body. Some of its fibers meet in the midline along a fibrous band (a raphé) forming a semirigid floor to the mouth, which is important in swallowing. Other fibers insert on the hyoid, and elevate it.

Innervated by *Mylohyoid Nerve* = branch of *Trigeminal Nerve (CN V3)*

Geniohyoid 4 runs from a small spine on the back of the mandibular symphysis to the hyoid. It elevates the hyoid during swallowing.

Innervated by *Hypoglossal Nerve (CN XII)*

Identify the four suprahyoid muscles in figure 5.10.

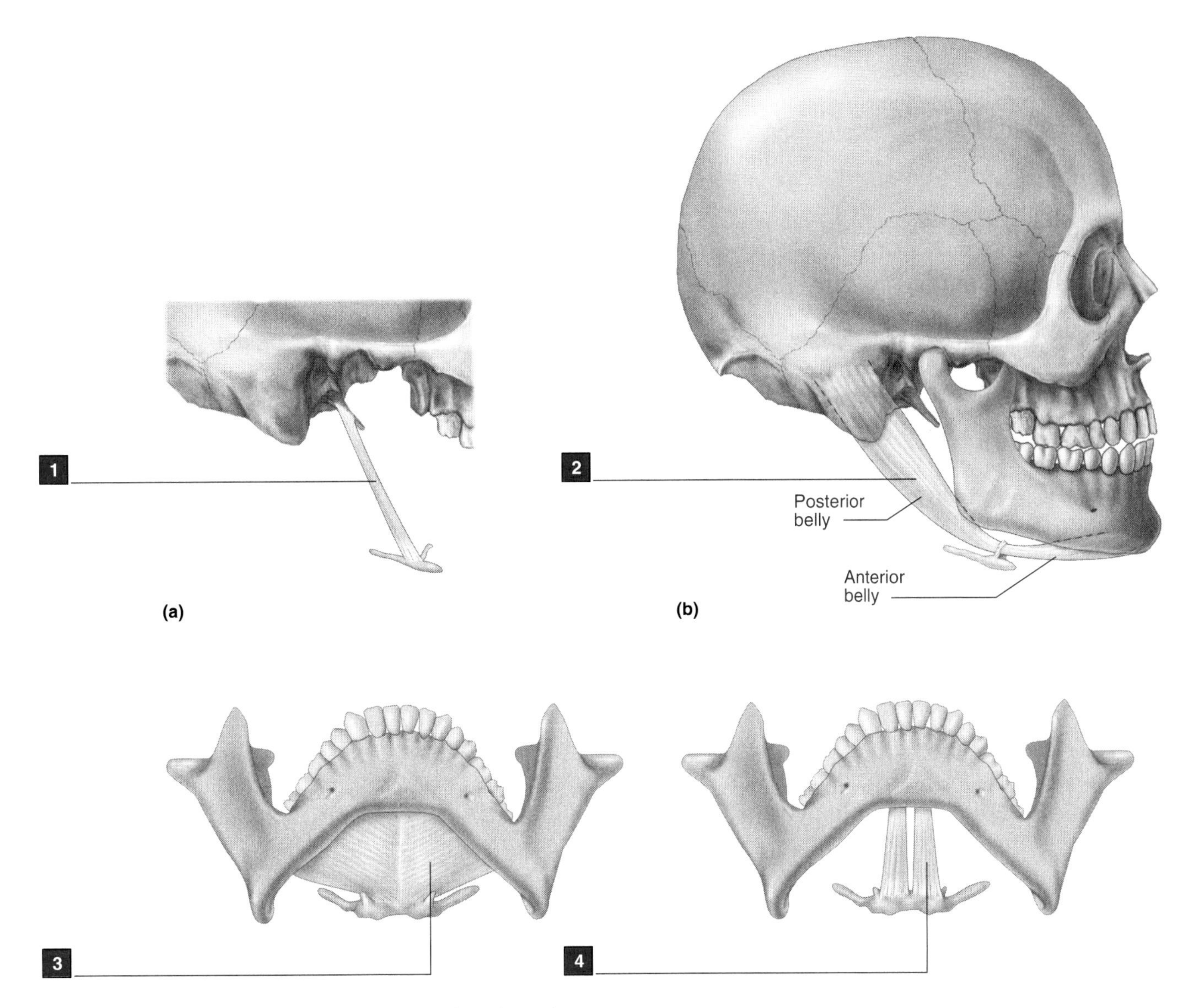

Figure 5.10 **Hyolaryngeal Muscles: Suprahyoid Muscles**
(a,b) lateral views; (c,d) inferior views.

Omohyoid 1 has two bellies connected by an intermediate tendon that is bound to the medial end of the clavicle by a fascial sling. The superior belly arises from the lower margin of the hyoid. The inferior belly arises from the superior border of the scapula medial to the suprascapular notch. It depresses the hyoid during vocalization and at the end of swallowing.

Innervated by *Ventral Rami of C1–C3* (via *Ansa Cervicalis*)

Sternohyoid 2 arises from the back of the manubrium and the medial end of the clavicle, and runs upward to insert onto the lower margin of the hyoid. It depresses the hyoid.

Innervated by *Ventral Rami of C1–C3* (via *Ansa Cervicalis*)

Thyrohyoid 3 runs from an oblique line along the front of the thyroid cartilage to the lower border of the hyoid. It can depress the hyoid, or elevate the larynx during swallowing.

Innervated by *Ventral Ramus of C1*

Sternothyroid 4 arises from the back of the manubrium and the medial part of the first costal cartilage (the cartilaginous rod that joins the rib to the manubrium). It runs upward, deep to the *Sternohyoid,* to insert onto the front of the thyroid cartilage. It depresses the larynx.

Innervated by *Ventral Rami of C2–C3* (via *Ansa Cervicalis*)

The Omohyoid, Sternohyoid, Thyrohyoid and Sternothyroid form the so-called Rectus cervicis. This is developmentally homologous with the Rectus abdominus muscle that runs longitudinally along the front of the abdomen.

Identify the infrahyoid muscles in figure 5.11, and the suprathyroid and infrathyroid muscles in figure 5.12.

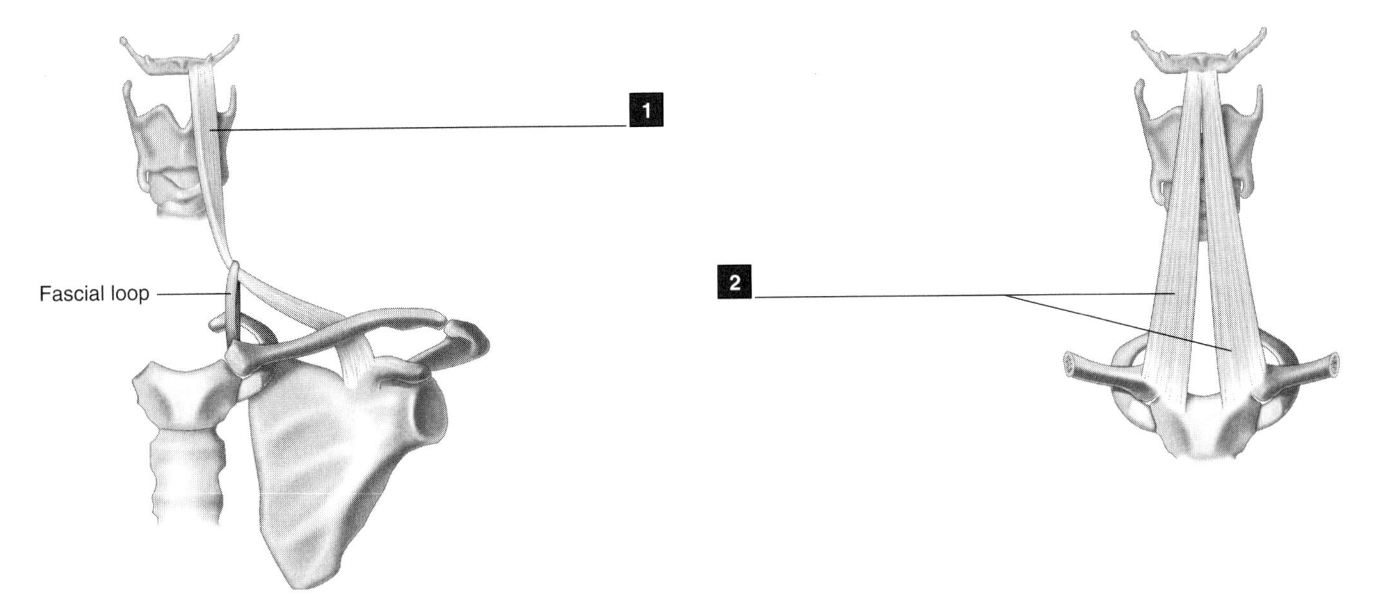

Figure 5.11 **Hyolaryngeal Muscles: Infrahyoid Muscles**

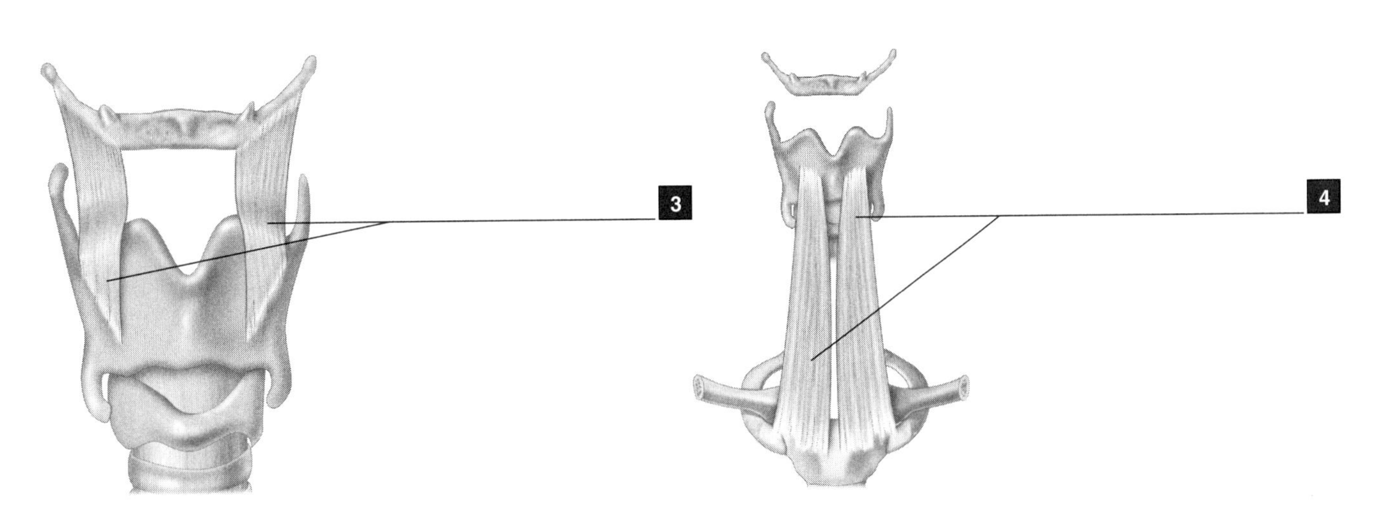

Figure 5.12 **Hyolaryngeal Muscles: Suprathyroid and Infrathyroid Muscles**

Intrinsic Laryngeal Muscles

Five muscles are capable of changing the tension of the vocal cords by moving the cricothyroid joint (one muscle) or the cricoarytenoid joint (four muscles).

Because vocal cord vibration is the immediate source of voice production, it is understandable that four of the intrinsic laryngeal muscles adduct (bring together) the cords. Only one abducts them. One tightens the cords and one slackens them.

Sound pitch results mainly from vocal cord tension. Two muscles tighten the cords and one causes them to slacken.

All five muscles are innervated by branches of the *Vagus Nerve (CN X)*.

Cricothyroid 1 runs from the front of the cricoid cartilage to the inferior border and horn of the thyroid cartilage. It rotates the thyroid forward and downward at the cricothyroid joint. This causes the vocal cords to tighten and adduct (slightly).

Innervated by *External Laryngeal nerve*

Lateral Cricoarytenoid 2 runs up and back from the upper rim of the cricoid to insert onto the lateral side of the base of the arytenoid cartilage. It rotates the arytenoid medially. This causes the vocal cords to adduct.

Innervated by *Recurrent laryngeal nerve*

Thyroarytenoid 3 arises from the back of the thyroid angle, and runs posteriorly to insert onto the anterolateral aspect of the arytenoid. Its most medial fibers are considered to comprise the *Vocalis Muscle*. Its uppermost fibers pass upward into the aryepiglottic fold to reach the epiglottis; these are referred to as the *Thyroepiglottic Muscle*. It rotates the arytenoid medially, which adducts the vocal cords and causes them to slacken. It also inverts the sensitive aryepiglottic folds during swallowing. This avoids the cough reflex, which would otherwise be elicited by the passage of food over the laryngeal opening.

Innervated by *Recurrent laryngeal nerve*

Arytenoideus 4 is composed of transverse and oblique fibers that run between the arytenoid cartilages. It causes the arytenoid cartilages to approximate one another. This adducts and tightens the vocal cords.

Innervated by *Recurrent laryngeal nerve*

Posterior Cricoarytenoid 5 runs from the back of the cricoid to insert onto the posterolateral side of the arytenoid cartilage. It laterally rotates the arytenoid cartilages. This causes the vocal cords to abduct. Because it is the only vocal cord abductor, it holds the glottis open during heavy breathing.

Innervated by *Recurrent laryngeal nerve*

Identify the five intrinsic laryngeal muscles in figure 5.13.

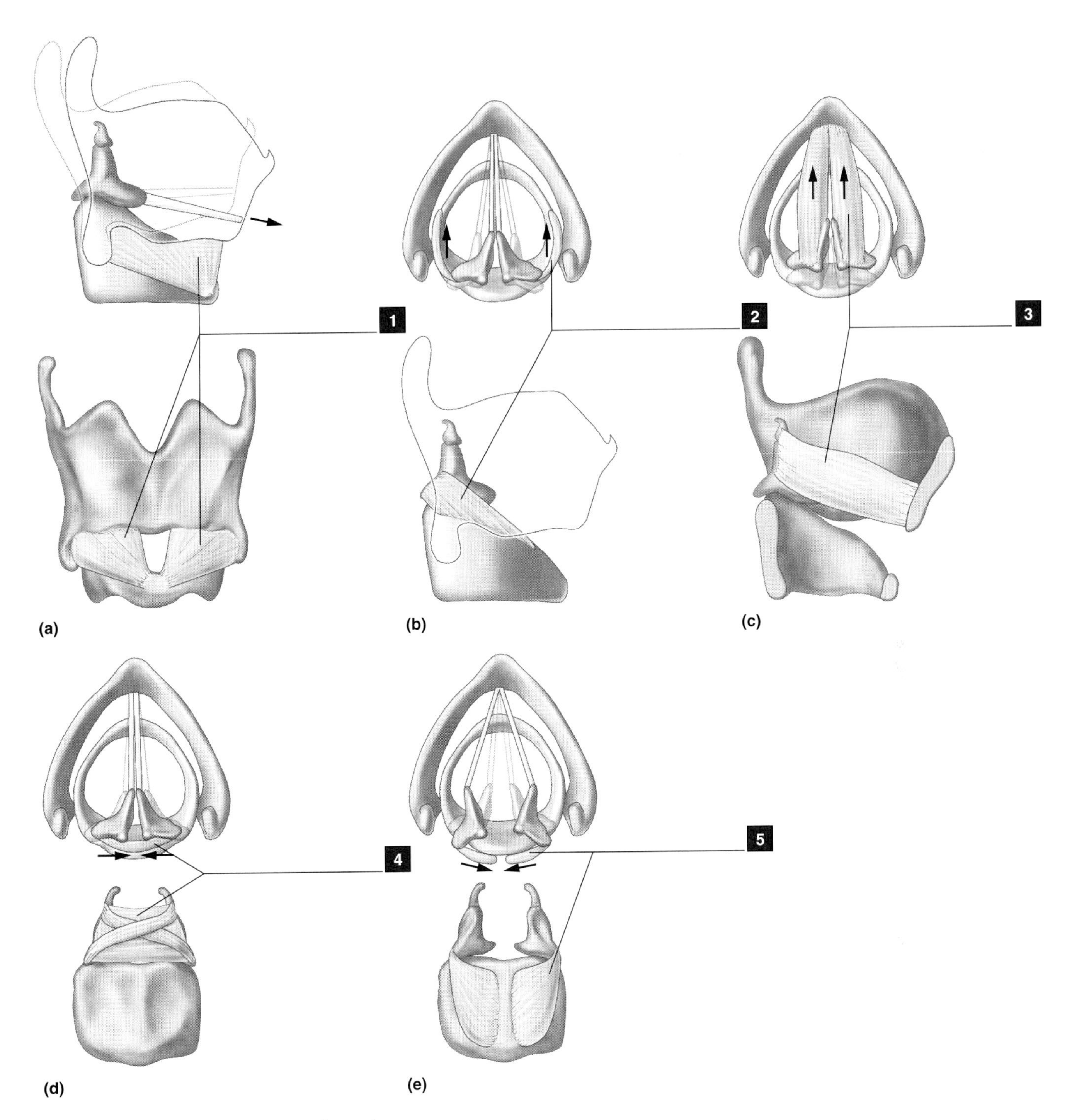

Figure 5.13 **Intrinsic Laryngeal Muscles**
(a, top) Lateral view; (a, bottom) anterior view; (b, top) superior view; (b, bottom) lateral view; (c, top) superior view; (c, bottom) medial view; (d, top) superior view; (d, bottom) posterior view; (e, top) superior view; (e, bottom) posterior view.

The Pharynx and Esophagus

The pharynx is a muscular tube that extends from the base of the skull to the level of the lower border of the larynx, where it becomes continuous with the esophagus. Its upper half is open anteriorly into the nasal and oral cavities.

The Pharynx has three parts. From top to bottom these are:

Nasopharynx
Oropharynx
Laryngopharynx

The **Nasopharynx** 1 opens into the **Nasal Cavity** 2 of the skull. The **Oropharynx** 3 lies below the **Palate** 4 and above the **Epiglottis** 5. It opens into the oral cavity. It is bounded around this opening by the **Uvula** 6, **Palatine Tonsils** 7, and **Tongue** 8. The **Laryngopharynx** 9 is continuous with the **Esophagus** 10.

The pharynx is composed of three constrictor muscles that are arranged like stacked funnels. From top to bottom these are

Superior Constrictor
Middle Constrictor
Inferior Constrictor

The **Superior Constrictor** 11 arises from the medial pterygoid plate of the skull and a narrow band of connective tissue (*Pterygomandibular Raphé*) that runs from this plate to the mandible behind the third molar. It has a tiny attachment to the base of the skull in the midline in front of the foramen magnum. The region between the muscle and the skull is filled by a sheet of tissue, the *Pharyngobasilar Fascia*. The fibers from each side curve around posteriorly to meet in the midline along a thin band of connective tissue (*Median Raphé*).

The **Middle Constrictor** 12 arises from the upper margin of the greater horn of the hyoid bone, deep to the *Hyoglossus* muscle. The fibers from each side curve around to meet at the median raphé.

The **Inferior Constrictor** 13 arises from the thyroid cartilage, the fascia over the *Cricothyroid* muscle, and the side of the cricoid cartilage. Its lowermost fibers intertwine with muscle fibers of the esophagus. The fibers from each side curve around to meet along a thin median raphé.

The pharynx is supplied by a plexus of nerves that lies on the back of the middle constrictor. This plexus is made up by branches from the *Glossopharyngeal Nerve (CN IX)* [sensory] and the *Vagus Nerve (CN X)* [motor]. It also receives sympathetic innervation by branches from the *Superior Cervical Ganglion*.

Identify the aforementioned structures in figure 5.14.

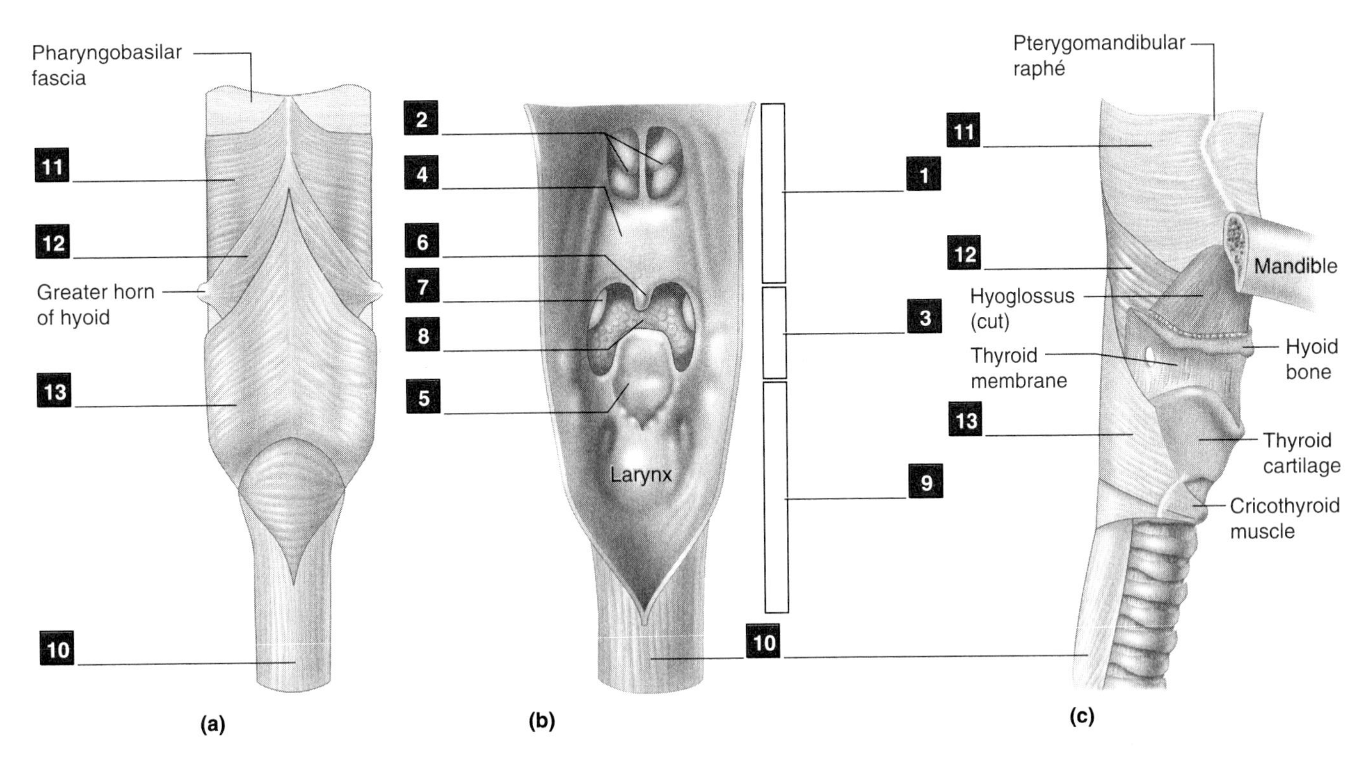

Figure 5.14 **The Pharynx and Esophagus**
(a) posterior view; (b) posterior view with pharynx cut open; (c) lateral view.

5.4 BLOOD VESSELS OF THE NECK

Arteries

Recall that on the *right side* of the body, a single artery—the **Brachiocephalic Trunk** 1—comes off the aortic arch. From the brachiocephalic trunk branch the **Common Carotid Artery** 2 and **Subclavian Artery** 3.

On the *left side* of the body, the **Subclavian Artery** and **Common Carotid Artery** arise directly from the aortic arch.

On *both sides* of the body, the **Common Carotid Artery** runs upward adjacent to the trachea and esophagus.

At the upper margin of the thyroid cartilage, the **Common Carotid Artery** divides into the **Internal Carotid Artery** 4 and **External Carotid Artery** 5.

Branches of the Internal Carotid Artery

The internal carotid artery has no branches in the neck. It runs behind the external carotid artery to enter the skull through the *Carotid Foramen*. It supplies the brain.

Branches of the External Carotid Artery

The external carotid artery has *seven* main branches.

Superior Thyroid Artery 6 comes off immediately to supply the thyroid and parathyroid glands.
Ascending Pharyngeal Artery 7 supplies the pharynx and soft palate.
Lingual Artery 8 comes off at the level of the hyoid to supply the tongue.
Facial Artery 9 curves around the lower edge of the mandible to ascend into the face.
Occipital Artery 10 supplies the back of the head.
Posterior Auricular Artery 11 supplies the scalp behind the ear.
Maxillary Artery 12 comes off just below the jaw joint and branches to the mandible, the flat bones of the cranium, and the maxilla.

After the maxillary artery comes off, the external carotid changes its name to **Superficial Temporal Artery** 13, which runs anterior to the ear.

Branches of the Subclavian Artery

The subclavian artery has *several* main branches.

Vertebral Artery 14 comes off first. It runs up the neck through the *Foramina Transversarium* of the cervical vertebrae. It enters the skull through the *Foramen Magnum* to supply the brain.
Internal Thoracic Artery 15 runs into the thorax behind the ribs, just lateral to the sternum.
Thyrocervical Trunk 16 is also a branch of the subclavian. It divides into *three* branches: (1) the **Suprascapular Artery** 17 supplies the scapular muscles; (2) the **Inferior Thyroid Artery** 18 supplies the thyroid and parathyroid glands; (3) the **Costocervical Trunk** is the third branch.

Identify the aforementioned arteries in figure 5.15.

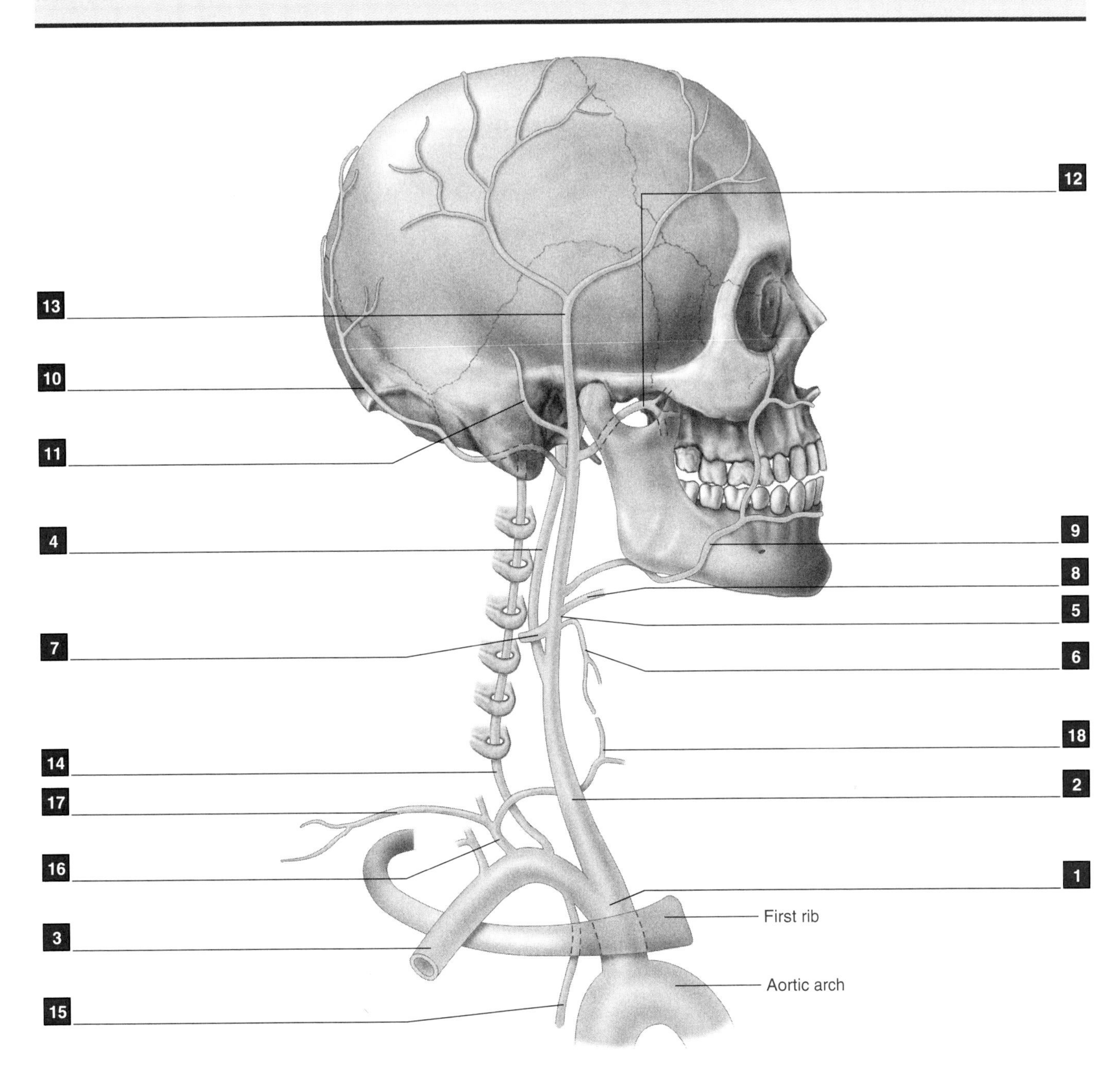

Figure 5.15 **Major Arteries of the Neck**
(Right side. Note that on the left side, the subclavian and common carotid arteries both arise directly from the aortic arch.)

Veins

The neck contains deep and superficial veins, with anastomotic connections between them. The superficial veins are quite variable in their drainage patterns.

Deep veins run together with some of the arteries, and carry their names. Thus, for example, the subclavian and vertebral arteries are accompanied by like-named veins.

The **Internal Jugular Vein** 1 exits the skull via the *Jugular Foramen*. It runs down the neck along the lateral surfaces of the internal and common carotid arteries. The vein and artery are surrounded by a tube of fascia, the *Carotid Sheath*, for most of their conjoint length.

The internal jugular vein drains into the **Subclavian Vein** 2. Their junction forms the **Brachiocephalic Vein** 3. The two brachiocephalic veins join to form the *Superior Vena Cava*.

The internal jugular vein receives blood from the **Superior Thyroid Vein** 4 and the **Middle Thyroid Vein** 5. The **Inferior Thyroid Vein** 6 drains into the left brachiocephalic vein. The internal jugular vein also receives blood via the **Common Facial Vein** 7 from the **Retromandibular Vein** 8 and the **Facial Vein** 9.

The retromandibular vein usually bifurcates. One branch, as we have seen, drains into the common facial vein. The other branch is joined by the **Posterior Auricular Vein** 10 and the **Occipital Vein** 11 to form the **External Jugular Vein** 12. The external jugular drains into the *Subclavian Vein*.

The retromandibular and facial veins join to form the **Anterior Communicating Vein** 13, which drains into the **Anterior Jugular Vein** 14. The anterior jugular vein drains into the *External Jugular Vein*.

The **Vertebral Vein** 15 exits the skull through the *Foramen Magnum*, and runs through the *Foramina Transversarium* of the cervical vertebrae. It drains into the junction of the *External Jugular Vein* and *Subclavian Vein*.

Identify the aforementioned veins in figure 5.16.

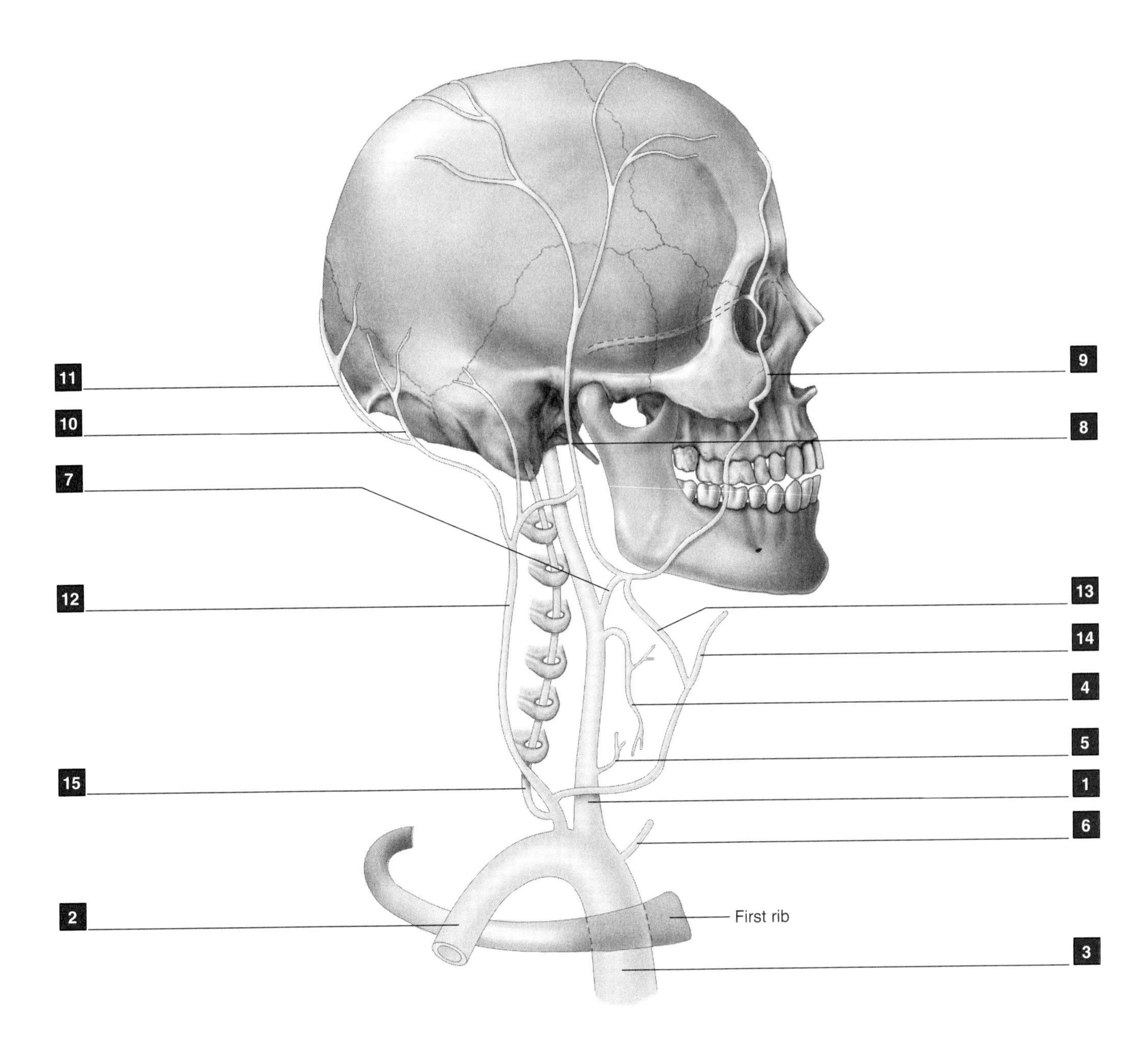

Figure 5.16 **Major Veins of the Neck**

Relationships of the Great Vessels in the Root of the Neck

A number of very important structures occupy the root of the neck. Their anatomical relationships are significant.

Anteriorly, the **Internal Jugular Veins** 1 merge with the **Subclavian Veins** 2 to form the **Brachiocephalic Veins** 3. The left brachiocephalic vein crosses the midline behind the manubrium, receiving the **Inferior Thyroid Vein** 4. It then joins the right brachiocephalic vein to form the **Superior Vena Cava** 5.

Posterior to the veins, the ***Aortic Arch*** gives off the **Brachiocephalic Trunk** 6 followed by the **Left Common Carotid Artery** 7 and the **Left Subclavian Artery.** 8 Behind the sternoclavicular joint, the brachiocephalic trunk divides into the **Right Common Carotid Artery** 9 and **Right Subclavian Artery.** 10 The subclavian artery gives rise to the *Vertebral Artery*, and the *Thyrocervical* and *Costocervical Trunks* before crossing over the first rib to become the *Axillary Artery*.

Between the veins and arteries run the **Vagus Nerves** 11 and **Phrenic Nerves** 12 on their way into the thoracic cavity. The phrenic nerve runs between the subclavian vein and subclavian artery. On the *left* side, the vagus nerve runs between the brachiocephalic vein and subclavian artery. On the *right* side, the vagus nerve runs between the brachiocephalic vein and brachiocephalic trunk.

The **Trachea** 13 runs behind and between the common carotid arteries. The **Esophagus** 14 lies behind the trachea.

The space bounded anteriorly by the carotid and subclavian arteries, medially by the trachea and esophagus, and posterolaterally by the first ribs is occupied by the apex of the **Pleural Dome** 15. The pleural sac lines that part of the thoracic cavity occupied by the lungs, and during deep inspiration the lungs will protrude into the pleural dome.

The **Thoracic Duct** 16 is the largest of the lymphatic channels. It enters the neck from the thorax on the *left* side of the esophagus. It ascends to the level of the thyroid gland, where it turns inferolaterally to drain into the brachiocephalic vein at its junction with the internal jugular vein. The **Right Lymphatic Duct** 17 drains into the right subclavian vein at its junction with the internal jugular vein.

Identify the aforementioned structures in figure 5.17.

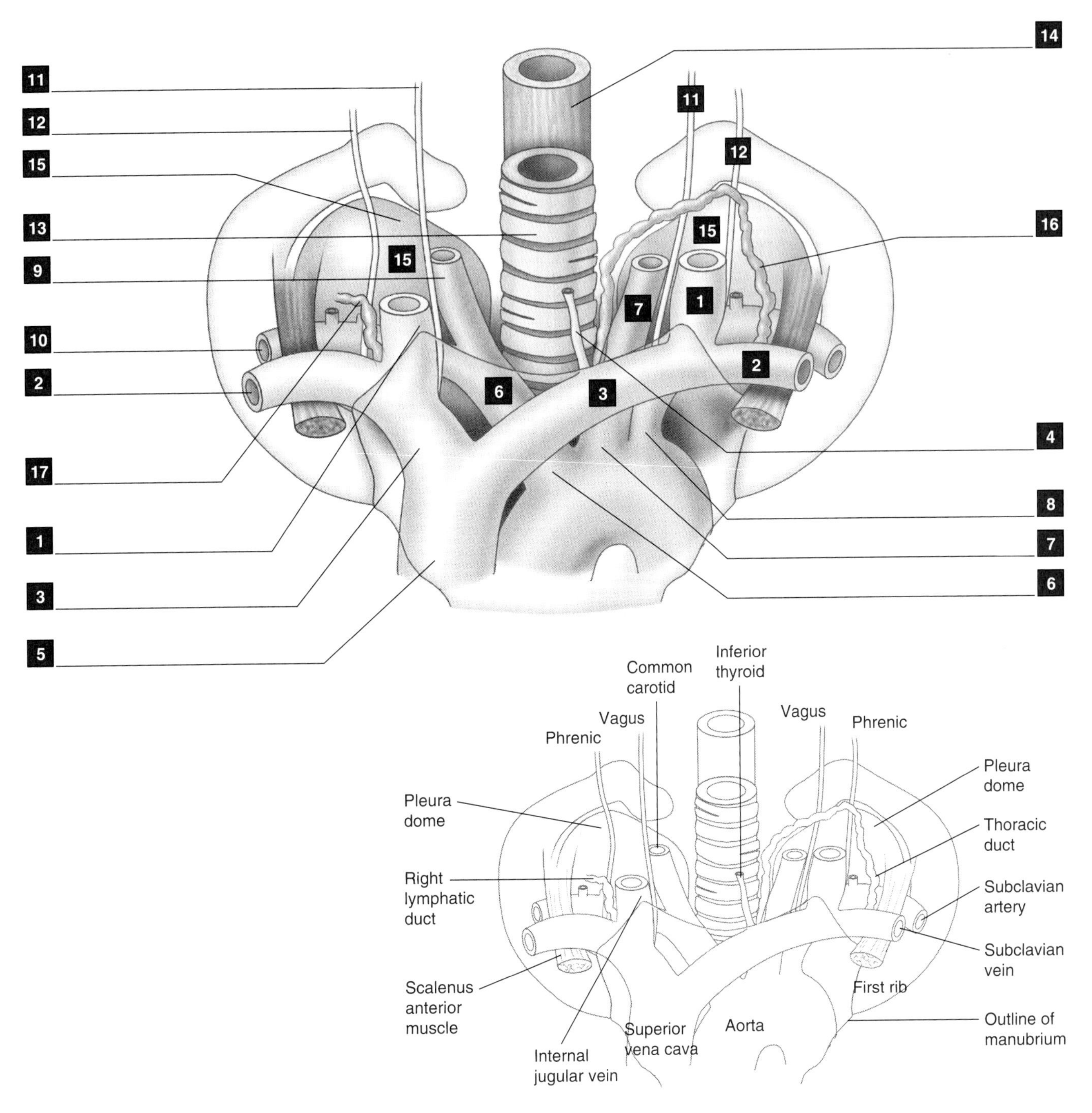

Figure 5.17 **The Root of the Neck**
Some of the structures discussed on page 162 are indicated in the inset illustration to assist you in orientation.

5.5 THYROID AND PARATHYROID GLANDS

There are two kinds of glands: *Exocrine* and *Endocrine*.

An ***Exocrine Gland*** secretes its product into a duct, which transports it to the outer surface of the body (turn back to laboratory 2, p. 22 for a review) or into the lumen of an organ (e.g., digestive and mucous glands).

An ***Endocrine Gland*** secretes its product (*hormone*) into the space around the secretory cells. The hormone diffuses into capillaries and is transported by the circulatory system throughout the body.

The thyroid and parathyroids are *Endocrine* glands.

Anatomy and Position of the Glands

The **Thyroid Gland** lies at the junction of the larynx and trachea. It is H-shaped, with two **Lateral Lobes** 1 connected by a horizontal **Isthmus** 2. The lobes cover the lower sides of the thyroid cartilage, the cricoid cartilage, and the upper six tracheal rings. The isthmus lies in front of the second to fourth tracheal rings. About 50% of people have a fingerlike **Pyramidal Lobe** 3 that projects upward onto the larynx from the middle of the isthmus.

The four lentil-sized **Parathyroid Glands** 4 are embedded on the back of the thyroid gland. The superior pair lie at the level of the lower margin of the pharynx; the position of the inferior pair is more variable.

Blood Supply of the Glands

Because they are endocrine glands, the thyroid and parathyroids are highly vascular.

The thyroid and parathyroid glands derive their blood supply from the superior and inferior thyroid arteries. The **Superior Thyroid Artery** 5 is a branch of the **External Carotid Artery** 6. It runs downward across the thyroid cartilage to the gland. The **Inferior Thyroid Artery** 7 is a branch of the Thyrocervical Trunk, which emerges from the **Subclavian Artery** 8.

Blood from the glands is drained by the **Superior Thyroid Vein** 9 and the **Middle Thyroid Vein** 10, which drain into the **Internal Jugular Vein** 11. They are also drained by the **Inferior Thyroid Veins** 12, which unite to drain into the **Left Brachiocephalic Vein** 13 behind the manubrium.

Identify the aforementioned structures in figure 5.18.

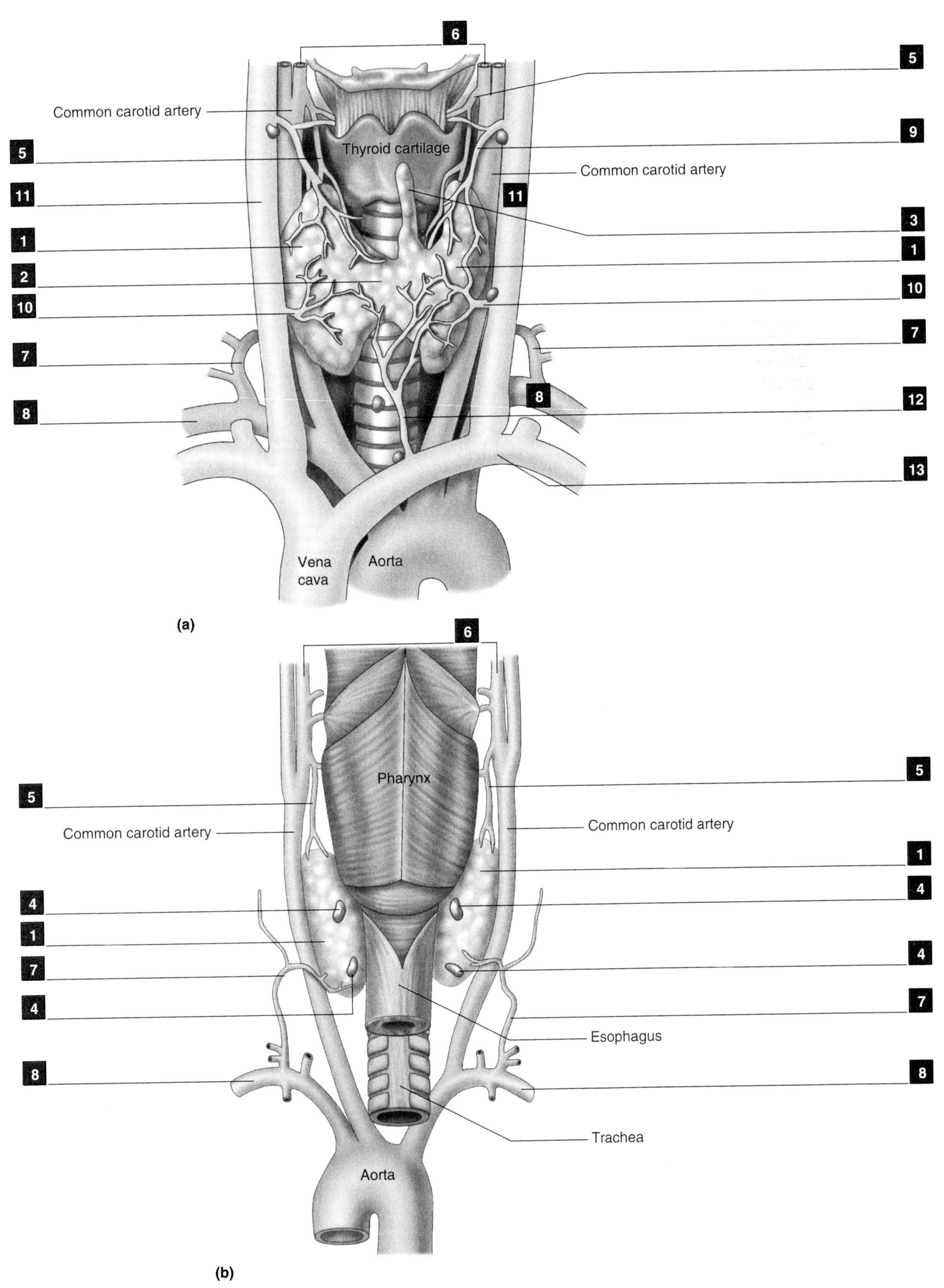

Figure 5.18 **Blood Vessels of the Endocrine Glands in the Neck**
(a) anterior view; (b) posterior view (veins not shown).

LABORATORY 6

The Head

6.1 THE SKULL

There are some 22 bones in the head. All but six are paired. These bones constitute the *Skull*. The skull can be divided into two parts: the *Cranium* houses the brain, and the *Face* surrounds the oral and nasal cavities.

Bones of the Cranium

Frontal

The **Frontal** 1 is a scoop-shaped bone that forms the forehead and the roof of the orbit (eye socket). It articulates posteriorly, along the *Coronal Suture*, with the two parietal bones. At birth, there are two frontals separated by a median sagittal suture. The region where the four incompletely ossified frontal and parietal bones will eventually meet is not yet ossified. In this region, the brain is covered only by membranous tissue. This "soft spot" is the *Bregmatic Fontanelle*. There are several other fontanelles in an infant's cranium; these are obliterated by bone growth usually by two years of age. The frontal contains a *Paranasal Sinus* of variable size just above the nose.

Parietal

The paired **Parietals** 2 are quadrilateral bones on the top and sides of the cranium. They articulate with one another along the midline *Sagittal Suture*, posteriorly with the occipital bone along the *Lambdoid Suture*, and inferiorly with the temporal bones along the *Squamosal Suture*. Internally, there is a gutter along the sagittal suture that is created by the *Superior Sagittal Sinus*.

Occipital

The **Occipital** 3 is a scoop-shaped bone at the back of the skull. It contains the large **Foramen Magnum** 4**,** through which passes the medulla oblongata of the brain stem, and the vertebral artery. On either side of the foramen magnum lie the convex *Occipital Condyles* by which the skull articulates with the atlas. Between each condyle and the foramen magnum is the **Hypoglossal Canal** 5**,** which transmits the hypoglossal nerve (CN XII). Neck muscles attach to the *Nuchal Lines* along the back of the bone. Internally, there is an inverted T-shaped sulcus formed by the *Superior Sagittal Sinus* and the left and right *Transverse Sinuses*. The occipital articulates anteriorly with the sphenoid bone by a synchondrosis.

Temporal

The **Temporal** 6 has three parts: Squamous, Petrous, and Tympanic.

The flat ***Squamous Portion*** forms part of the lateral wall of the braincase. Its anterolateral process constitutes the back of the *Zygomatic Arch*. At its base, the *Glenoid Fossa* forms the jaw (temporomandibular) joint with the condyle of the mandible. The *Mastoid Process*, which contains numerous air cells, projects inferiorly from the back of the squamous temporal.

The pyramid-shaped **Petrous Portion** houses the *Middle* and *Inner Ear Cavities*. Endocranially, it separates the *Middle* and *Posterior Cranial Fossae*. There is a deep, S-shaped sulcus along its border with the occipital that is formed by the *Sigmoid Sinus*, which becomes the internal jugular vein as it drains through the **Jugular Foramen** 7**.** This foramen also transmits the glossopharyngeal (CN IX), vagus (CN X), and accessory (CN XI) nerves. Endocranially, the **Internal Auditory (Acoustic) Meatus** 8 pierces the back wall of the bone; it transmits the facial (CN VII) and vestibulocochlear (CN VIII) nerves into the inner and middle ear cavities. The under side of the bone has the **Carotid Foramen** 9**,** through which the internal carotid artery enters the *Carotid Canal* to emerge endocranially where the petrous temporal meets the body of the sphenoid bone. The **Foramen Lacerum** 10 is located at this junction. The *Styloid Process* projects inferiorly just medial to the mastoid process, and next to it is the **Stylomastoid Foramen** 11**,** by which the facial nerve (CN VII) exits the skull.

The ***Tympanic*** bridges the gap between the glenoid fossa and mastoid process, creating the *Auditory Canal* and the *External Auditory (Acoustic) Meatus*.

Sphenoid

The **Sphenoid** 12 is a complex bone. The boxlike body of the sphenoid articulates posteriorly with the occipital, and anteriorly with the ethmoid and vomer, forming the back roof of the nasal cavity. The body has a large sinus. Its endocranial aspect resembles a saddle (the *Sella Turcica*), the seat of which houses the pituitary gland (*Hypophyseal Fossa*). In front of the hypophyseal fossa, are the **Optic Canals** 13, which transmit the optic nerves (CN II). The *Lesser Wings* splay laterally to form the back part of the floor of the *Anterior Cranial Fossa*. The lesser wing is separated from the *Greater Wing* by the **Superior Orbital Fissure** 14, which transmits the oculomotor (CN III), trochlear (CN IV), abducent (CN VI), and the ophthalmic division of the trigeminal nerve (CN V1). Just inferior to the medial end of the superior orbital fissure is the **Foramen Rotundum** 15; it transmits the maxillary division of the trigeminal nerve (CN V2). Behind it is the **Foramen Ovale** 16, which transmits the mandibular division of the trigeminal nerve (CN V3). Just lateral to the foramen ovale is the small **Foramen Spinosum** 17, through which the middle meningeal artery enters the cranial cavity. The greater wing splays out laterally from the body to form part of the lateral wall of the braincase and the back wall of the orbit. The *Medial* and *Lateral Pterygoid Plates* project from the bottom of the greater wing where it joins the body.

Bones of the Face

Maxilla

The **Maxillae** 18 are paired bones that constitute most of the hard palate and the face below the orbits. They form the floors of the orbits, and each has a very large sinus. They articulate with one another along the median palatine suture, which is interrupted anteriorly by the **Incisive Foramen** 19. The *Alveolar Process* houses the roots of the teeth around the front and side of the palate. The maxilla articulates laterally with the zygomatic bone. Immediately below the inferior margin of the orbit is the **Infraorbital Foramen** 20. The maxillae form the inferior and lateral margins of the *Nasal (Pyriform) Aperture*, and the *Anterior Nasal Spine*, to which is attached the septal cartilage of the nose. The perpendicular plate of the vomer articulates along the length of the maxillae in the midline. This partly divides the nasal cavity into left and right halves. The maxilla articulates with the frontal, lacrimal, and nasal bones. The *Nasolacrimal Duct* traverses the maxilla from the orbit to drain into the nasal cavity just below the inferior nasal concha. Because of this duct, your nose runs when you cry.

Nasal

The **Nasals** 21 are paired small bones that articulate superiorly with the frontal and laterally with the maxillae. Their free inferior margins form the superior border of the *Pyriform Aperture*.

Lacrimal

The **Lacrimals** 22 are paired little bones that articulate with the maxillae, forming the *Lacrimal Duct* along the medial side of the orbit.

Zygomatic

The **Zygomatics** 23 are paired bones that form the cheeks. A thin plate extends posteriorly to contact the fingerlike projection from the temporal, forming the *Zygomatic Arch*.

Ethmoid

The **Ethmoid** 24 is an unpaired, box-shaped bone. It has a top side and lateral sides, and the interior of the box contains several thin bony sheets and some air sinuses. Its top, which fills a small midline gap in the frontal bone in the *Anterior Cranial Fossa*, is perforated by numerous holes. This is known as the *Cribriform Plate*, and it transmits the olfactory nerves (CN I) from the nasal cavity. A perpendicular plate, the *Crista Galli*, rises between the two olfactory bulbs to anchor a fold of dura mater. The sides of the box, which form the medial wall of the orbit, balloon inward to form the *Superior Nasal Concha*. Inferior to this, another thin bony plate, the *Middle Nasal Concha*, extends downward into the nasal cavity. In the midline, the thin *Perpendicular Plate* of the ethmoid articulates with the vomer to form the *Nasal Septum*.

Inferior Nasal Concha

The **Inferior Nasal Conchae** 25 are paired small bones that extend along the lateral walls of the nasal cavity just below the middle nasal conchae. They articulate with the maxillae.

Vomer

The **Vomer** 26 is a thin, unpaired bone that forms part of the midline *Nasal Septum*. It articulates inferiorly with the maxillae and superiorly with the ethmoid and sphenoid.

Palatine

The **Palatines** 27 are paired, L-shaped small bones, with vertical and horizontal plates. The horizontal plates articulate with the maxillae, forming the back of the palate, and with the pterygoid plates of the sphenoid. The horizontal plate is perforated by a *Greater Palatine Foramen*. The vertical plate extends upward, forming a tiny part of the lateral wall of the nasal cavity and the *Sphenopalatine Foramen*.

Mandible

The **Mandible** 28 constitutes the lower jaw. It has a *Body* that contains the roots of the teeth within its *Alveolar Process*. The body projects forward in the midline as the *Mental Protuberance* (chin); its posterior surface has small *Genial Spines* for attachment of tongue muscles. The lateral surface of the body is pierced by the **Mental Foramen** 29, which transmits one of the terminal branches (mental nerve) of the mandibular division of trigeminal (CN V). The vertical *Ramus* has two process that extend superiorly. The anterior is the *Coronoid Process;* the posterior has a *Condyle* that articulates with the temporal. The internal aspect of the ramus is pierced by the **Mandibular Foramen** 30, by which the inferior alveolar nerve enters the mandible.

Identify the aforementioned structures in figures 6.1, 6.2, 6.3, and 6.4.

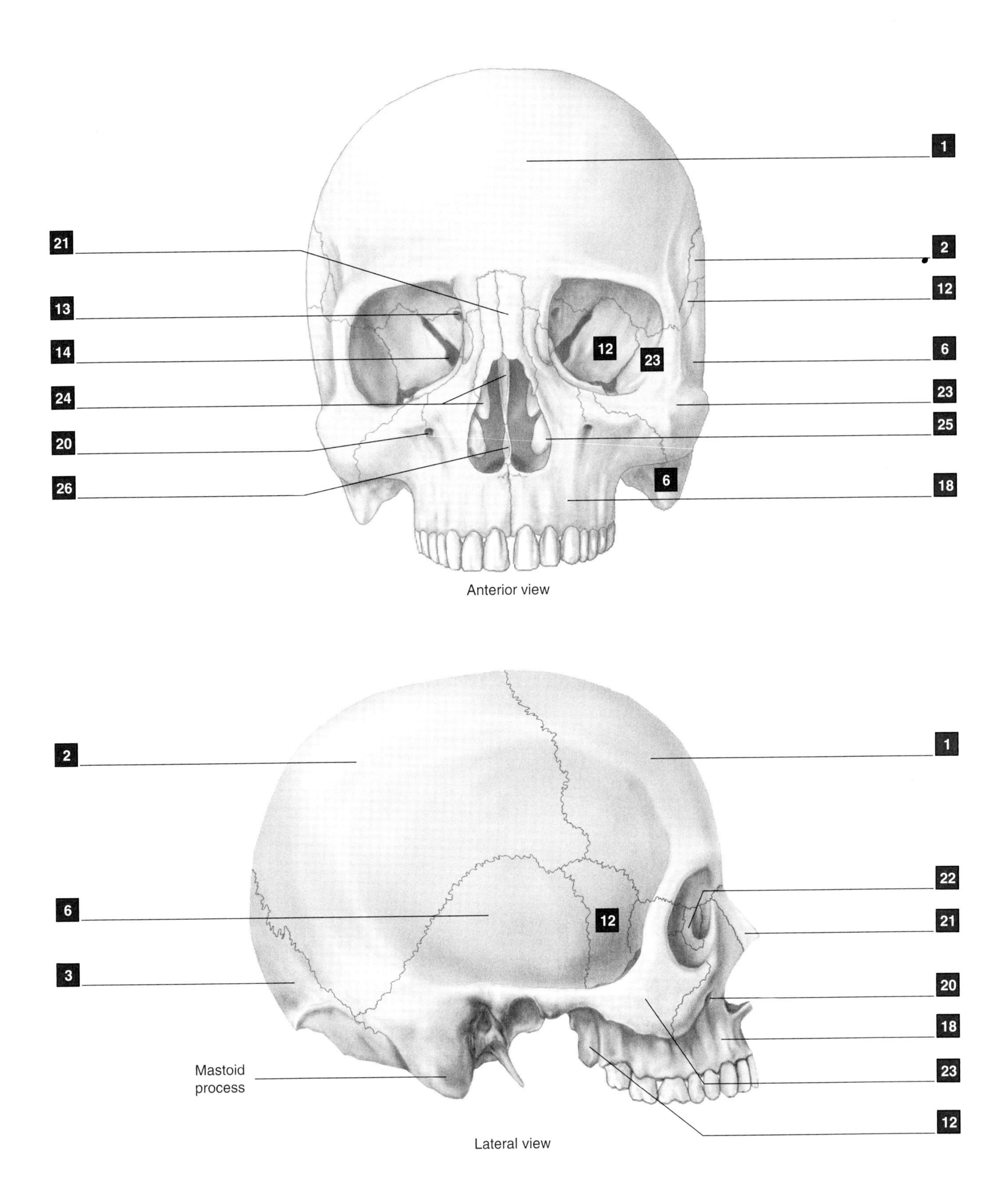

Figure 6.1 **Bones of the Skull**

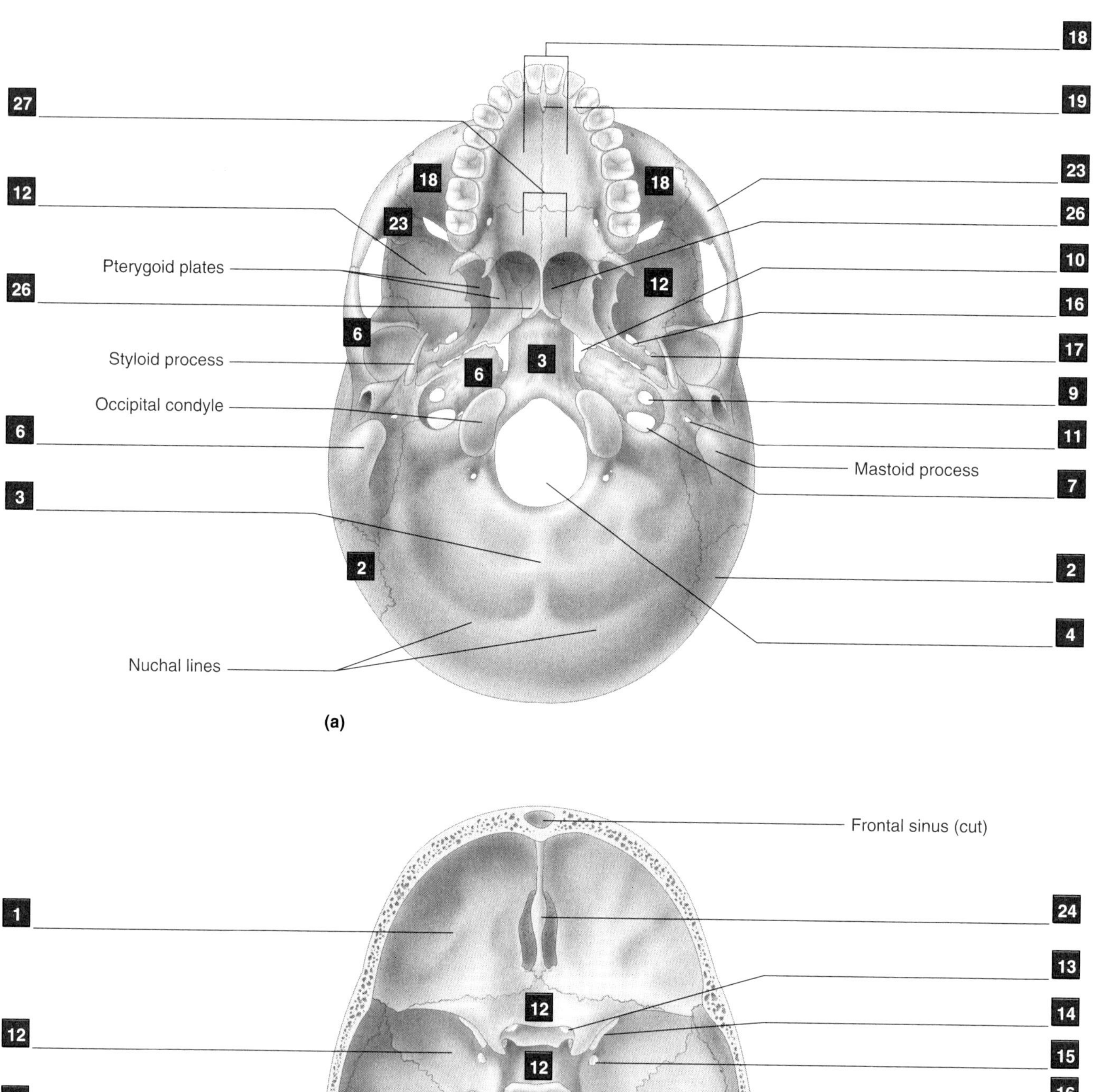

Figure 6.2 **Bones of the Skull**
(a) basal view; (b) internal view of base.

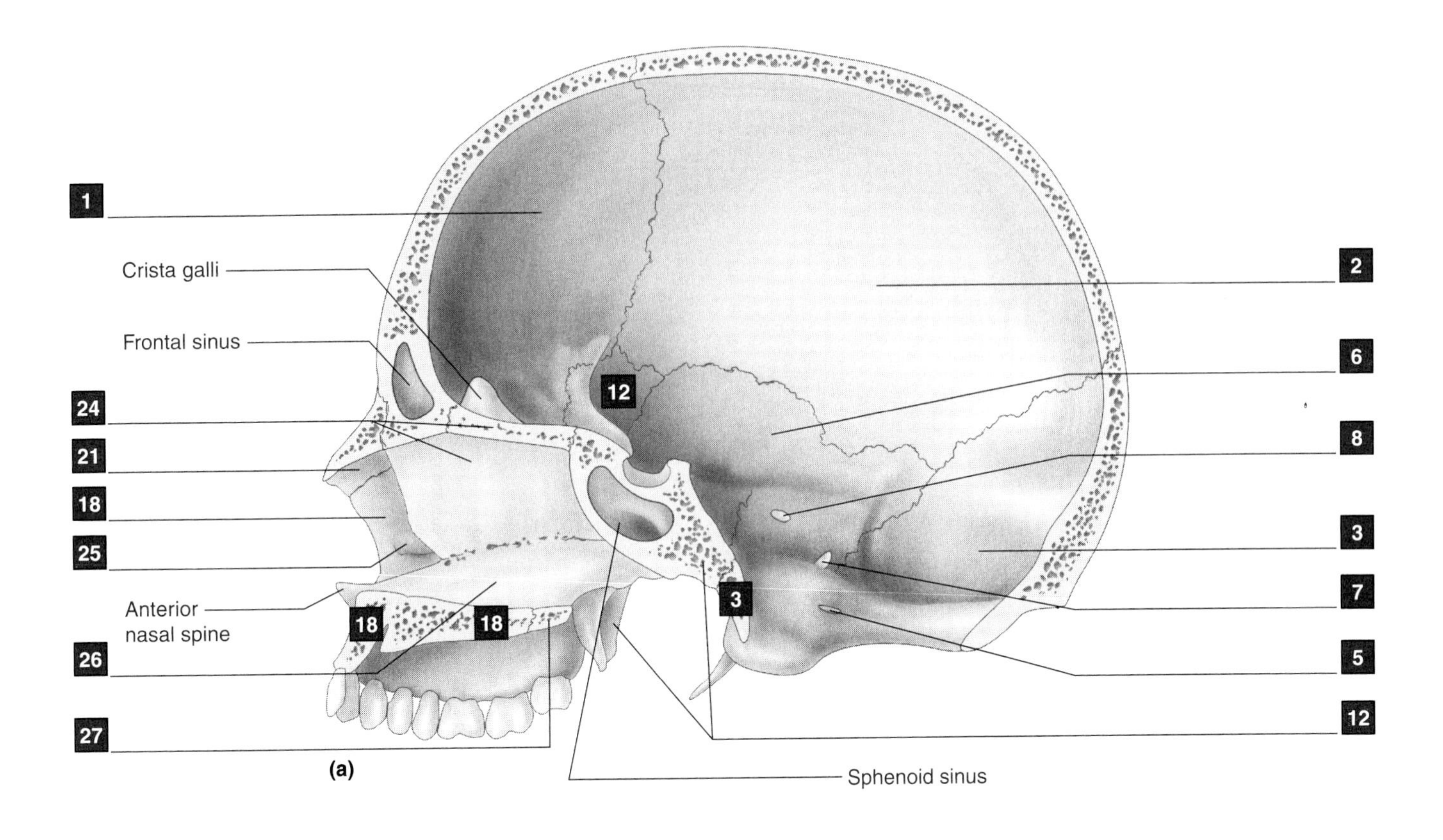

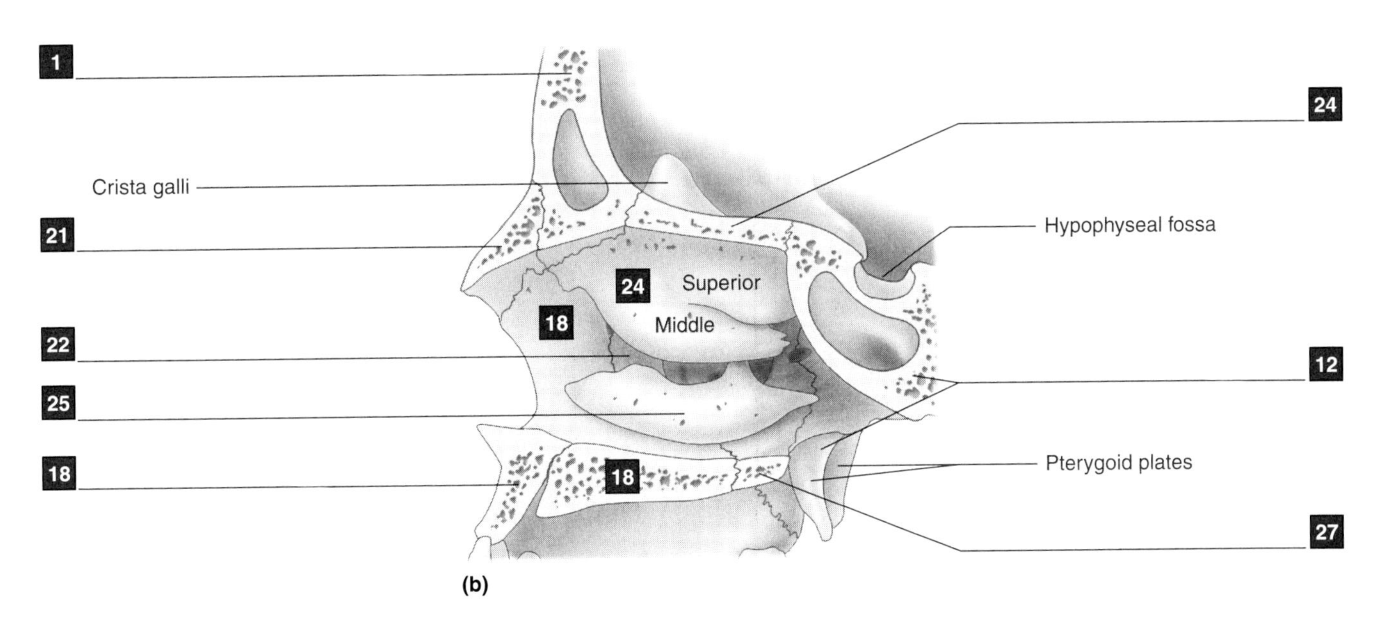

Figure 6.3 **Bones of the Skull**
(a) internal view of right median sagittal section; (b) internal view of right side of nasal cavity (the perpendicular plates of the ethmoid and vomer have been removed).

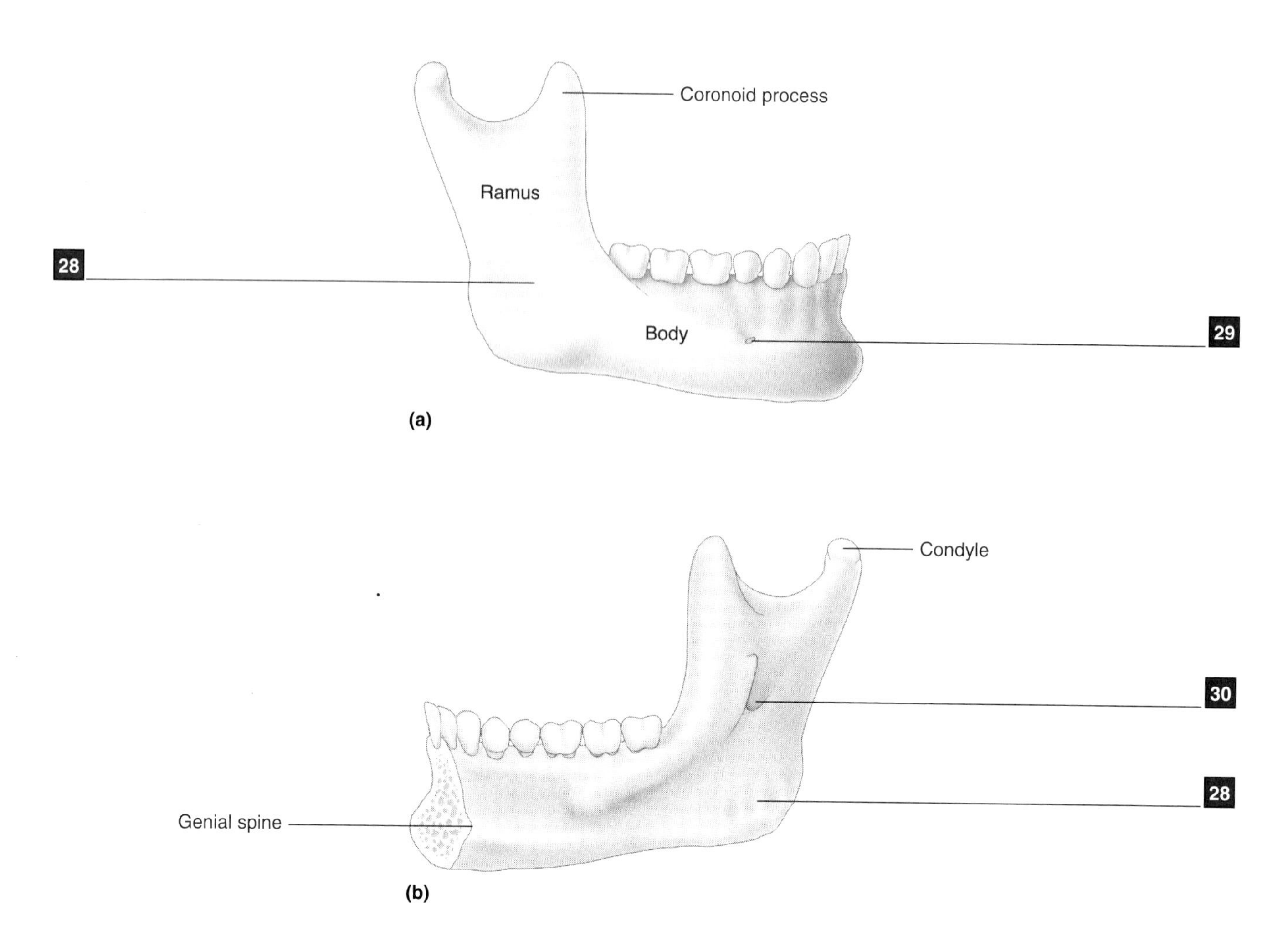

Figure 6.4 **Bones of the Skull: the Mandible**
(a) lateral view; (b) internal view of right side (median sagittal section).

6.2 THE DENTITION

Humans, like almost all other mammals, have two sets of teeth during life. The first set erupts into the mouth during early childhood. This is known as the *Deciduous Dentition*, because it is shed and replaced by a second set. The second set is retained (hopefully) throughout the rest of life, and is therefore known as the *Permanent Dentition*.

Permanent Dentition

The 32 (16 maxillary and 16 mandibular) teeth of an adult comprise the permanent dentition. Each maxilla has eight and each side of the mandible has eight teeth. Each of these four quadrants contains four types of teeth that differ morphologically: **Incisors (I = *two teeth*)**, **Canine (C = *one tooth*)**, **Premolars (P = *two teeth*)**, and **Molars (M = *three teeth*)**. The maxillary canine is sometimes called the "eye tooth" because it has a long root that extends upward toward the orbit, the premolars may be referred to as "bicuspids" because they usually have two cusps. The last (third) molars are known as "wisdom teeth" because they generally erupt between 18 and 21 years, the age of wisdom!

Note the positions of the four types of teeth found in the permanent dentition as shown in figure 6.5. Color in the teeth, using a different color to represent each type of tooth.

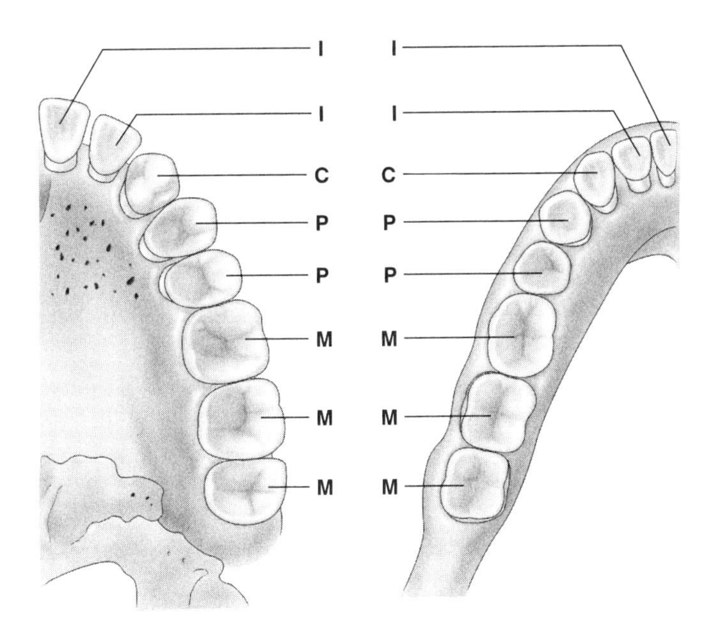

Figure 6.5 **The Permanent Dentition**
Maxillary teeth (left) and mandibular teeth (right).

Deciduous Dentition

The deciduous dentition consists of 20 teeth, with 5 in each quadrant. These are divided into three types according to their morphology: **Incisors (*two teeth*), Canine (*one tooth*),** and **Premolars (*two teeth*).** The deciduous premolars are commonly referred to as deciduous molars because they more closely resemble permanent molars than premolars in morphology, but they occupy premolar positions.

Dental Development and Eruption

Teeth develop (calcify) within the jawbones before erupting into the oral cavity. The timing of tooth calcification and eruption is reasonably well correlated with ontogenetic age. Because of this, dental development is useful in *Forensic* investigations when age at death of a subadult individual must be determined.

At birth, usually no teeth have erupted. The last permanent tooth to emerge generally erupts at about 18 years. Within this time span, there are several "eruptive stages" with which you should be familiar.

Age	**Teeth Erupted**
Birth	No teeth have erupted
Six Months	Deciduous Central Incisor
Three Years	Deciduous Dentition fully erupted
Six Years	Permanent First Molars and/or Central Incisors
Ten Years	Permanent Premolars and/or Canines
Twelve Years	Permanent Second Molars
Nineteen Years	Permanent Dentition fully erupted

Figure 6.6 shows the state of development of the crowns and roots of deciduous (*dark*) and permanent (*light*) teeth at different ages. Note that the crown erupts into the mouth before the root is completely formed. Color the permanent teeth with the same colors you used on figure 6.5 in order to identify each type.

3 years

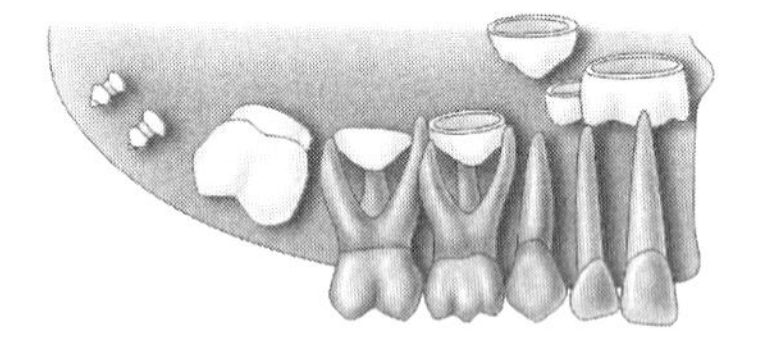

6 years

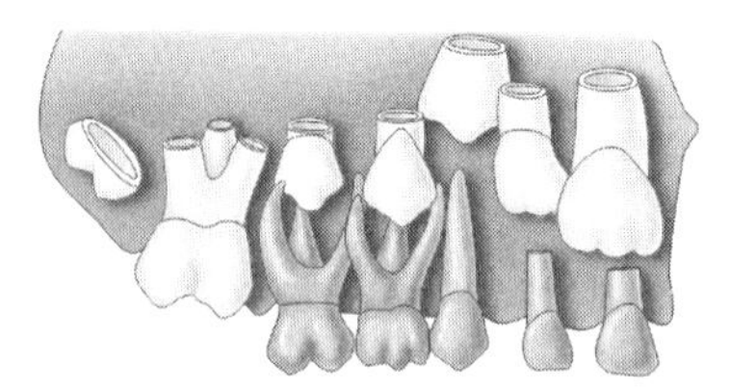

10 years

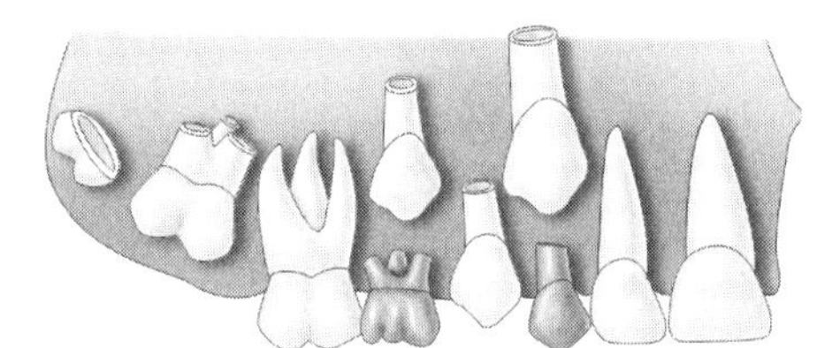

12 years

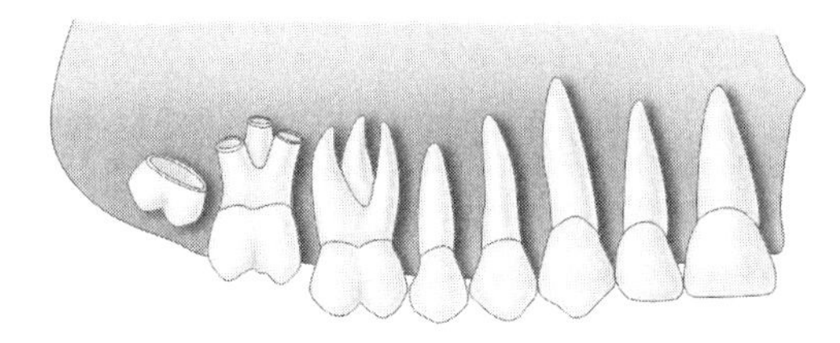

21 years

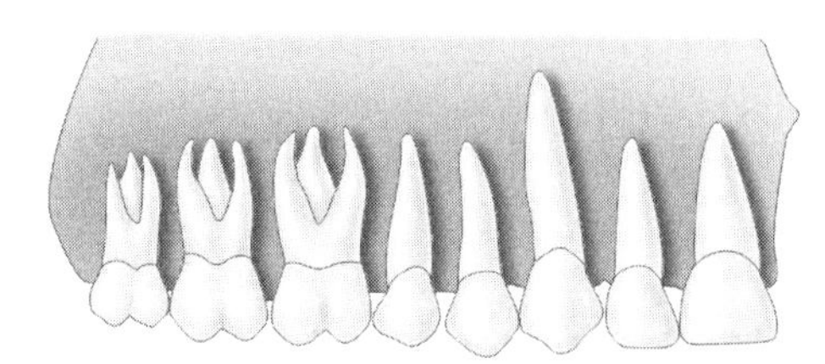

Figure 6.6 Development and Eruption of Permanent Teeth

Dental Anatomy

Each tooth has two parts: the **Crown** 1 and **Root** 2. The crown is that part enclosed by the white, shiny **Enamel Cap** 3. This projects above the **Gingiva** 4 into the oral cavity. Deep to the enamel cap is the **Dentine** 5, which forms the bulk of the tooth root. The root is embedded in **Alveolar Bone** 6. The external surface of the root has a thin layer of bonelike **Cementum** 7, which serves to anchor the root in the alveolar bone by numerous short **Periodontal Ligaments** 8. The dentine surrounds the **Pulp Chamber** 9, which is the nutritive portion of the tooth, containing blood vessels and nerves. These are branches of the **Alveolar Nerves and Vessels** 10, and they make their way into the pulp chamber by means of canals in the roots.

Identify the components of a tooth and its supporting structure in figure 6.7.

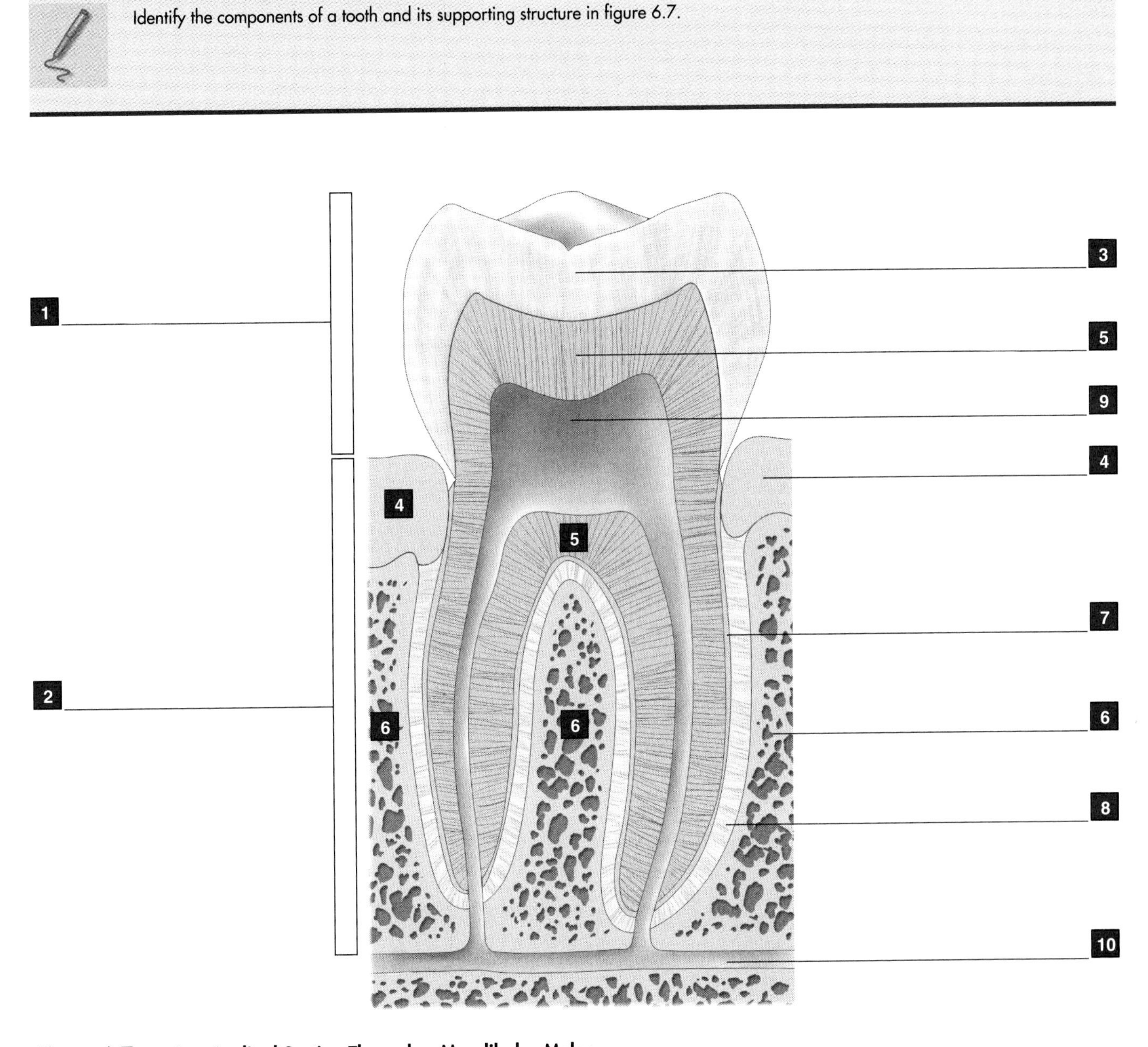

Figure 6.7 **Longitudinal Section Through a Mandibular Molar**

6.3 MUSCLES OF THE HEAD

We have already studied the muscles that move the head in laboratory 5 (pp. 145–146). Here we will examine the muscles of facial expression and mastication, and those that move the soft palate and tongue. The other muscles of the head—those that move the eyeball—will be examined in laboratory 8 (pp. 232–233).

Muscles of Facial Expression

There are about 25 muscles that move the skin, lips, nostrils, and eyelids in different facial expressions. We will concern ourselves with the more important ones. One other muscle of "facial" expression, the *Platysma*, was studied previously (laboratory 5, p. 149).

All muscles of facial expression are innervated by the *Facial Nerve (CN VII)*.

Frontalis 1 arises superiorly from a flat tendon (the *Epicranial Aponeurosis*) that stretches across the parietals to the occipital bone. It inserts into the skin of the eyebrows, and draws them upward, thus wrinkling the skin of the forehead.

Orbicularis Oculi 2 encircles the orbit. It consists of three portions; each has a different function. The ***Palpebral Portion*** runs in the eyelids, lowering them during blinking and voluntary closing of the eyes. The ***Orbital Portion*** runs around the periphery of the eyelids, and acts during forceful closure of the eyes. The ***Lacrimal Portion*** runs from the crest of the lacrimal bone to the medial part of the eyelids; it compresses the lacrimal sac.

Zygomaticus major 3 arises from the front of the zygomatic bone and inserts into the corners of the mouth. It draws the corners of the mouth upward, as in smiling.

Buccinator 4 arises from the lateral surfaces of the alveolar processes of the maxilla and mandible opposite the molar teeth. It inserts into the corners of the mouth. It increases the rigidity of the cheek, and is, therefore, important in mastication. It is pierced by the *Parotid Duct* from the *Parotid (Salivary) Gland*, which opens into the oral cavity opposite the second molar.

Orbicularis Oris 5 encircles the mouth within the lips. It closes the lips, compresses them against the front teeth, and purses them, as in pouting or kissing.

Identify the muscles of facial expression in figure 6.8.

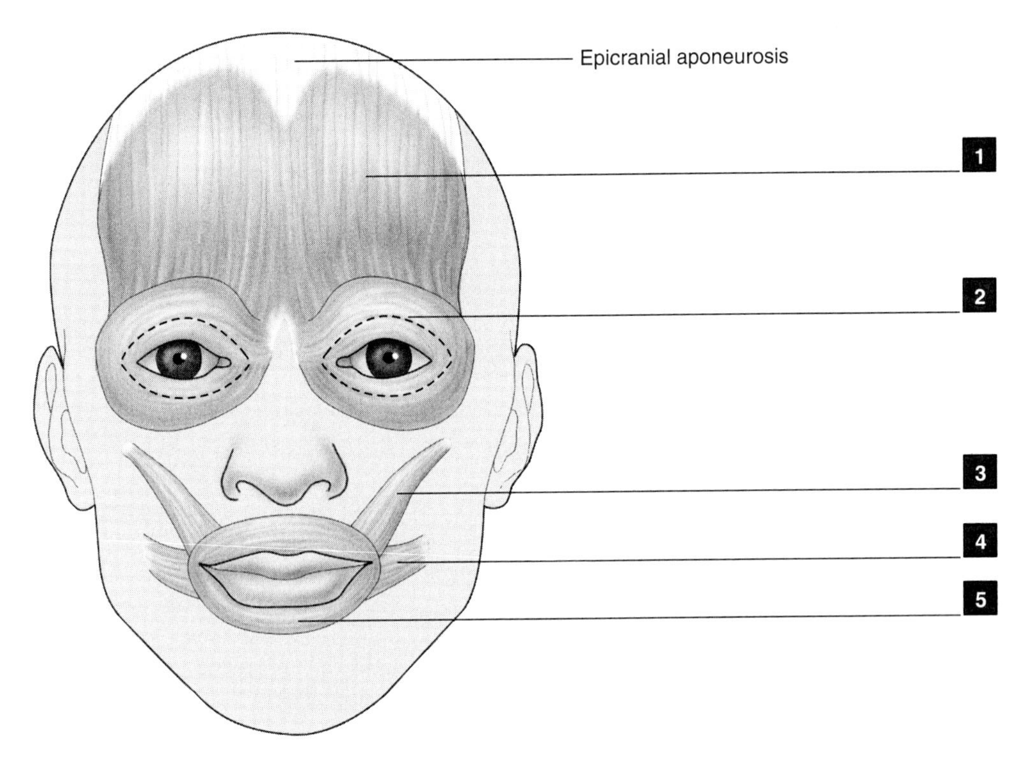

Figure 6.8 **Muscles of Facial Expression**

Muscles of Mastication

Five muscles move the mandible during chewing. One of these, the *Digastric*, opens the mouth. It was studied earlier in laboratory 5 (pp. 150–151) because it extends into the neck. The other four elevate, protrude, retract, and/or laterally translate the mandible.

Recall that the anterior and posterior bellies of *Digastric* are innervated independently by the mandibular division of the *Trigeminal nerve (CN V3)* and the *Facial nerve (CN VII)*, respectively.

The other four muscles of mastication—masseter, temporalis, and the medial and lateral pterygoids—are supplied by the mandibular division of the *Trigeminal nerve (CN V3)*.

Masseter **1** arises from the inferior edge of the zygomatic arch and inserts on the lateral surface of the ramus and angle of the mandible. It elevates the mandible, and can also pull it forward (protraction).

Temporalis **2** is a fan-shaped muscle that arises from the lateral surface of the cranial vault along a curved line that runs from behind the orbit on the frontal bone, across the parietal bone, and onto the temporal bone just above the mastoid process. It passes deep to the zygomatic arch (in the *Temporal Fossa*) and inserts onto the tip of the coronoid process of the mandible. It elevates the mandible.

Medial Pterygoid **3** runs from the medial surface of the lateral pterygoid plate of the sphenoid, and inserts onto the medial surface of the angle of the mandible. It elevates and protrudes the mandible.

Lateral Pterygoid **4** has two parts that arise from the lateral surface of the lateral pterygoid plate and base of the sphenoid. One (*Superior Head*) inserts onto the articular disc that sits between the glenoid fossa and the mandibular condyle. The other (*Inferior Head*) inserts onto the front of the neck of the mandibular condyle. Together they pull the condyle of the mandible downward and forward, which causes the mandible to protrude and move from side to side.

Identify the muscles of mastication in figure 6.9.

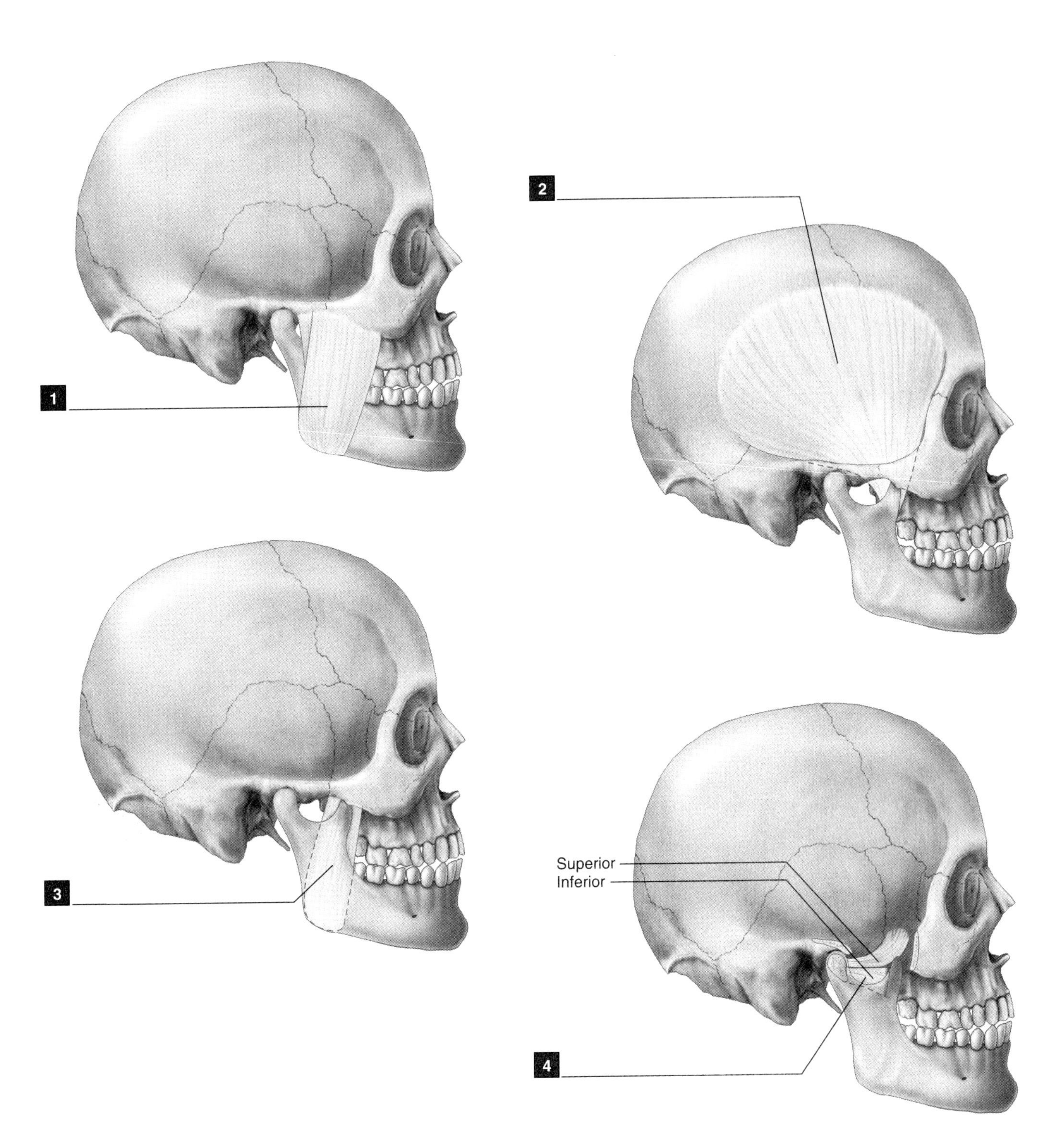

Figure 6.9 **Muscles of Mastication**

Muscles That Move the Soft Palate

Two muscles move the soft palate. One (*Tensor Palatini*) tightens it and one (*Levator Palatini*) elevates it during swallowing. Working in conjunction with the *Superior Pharyngeal Constrictor,* they close off the nasopharynx from the oropharynx, which prevents food and drink from passing upward into the nasopharynx and nasal cavity.

Tensor palatini and Levator palatini arise from the *Auditory (Eustachian) tube*. Thus, they also open the pharyngeal end of the tube during swallowing, and this equalizes pressure in the middle ear and pharynx.

Tensor (Veli) Palatini 1 arises from the hollow (*Scaphoid Fossa*) between the medial and lateral pterygoid plates, and from the edge of the auditory tube. It curves medially around the little hook at the inferior edge of the medial pterygoid plate (*Pterygoid Hamulus*) to insert into the soft palate. It tightens the soft palate.

Innervated by mandibular division of *Trigeminal nerve (CN V3)*

Levator (Veli) Palatini 2 arises from the petrous portion of the temporal bone just anterior to the carotid foramen, and from the edge of the auditory tube. It runs along the tube to insert into the upper surface of the soft palate. It elevates the soft palate.

Innervated by *Pharyngeal Branch* of *Vagus nerve (CN X)*

Identify the muscles of the soft palate in figure 6.10.

Swallowing

Swallowing involves elevation of the tongue and sealing the top of the oropharynx. The following diagrams illustrate this. (a) The tongue pushes food backward against the palate, from the oral cavity into the oropharynx. (b) The oropharynx is sealed off from the nasopharynx and nasal cavity through the elevation and tightening of the soft palate by the levator palatini and tensor palatini in conjunction with the forward contraction of the superior pharyngeal constrictor.

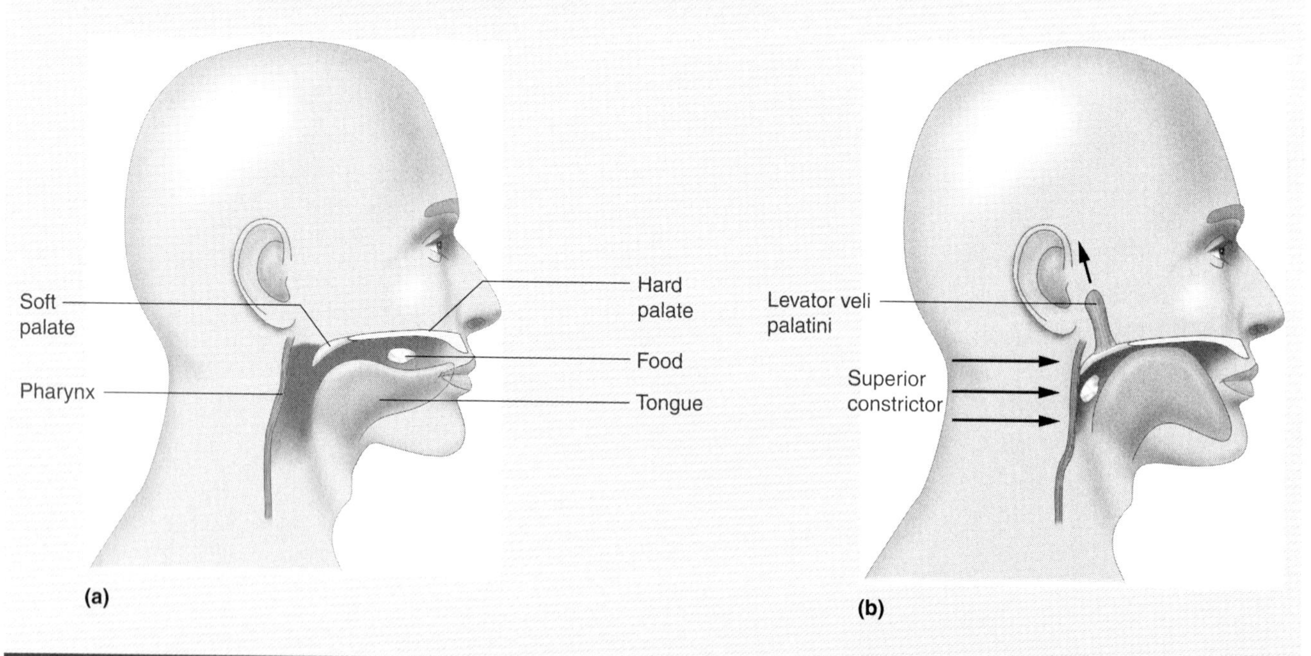

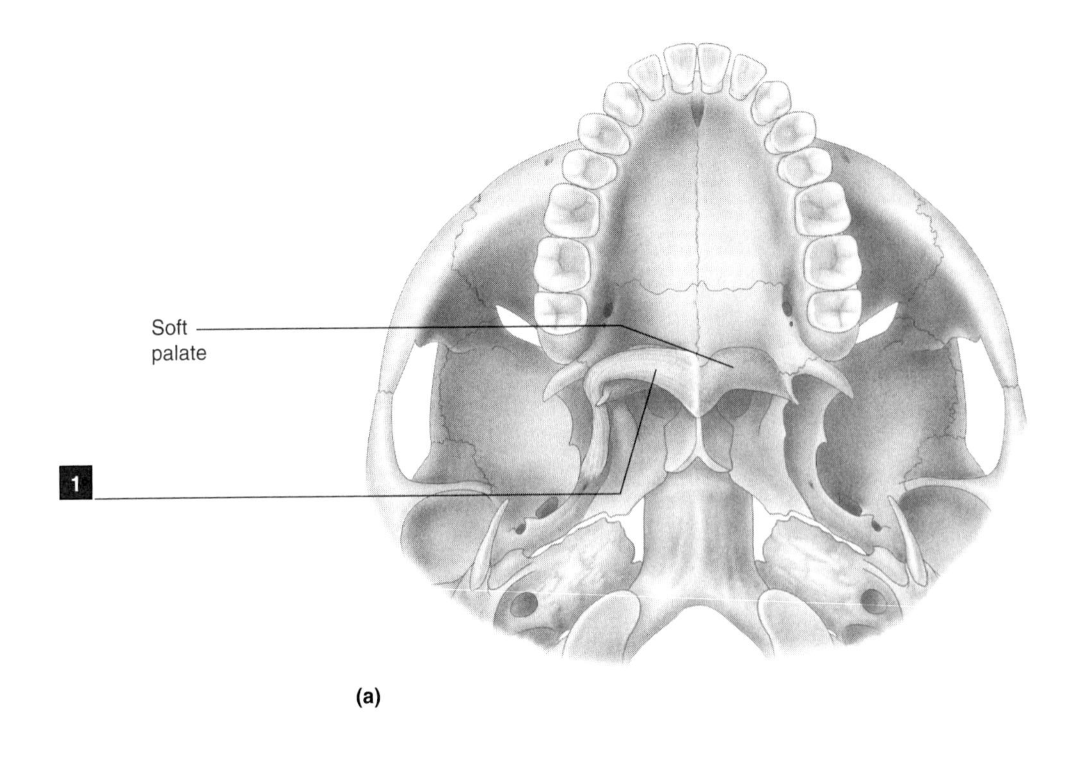

(a)

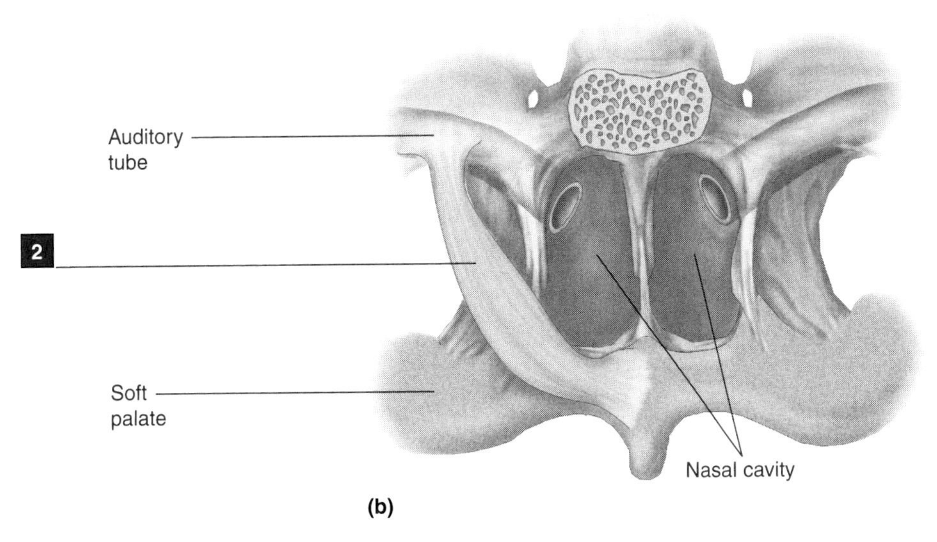

(b)

Figure 6.10 **Muscles That Move the Soft Palate**
(a) inferior view of skull; (b) posterior view of nasopharynx

Muscles That Move the Tongue

Six muscles move the tongue. They comprise two groups:

Intrinsic Muscles alter the shape of the tongue. They are confined to the tongue.

Extrinsic Muscles protrude, retract, elevate, and depress the tongue. They attach the tongue to the bones of the skull and hyoid.

The body of the tongue is wholly muscular. It is divided into left and right halves by a fibrous *Median Septum*. The upper part of the tongue is made up of intrinsic muscle fibers that attach to the septum. The lower part is comprised mostly by an extrinsic muscle (*Genioglossus*).

Intrinsic Tongue Muscles

Transverse and Longitudinal Fibers 1 attach to the fibrous median septum and to the mucous membrane of the tongue.

Supplied by *Hypoglossal Nerve (CN XII)*

Extrinsic Tongue Muscles

These four muscles attach the tongue to the bones of the skull, the hyoid, and the soft palate. All have the suffix "glossus" in their names. One is innervated by the *Vagus nerve (CN X)*; the others are supplied by the *Hypoglossal nerve (CN XII)*.

Genioglossus 2 forms the greater part of the bulk of the tongue. It arises from the genial spines on the back of the mandibular symphysis. Its fibers fan posteriorly into the substance of the tongue along the median septum from its base to the tip. Some fibers reach the hyoid bone. It protrudes the tongue, and prevents it from being sucked into the oropharynx during inspiration.

Supplied by *Hypoglossal nerve (CN XII)*

Hyoglossus 3 arises from the hyoid bone, and inserts into the fibrous tissue of the tongue near its dorsum. It flattens (depresses) the tongue.

Supplied by *Hypoglossal nerve (CN XII)*

Styloglossus 4 arises from the styloid process, and inserts into the edge of the tongue as far as its tip. It retracts the tongue upward and backward, important for swallowing.

Supplied by *Hypoglossal nerve (CN XII)*

Palatoglossus 5 arises in the tissue of the soft palate, and descends in the *Palatoglossal Arch* in the back of the oral cavity to the tongue. It retracts the tongue upward and backward, and approximates one arch to the other.

Supplied by *Vagus nerve (CN X)*

Identify the muscles of the tongue in figure 6.11.

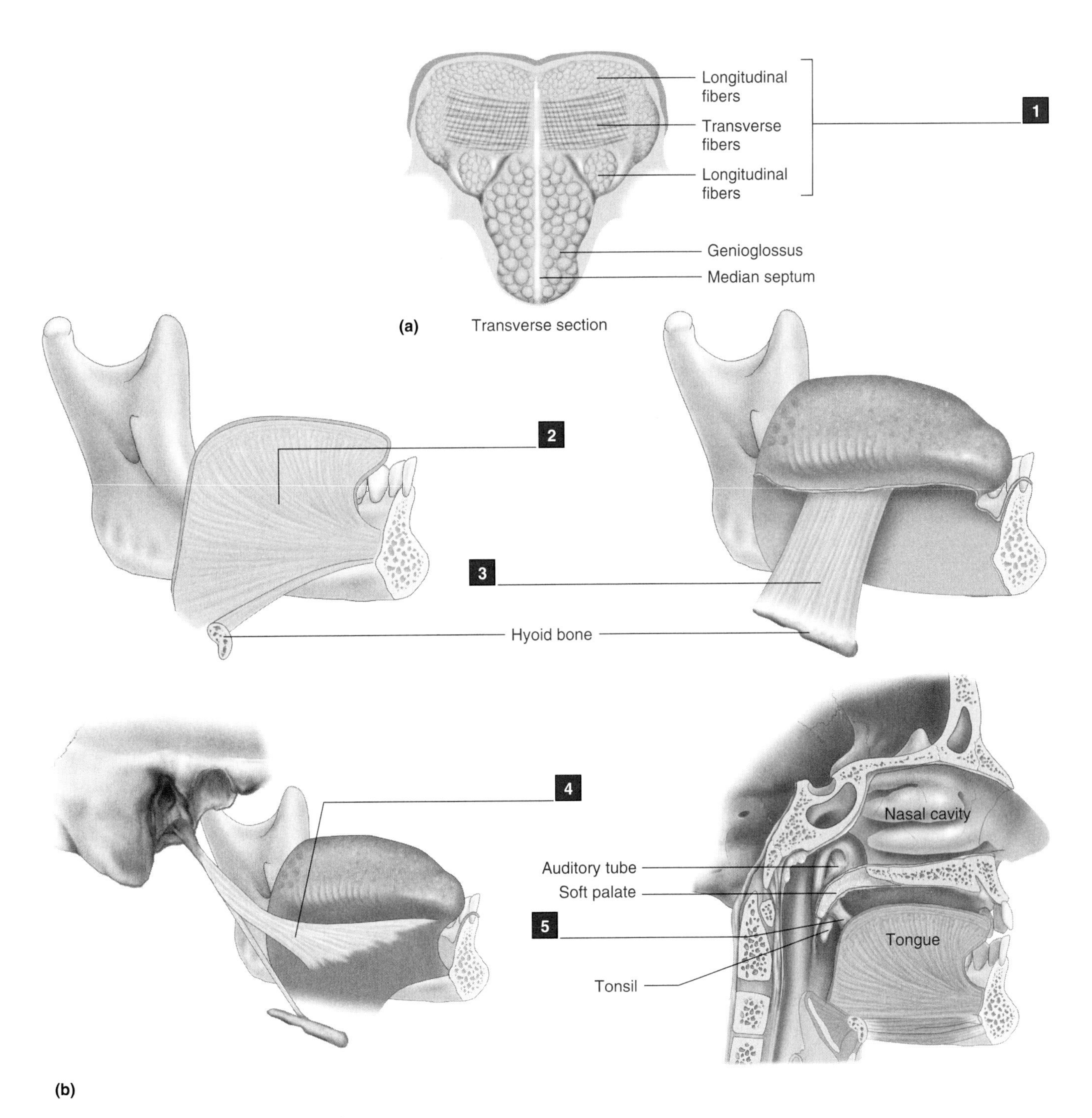

Figure 6.11 **Muscles That Move the Tongue**
(a) intrinsic tongue muscles; (b) extrinsic tongue muscles.

6.4 THE NASAL AND ORAL CAVITIES

Nasal Cavity and Nasopharynx

The nasal cavity opens anteriorly through the *Nostrils* to the external environment, and posteriorly through the *Internal Nasal Choana* to the *Nasopharynx*.

The **Ethmoid** 1 forms the roof of the nasal cavity; the **Sphenoid** 2 forms the roof of the nasopharynx. The ethmoid also forms the upper part of the lateral wall of the nasal cavity, and has two platelike projections—the **Superior Concha** 3 and **Middle Concha** 4—that are covered with mucous membrane. The lower third of this wall is formed by the **Inferior Nasal Concha** 5 and the **Maxilla** 6. The maxilla together with the **Palatine** 7 form the floor of the nasal cavity, and the roof of the oral cavity. The nasal cavity is divided into left and right halves by a septum, made up of the perpendicular plates of the ethmoid (superiorly) and **Vomer** 8 (inferiorly).

The **Maxillary Sinus** 9 is continuous with the nasal cavity via an opening underneath the middle concha.

The **Auditory Hiatus** 10 in the lateral wall of the nasopharynx opens into the *Auditory Tube*. The back wall of the hiatus is raised as a low ridge called the *Torus Tubarius*.

Oral Cavity and Oropharynx

The oral cavity opens anteriorly through the mouth to the external environment, and posteriorly into the *Oropharynx* through the *Fauces*. The roof of the oral cavity is formed by the **Maxilla** and **Palatine;** which constitute the ***Hard Palate.*** The **Soft Palate** 11 separates the oropharynx and nasopharynx. The lateral wall of the oral cavity is lined by mucous membrane that covers the deep surface of the *Buccinator* muscle.

Look into your mouth using a mirror. Note that the dorsum of your tongue is divided by a V-shaped *Sulcus Terminalis* into an anterior two-thirds and a posterior one-third. Just in front of the sulcus is a row of **Vallate Papillae** 12 that are studded with taste buds. Note that your tongue's free underside is attached to the floor of the mouth by a band of tissue called a *Frenulum*.

Note also that a fingerlike projection, the **Uvula** 13 hangs from the midline of the soft palate. On either side of it, making up the arches (*Fauces*) at the back of the mouth are the **Palatopharyngeal Folds** 14. In front of them are the **Palatoglossal Folds** 15, within which run the *Palatoglossus* muscles. The **Palatine Tonsils** 16 lie between these two folds.

Identify the aforementioned structures in figure 6.12.

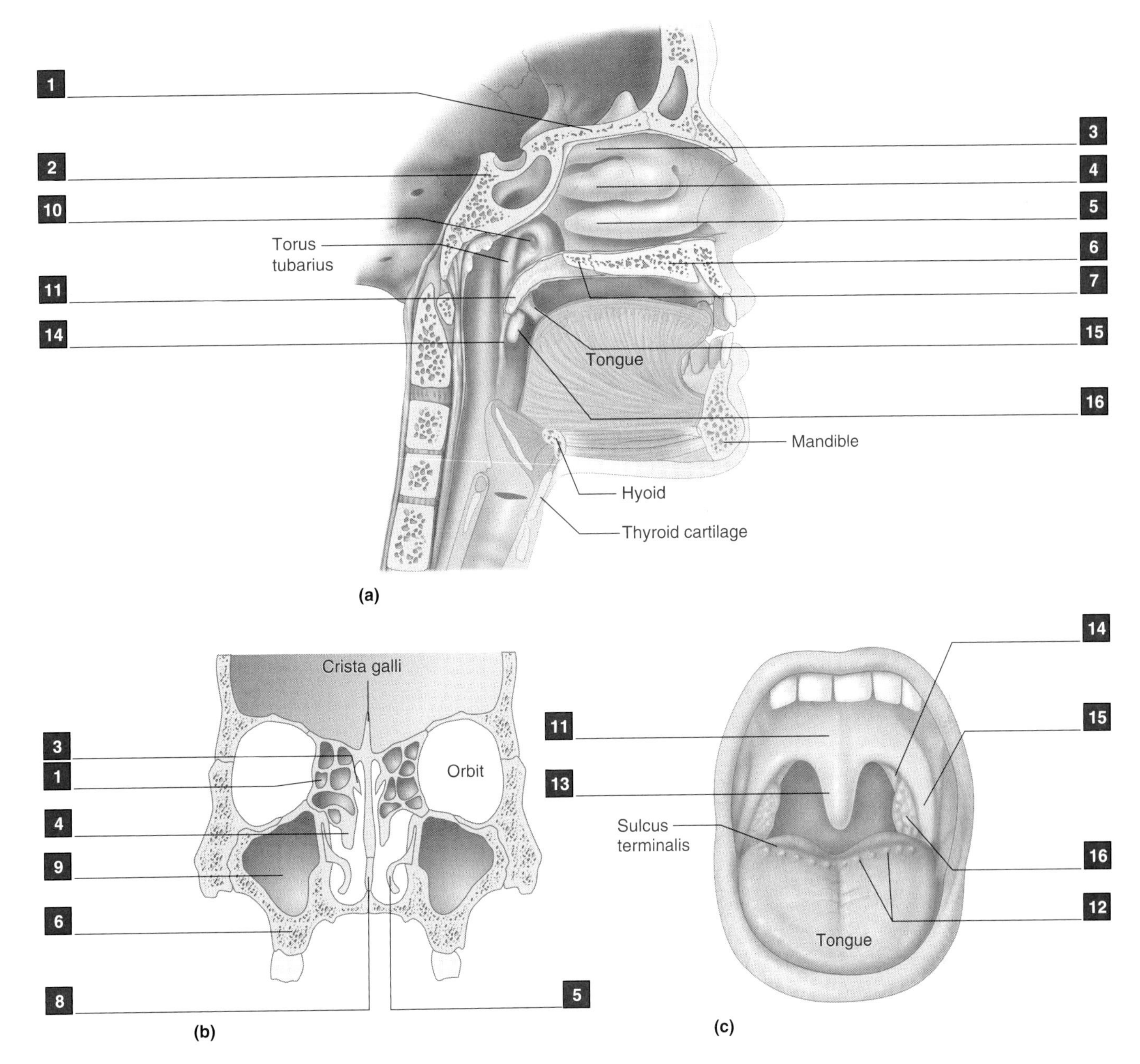

Figure 6.12 **Nasal and Oral Cavities**
(a) median sagittal section; (b) schematic coronal section of nasal cavity; (c) anterior view of mouth.

6.5 BLOOD VESSELS OF THE HEAD

Arteries

Blood is supplied to the head by three arteries. The vertebral artery supplies the cerebellum and the back of the cerebrum. The internal carotid artery supplies the rest of the brain, and the eyes. The external carotid artery supplies the rest of the head.

External Carotid Artery

Recall that the main branches of the external carotid artery are the ***Facial, Occipital, Posterior Auricular, Maxillary,*** and ***Superficial Temporal*** arteries. These were studied in laboratory 5 (pp. 158–159).

The **Maxillary Artery** gives off the **Middle Meningeal Artery,** which enters the cranium via the *Foramen Spinosum* to supply the flat bones of the neurocranium.

Vertebral and Internal Carotid Arteries: Blood Supply of the Brain

The **Vertebral Arteries** 1 enter the cranial cavity via the *Foramen Magnum*. Each gives off a **Posterior Inferior Cerebellar Artery** 2 to the *Cerebellum*. The vertebral arteries then merge anteriorly to form the median **Basilar Artery** 3. It gives off two additional branches to the *Cerebellum*—the **Anterior Inferior Cerebellar** 4 and the **Superior Cerebellar Arteries** 5.

The basilar artery then bifurcates to form the left and right **Posterior Cerebral Arteries** 6. They supply the *Occipital Lobes of the Cerebrum*. Each posterior cerebral artery sends a branch forward—the **Posterior Communicating Artery** 7—that connects to the internal carotid artery.

The **Internal Carotid Arteries** 8 enter the skull via the *Carotid Canals* to emerge on either side of the body of the sphenoid. From here they travel forward along the sides of the hypophyseal fossa, within the *Cavernous Sinus*, to a point just below the optic canals, where each gives off an **Ophthalmic Artery** 9. This travels into the *Orbit* with the optic nerve.

The internal carotid then turns backward and bifurcates to form two branches. One—the **Middle Cerebral Artery** 10—runs into the *Lateral (Sylvian) Fissure* between the frontal and temporal lobes of the brain. The second—the **Anterior Cerebral Artery** 11—runs forward into the *Longitudinal Fissure* between the left and right cerebral hemispheres, and then up and back over the *Corpus Callosum* to supply the frontal and parietal lobes of the brain. The anterior cerebral arteries are connected by a short **Anterior Communicating Artery** 12.

The Circle of Willis

The anterior and posterior communicating arteries permit blood from one side to reach the other, and blood from the vertebral network to reach the carotid network (and vice versa). This arterial circle (of Willis) provides for collateral circulation, which becomes potentially very important if one of the channels becomes occluded.

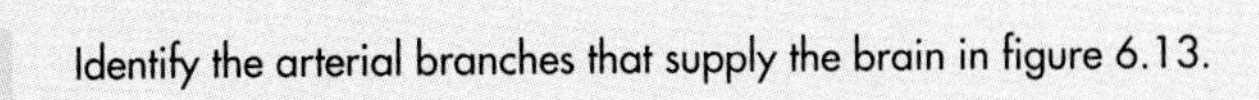

Identify the arterial branches that supply the brain in figure 6.13.

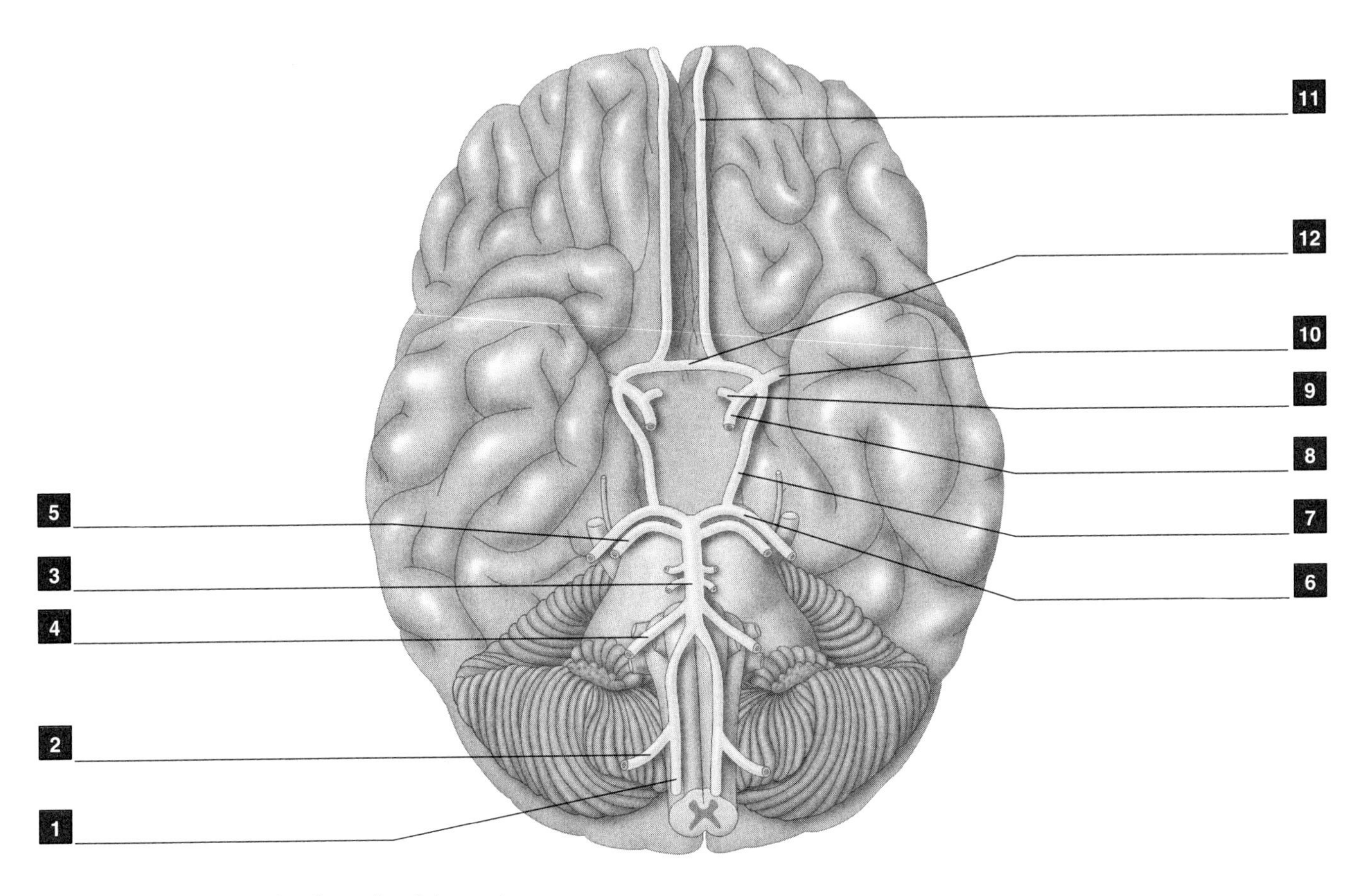

Figure 6.13 Blood Supply of the Brain

Veins and Venous Sinuses

Blood is drained from the brain by numerous small veins that empty into the *Venous Sinuses*. The sinuses empty into the *Internal Jugular Vein*, which also receives blood from the *Facial Vein*. The facial vein drains part of the face, and is a secondary channel for blood from the endocranial cavity via the orbit. The *External Jugular Vein* drains blood from the scalp and part of the face.

External Jugular Vein

The external jugular vein and its main tributaries—***Retromandibular, Posterior Auricular,*** and ***Occipital Veins***-were examined in laboratory 5 (pp. 160–161).

Endocranial Venous Sinuses: Blood Drainage from the Brain

Within the cranial cavity, neither the vertebral arteries, nor internal carotid arteries, nor their principal branches have *Venae comitantes*. Instead, the veins that drain the brain empty into cranial venous sinuses.

Venous Sinus Structure There are two major layers of tissue between the scalp and the surface of the brain: **Cranial bone 1**, and the **Meninges 2**. Recall that the outer surface of a bone is covered with *Periosteum*. The trabecular "cavity" of the cranial bones—the **Diploë 3** with its layer of endosteum—is sandwiched between external and internal layers of compact bone. The superficial and deep layers of cranial bone periosteum are called **Pericranium 4** and **Endocranium 5** respectively.

The outermost layer of the meninges is the **Dura Mater 6**. It is closely attached to the endocranium over the entire inner surface of the cranial cavity except where it and the other meningeal layers—**Arachnoid Mater 7** and **Pia Mater 8**—follow the major fissures of the brain. Between the left and right cerebral hemispheres, between the frontal and temporal lobes, and between the cerebrum and cerebellum the *Dura Mater* pulls away from the *Endocranium*, creating a **Venous Sinus 9**.

Identify the components of a cranial venous sinus in figure 6.14.

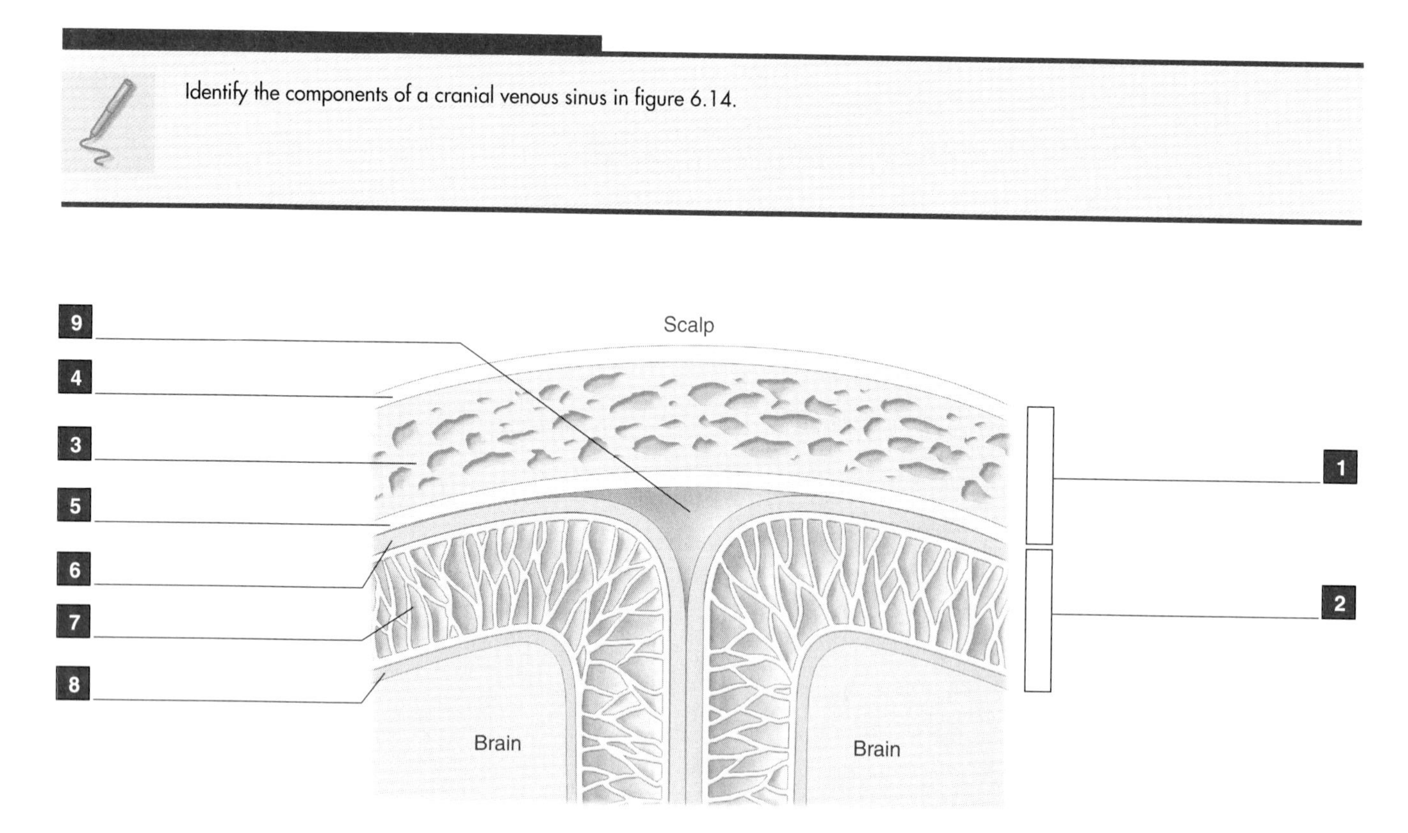

Figure 6.14 **Structure of a Cranial Venous Sinus**
Schematic coronal section through the superior sagittal sinus.

Dural Folds Where the *Dura Mater* pulls away from the endocranium it will lie against itself if the fissure into which it dives is deep enough. Here the double dural layer forms a thick, fibrous sheet, the free edge of which may contain a venous sinus where the sheet of dura curves back around. There are two fissures deep enough to create a dural fold.

One dural fold is *between the left and right cerebral hemispheres*. This fold extends downward vertically in the median sagittal plane. Its free inferior margin is curved so that the sheet resembles a sickle. It is known as the ***Falx Cerebri.*** The second dural fold is *between the cerebrum and the cerebellum*. This dural fold extends in the horizontal plane from the occipital bone and the superior margin of the petrous portion of the temporal bone. Its free edge is curved around the brain stem. It forms a tent over the cerebellum and is known as the ***Tentorium Cerebelli.***

Venous Sinus Drainage Beginning anteriorly at the crista galli of the ethmoid bone, the dura separates from the endocranium in the midline to create the **Superior Sagittal Sinus 1.** It forms the root of the *Falx Cerebri*. In the middle of the occipital bone, the superior sagittal sinus divides into the left and right **Transverse Sinuses 2,** which run laterally in the root of the *Tentorium Cerebelli* across the occipital to the back of the petrous pyramid. At this point, the sinus changes its name to the **Sigmoid Sinus 3** because it makes an S-shaped curve around the temporal bone to exit via the *Jugular Foramen* into the *Internal Jugular Vein*.

The inferior edge of the *Falx Cerebri* contains the **Inferior Sagittal Sinus 4,** which runs posteriorly to the junction of the falx cerebri and the tentorium cerebelli. Here it is joined by the **Great Cerebral Vein 5** to form the **Straight Sinus 6.** The straight sinus runs posteriorly between the two dural folds to the confluence of the superior sagittal and transverse sinuses.

The **Cavernous Sinus 7** lies on either side of the *Hypophyseal fossa* of the sphenoid bone. This sinus is drained by the **Superior Petrosal Sinus 8,** which runs along the superior margin of the petrous pyramid of the temporal bone to the sigmoid sinus. The cavernous sinuses communicate with one another via channels across the front and back of the hypophyseal fossa. The cavernous sinus receives blood from the **Sphenoparietal Sinus 9,** which runs along the back of the lesser wing of the sphenoid. The cavernous sinus also receives blood from the **Ophthalmic Vein 10,** which drains posteriorly from the orbit through the *Superior Orbital Fissure*. The ophthalmic vein anastomoses with the **Facial Vein 11,** from which it receives blood. Recall that the facial vein is one of the principal tributaries of the *External jugular vein* (laboratory 5: p. 160; fig. 5.16).

Because of the anastomosis between the facial vein and the ophthalmic vein, blood from the facial vein may make its way into the cavernous sinus. This is of potential clinical significance because it will permit a facial infection to spread to the endocranial cavity.

Identify the cranial venous sinuses in figure 6.15

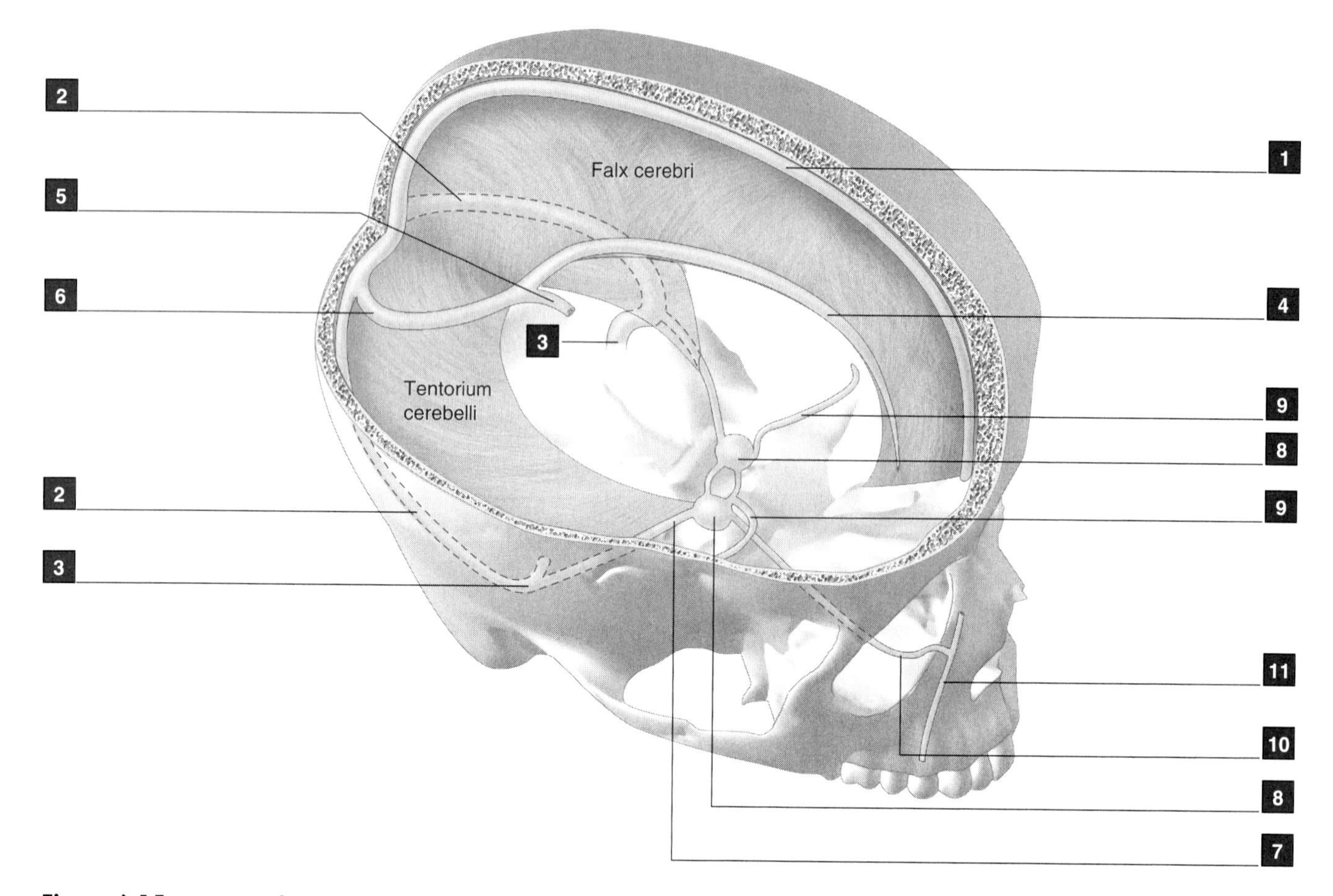

Figure 6.15 **Cranial Venous Sinuses**

LABORATORY 7

The Brain and Cranial Nerves

7.1 THE BRAIN 194
Rhombencephalon 196
Medulla Oblongata 196
Pons 196
Cerebellum 196
Mesencephalon 198
Midbrain 198
Prosencephalon 200
Diencephalon 200
Cerebrum 202
Cerebral Cortex 202
White Matter 204
Basal Nuclei 205
Ventricles and Cerebrospinal Fluid 207
Ventricular System 207
CSF Circulation and Resorption 207

7.2 CRANIAL NERVES 209
Olfactory (I) 210
Optic (II) 211
Oculomotor (III) 211
Trochlear (IV) 211
Abducens (VI) 211
Trigeminal (V) 213
Facial (VII) 215
Vestibulocochlear (VIII) 217
Glossopharyngeal (IX) 218
Vagus (X) 220
Accessory (XI) 222
Hypoglossal (XII) 223

7.1 THE BRAIN

The brain is the control center of the central nervous system. It occupies the endocranial cavity of the skull, and is continuous through the foramen magnum with the spinal cord. Just as the spinal cord has peripheral nerves, so too does the brain. The peripheral nerves of the brain are known as *Cranial Nerves*.

The brain is enclosed by the same three layers of meninges that are continuous around the spinal cord—*Dura Mater, Arachnoid Mater*, and *Pia Mater*. The *Cerebrospinal Fluid* that occupies the subarachnoid space around the brain and spinal cord is produced and circulates in hollow, interconnected *Ventricles* within the brain.

The brain is composed of both myelinated and unmyelinated neurons, clusters of nerve cell bodies, and support cells (neuroglia). The *Gray Matter* consists of neuron cell bodies and unmyelinated fibers that occupy the outer, convoluted layer of the cerebrum and cerebellum. The *White Matter* comprises myelinated axons that form fiber tracts deep to the cortex. Clusters of gray matter—the *Basal Nuclei*—that are formed by nerve cell bodies lie deep within the white matter.

The brain is traditionally divided into three components: the Rhombencephalon, Mesencephalon, and Prosencephalon.

Rhombencephalon 1, also known as the hindbrain, comprises the *Medulla Oblongata, Pons*, and *Cerebellum*. It houses the *Fourth Ventricle*.

Mesencephalon 2, also known as the midbrain, is a stalk of white matter connecting the rhombencephalon and prosencephalon. It contains the *Cerebral Aqueduct*, which connects the third and fourth ventricles.

Prosencephalon 3, also known as the forebrain, comprises the *Diencephalon* and *Cerebrum*, or *Telencephalon*. The diencephalon comprises the *Thalami*, and surrounds the *Third Ventricle*. The cerebrum is composed of two large hemispheres connected to one another by the *Corpus Callosum*. Each hemisphere has a cortex of gray matter, abundant white matter, a core of basal nuclei, and a lateral ventricle.

Identify and color the three components of the brain in figure 7.1

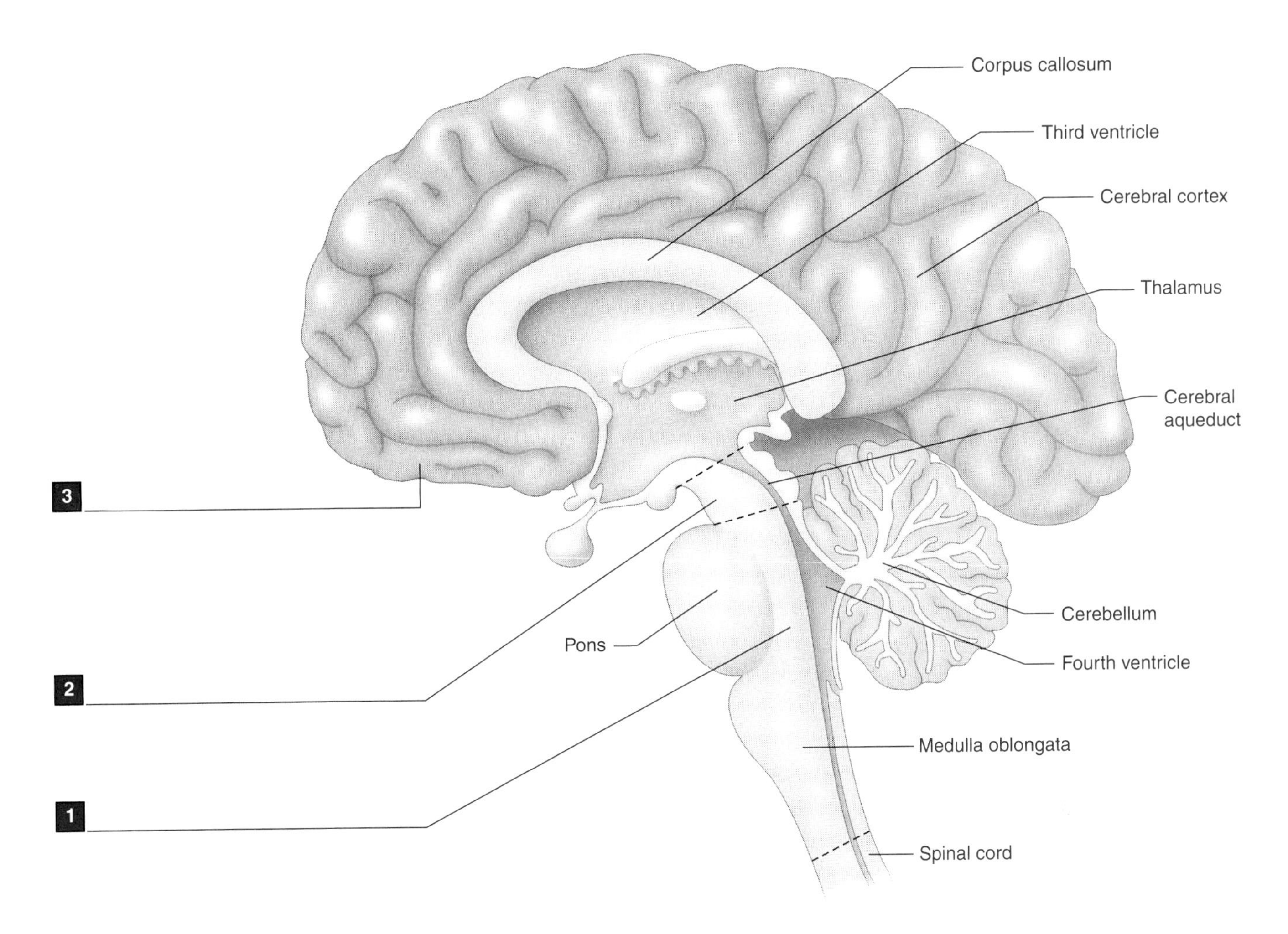

Figure 7.1 **Divisions of the Brain**
Median sagittal section.

Rhombencephalon

Medulla Oblongata

The **Medulla Oblongata** 1 is one of the three parts of the *brainstem*—the other two are the *Pons* and *Midbrain*. All cranial nerves except I and II emerge from the brainstem. The medulla oblongata is continuous with the spinal cord, and contains the cardiac, vasomotor and respiratory nerve centers. It regulates vomiting, breathing, sneezing, coughing and swallowing, and monitors the level of carbon dioxide in the blood.

The medulla is continuous with the *Pons* superiorly. It is connected to the *Cerebellum* by the **Inferior Cerebellar Peduncle** 2.

The lateral surfaces of the medulla are swollen to form the **Olives** 3. The olives mediate impulses that pass from the forebrain and midbrain to the cerebellum by way of the inferior cerebellar peduncles. The ventral surface is elevated into two parallel ridges, the **Pyramids** 4, that carry motor nerve fibers from the prosencephalon to the spinal cord. Almost all of these fibers cross over (decussate) just distal to the pyramids at the **Pyramidal Decussation** 5. This decussation is responsible for the fact that the left side of the brain controls body movements on the right side and vice versa.

Pons

The **Pons** 6 is the bridge between the *Medulla Oblongata* and the *Midbrain*. Its dorsal surface forms the floor of a space known as the fourth ventricle of the brain. It is connected to the *Cerebellum* by the **Middle Cerebellar Peduncle** 7.

Cerebellum

The **Cerebellum** 8 makes up a large part of the brain. It is wedged between the brainstem and the back of the cerebrum. Its functions relate to balance, the timing and precision (coordination) of movements, and body posture.

The cerebellum is divisible into three parts. There is a midline *Vermis*, two small *Flocculonodular Lobes* that are involved in the maintenance of skeletal muscle tone, and two large *Lateral Lobes* (*Corpus Cerebelli*) that synchronize the precise timing of skeletal muscle contractions.

The **Cerebellar Cortex** 9 consists of a surface layer of gray matter. It is corrugated, with numerous parallel ridges, the *Folia Cerebelli*, separated by shallow fissures. The underlying white matter has a branching arrangement known as the **Arbor Vitae** 10.

The cerebellum is connected to the

- *Medulla Oblongata* by the **Inferior Cerebellar Peduncle** 2
- *Pons* by the **Middle Cerebellar Peduncle** 7
- *Midbrain* by the **Superior Cerebellar Peduncle** 11.

Identify the components of the rhombencephalon in figure 7.2.

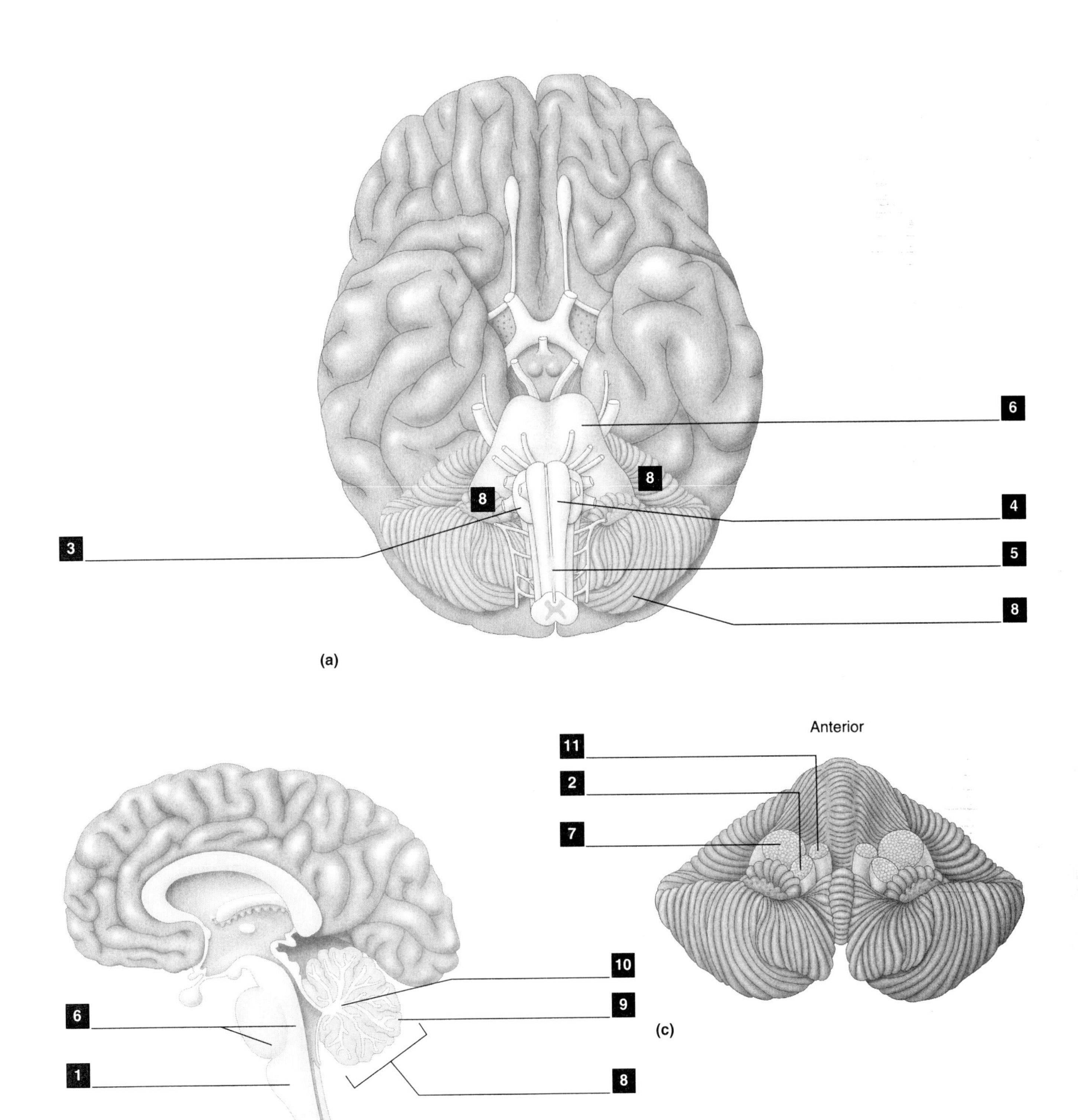

Figure 7.2 **Rhombencephalon: Medulla Oblongata, Pons, Cerebellum**
(a) ventral surface of brain; (b) median sagittal section; (c) ventral surface of cerebellum.

Mesencephalon

Midbrain

The midbrain is a short section of the brainstem that connects the rhombencephalon and prosencephalon. It passes through the notch in the dural sheet known as the *Tentorium Cerebelli*.

The midbrain comprises two ventrolateral halves called the **Crus Cerebri** **1**. The ventral portions of the crura are known as the *Cerebral Peduncles*. These masses of white matter serve as the fiber tracts that convey all somatic motor impulses between the cerebral cortex and the spinal cord.

Between and slightly posterior to the cerebral peduncles runs a hollow, fluid-filled canal called the **Cerebral Aqueduct** **2**. It connects the third and fourth ventricles.

The dorsal surface of the midbrain is known as the *Tectum*. It has four small bumps—the *Colliculi*—that are nerve reflex centers. The two **Superior Colliculi** **3** are related to the *eyes*. They are involved in the coordination of eye movements as well as focusing and pupillary response. The two **Inferior Colliculi** **4** are related to the *ears*. More particularly, they are related to auditory reflexes. The four colliculi together constitute the *Corpora Quadrigemina*.

The midbrain contains the *Substantia Nigra*, a nucleus that is functionally connected to the *basal nuclei*.

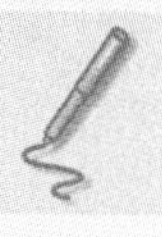

Identify the components of the mesencephalon in figure 7.3.

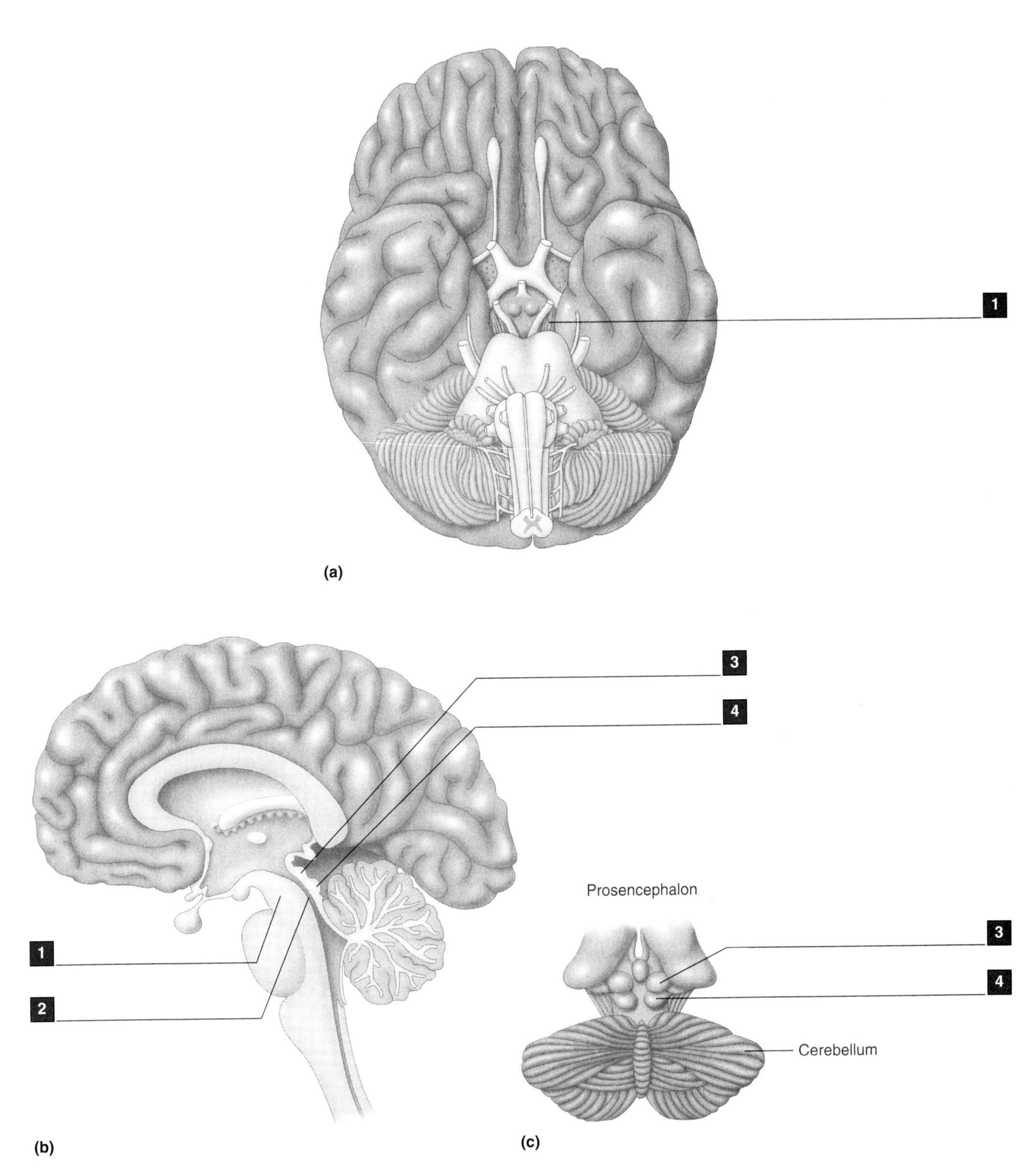

Figure 7.3 **Mesencephalon: Midbrain**
(a) ventral aspect; (b) median sagittal section; (c) dorsal surface.

Prosencephalon

Diencephalon

The diencephalon lies deep within the prosencephalon, forming the walls and floor of the third ventricle. It is comprised of the *Thalamus*, *Hypothalamus*, *Subthalamus*, and *Pineal Gland*.

The **Thalamus** 1 is an egg-shaped mass of gray matter that constitutes most of the lateral wall of the third ventricle. The left and right thalami are joined by the *Interthalamic Adhesion*. The thalami serve as relay points and processing centers for all sensory impulses (except olfaction).

The **Hypothalamus** 2 lies directly below the thalamus, forming the floor and lower walls of the third ventricle. It has a number of nuclei, including the two **Mammillary Bodies** 3. They project inferiorly between the cerebral peduncles of the midbrain and the *Optic Chiasm*. The **Hypophysis (Pituitary Gland)** 4 extends inferiorly from the hypothalamus by the **Infundibulum** 5. The pituitary gland functions largely in the regulation of other endocrine glands. The hypothalamus is important in a number of functions, including the modification of autonomic responses (e.g., blood pressure and heart beat), body temperature regulation, the maintenance of electrolyte balance, and the expression of emotional behaviors.

The ***Subthalamus*** is located below the thalamus and behind the mammillary body. It regulates and modulates the output of the *Basal Nuclei*.

The **Pineal Gland** 6 occupies the caudal part of the roof of the third ventricle. This is an endocrine gland of regulatory (generally inhibitory) importance.

Identify the components of the diencephalon in figure 7.4.

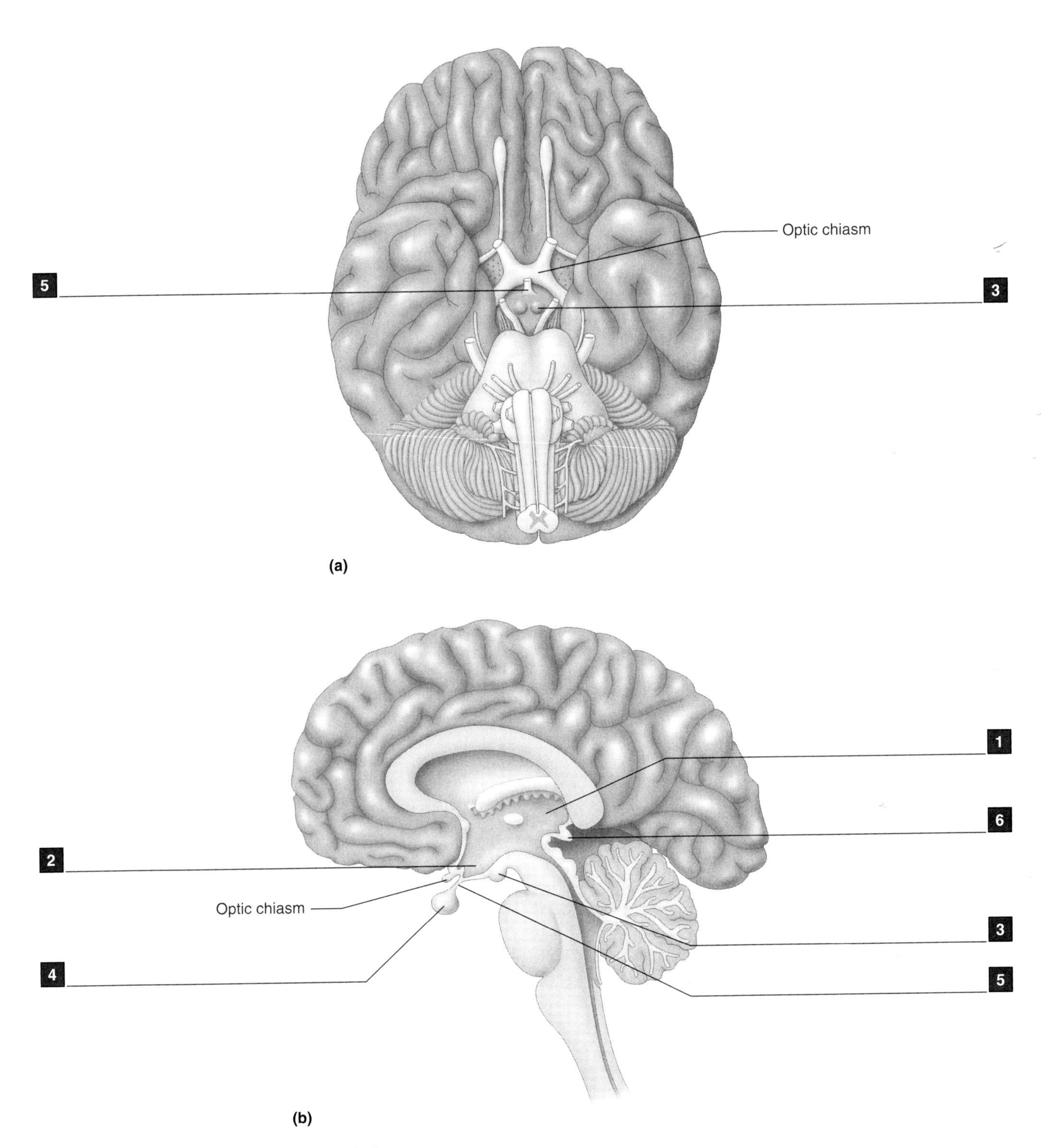

Figure 7.4 **Prosencephalon: Diencephalon**
(a) ventral view; (b) median sagittal section.

Cerebrum

The cerebrum, or telencephalon, is the largest part of the brain, accounting for about 85% of its tissue. It comprises two *Cerebral Hemispheres* that are connected to one another by the *Corpus Callosum*, a narrow band of transverse fibers. Each hemisphere has a *Cortex* of gray matter, abundant *White Matter* comprising the fiber tracts, and a deep core of gray matter called the *Basal Nuclei*.

Cerebral Cortex

The cerebral cortex is a thin layer of gray matter that is convoluted by ridges (*Gyri*) and shallow grooves (*Sulci*). The cerebral cortex is divided into six lobes; four are named according to the skull bone that covers them. Several gyri and sulci are important enough to be named.

The **Frontal Lobe** 1 is separated from the **Parietal Lobe** 2 by the *Central Sulcus*. The frontal and parietal lobes are separated from the **Temporal Lobe** 3 by the *Lateral* (*Sylvian*) *Sulcus*. The **Occipital Lobe** 4 is separated from the parietal and temporal by arbitrary lines that meet dorsally at the parietooccipital sulcus.

The **Precentral Gyrus** 5 of the *Frontal Lobe* is located immediately anterior to the *Central Sulcus*. It contains the ***Primary Motor Cortex,*** which controls all voluntary skeletal muscle movement (figure 7.5c). The frontal lobe is also involved in speech.

The **Postcentral Gyrus** 6 of the *Parietal Lobe* is located immediately behind the *Central Sulcus*. It contains the ***Primary Sensory Cortex,*** which receives sensory information from receptors throughout the body (figure 7.5c). The parietal lobe is also involved in the sensation of taste.

The temporal lobe is involved in hearing and equilibrium. The occipital lobe processes visual information and is related to our understanding of the written word.

The *Insula* is a small region of cortex located within the lateral sulcus.

The so-called *Limbic Lobe* comprises several gyri surrounding the corpus callosum and third ventricle on the medial surface of the cerebral hemisphere. These gyri are the **Cingulate Gyrus** 7**, Isthmus** 8**, Parahippocampal Gyrus** 9**,** and **Uncus** 10 **.**

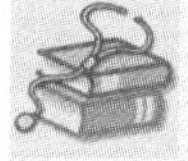

Limbic System

The limbic system constitutes neurons that act together as a functional unit. These neurons are widespread. Some occupy the cerebral cortex; others are in the diencephalon. The limbic system is involved in memory and emotions, and in basic "survival" drives and behaviors such as hunger, thirst, and sex.

The limbic system is made up by the "limbic lobe" of the cerebral cortex, the mammillary bodies of the hypothalamus, the thalamus, the hippocampus, the amygdaloid, the fornix, and the olfactory bulbs.

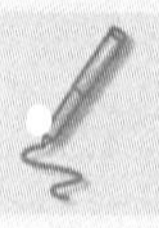

Color and label the components of the cerebrum figure 7.5.

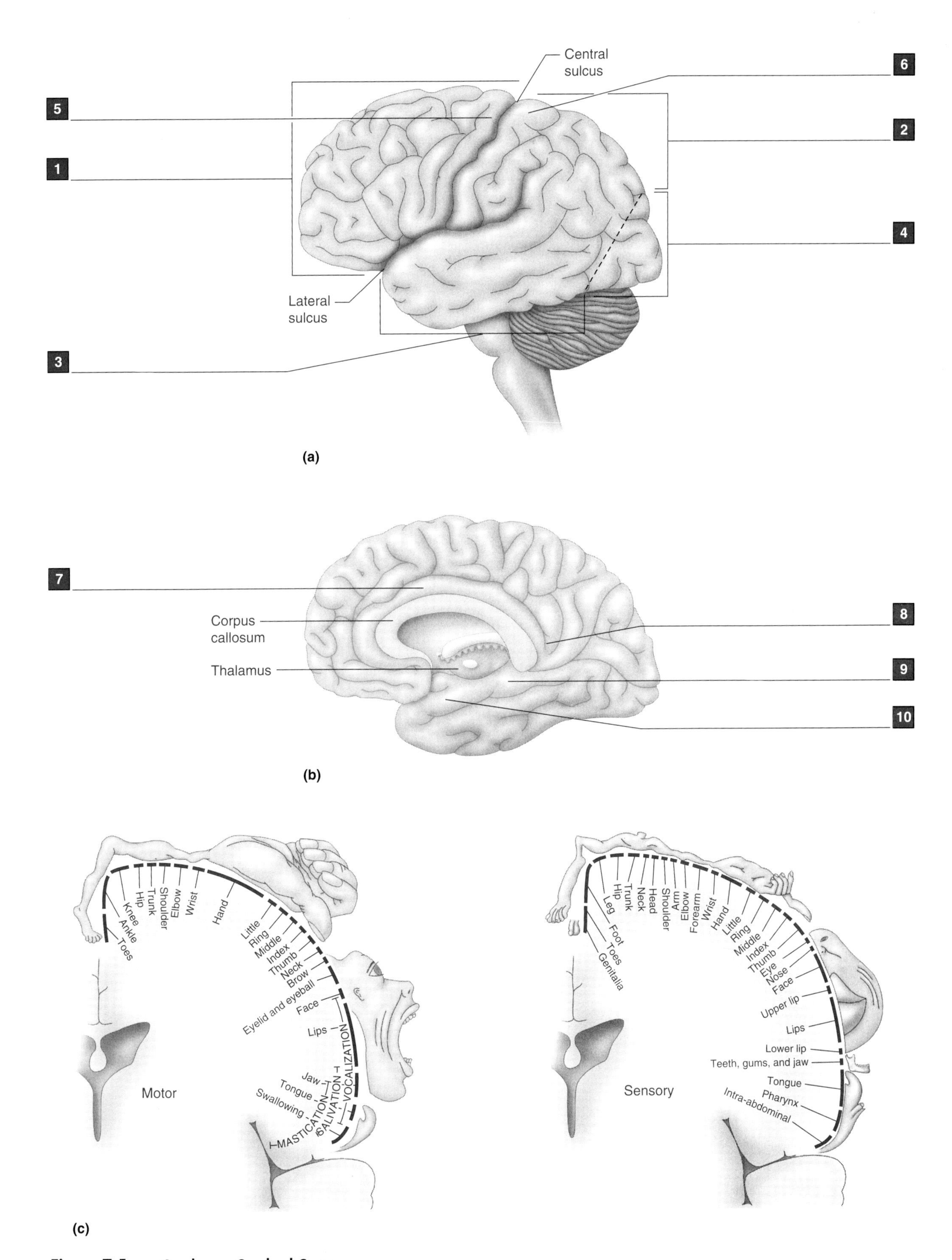

Figure 7.5 **Cerebrum: Cerebral Cortex**
(a) lateral view; (b) median sagittal section—brainstem removed; (c) these "somatotopical" diagrams depict the proportional representation of different body parts along coronal sections through the precentral gyrus (left) and postcentral gyrus (right).

White Matter

The mass of the cerebrum is composed of myelinated fibers that create a core of white matter. These nerve fibers follow numerous tracts that lead from one place to another. There are three types of fiber tracts:

Association Fibers 1 connect areas of the cerebral cortex to one another within the same hemisphere.

Commissural Fibers 2 connect corresponding areas of the cerebral cortex between the two hemispheres. These traverse the *Corpus Callosum*.

Projection Fibers 3 connect areas of the cerebral cortex to other regions of the brain and to the spinal cord. The *Corticospinal Tract*, which runs from the precentral gyrus through the crus cerebri of the midbrain and medullary pyramids of the medulla oblongata into the spinal cord, is composed of projection fibers.

Identify the courses of the fiber tracts of the cerebrum in figure 7.6.

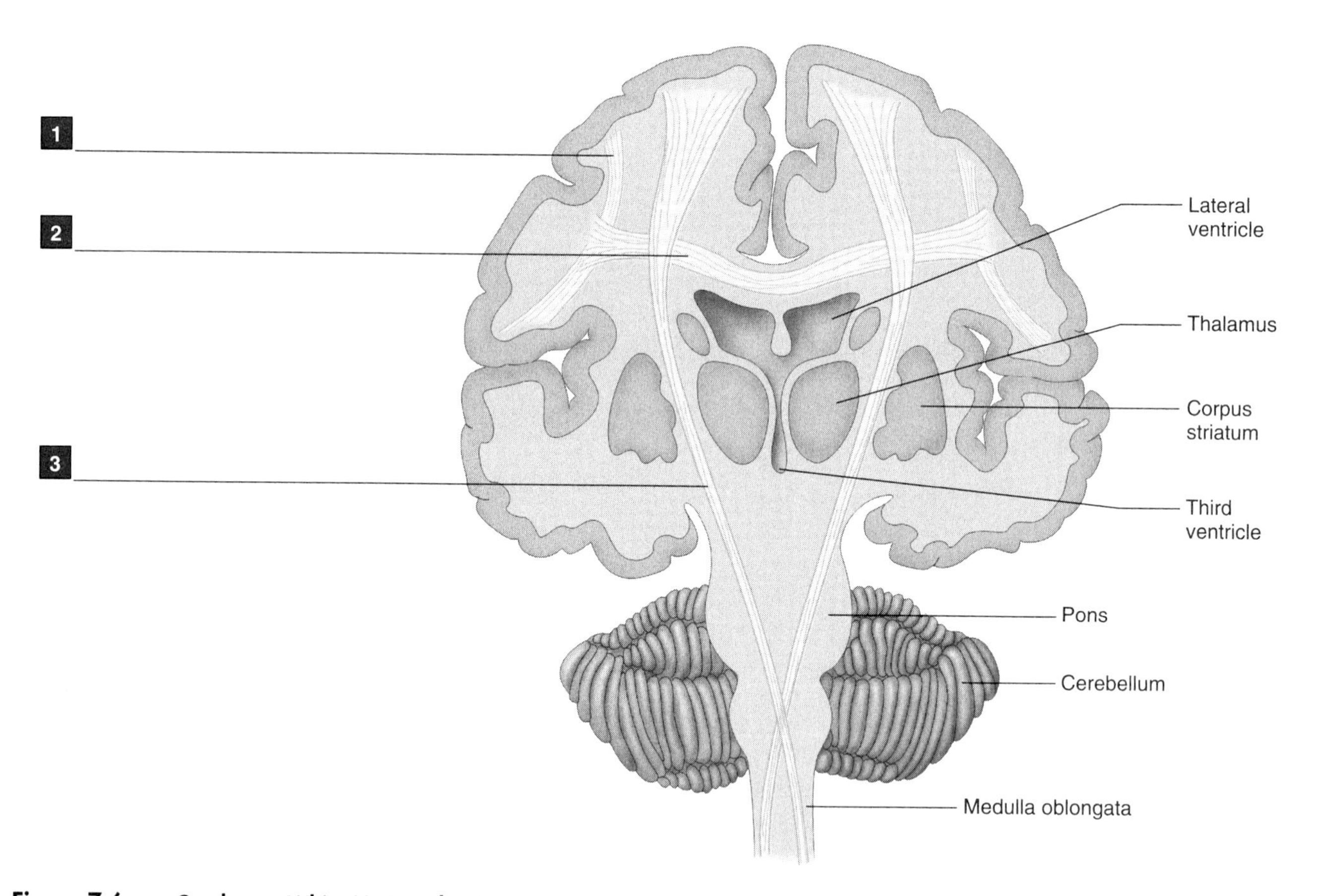

Figure 7.6 **Cerebrum: White Matter Fiber Tracts**
This schematic ("oblique coronal") section through the brain shows the courses of the three types of myelinated fiber tracts.

Basal Nuclei

The basal nuclei are aggregations of neuron cell bodies within the white matter of the cerebrum. (Older textbooks erroneously refer to the basal nuclei as "***Basal Ganglia.***" Recall that ganglia are collections of nerve cell bodies outside the CNS.)

The basal nuclei are functionally associated with other parts of the cerebrum in the production and control of motor responses. These nuclei initiate voluntary movements, and coordination of slow skeletal muscle contractions, such as those employed in posture and balance (e.g., arm swinging while walking). Some of the basal nuclei also seem to be involved in cognitive functions. Lesions to these nuclei result in a variety of motor dysfunctions, such as *Parkinson's Disease* and *Huntington's Disease*.

The caudate nucleus, putamen, and globus pallidus are the three main components of the basal nuclei. They comprise what is called the *Corpus Striatum*.

The **Caudate Nucleus** 1 is curved like a C; its superior part is known as the head, and the inferior part is the tail.

The **Putamen** 2 and **Globus Pallidus** 3 constitute what is known as the **Lentiform Nucleus** 4**.** The putamen comprises its lateral portion, and the globus pallidus makes up its medial portion. The lentiform nucleus lies along the dorsolateral side of the tail of the caudate nucleus.

Other structures that are functionally linked to the basal nuclei are the *Substantia Nigra* (a midbrain nucleus), the *Subthalamic Nucleus*, and the *Red Nucleus*.

Identify the basal nuclei of the cerebrum in figure 7.7.

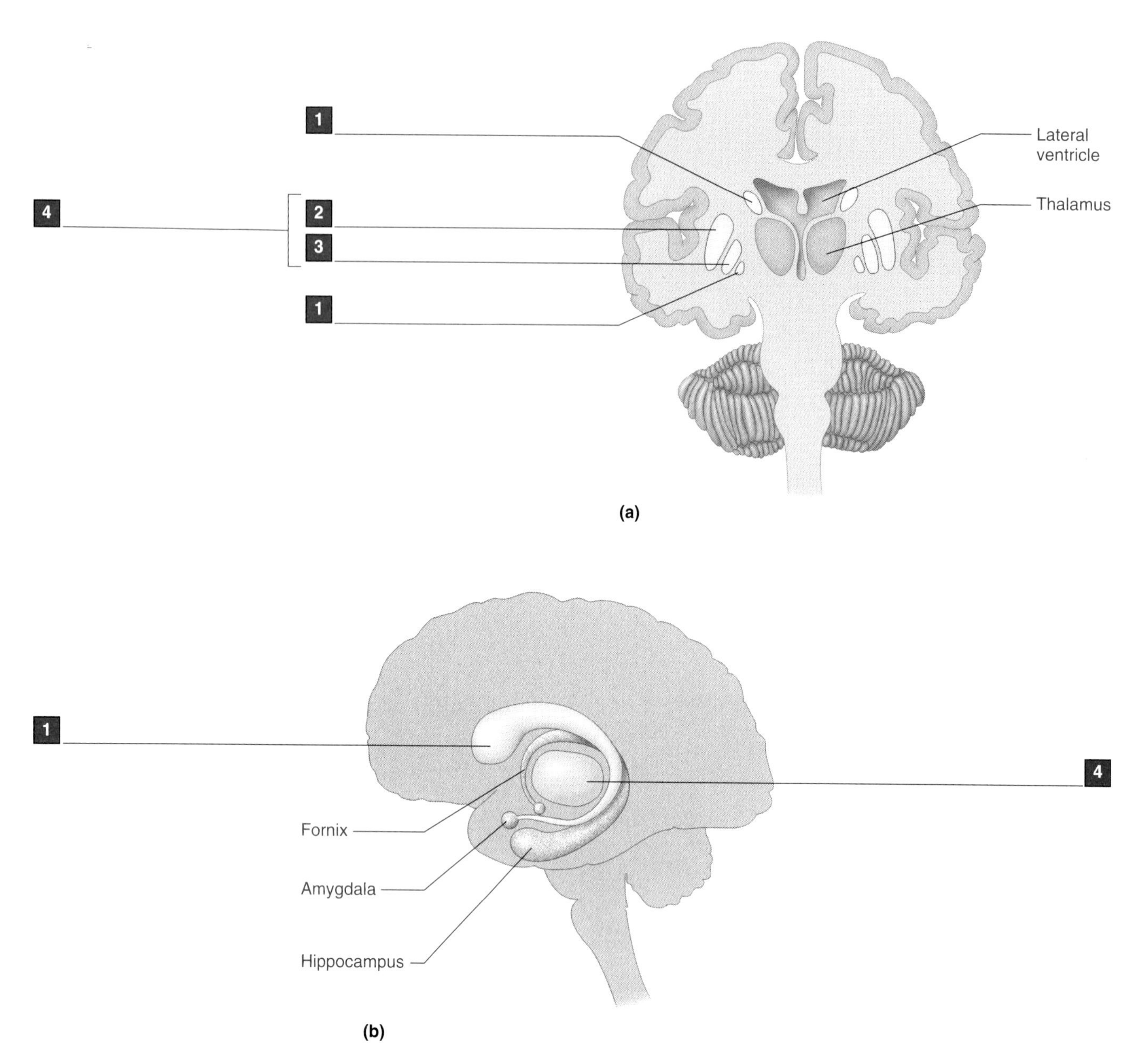

Figure 7.7 **Cerebrum: Basal Nuclei**

The components of the *Corpus Striatum* are shown in these illustrations.
(a) a schematic ("oblique coronal") section through the brain showing the relationship of the basal nuclei to the ventricles and thalamus; (b) the relationship of the basal nuclei to components of the *Limbic System* (stippled), as they would be seen if the cerebrum was transparent.

Ventricles and Cerebrospinal Fluid

The brain and spinal cord are protected against physical injury by the ***Cerebrospinal Fluid (CSF),*** which serves as a shock absorber. Recall that the CSF circulates in the *Subarachnoid Space*—that is, the space between the arachnoid and pia mater—around the brain and spinal cord, and through hollow, interconnected *Ventricles* within the brain.

CSF is produced by the *Choroid Plexus*. This is a network of capillaries and epithelial tissue in the walls of the ventricles that filter blood plasma.

Ventricular System

There are four ventricles. The two **Lateral Ventricles** 1 occupy space within the cerebrum. Each is C-shaped. The anterior part is separated from its opposite by a thin vertical partition, the *Septum Pellucidum*. The upper part runs below the corpus callosum and above the thalamus, and it extends as a posterior horn into the occipital lobe of the cerebrum. The inferior part of the C extends anterolaterally into the temporal lobe. The caudate nucleus forms the lateral wall of the lateral ventricle.

The two lateral ventricles communicate by the **Interventricular Foramen** 2 with the midline **Third Ventricle** 3**.** This occupies a narrow cleft between the left and right thalami. The choroid plexus on the roof of the third ventricle is continuous with that on the floor of the lateral ventricles.

The third ventricle is connected by the narrow **Cerebral Aqueduct** 4 to the midline **Fourth Ventricle** 5**.** This ventricle occupies the space between the dorsum of the pons and medulla and the overlying cerebellum.

At the junction of the pons and medulla there are two **Lateral Aperatures** 6**,** and below the cerebellum there is a midline **Median Aperture** 7 in the roof of the fourth ventricle. It is through these apertures that the ventricular system becomes continuous with the **Subarachnoid Space** 8**.** The fourth ventricle is continuous distally with the **Central Canal** 9 of the medulla oblongata and spinal cord.

CSF Circulation and Resorption

The CSF circulates around the brain and spinal cord in the subarachnoid space. It is gradually reabsorbed from this space into the vascular system by means of the **Arachnoid Villi** 10**.** These are fingerlike extensions of the arachnoid mater that project into the **Superior Sagittal Sinus** 11**.**

Identify the components of the ventricular system in figure 7.8.

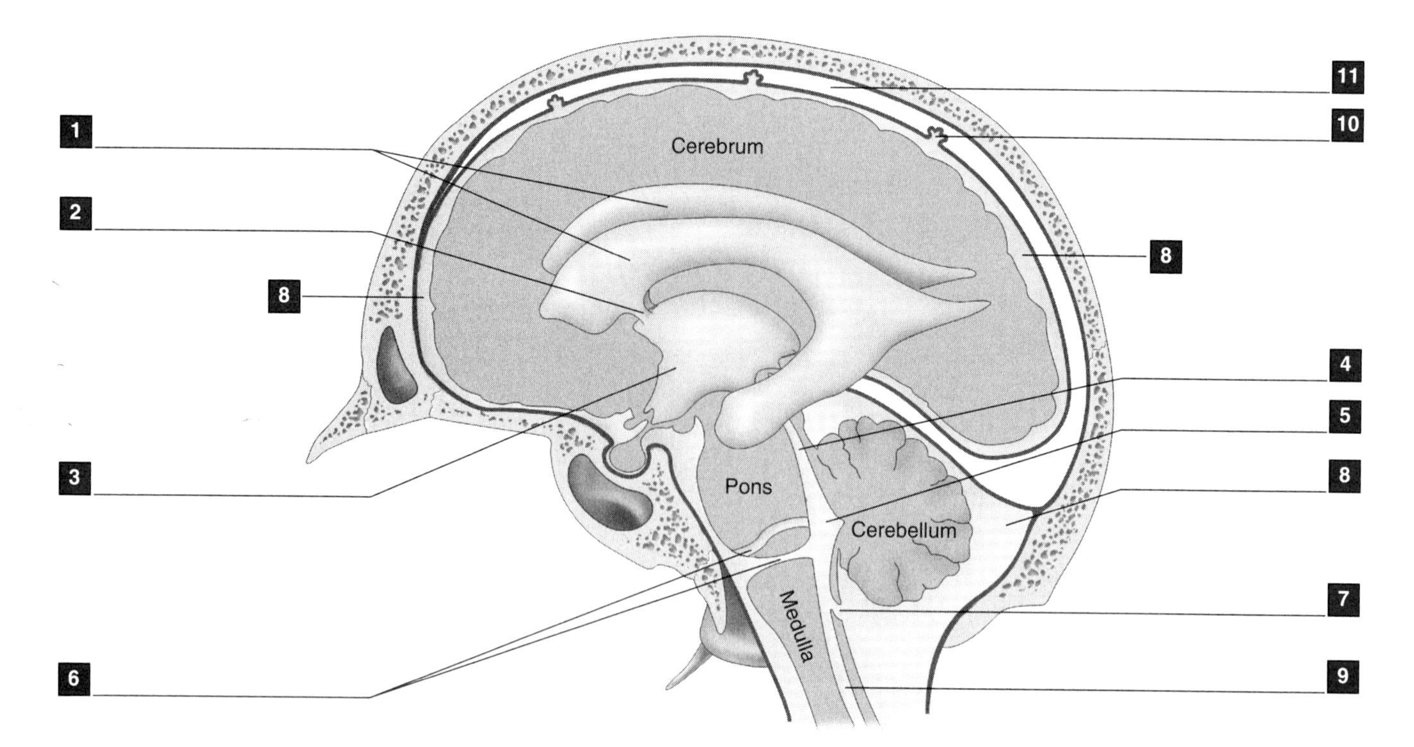

Figure 7.8 **Ventricles and CSF Circulation**

7.2 CRANIAL NERVES

The 12 cranial nerves can be thought of as peripheral nerves that emanate directly from the brain. Recall that they are numbered with Roman numerals (I–XII) according to the order in which they exit the brain from anterior to posterior. All emerge from the brainstem except I (olfactory) and II (optic), which derive from the cerebrum.

The 12 cranial nerves are:

I	**Olfactory**	**VII**	**Facial**
II	**Optic**	**VIII**	**Vestibulocochlear**
III	**Oculomotor**	**IX**	**Glossopharyngeal**
IV	**Trochlear**	**X**	**Vagus**
V	**Trigeminal**	**XI**	**Accessory**
VI	**Abducens**	**XII**	**Hypoglossal**

Cranial nerves may be classified into one of three categories according to their function: (1) *Sensory*, (2) *Motor*, and (3) *Mixed* (*Sensory and Motor*). In addition to conveying somatic motor impulses, cranial nerves may also transmit autonomic (*Parasympathetic*) motor impulses.

Three are purely ***Sensory:***

I	Olfactory
II	Optic
VIII	Vestibulocochlear

Five are ***Primarily Motor:***

III	Oculomotor	***(P)***
IV	Trochlear	
VI	Abducens	
XI	Accessory	
XII	Hypoglossal	

Four are ***Mixed:***

V	Trigeminal	
VII	Facial	***(P)***
IX	Glossopharyngeal	***(P)***
X	Vagus	***(P)***

Of the nine that are primarily motor or mixed, four also convey ***Parasympathetic*** fibers to peripheral ganglia. These are indicated in the preceding list with a ***(P)***.

Let us examine the course and distribution of each cranial nerve.

I Olfactory

The **Olfactory Nerve** 1 carries sensory (smell) information from receptors in the olfactory epithelium in the upper part of the nasal cavity. Axons from these cells form numerous bundles that pierce the *Cribriform Plate* of the ethmoid bone. They then enter the **Olfactory Bulb** 2, where they synapse. The postsynaptic fibers travel into the brain along the ***Olfactory Tracts.*** The olfactory tracts and bulbs are part of the cerebrum.

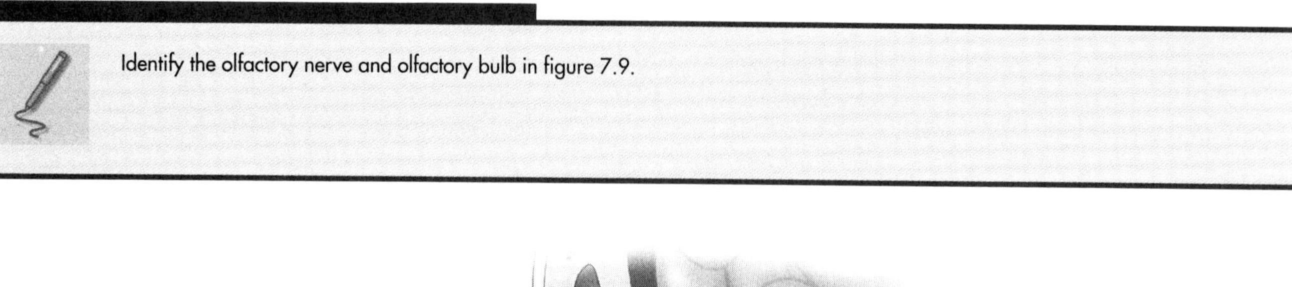

Identify the olfactory nerve and olfactory bulb in figure 7.9.

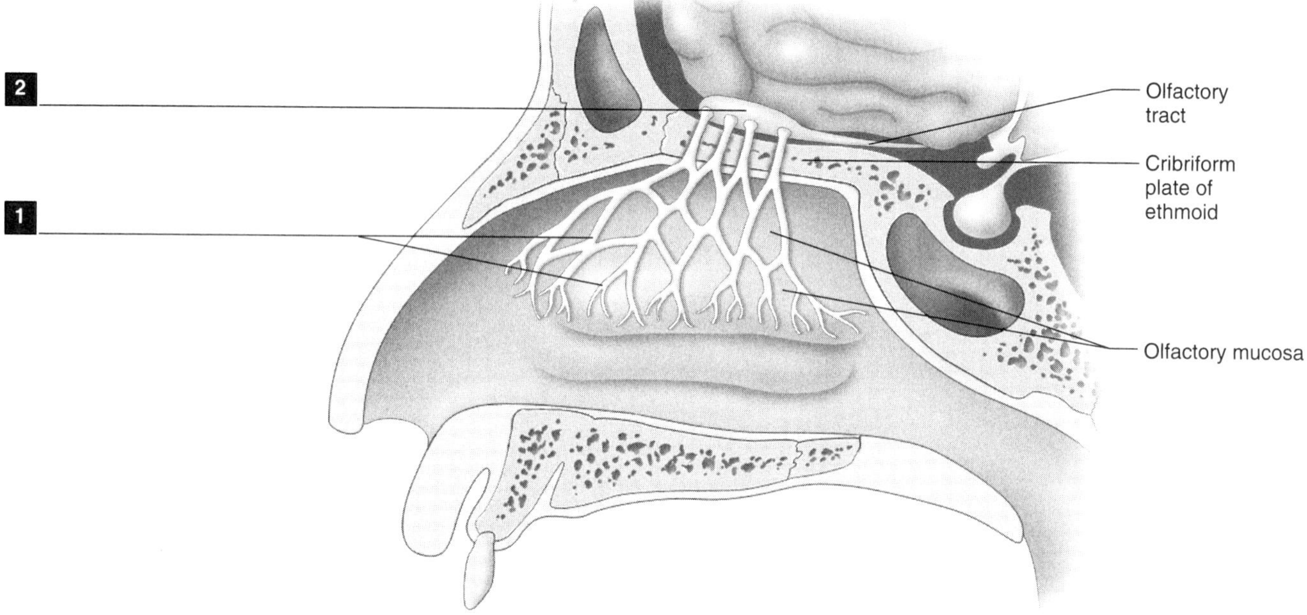

Figure 7.9 **Olfactory Nerve**

II Optic

The **Optic Nerve** 1 carries sensory (vision) information from receptors in the retina of the eye. The nerves pass back through the *Optic Canal* of the sphenoid bone and then converge in the midline at the **Optic Chiasm** 2 immediately anterior to the hypophyseal fossa. At the optic chiasm, some of the fibers from each nerve cross over to join the **Optic Tract** 3 on the opposite side to pass back to the thalamus.

Receptors in the *Medial Half* of each retina receive information from the **Temporal Visual Field** 4. Axons from these receptors in each eye *cross over* to the opposite side at the optic chiasm. Thus information from the temporal visual field is relayed to the opposite side of the brain.

Receptors in the *Lateral Half* of each retina receive information from the **Nasal Visual Field** 5. Axons from these receptors *do not cross over* at the chiasm. Thus information from the nasal visual field is relayed to the same side of the brain. Perception of visual stimuli occurs in the occipital lobe of the cerebrum.

Identify the aforementioned structures in figure 7.10. Color each temporal visual field, the medial side of the eyeball, and the appropriate axons red. Color each nasal visual field, the lateral side of the eyeball, and the appropriate axons blue.

III Oculomotor

The **Oculomotor Nerve** 1 carries motor information to the muscle that raises the upper eyelid (*Levator Palpebrae Superioris*). It also carries motor information to four of the six muscles that move the eyeball (*Superior Rectus*, *Medial Rectus*, *Inferior Rectus*, *Inferior Oblique*).

This nerve also conveys ***parasympathetic*** fibers to two of the intrinsic muscles of the eye: the *Ciliary Muscle* controls focus of the lens, and the *Constrictor Pupillae* of the iris contracts to constrict the pupil under bright light. The parasympathetic axons leave the oculomotor nerve and synapse in the **Ciliary Ganglion** 2, which lies next to the optic nerve.

The oculomotor nerve enters the orbit via the *Superior Orbital Fissure*.

IV Trochlear

The **Trochlear Nerve** 3 carries motor information to one of the six muscles that move the eyeball (*Superior Oblique*).

It enters the orbit with the oculomotor nerve via the *Superior Orbital Fissure*.

The ***Trigeminal Nerve*** **(CN V)** will be discussed on the following page of text. Let us consider the abducens nerve (CN VI) next because, like the oculomotor and trochlear, it conveys motor information to extrinsic eye muscles.

VI Abducens

The **Abducens Nerve** 4 carries motor fibers to one of the six muscles that move the eyeball (*Lateral Rectus*).

It enters the orbit through the *Superior Orbital Fissure*.

Identify cranial nerves III, IV, and VI in figure 7.11.

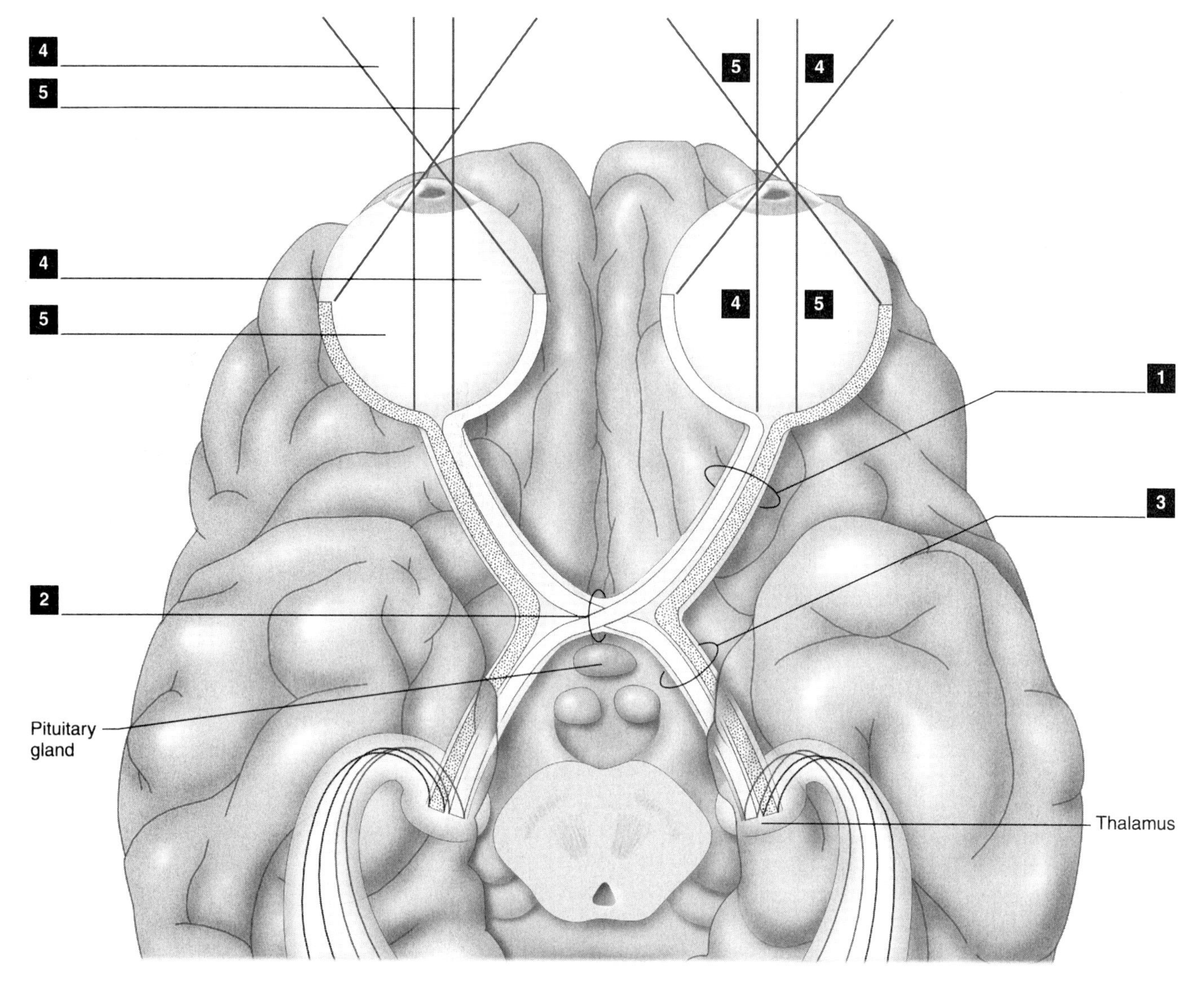

Figure 7.10 **Optic Nerve**

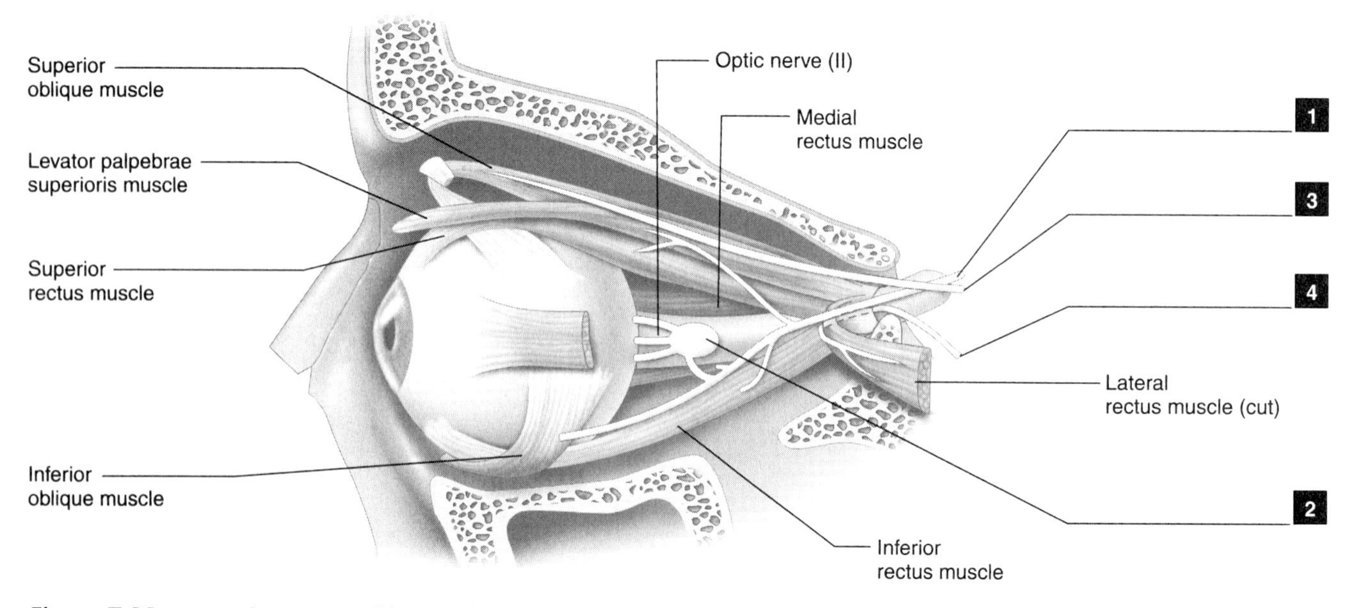

Figure 7.11 **Oculomotor, Trochlear, and Abducens Nerves**

V Trigeminal

The **Trigeminal Nerve** 1 is mixed, conveying both sensory and motor fibers. The sensory neurons have their cell bodies in a very large ganglion (the *Trigeminal Ganglion*) that lies on the floor of the greater wing of the sphenoid. It is equivalent to the dorsal root ganglion of a spinal nerve. Both sensory and motor fibers pass through this ganglion.

The trigeminal nerve has three divisions (designated by Arabic numerals following the Roman numeral):

V1	**Ophthalmic** (purely sensory)
V2	**Maxillary** (purely sensory)
V3	**Mandibular** (mixed sensory and motor)

The **Ophthalmic Division** 2 exits the cranial cavity through the *Superior Orbital Fissure*. It supplies the cornea of the eye, the nasal cavity and sinuses, and the skin of the forehead, eyelids, and middle of the nose. One of its branches exits the orbit via the *Supraorbital Foramen* (*Notch*) to reach the forehead.

The **Maxillary Division** 3 exits the cranial cavity through the *Foramen Rotundum*. It supplies the skin of the cheek and upper lip, the nasal cavity, the maxillary sinus, and the upper teeth. One of its branches reaches the face by exiting the maxilla through the *Infraorbital Foramen*.

The **Mandibular Division** 4 exits the cranial cavity through the *Foramen Ovale*. It carries motor fibers that serve the muscles of mastication, as well as the *Mylohyoid*, *Tensor Palatini*, and *Tensor Tympani* muscles. It conveys sensation, but not taste, from the mucosa of the mouth and tongue via one of its principal branches, the **Lingual Nerve** 5**.** It conveys sensation from the teeth, lower lip, and skin over the mandible via the **Inferior Alveolar Nerve** 6**,** which enters the mandible through the *Mandibular Foramen*, and exits it through the *Mental Foramen*. It supplies sensation from the ear and the skin above it by the **Auriculotemporal Nerve** 7**,** which emerges behind the neck of the mandible.

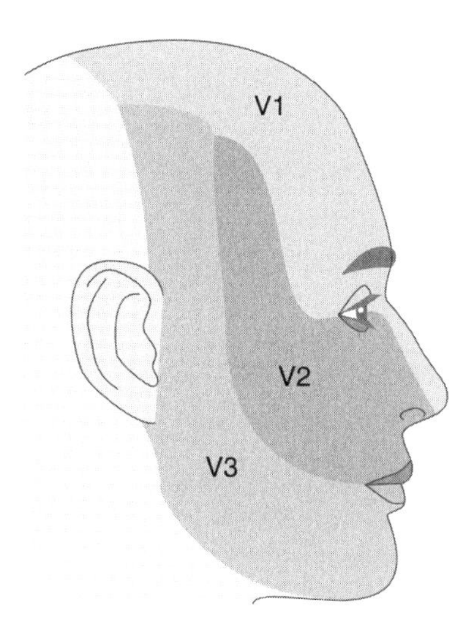

Cutaneous distribution of trigeminal nerve divisions.

Identify the branches of the trigeminal nerve in figure 7.12.

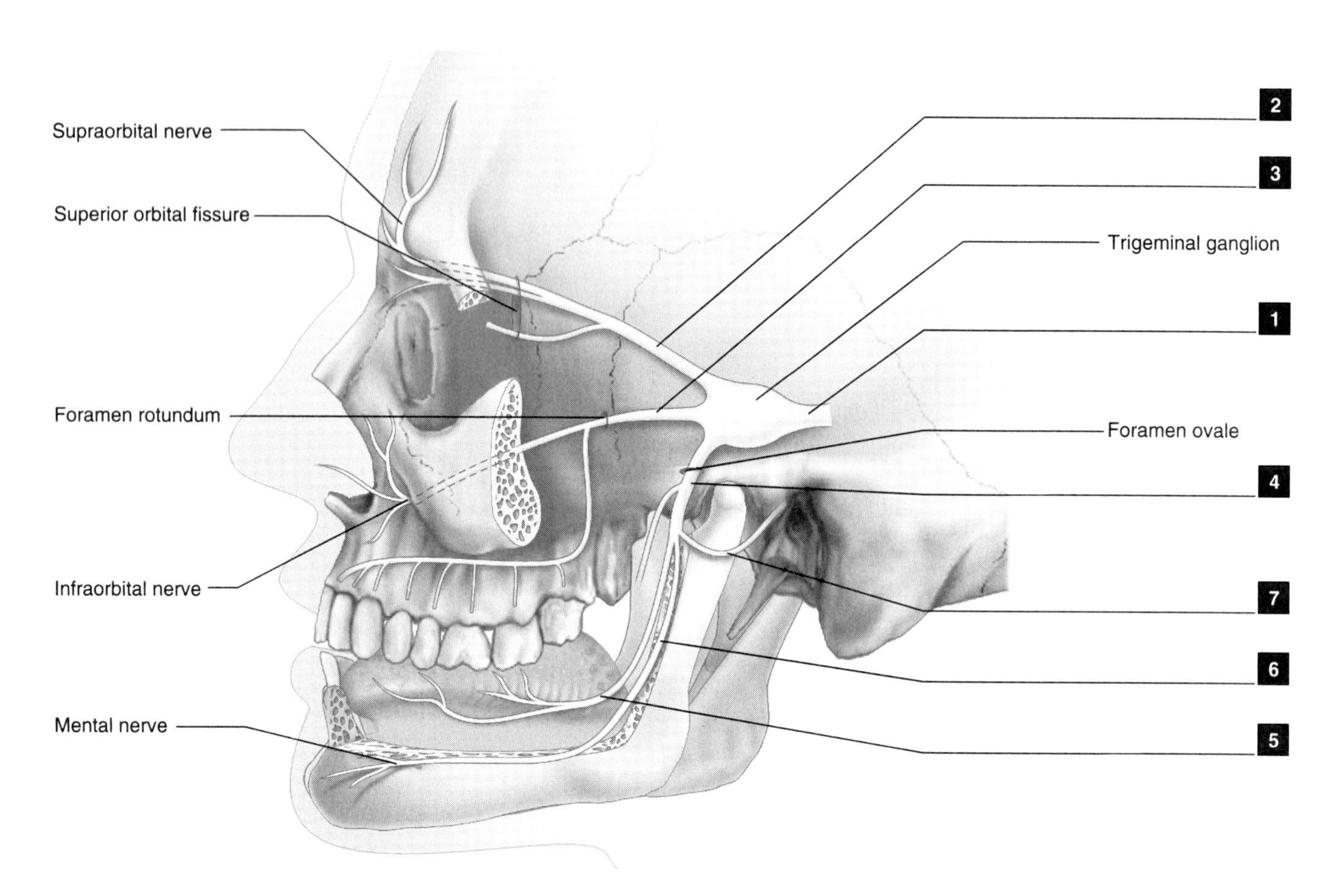

Figure 7.12 **Trigeminal Nerve**

The ***Abducens Nerve* (CN VI)** was discussed previously, together with the oculomotor and trochlear, which also convey motor information to the extrinsic eye muscles.

VII Facial

The **Facial Nerve** 1 is mixed, conveying both sensory and motor information. It carries sensory (taste) fibers from the anterior two-thirds of the tongue, and motor fibers to the muscles of facial expression. It also conveys ***parasympathetic*** fibers to the lacrimal and salivary glands.

It enters the petrous portion of the temporal bone via the *Internal Acoustic Meatus*. It travels through the bone for a short distance, where it encounters the *Geniculate Ganglion*. This is the location of the cell bodies of its sensory neurons. At the geniculate ganglion, the facial nerve gives off a branch, the *Greater Petrosal Nerve*. Soon after the geniculate ganglion, the facial nerve gives off a second branch, the *Chorda Tympani*.

The **Greater Petrosal Nerve** 2 carries ***parasympathetic*** fibers. It emerges from the front of the petrous bone into the middle cranial fossa, which it then exits through the *Foramen Lacerum*. It then runs anteriorly through the sphenoid at the root of the medial pterygoid plate to synapse in the **Pterygopalatine Ganglion** 3**.** Postganglionic fibers then follow a rather tortuous path to innervate the lacrimal gland. First, they jump onto the maxillary division of the trigeminal nerve (V2), they then turn and travel backward along it to the trigeminal ganglion, where they then turn again and travel in company with the ophthalmic division of the trigeminal nerve (V1) into the orbit to supply the *Lacrimal Gland*.

The **Chorda Tympani** 4 carries sensory (taste) and ***parasympathetic*** fibers. It leaves the temporal bone through a narrow slit between the petrosal and tympanic plate next to the jaw joint. It then travels in company with the lingual nerve, which is a branch of the mandibular division of the trigeminal (V3). Its sensory fibers run with V3 into the tongue, conveying taste from its anterior two-thirds. Its parasympathetic fibers leave V3 to synapse in the **Submandibular Ganglion** 5**.** Postganglionic fibers innervate the *Submandibular and Sublingual Salivary Glands*.

The **Facial Nerve** itself exits through the *Stylomastoid Foramen* to provide motor fibers to the muscles of facial expression via some six named branches (*Posterior Auricular*, *Cervical*, *Mandibular*, *Buccal*, *Zygomatic*, and *Temporal*).

Identify the branches of the facial nerve in figure 7.13.

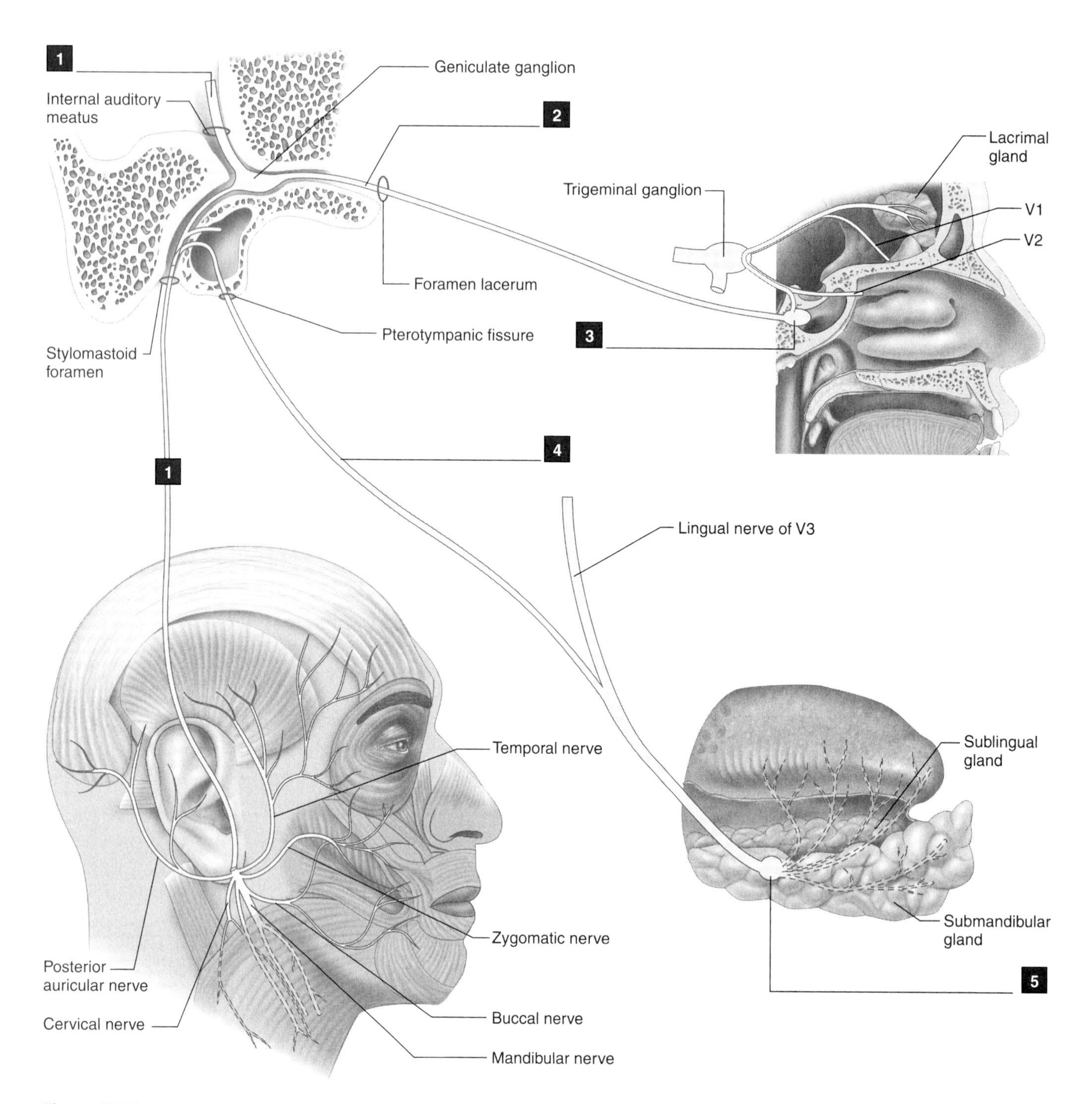

Figure 7.13 **Facial Nerve: Facial, Greater Petrosal, and Chorda Tympani**

VIII Vestibulocochlear

The **Vestibulocochlear Nerve** 1 carries sensory information from the inner ear. It conveys impulses related to both *Balance* and *Hearing*. This nerve is sometimes referred to as the Stato-acoustic nerve.

Impulses related to hearing originate in the snail-shaped *Cochlea* and are conveyed by the **Cochlear Branch** 2 of this nerve.

Information relating to balance and equilibrium derives from the *Semicircular Canals*, *Utricle*, and *Saccule* and is conveyed by the **Vestibular Branch** 3.

The cochlear and vestibular branches join together, and enter the endocranial cavity through the *Internal Acoustic Meatus*, in company with the facial nerve.

Identify the components of the vestibulocochlear nerve in figure 7.14.

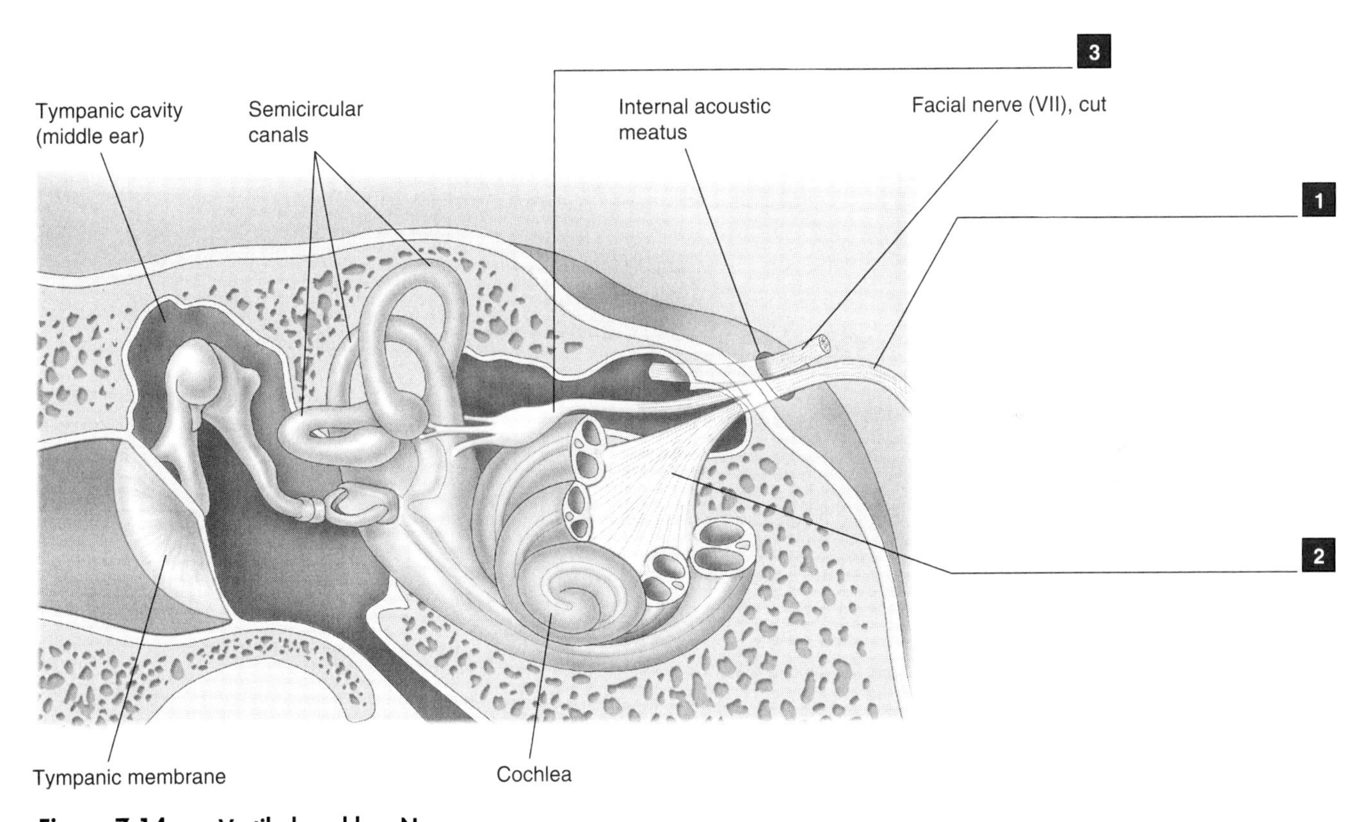

Figure 7.14 **Vestibulocochlear Nerve**

IX Glossopharyngeal

The **Glossopharyngeal Nerve** 1 is mixed. It is mostly sensory, relaying somatic and visceral information, as well as that related to taste. It also conveys somatic and ***parasympathetic*** motor impulses. It has several important branches.

It exits the skull through the *Jugular Foramen*. Within the foramen, the glossopharyngeal gives off a branch, the *Tympanic Nerve*.

The **Tympanic Nerve** 2 carries ***parasympathetic*** fibers. It runs to a plexus atop the petrous temporal. It emerges from this plexus as the **Lesser Petrosal Nerve** 3, which exits the skull through the *Foramen Ovale* to synapse in the **Otic Ganglion** 4 immediately below the foramen. Postganglionic fibers then jump onto the auriculotemporal nerve (a branch of the mandibular division of trigeminal, which also traverses the foramen ovale), and travel with it to innervate the *Parotid Gland*.

The Glossopharyngeal nerve runs into the neck, giving **Pharyngeal Branches** 5 to the pharyngeal plexus on the back of the *Middle Constrictor* muscle. These branches convey sensory information from the pharynx, and are partially responsible for the *gag reflex*.

The Gag Reflex

The reflex consists of constriction of the pharynx, through contraction of the three pharyngeal constrictors, when the back of the oropharynx (uvula) is touched. The glossopharyngeal nerve is the sensory limb of this reflex because it conveys somatic sensory information from the pharynx. The vagus nerve is the motor limb of the reflex, because it provides these impulses to the pharyngeal constrictors.

The glossopharyngeal nerve runs along the *Stylopharyngeus* muscle, supplying it with motor fibers, and then turns forward to reach the back of the tongue. It conveys taste from the posterior one-third of the tongue.

The glossopharyngeal sends off a branch that runs down the *Internal Carotid Artery* to its bifurcation with the external carotid artery. This **Carotid Branch** 6 provides visceral sensory information from chemoreceptors in the *Carotid Body* (measuring levels of CO_2) and stretch receptors in the *Carotid Sinus* (monitoring blood pressure).

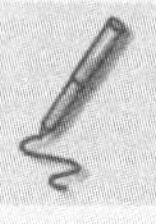

Identify the branches of the glossopharyngeal nerve in figure 7.15.

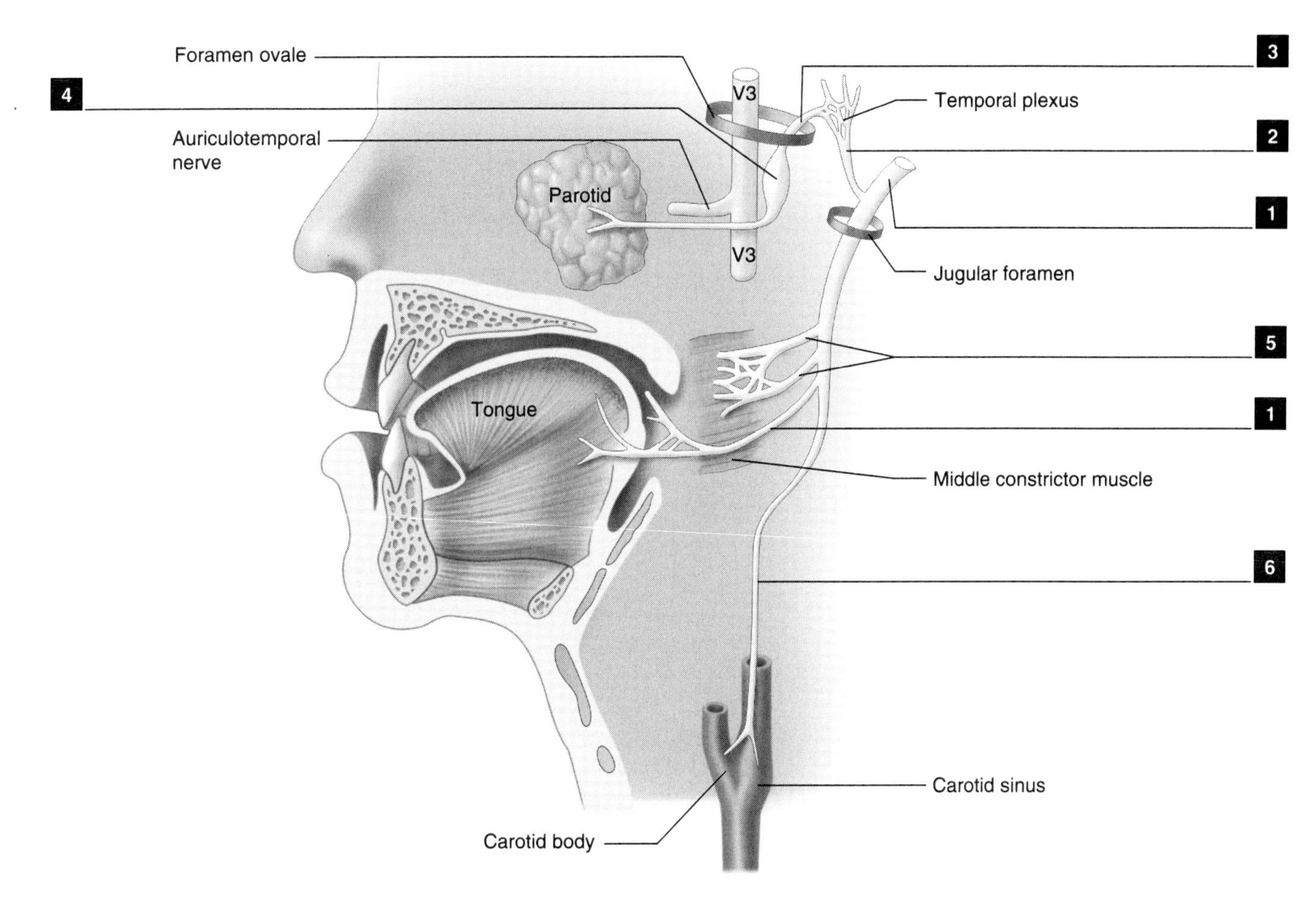

Figure 7.15 Glossopharyngeal Nerve

X Vagus

The **Vagus Nerve** 1 is mixed. It has the most extensive distribution of any cranial nerve, providing structures in the head, neck, thorax, and abdomen with innervation.

It gives somatic motor fibers to the pharyngeal constrictors and the laryngeal muscles (swallowing and speaking). It provides ***parasympathetic*** innervation to cardiac (heart) muscle, and to the smooth muscle and glands of the digestive system. It conveys sensory information from the skeletal muscles it innervates, the external auditory meatus, and the tracheobronchial tree of the lung. It also conveys taste from the epiglottis.

It exits the skull through the *Jugular Foramen*. Cell bodies of the sensory neurons are located within ganglia in (*Superior Jugular Ganglion*) and just below (*Inferior Nodose Ganglion*) this foramen. The ***Auricular Branch*** runs from the superior ganglion to the external ear.

The vagus runs through the neck between the internal carotid artery and jugular vein in the *Carotid Sheath*. It sends off two principal branches in the neck. The **Pharyngeal Nerve** 2 supplies the pharyngeal constrictors and most of the palatal muscles, and the **Superior Laryngeal Nerve** 3 supplies the *Cricothyroid* muscle.

Below the neck, the *Right Vagus Nerve* sends off the *Right Recurrent Laryngeal Nerve* beneath the subclavian artery. The *Left Vagus Nerve* runs along the aortic arch, sending off the **Left Recurrent Laryngeal Nerve** 4 under the arch. The right recurrent laryngeal nerves then run back up into the neck. The recurrent laryngeal nerves supply all of the intrinsic laryngeal muscles with the exception of the *Cricothyroid*.

The vagus provides fibers to the *Cardiac Plexus* and the *Pulmonary Plexus*. Branches destined for the heart emerge from the former; branches destined for the lung come from the latter.

The vagus nerves run down the esophagus as the *Anterior and Posterior Vagal Trunks* into the abdomen.

The left vagus contributes primarily to the **Anterior Vagal Trunk** 5. It supplies the stomach, liver, pancreas, and parts of the intestines.

The right vagus constitutes the bulk of the ***Posterior Vagal Trunk.*** It contributes to the **Celiac Ganglion** on the front of the aorta, from which the majority of abdominal structures receive innervation.

Identify the branches of the vagus nerve in figure 7.16.

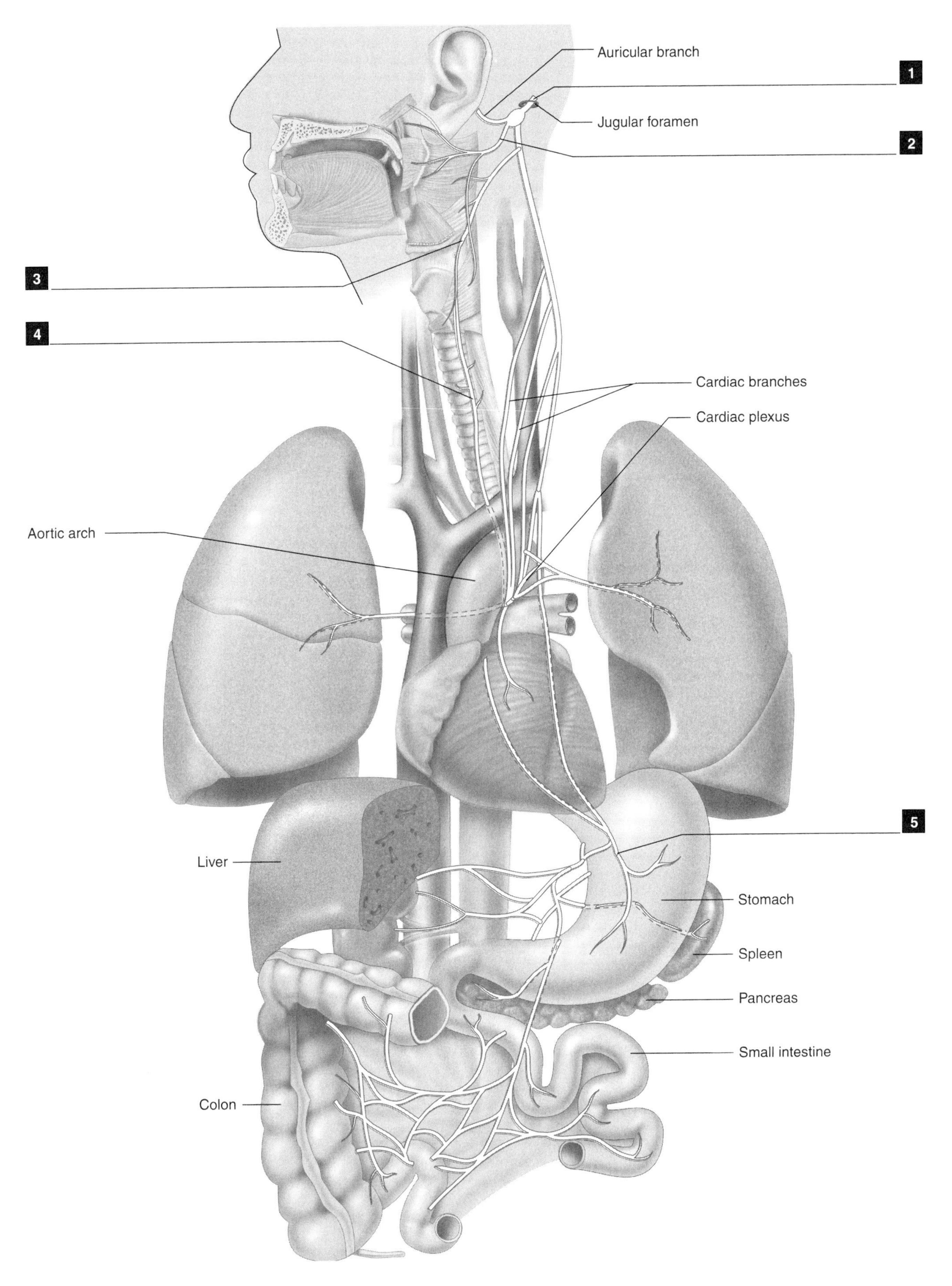

Figure 7.16 **Vagus Nerve**

XI Accessory

The **Accessory Nerve** 1 conveys motor impulses to skeletal muscles. It is really two nerves, and is sometimes referred to as the *Spinal Accessory Nerve*, a name that is anatomically appropriate.

Because it comprises two nerves that are conjoined for a short distance, it has *two roots* and *two branches*.

The two roots of the accessory nerve are: spinal and cranial (medullary).

The **Spinal Root** 2 derives from a column of gray matter in the lateral horn of the *Spinal Cord from C1–C5*. The axons from these neural cell bodies join together as they travel upward through the vertebral canal forming a single bundle that passes upward through the *Foramen Magnum*.

The **Cranial Root** 3 derives from motor nuclei in the *Medulla Oblongata*.

The spinal and cranial roots join together to form the accessory nerve, which exits the cranium through the *Jugular Foramen* in company with the vagus nerve.

The two branches of the accessory nerve are: spinal and cranial (medullary).

Upon emerging from the jugular foramen, the *Spinal Root* fibers of the accessory nerve split off as the **Spinal Branch** 4 to innervate *Sternocleidomastoid* and *Trapezius*.

The *Cranial Root* fibers of the accessory nerve form the **Cranial Branch** 5. The cranial branch is an *accessory* to the vagus.

Identify the roots and branches of the accessory nerve in figure 7.17.

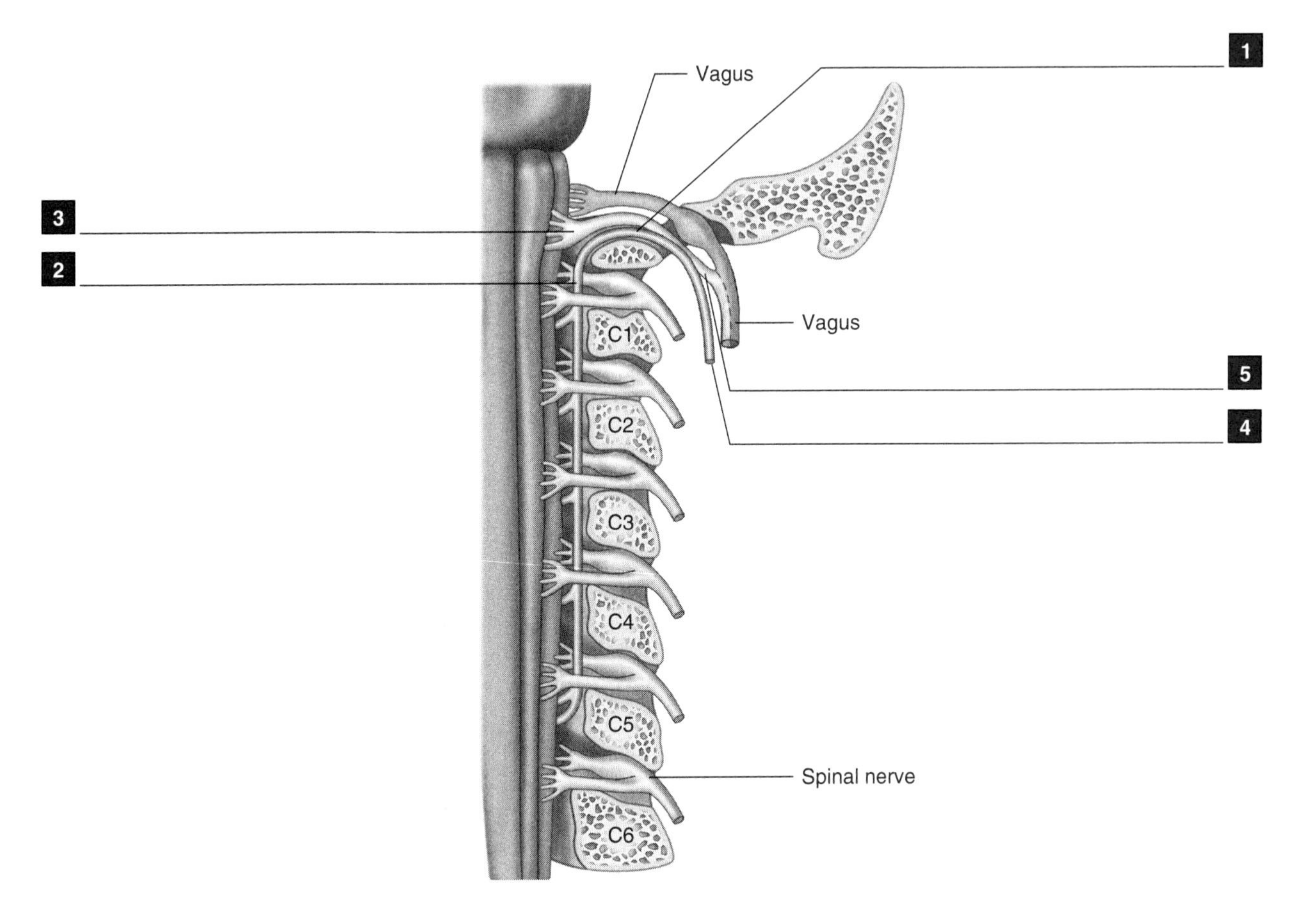

Figure 7.17 **Accessory Nerve**

XII Hypoglossal

The **Hypoglossal Nerve** 1 conveys somatic motor impulses to the tongue musculature.

It exits the skull through the *Hypoglossal Canal*, and supplies the two intrinsic tongue muscles and three extrinsic tongue muscles: *Hyoglossus*, *Genioglossus*, and *Styloglossus*.

Color and label the hypoglossal nerve in figure 7.18.

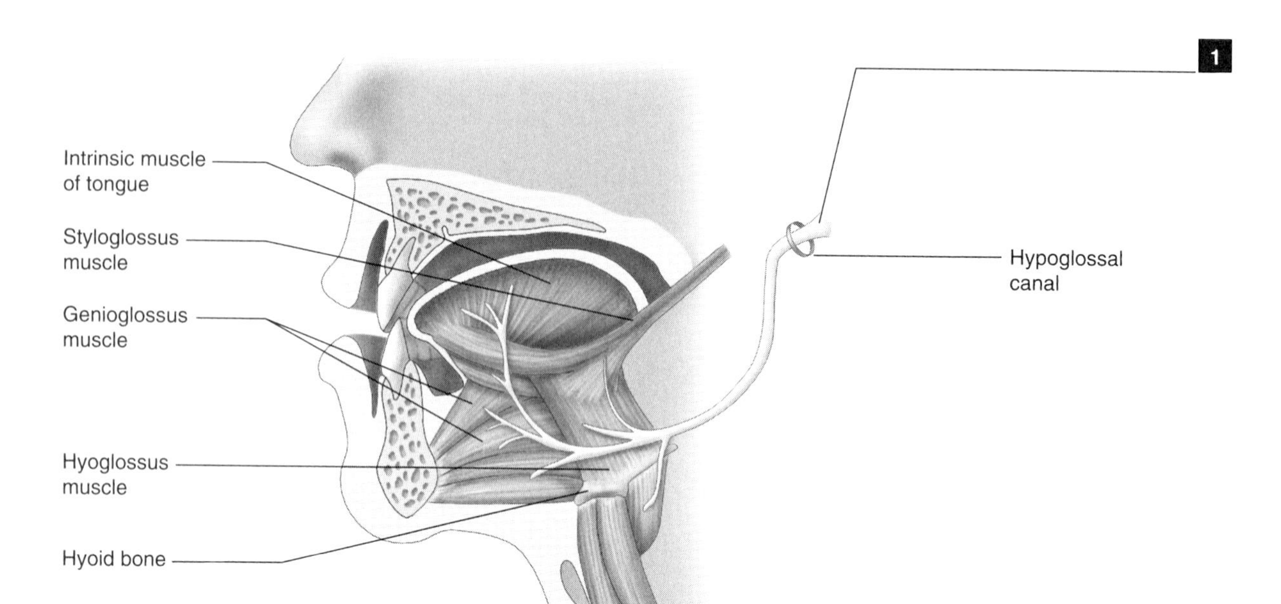

Figure 7.18 **Hypoglossal Nerve**

LABORATORY

8

The Eye and The Ear

8.1 THE EYE

The eye is a sensory organ that comprises a fluid-filled globe (eyeball) with light-sensitive receptors which is contained within a bony cavity (orbit) of the skull. The orbit contains muscles that move the eyeball, as well as fat, nerves, blood vessels, and the lacrimal apparatus. The orbit and eyeball are guarded anteriorly by two thin, moveable folds, the eyelids.

External Features

Below the eyebrow are two eyelids, the **Palpebrae** [1]. They help protect the eye from injury. The elliptical opening between them is the *Palpebral Fissure*. Where the eyelids meet, they form a rounded **Medial Canthus** [2] and a more acute **Lateral Canthus** [3]. The lateral canthus lies in direct contact with the eyeball; the medial does not.

The margin of each eyelid along the palpebral fissure is lined with hairs. Specialized, elongate sebaceous glands exude oily secretions through openings behind the eyelashes; this material helps make the closed eyelids both airtight and watertight.

The lids are covered externally by skin, and lined internally by a clear, thin mucous membrane known as **Conjunctiva** [4]. The palpebral conjunctiva that lines the eyelids is continuous with the bulbar conjunctiva that covers the eyeball.

The medial canthus is separated from the eyeball by a small, pink space, in the center of which is a small bump, the **Lacrimal Caruncle** [5]. Just lateral to the lacrimal caruncle is a short, curved fold of tissue, the **Plica Semilunaris** [6]. The medial end of each eyelid has a small black dot known as a **Lacrimal Punctum** [7]. This is the opening of the *Lacrimal Canal*, which leads to the lacrimal sac.

The *Epicanthic Fold* is a fold of skin of varying extent that runs from the upper eyelid to inferior corner of the medial canthus. Its presence accounts for some of the differences in facial appearance among different populations. It is most common in individuals of Asian heritage, although it is not as frequent in people from South Asia and the offshore Asian islands.

The white of the eyeball that is covered by transparent bulbar conjunctiva is known as the **Sclera** [8]. The ***Cornea*** forms a transparent bulge on the front of the eyeball. The cornea is avascular, and has no lymphatic drainage. Deep to the cornea is the variably pigmented **Iris** [9]. In the center of the iris is the **Pupil** [10].

Identify the aforementioned features of the eye in figure 8.1. Identify these features through self-examination with a mirror, and on fellow students in the lab.

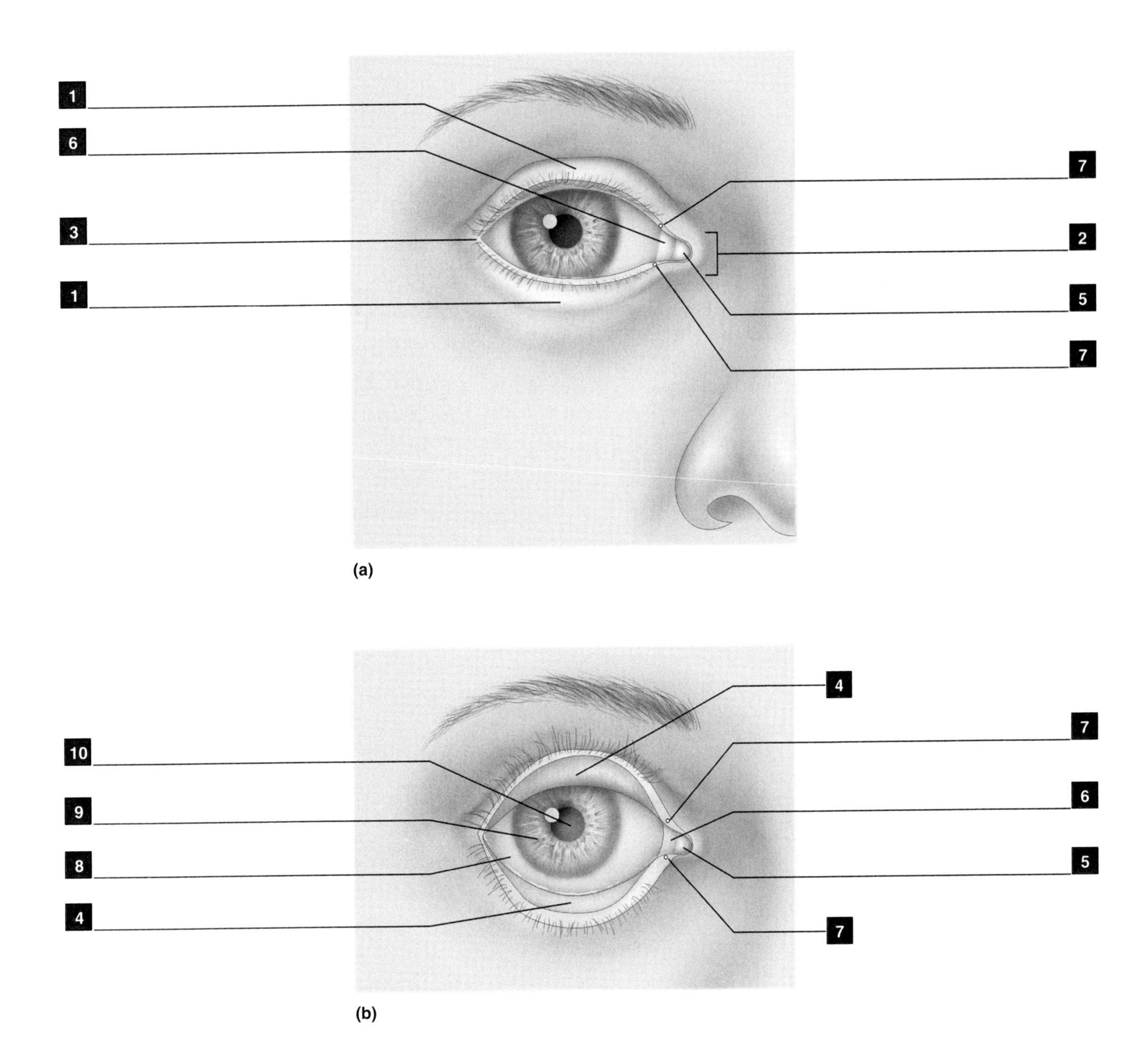

Figure 8.1 **External Features of the Eye**
(a) anterior view; (b) anterior view with eyelids reflected.

The Bony Orbit

Six bones comprise the orbit. We briefly examined them when we studied the skull (laboratory 6, pp. 172–173).

The **Frontal** 1 makes up the superior margin and roof of the orbit. The **Zygomatic** 2 makes up the lateral margin and part of the lateral wall. The **Maxilla** 3 comprises the floor and much of the inferior and medial margins. The medial wall is comprised by the **Lacrimal** 4 anteriorly, and the **Ethmoid** 5 posteriorly. The **Sphenoid** 6 constitutes the back of the cavity and much of its lateral wall.

The **Optic Canal** 7 pierces the lesser wing of the sphenoid. It conveys the *Optic Nerve*. The greater and lesser wings of the sphenoid are separated by the **Superior Orbital Fissure** 8. It conveys the *Oculomotor, Trochlear, ophthalmic division of Trigeminal*, and *Abducens Nerves*. A thin **Inferior Orbital Fissure** 9 separates the sphenoid and maxilla.

The lacrimal bone has a sharp crest. Together with a crest along the frontal process of the maxilla, it defines a hollow known as the **Lacrimal Fossa** 10. This is continuous inferiorly with the *Nasolacrimal Canal*, which opens into the nasal cavity below the inferior nasal concha.

Identify the aforementioned features of the bony orbit in figure 8.2.

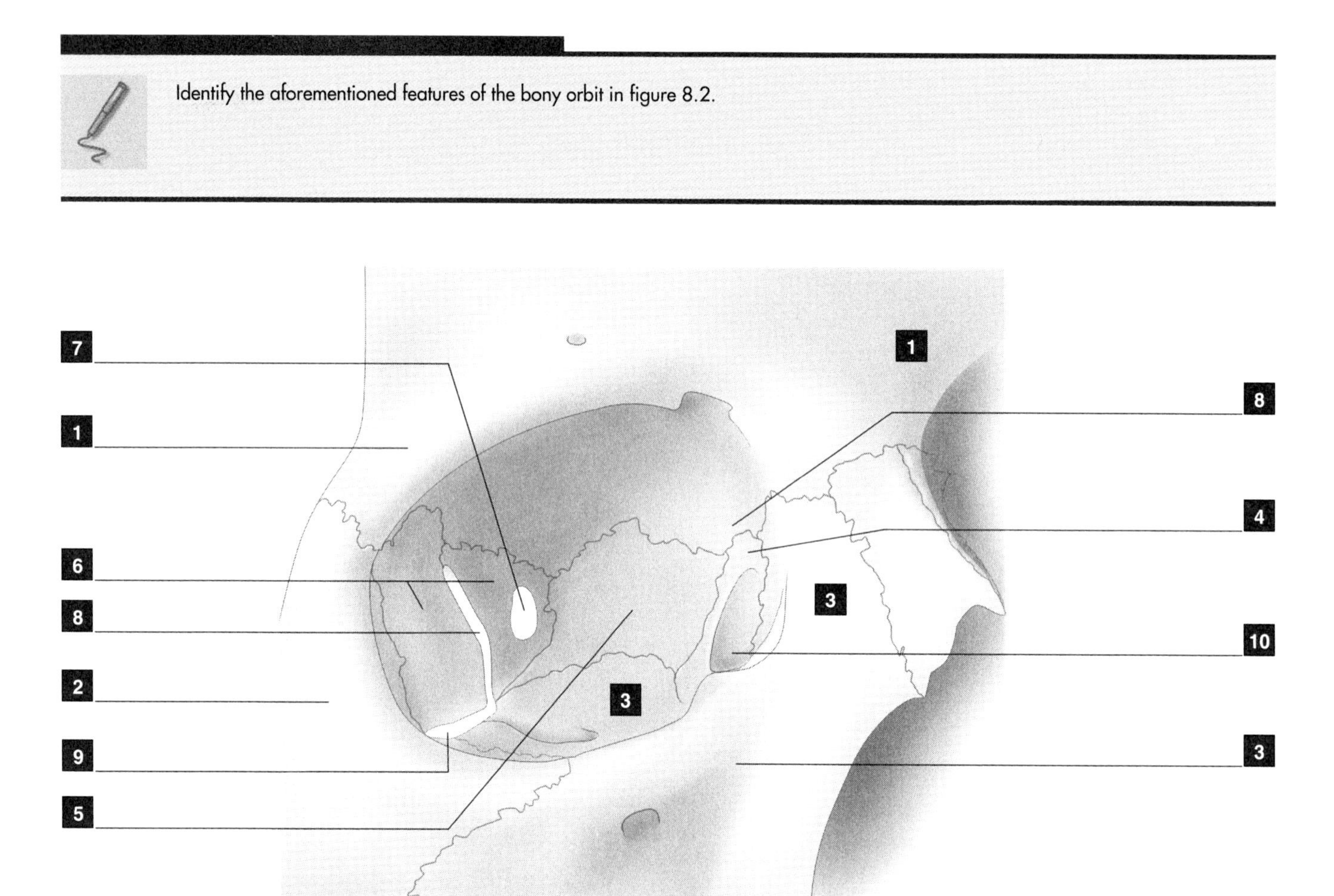

Figure 8.2 **The Bony Orbit**

The Eyelids

Each eyelid has a fibrous "skeleton" that is lined on the outside by muscle, superficial fascia, and skin, and on the inside by modified sebaceous glands and a thin, transparent mucous membrane. The upper lid has a muscle that retracts it.

The fibrous "skeleton" of each eyelid is called the **Tarsal Plate** 1. Each plate is connected to the medial and lateral orbital walls by a ligament, and to the orbital margin by a thin sheet of connective tissue called the **Orbital Septum** 2.

External to the tarsal plate and orbital septum, the fibers of the **Orbicularis Oculi** 3 muscle arise from the medial tarsal ligament and sweep laterally in both lids to insert into the lateral tarsal ligament. This muscle closes the eyelids. The fibers near the margin of the lid are employed in blinking; the remaining fibers are used to voluntarily close the lids.

The fibers of the **Levator Palpebrae Superioris** 4 muscle insert into the upper margin of the superior tarsal plate. Recall that this muscle elevates the lid. It is, therefore, active while the eyes are open.

Recall that the **Palpebral Conjunctiva** 5 lining the eyelids is continuous with the **Bulbar Conjunctiva** 6 over the sclera of the eyeball.

Identify the features of the eyelids in figure 8.3.

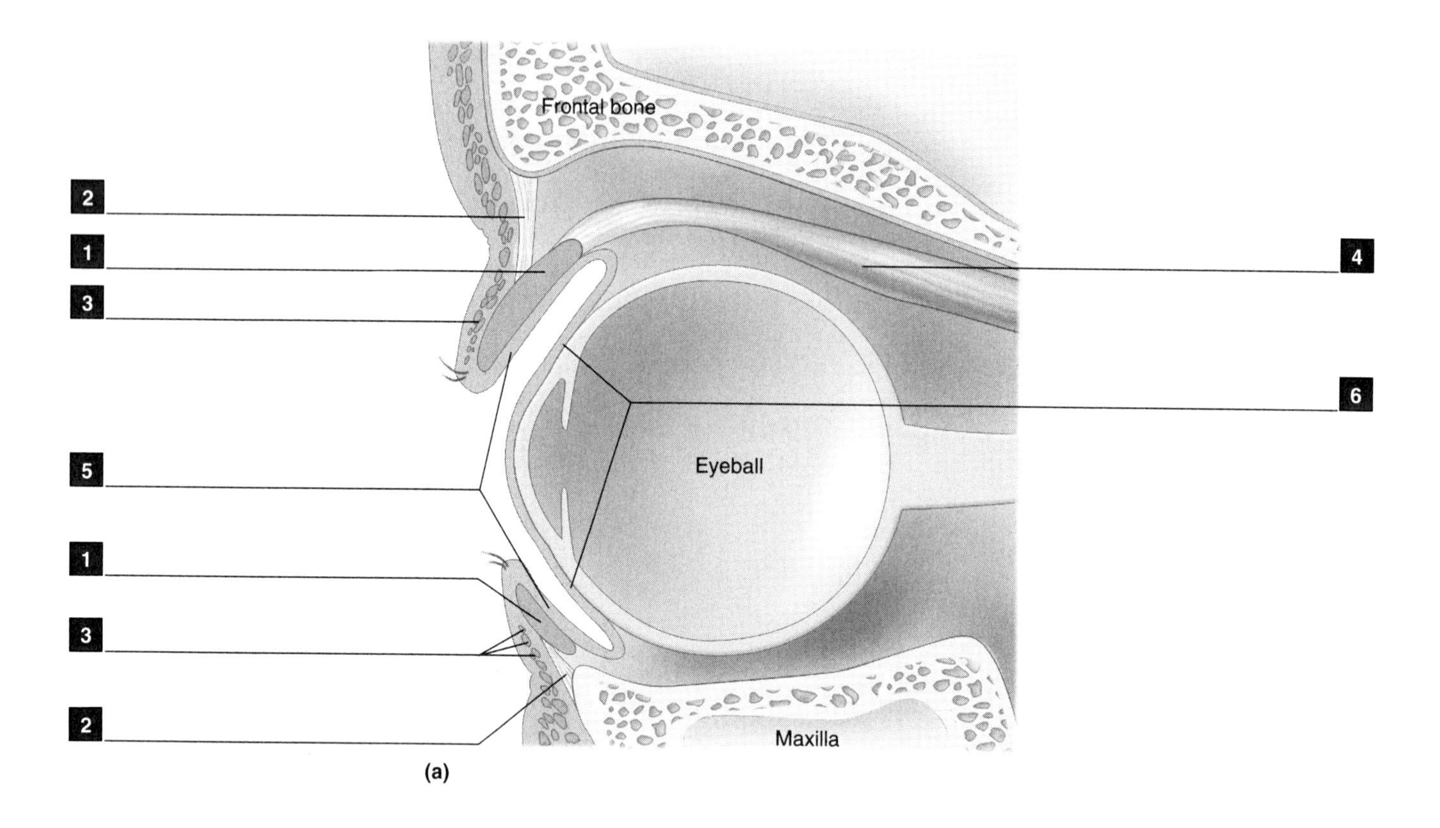

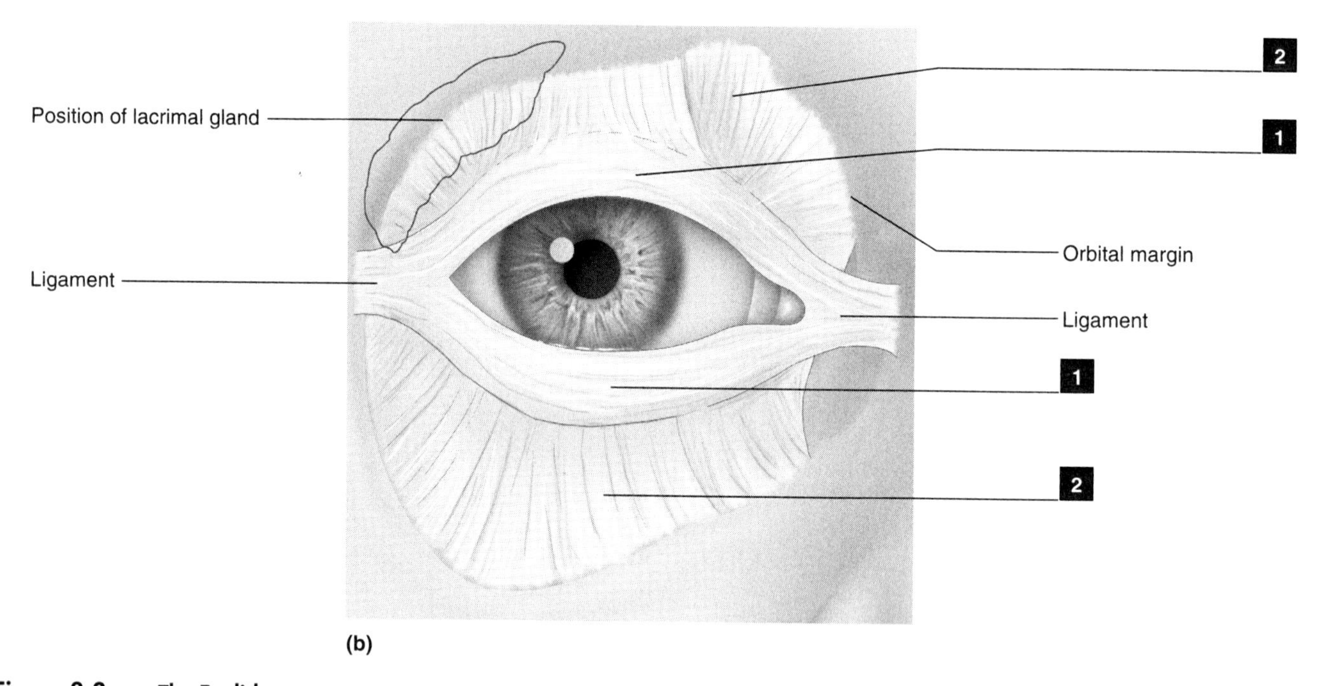

Figure 8.3 **The Eyelids**
(a) sagittal section; (b) anterior view of right eye.

The Lacrimal Apparatus

The lacrimal apparatus consists of a gland that produces tears, and the tubes that drain this fluid from the surface of the eyeball.

The **Lacrimal Gland** **1** is about the size of an almond. It lies under cover of the superolateral margin of the bony orbit. It has numerous fine excretory ducts that open into the space between the palpebral and bulbar conjunctiva. The lacrimal gland is innervated by the *Greater Petrosal Nerve*, which is a branch of the *Facial Nerve* (*CN VII*).

Tears from the gland flow across the front of the eyeball and are collected by a small opening, the **Lacrimal Punctum** **2**, at the medial end of each eyelid. Each lacrimal punctum opens into a small tube called the **Lacrimal Canaliculus** **3** that empties into the **Lacrimal Sac** **4**.

The lacrimal sac lies in the sulcus formed by the lacrimal bone and maxilla on the inferomedial corner of the orbit. It drains into the **Nasolacrimal Duct** **5**. This duct runs in the bony *Nasolacrimal Canal*, which opens below the inferior concha into the side of the nasal cavity.

The presence of the lacrimal canaliculi, sacs, and ducts explains why your nose runs when you cry.

Identify the components of the lacrimal apparatus in figure 8.4.

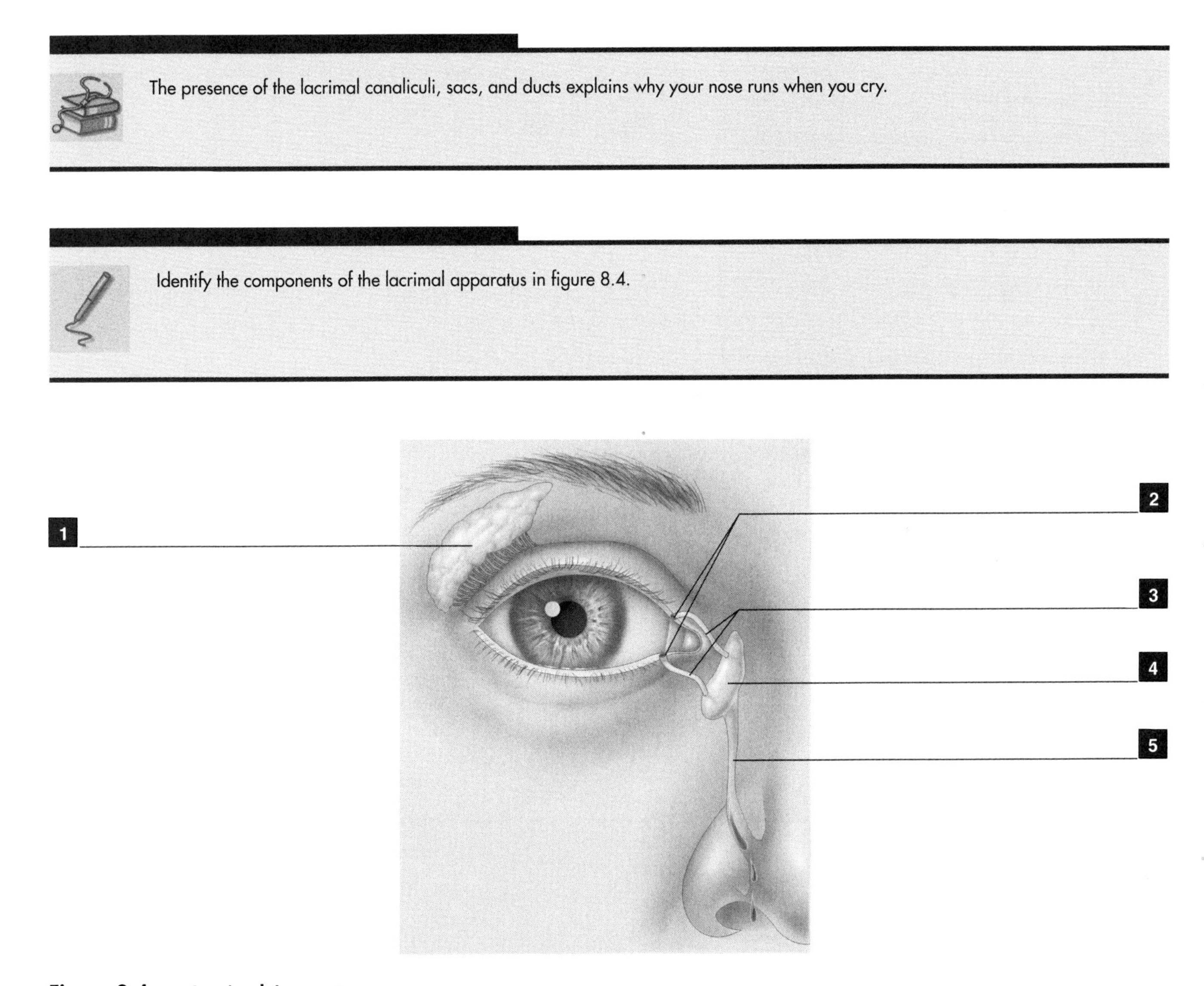

Figure 8.4 **Lacrimal Apparatus**

Extrinsic Eye Muscles

The orbit contains seven extrinsic (extraocular) eye muscles. Six of these move the eyeball; one elevates the upper eyelid. At the back of the orbit a tendinous ring spans the optic canal and the middle of the superior orbital fissure. This common tendinous ring is the origin of four of the six muscles that move the eyeball.

A Muscle That Moves the Eyelid

Levator Palpebrae Superioris 1 arises from the lesser wing of the sphenoid bone and runs across the roof of the orbit to insert into the upper eyelid. It elevates the lid, and is continuously active when the eyes are open.

Innervated by *Oculomotor Nerve (CN III)*

Muscles that Move the Eyeball

Superior Rectus 2 arises from the tendinous ring and runs forward to insert into the sclera of the eyeball above the cornea. It moves the cornea upward and medially.

Innervated by *Oculomotor Nerve (CN III)*

Inferior Rectus 3 arises from the tendinous ring and runs forward to insert into the sclera below the cornea. It moves the cornea downward and medially.

Innervated by *Oculomotor Nerve (CN III)*

Medial Rectus 4 arises from the tendinous ring and runs forward to insert into the sclera medial to the cornea. It moves the cornea medially.

Innervated by *Oculomotor Nerve (CN III)*

Inferior Oblique 5 arises from the front of the orbit floor, next to the nasolacrimal fossa. It runs laterally and then turns upward to insert into the sclera. It turns the cornea upward and laterally.

Innervated by *Oculomotor Nerve (CN III)*

Superior Oblique 6 arises from the sphenoid bone and runs along the orbital roof. Its tendon passes through a fibrous pully (*Trochlea*) attached to the superomedial corner of the orbital rim, and then it turns backward and laterally to insert into the sclera. It turns the cornea downward and laterally.

Innervated by *Trochlear Nerve (CN IV)*

Lateral Rectus 7 arises from the tendinous ring and runs forward to insert into the sclera lateral to the cornea. It moves the cornea laterally.

Innervated by *Abducens Nerve (CN VI)*

Identify the aforementioned extraocular eye muscles in figure 8.5.

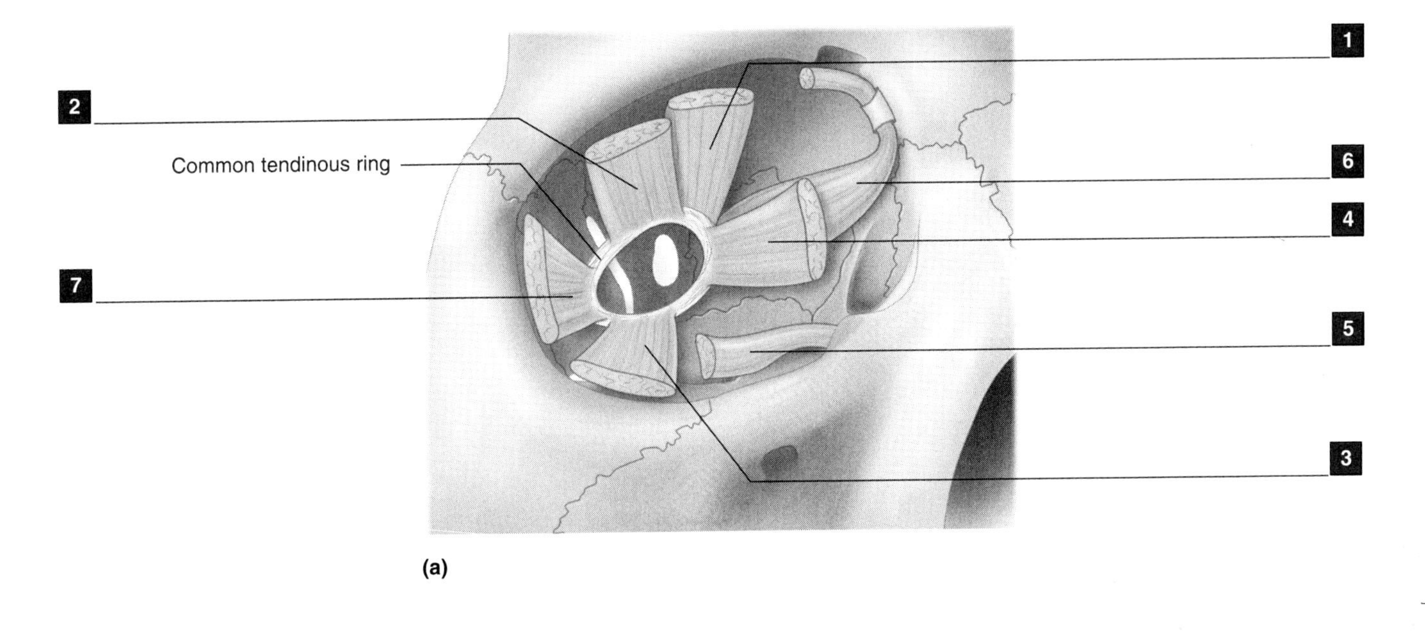

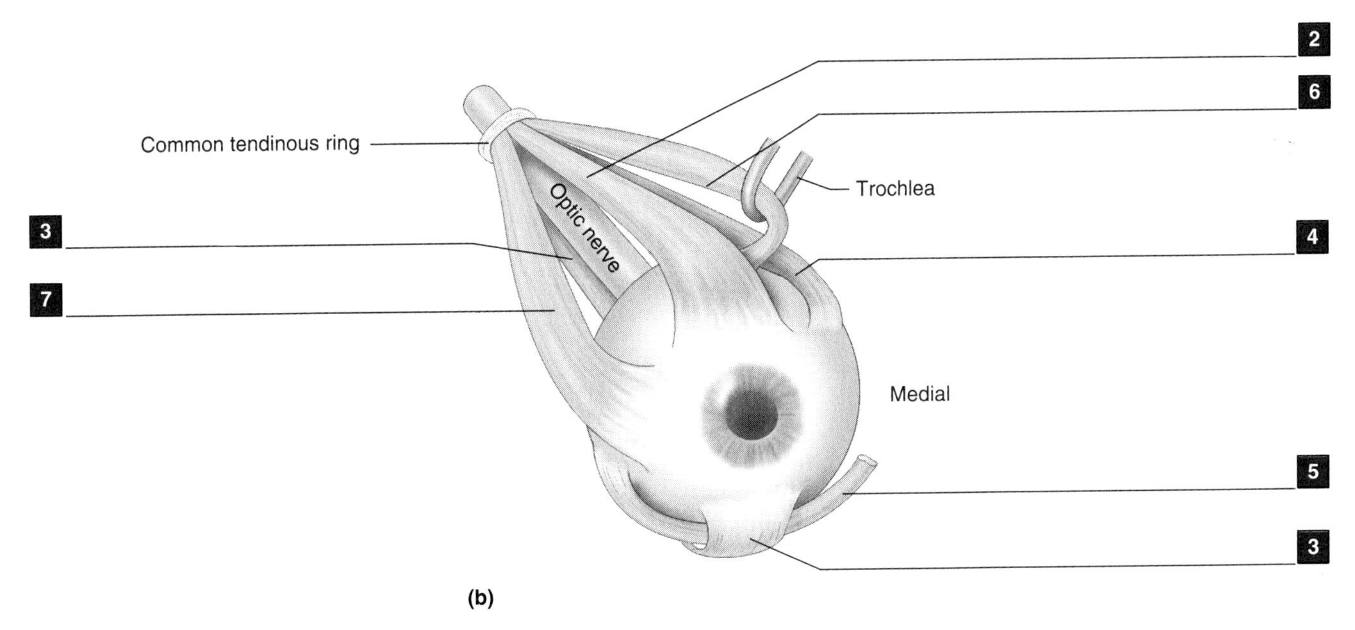

Figure 8.5 **Extrinsic Eye Muscles**
(a) origin of extrinsic muscles of the eye; (b) muscles that move the eye (right side).

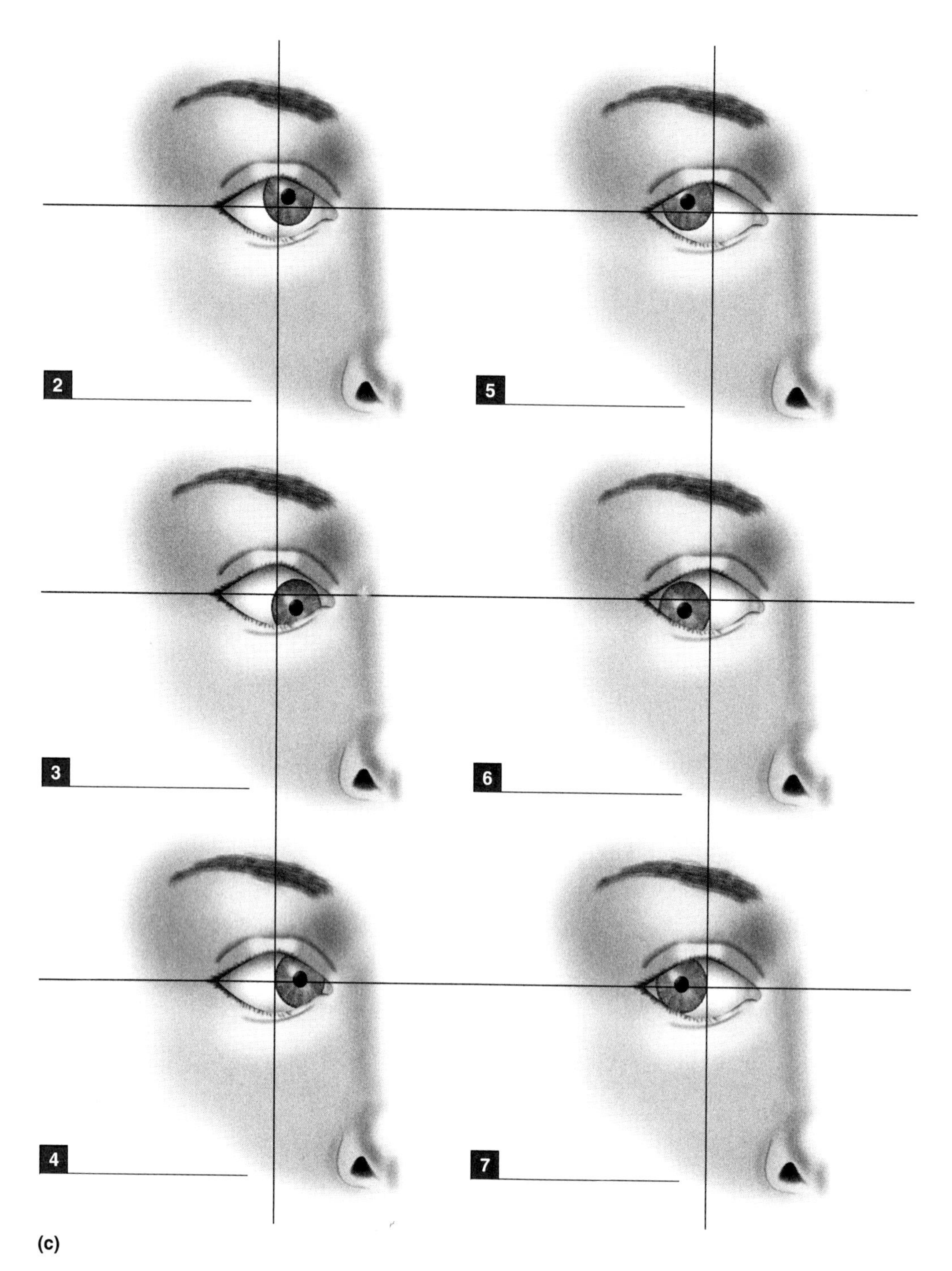

Figure 8.5 **Extrinsic Eye Muscles (Continued)**
(c) actions of the four recti and two oblique muscles of the right eye if each muscle is acting alone. Note the position of the pupil in relation to the horizontal and vertical axes.

The Eyeball

The eyeball is a fluid-filled structure that contains light-sensitive receptors.

The light sensitive layer (*Retina*) of the eyeball is surrounded by two concentric layers of tissue. The outer is a thick fibrous tunic. The muscles that move the eyeball insert into it. It is opaque white around most of the eyeball, where it is known as the **Sclera 1.** Anteriorly this layer bulges outward in front of the pupil and becomes the transparent **Cornea 2.**

Deep to the fibrous tunic is the thinner, vascular **Choroid 3.** Anteriorly, this layer becomes thickened by smooth muscle (the *Ciliary Muscle*) to form the **Ciliary Body 4.** In front of the ciliary body, the choroid layer thins to form the **Iris 5,** which also contains smooth muscle fibers. The iris may be pigmented by melanin; if none is present, it will appear blue. The iris surrounds an opening known as the **Pupil 6.**

The iris contains smooth muscles that cause the pupil to dilate or constrict. Under bright light, the circular fibers of the *Sphincter Pupillae* contract to constrict the pupil. The *Sphincter Pupillae* receives *parasympathetic* innervation from the *Oculomotor Nerve* (*CN III*) via the *Ciliary Ganglion*. Under dim light, the radial fibers of the *Dilator Pupillae* contract to dilate the pupil. The *Dilator Pupillae* is supplied by *sympathetic* fibers.

The *Ciliary Muscle* of the ciliary body gives rise to elastic fibers, the **Suspensory Ligaments 7,** that attach to the periphery of the **Lens 8.**

The function of the lens is to focus light entering the eye. This is accomplished by the smooth *Ciliary Muscle,* which may change the tension on the *Suspensory Ligaments.* This forces the lens to assume an ovoid (distance focus) or rounded (nearby focus) shape. The *Ciliary Muscle* receives *parasympathetic* innervation from the *Oculomotor Nerve* (*CN III*) via the *Ciliary Ganglion.*

The innermost of the three layers of the eyeball is the **Retina 9.** It contains the photoreceptor cells, the *Rods* (black and white shades) and *Cones* (colors). In the center of the back of the retina is a yellow spot known as the **Macula Lutea 10,** within which most of the cones are concentrated. The **Fovea Centralis 11** is a depression in the center of the macula lutea that contains cones, but no rods.

Where the **Optic Nerve 12** exits the eyeball the retina contains no photoreceptor cells. This spot is called the **Optic Disc(*Blind Spot*) 13.**

The retina is supplied with blood by the **Central Retinal Artery 14,** which runs into the optic nerve to enter the back of the eyeball. This is a branch of the *Ophthalmic Artery,* which comes from the internal carotid artery and runs through the superior orbital fissure to supply the contents of the orbit.

The cavity of the eyeball is divided into two parts by the lens. Behind the lens is the **Vitreous Chamber 15.** It is filled with a jellylike substance, the *Vitreous Body.* The space in front of the lens is filled with a clear fluid called the *Aqueous Humor.* This space is divided into two chambers by the iris, the free margin of which rests upon the lens. Behind the iris is the **Posterior Chamber 16;** in front of the iris is the **Anterior Chamber 17.**

Aqueous humor is produced in the posterior chamber, and flows through the pupil into the anterior chamber, where it is drained by a vein that encircles the cornea. This vein is known as the **Canal of Schlemm 18.**

Sufficient aqueous humor pressure is necessary to maintain proper eyeball shape. However, if the pressure rises because of inadequate drainage, the retina may be damaged. This condition is known as **Glaucoma.**

Recall that the anterior surface of the sclera is covered by a thin mucous membrane, the bulbar conjunctiva, which is continuous with that on the deep surface of the eyelid.

Identify the components of the eyeball in figure 8.6.

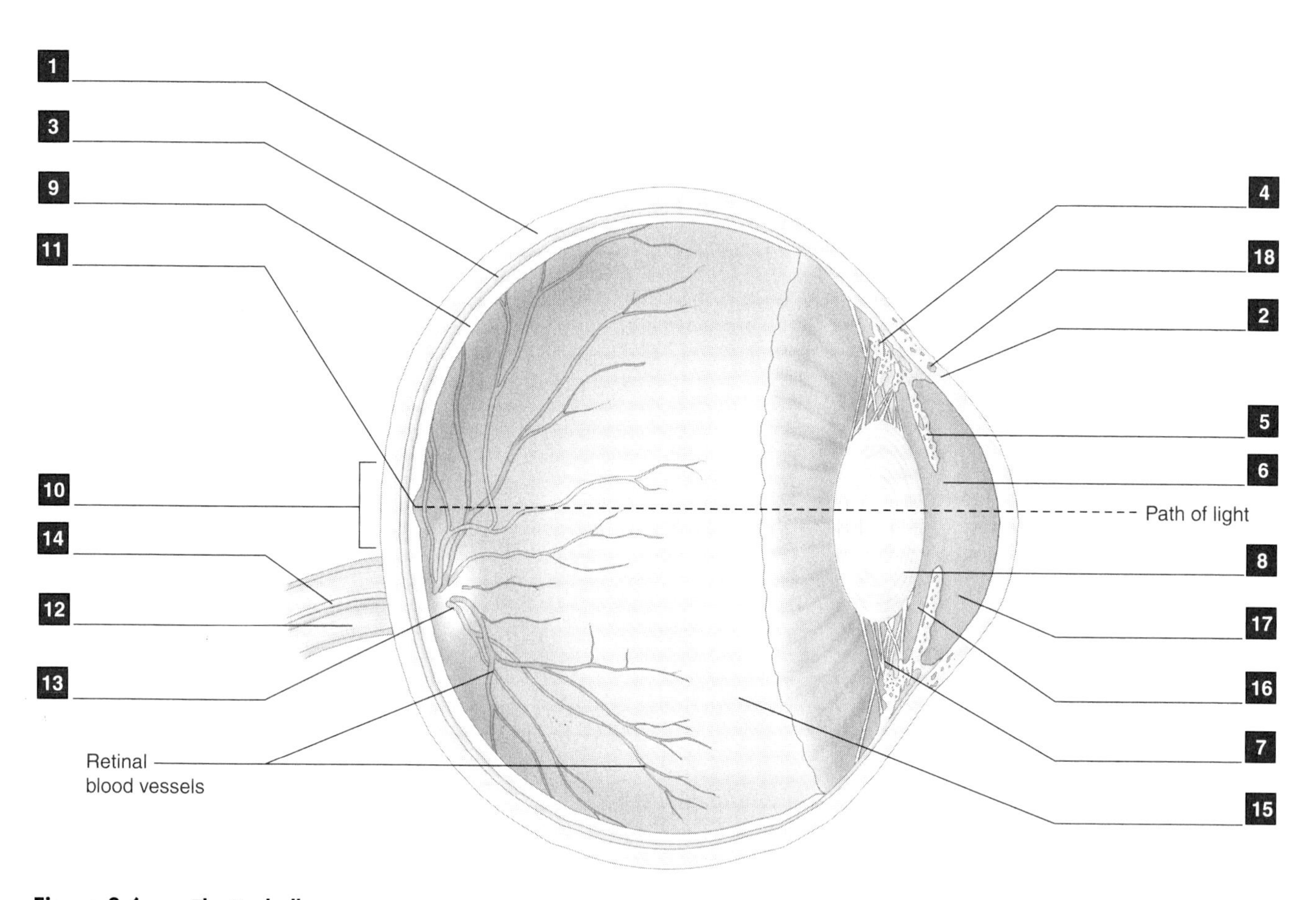

Figure 8.6 The Eyeball

8.2 THE EAR

The ear conveys special sensory information related to hearing and equilibrium (balance). Anatomically, the ear is divided into three regions: External Ear, Middle Ear, and Inner Ear. All three regions are involved in the sense of hearing. Only the inner ear is involved in the sense of equilibrium.

The components of the three anatomical regions of the ear follow.

External Ear

The **External Ear** 1 is designed to collect sound waves. It comprises the **Auricle** 2 and the **External Auditory (Acoustic) Canal** 3.

Middle Ear

The **Middle Ear** 4 is designed to amplify sound waves by turning them into mechanical vibrations. It comprises an air-filled **Tympanic Cavity** 5 in the temporal bone. It is bounded laterally by the **Tympanic Membrane** 6, and is continuous anteromedially with the **Auditory (Eustachian) Tube** 7. The middle ear contains the three **Auditory Ossicles** 8.

Inner Ear

The **Inner Ear** 9 is designed to translate mechanical vibrations from the middle ear into nerve impulses that result in hearing. It also translates head posture and movement into neural impulses that relate to the sense of balance. It is comprised of a series of hollow tubes within the *petrous temporal* bone. These bony tubes contain membranous sacs and are filled with fluids. Because of the complicated arrangement of the tubes, this structure is known as the **Labyrinth** 10. The **Vestibulocochlear Nerve (CN VIII)** 11 transmits information from the labyrinth to the brain.

Identify the principal components of the ear in figure 8.7.

Let us now examine the anatomical structures that comprise each of these three regions more closely.

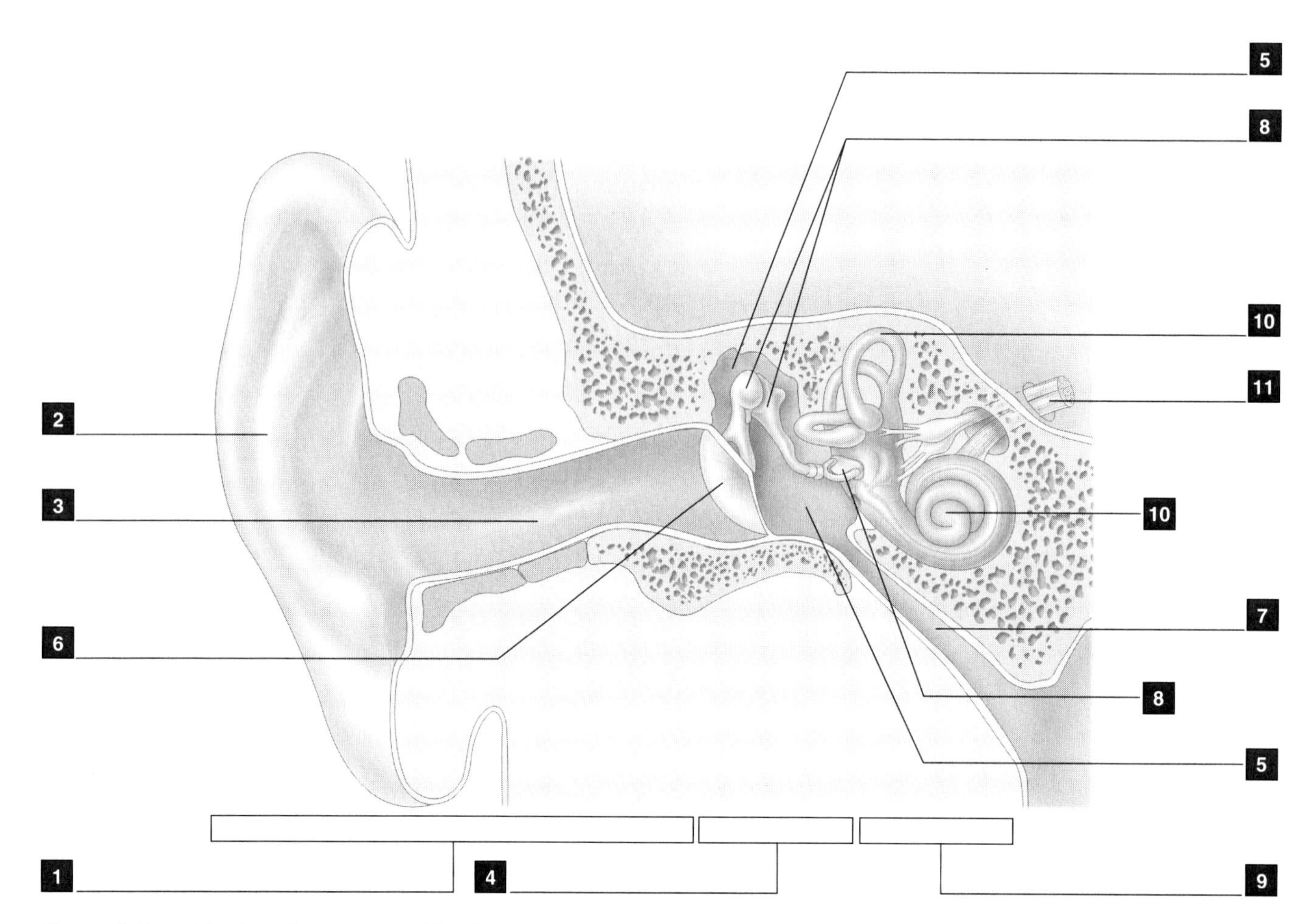

Figure 8.7 **The Ear: Components of the Three Regions**

The External Ear

The external ear comprises the structures lateral to the eardrum, or **Tympanic Membrane** 1. The largest of these is the prominent **Auricle** 2. It funnels sound waves into the **External Acoustic Meatus** 3, which is the lateral opening of the **External Auditory (Acoustic) Canal** 4. The outer half of the canal is supported by **Cartilage** 5. The inner half is formed by the **Temporal Bone** 6 above and the **Tympanic Bone** 7 below.

The tympanic membrane is a thin sheet of tissue that is concave laterally, and convex medially where it is adherent to the *Malleus* (an ear ossicle). It is innervated by the *Auriculotemporal Nerve* (branch of the mandibular division of the Trigeminal) and the *Auricular Nerve* (branch of the Vagus).

The auricle has a framework of elastic cartilage. Its curved outer rim, known as the **Helix** 8, originates just above the external auditory meatus, and terminates in a fleshy lobe. The **Antihelix** 9, a second curve internal to the helix, surrounds a hollow that leads into the external auditory meatus. The **Tragus** 10 is a small flap that projects immediately anterior to the meatus.

Identify the features of the external ear in figure 8.8.

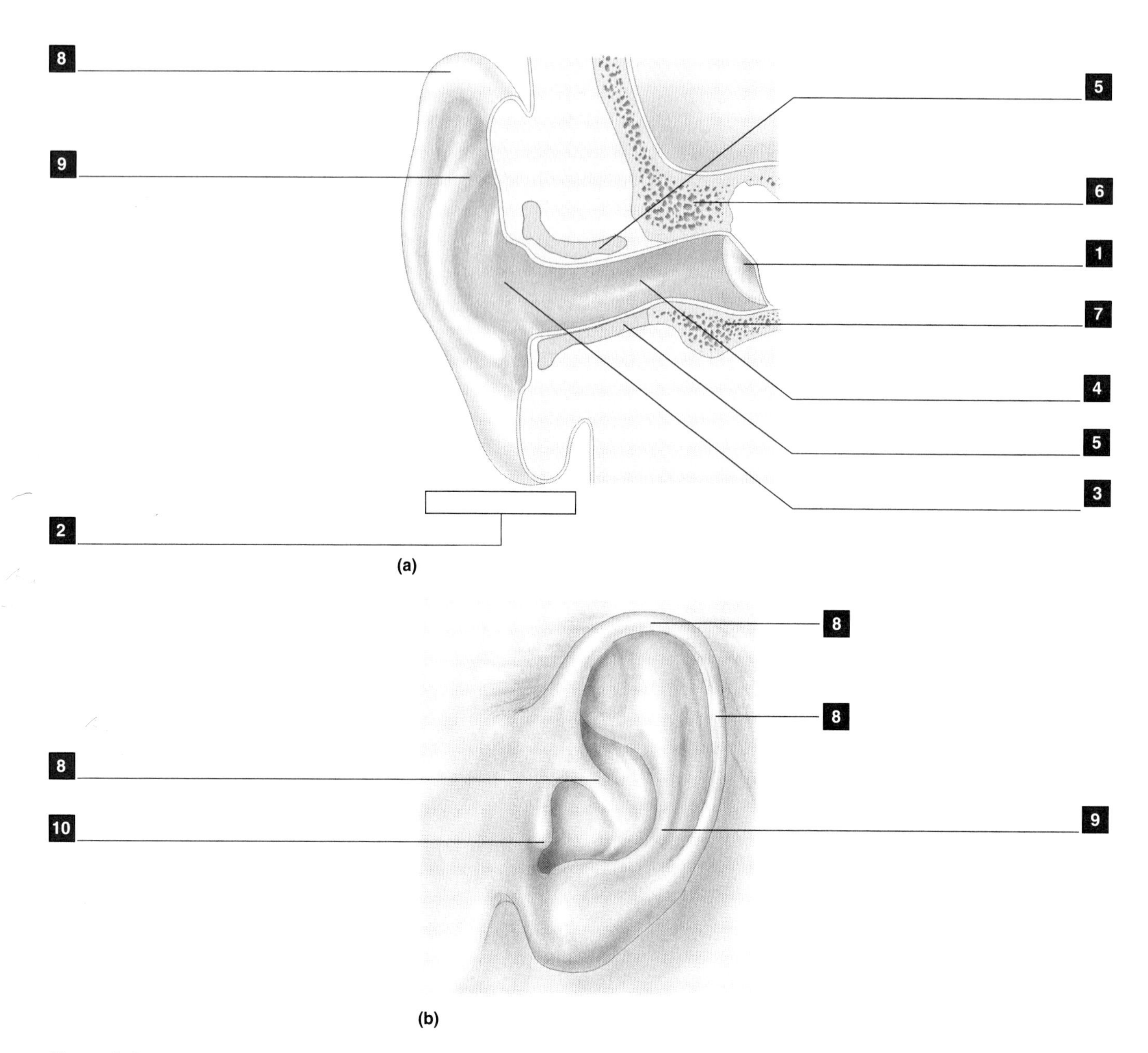

Figure 8.8 **External Ear**
(a) schematic-coronal section through right ear; (b) lateral view of left ear.

The Middle Ear

The middle ear comprises an air-filled **Tympanic Cavity** 1 that is almost completely surrounded by the petrous temporal. This cavity contains three tiny bones that transmit vibrations from the outer to the inner ear.

The tympanic cavity is bounded laterally by the **Tympanic Membrane** 2. It is connected posteriorly with the air cells of the *Mastoid Process* of the temporal bone. The tympanic cavity is continuous anteromedially with the **Auditory (Eustachian) Tube** 3, which opens into the nasopharynx.

The *Auditory Tube* is usually closed. It opens during swallowing and yawning to equalize the external air pressure with that in the tympanic cavity. Its nasopharyngeal aperture is opened by the actions of *Tensor Palatini* and *Levator Palatini*. Free vibration of the auditory ossicles requires that air pressure be the same on both sides of the tympanic membrane.

The bony medial wall of the tympanic cavity has two small openings—the **Oval Window** 4 and the **Round Window** 5—by which it communicates with the inner ear.

Bones of the Middle Ear

The *Auditory Ossicles* are three tiny bones that form a chain from the *Tympanic Membrane* to the *Oval Window*. From lateral to medial, the three bones are: Malleus - Incus - Stapes

The **Malleus** 6 is attached laterally to the *Tympanic Membrane*, and medially to the **Incus** 7. The incus, in turn, is connected to the **Stapes** 8, which communicates with the labyrinth of the inner ear by way of the *Oval Window*. The auditory ossicles form a series of levers that serves to increase the force of sound wave vibrations 20-fold from the tympanic membrane to the oval window.

Muscles of the Middle Ear

If sound waves that impact on the tympanic membrane are too strong, its tension may be increased to dampen the amplitude of vibration. This is the job of the **Tensor Tympani** 9 muscle. It runs in a canal through the petrous bone adjacent to the auditory tube, to which it is attached. It inserts on the malleus. It is innervated by the *mandibular* division of the *Trigeminal Nerve* (*CN V3*).

Similarly, if the level of vibration that eventually reaches the stapes is too strong, its pistonlike movement in the oval window may be dampened by the **Stapedius** 10 muscle. It runs from the mastoid wall of the tympanic cavity to insert on the stapes. It is innervated by the *Facial Nerve* (*CN VII*).

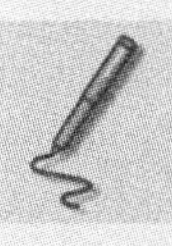

Identify the components of the middle ear in figure 8.9.

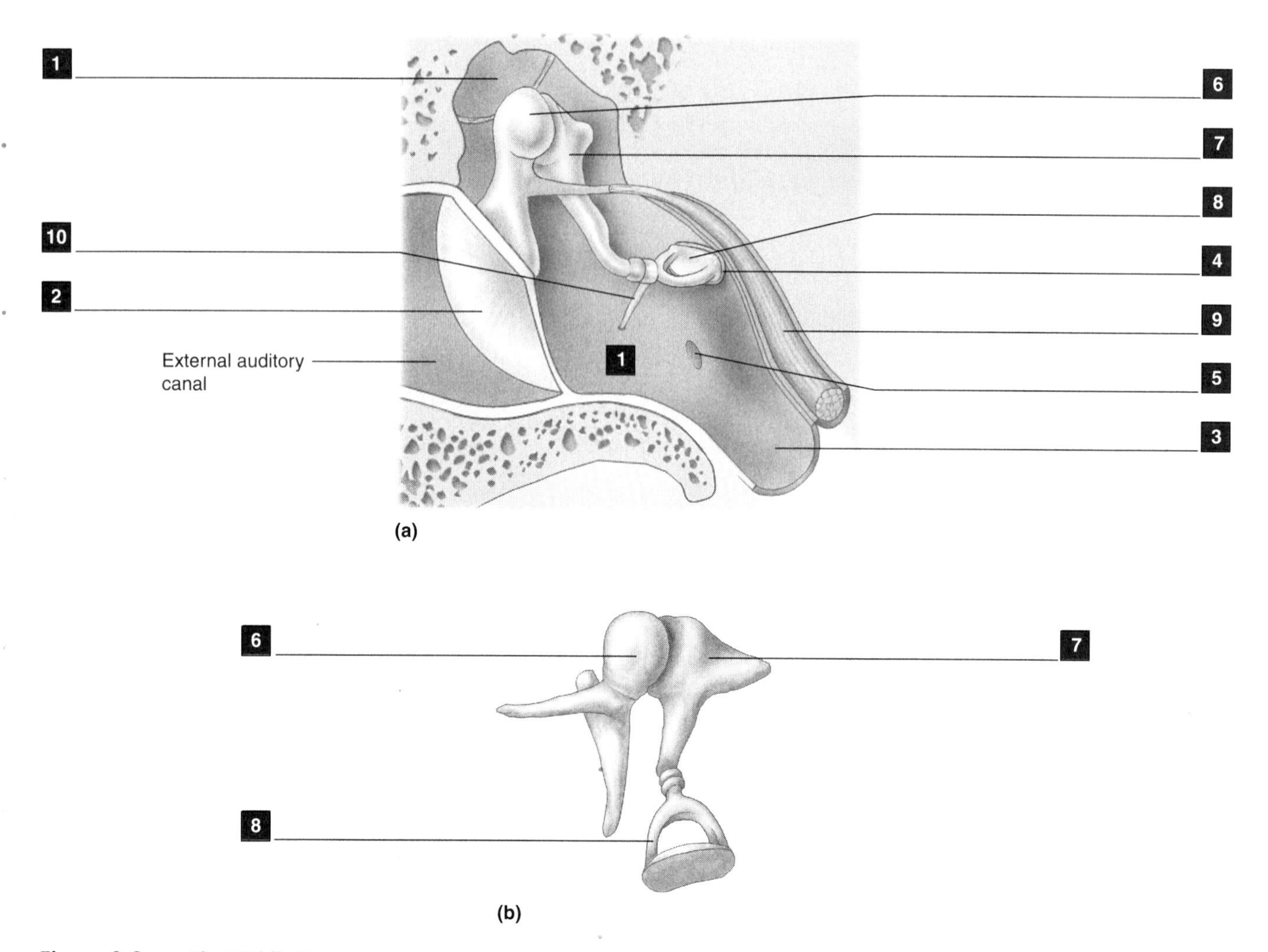

Figure 8.9 **The Middle Ear**
(a) anterior cutaway view of middle ear; (b) ossicles articulated: medial view.

The Inner Ear

Fluid-filled cavities and tubes within the temporal bone comprise the inner ear. Because of its complicated arrangement, this structure is referred to as the *Labyrinth*. The labyrinth is supplied by the *Vestibulocochlear Nerve* (CN *VIII*).

The cavity within the **Petrous Temporal** 1 constitutes the **Bony Labyrinth** 2. It is filled with a fluid called *Perilymph*. Housed within the bony labyrinth and surrounded by the perilymph are several membranous sacs and tubes that constitute the **Membranous Labyrinth** 3. The membranous labyrinth is filled with a fluid called *Endolymph*.

The labyrinth is divided into three parts: **Vestibule** 4, **Semicircular Canals** 5, and **Cochlea** 6.

The vestibule communicates with the footplate of the stapes via the **Oval Window** 7. Within the vestibule are two membranous sacs, the **Utricle** 8 and **Saccule** 9.

Within the semicircular canals are the **Semicircular Ducts** 10. They are continuous with the utricle.

The cochlea "communicates" with the middle ear via the **Round Window** 11. The cochlea contains the **Cochlear Duct** 12.

The vestibule and semicircular canals are concerned with *Equilibrium*. The vestibule and cochlea are involved in *Hearing*.

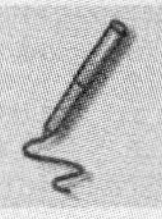

Identify the components of the inner ear in figure 8.10. Use shades of blue for the bony labyrinth, and shades of red for the parts of the membranous labyrinth.

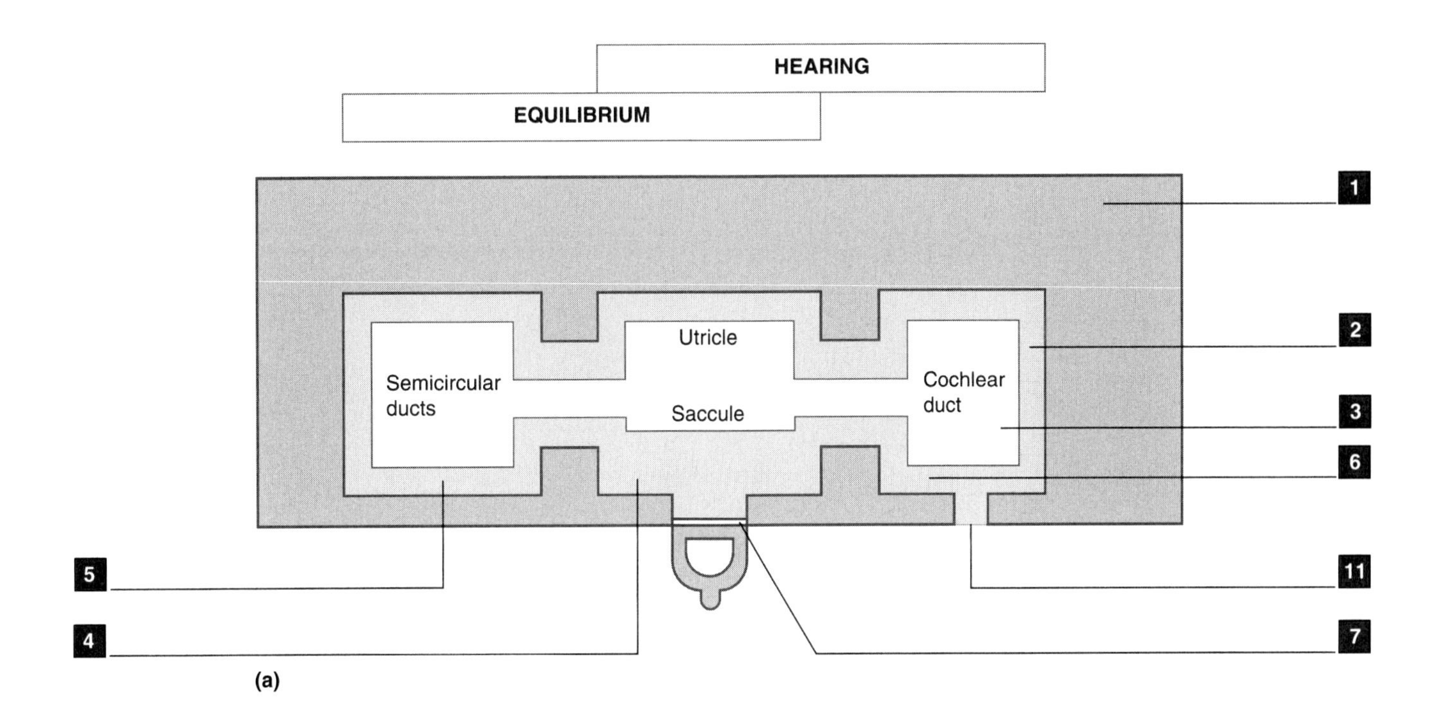

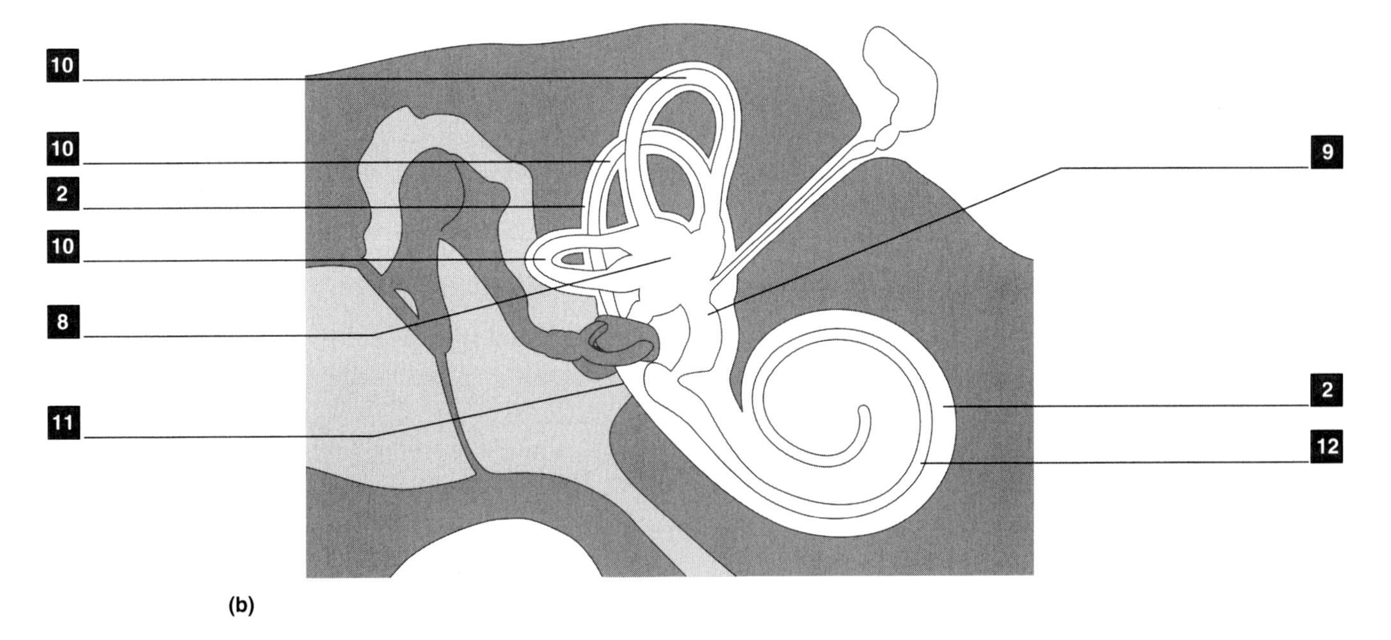

Figure 8.10 **Bony and Membranous Labyrinth**
(a) schematic of functional and anatomical components; (b) schematic of anatomical relationships.

Structures Related to Equilibrium

The membranous labyrinth in the vestibule and semicircular canals is concerned with the detection of static and dynamic equilibrium.

Within the **Vestibule** 1 are two membranous sacs: the **Utricle** 2 and **Saccule** 3. They are concerned with the maintenance of *Static Equilibrium*. The utricle is the more important of the two. Each has a receptor region known as the *Macula*. Within the macula, sensitive hair cells are embedded in a jellylike *Otolithic Membrane*. Calcium carbonate crystals known as *Otoliths* rest on this membrane.

When the head is tilted forward, gravity pulls the heavy otoliths and adherent otolithic membrane downward over the hair cells. This triggers them to generate an impulse that synapses with the *Vestibular Branch of CN VIII.*

Emerging from the utricle are three membranous **Semicircular Ducts** 4 contained within the bony **Semicircular Canal** 5. They are concerned with the detection of *Dynamic Equilibrium*. The canals occupy different planes:

The **Anterior Canal** 6 is in the sagittal plane

The **Posterior Canal** 7 is in the coronal plane

The **Lateral Canal** 8 is in the horizontal plane

Near their origins at the utricle, each semicircular duct is enlarged to form an **Ampulla** 9. This contains sensitive hair cells embedded in a mass of gelatinous material called the *Cupula*.

Movement of the head causes the endolymph in one (or more) of the semicircular ducts to move. This pushes the cupula over the hair cells, triggering them to generate an impulse that synapses with the *Vestibular Branch of CN VIII.*

Extending from the utricle and saccule is an accessory endolymphatic duct. It runs posteromedially through the petrous bone to emerge on its posterior surface under cover of the dura mater. The end of the duct is known as the **Endolymphatic Sac** 10. It has no known function.

The vestibule communicates with the middle ear via the **Oval Window** 11, which houses the footplate of the stapes.

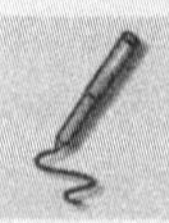

Identify the structures of the inner ear that are related to equilibrium in figure 8.11. Again, use shades of blue for the bony labyrinth, and shades of red for the parts of the membranous labyrinth.

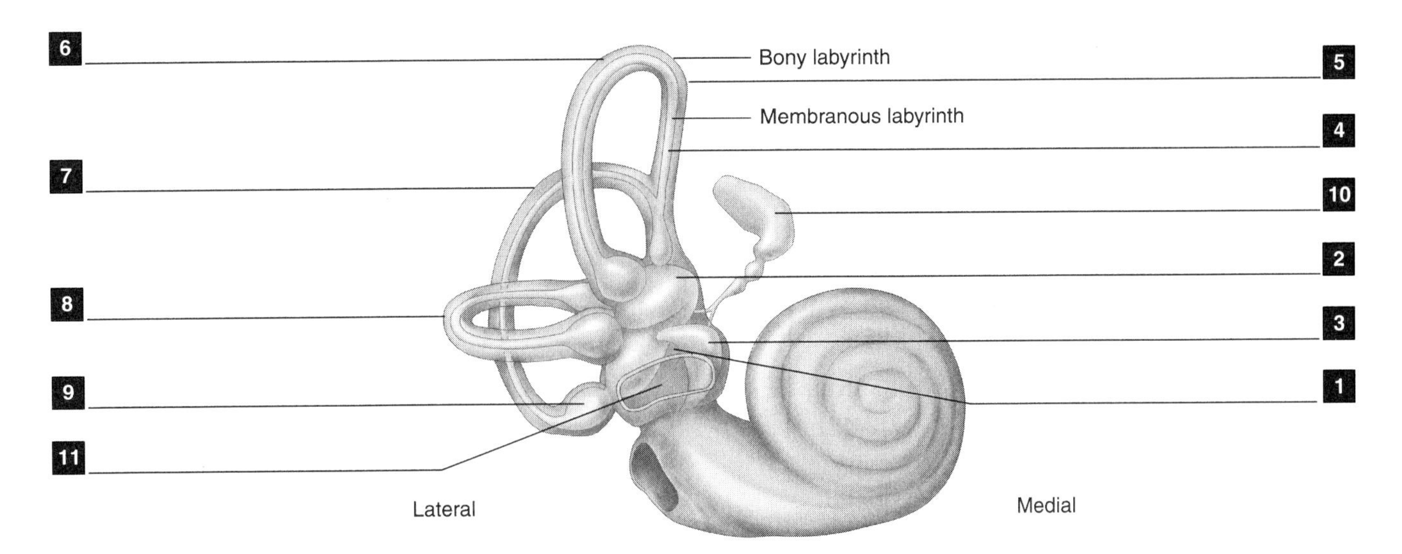

Figure 8.11 **Structures Related to Equilibrium**

Structures Related to Hearing

The vestibule and cochlea are concerned with hearing.

That part of the **Bony Labyrinth** 1 known as the **Cochlea** 2 is a coiled tube. Like the other bony labyrinthine structures, it is filled with *Perilymph*. Housed within the cochlea is a **Membranous Labyrinth** 3 tube known as the **Cochlear Duct** 4. Like the other membranous labyrinthine structures, it is filled with *Endolymph*. The cochlear duct is triangular in cross section.

The bony cochlea has a spur of bone that projects into the cochlear tube along its entire length. This spur joins the apex of the triangular membranous duct, thus dividing the bony labyrinth into two parts: the *scala vestibuli* and the *scala tympani*.

The **Scala Vestibuli** 5 is open to the *Vestibule* at its one end. The membranous sheet that separates the perilymph of the scala vestibuli from the endolymph of the cochlear duct is known as the **Vestibular Membrane** 6.

The **Scala Tympani** 7 "communicates" at its one end with the *Middle Ear* cavity via the **Round Window** 8. The round window is covered with a thin membrane. The membranous sheet that separates the perilymph of the scala tympani from the endolymph of the cochlear duct is known as the **Basilar Membrane** 9.

At the apex of the cochlea, the scala vestibuli and scala tympani are continuous with one another through a small opening known as the **Helicotrema** 10.

Within the cochlear duct is a sheet of hair cells that are the receptors for auditory sensations. These cells form the **Spiral Organ of Corti** 11, which lies against the *Basilar Membrane*. Projecting over the hair cells as a tent is the gelatinous **Tectorial Membrane** 12.

Pressure transmitted to the oval window from movement of the stapes travels through the perilymph along the scala vestibuli, through the helicotrema and into the perilymph of the scala tympani, which ends at the round window.

Pressure waves in the scala vestibuli deform the vestibular membrane, which causes pressure changes in the endolymph of the cochlear duct. These, in turn, cause the basilar membrane to move, which causes the hair cells of the spiral organ to move against the tectorial membrane.

Bending the hairs of the spiral organ cells triggers them to generate impulses that synapse with the *Cochlear Branch of CN VIII*.

Pressure waves that are generated in the scala tympani through the helicotrema and movement of the basilar membrane are transmitted to the round window, causing it to bulge outward into the middle ear.

Identify the structures related to hearing in figure 8.12.

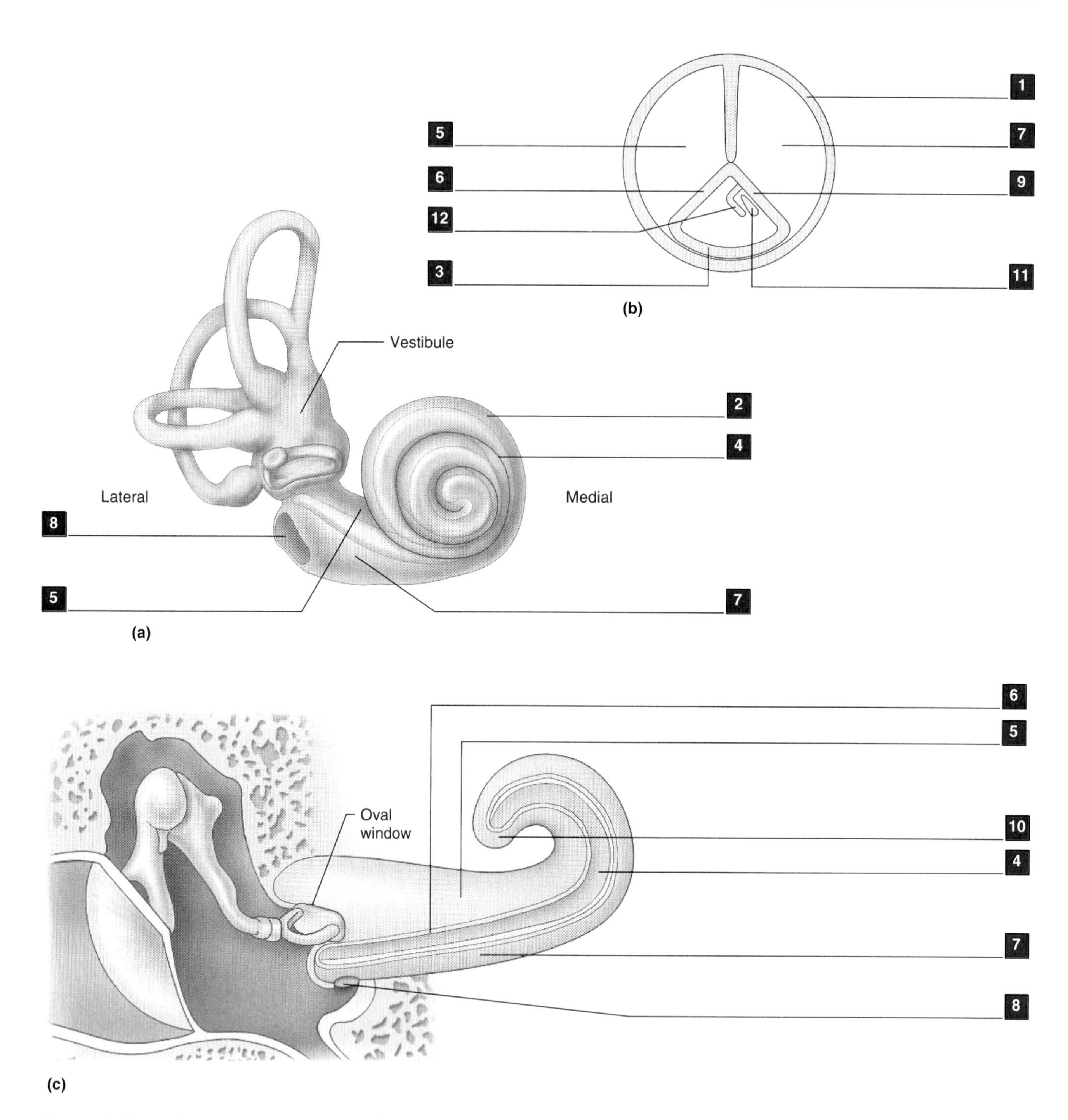

Figure 8.12 **Structures Related to Hearing**
(a) anterior view of inner ear with stapes in the oval window of the vestibule; (b) cochlea in cross-section; (c) cochlea in longitudinal section.

LABORATORY

9

The Thorax

9.1 THE BREAST

In both sexes, the tissues around the nipple and areola may become slightly swollen immediately after birth, and the nipples may even exude a small amount of secretion. Breast activity rapidly subsides, and its growth follows that of other subcutaneous tissues until puberty.

In females, the breast at puberty increases in size and the glandular tissue becomes surrounded with fatty superficial fascia. The female breast is known as the ***Mammary Gland.*** It is structurally similar to an apocrine sweat gland. Because the glandular and fatty tissue fails to hypertrophy in males, the male breast is a rudimentary structure consisting only of a nipple and areola.

The **Breast** 1 rests on the fascia covering the **Pectoralis Major** 2 and a bit of **Serratus Anterior** 3. An extension of the adipose tissue of the breast, known as the **Axillary Tail** 4 runs upward and laterally into the anterior fold of the axilla (armpit).

Lymphatic drainage of the breast is primarily into nodes at the end of the axillary tail. The blood supply of the breast derives from the internal and lateral thoracic arteries. These aspects will be examined a bit later.

The **Nipple** 5 is usually located over the fourth intercostal space (i.e., between the fourth and fifth ribs) in males and nulliparous females. It is surrounded by a pigmented ring of skin known as the **Areola** 6. The areola has a number of small elevations, the *Areolar Glands (=Montgomery Tubercles)*, that lubricate the nipple during lactation (milk secretion).

The nipple is pierced by some 15 to 20 **Lactiferous Ducts** 7. These enlarge immediately below the areola to form **Lactiferous Sinuses** 8, which constitute milk reservoirs during lactation. Each lactiferous duct drains a glandular **Lobe** 9, which is comprised of a number of smaller lobules. Each lobule has a number of simple secretory sacs called alveoli.

The glandular lobes are surrounded by **Adipose Tissue** 10, and are supported from the skin and the deep fascia overlying the Pectoralis major muscle by a series of **Suspensory Ligaments** (= *Cooper's Ligaments*) 11. These form fibrous septa within the breast.

Lactation

During pregnancy, the mammary gland develops the capacity to produce milk through high levels of progesterone and estrogen that stimulate growth of the glandular tissues and ducts. After birth of the baby, the levels of progesterone and estrogen fall, which enables prolactin to trigger milk production and secretion by the alveoli.

Identify the aforementioned features of the breast in figure 9.1.

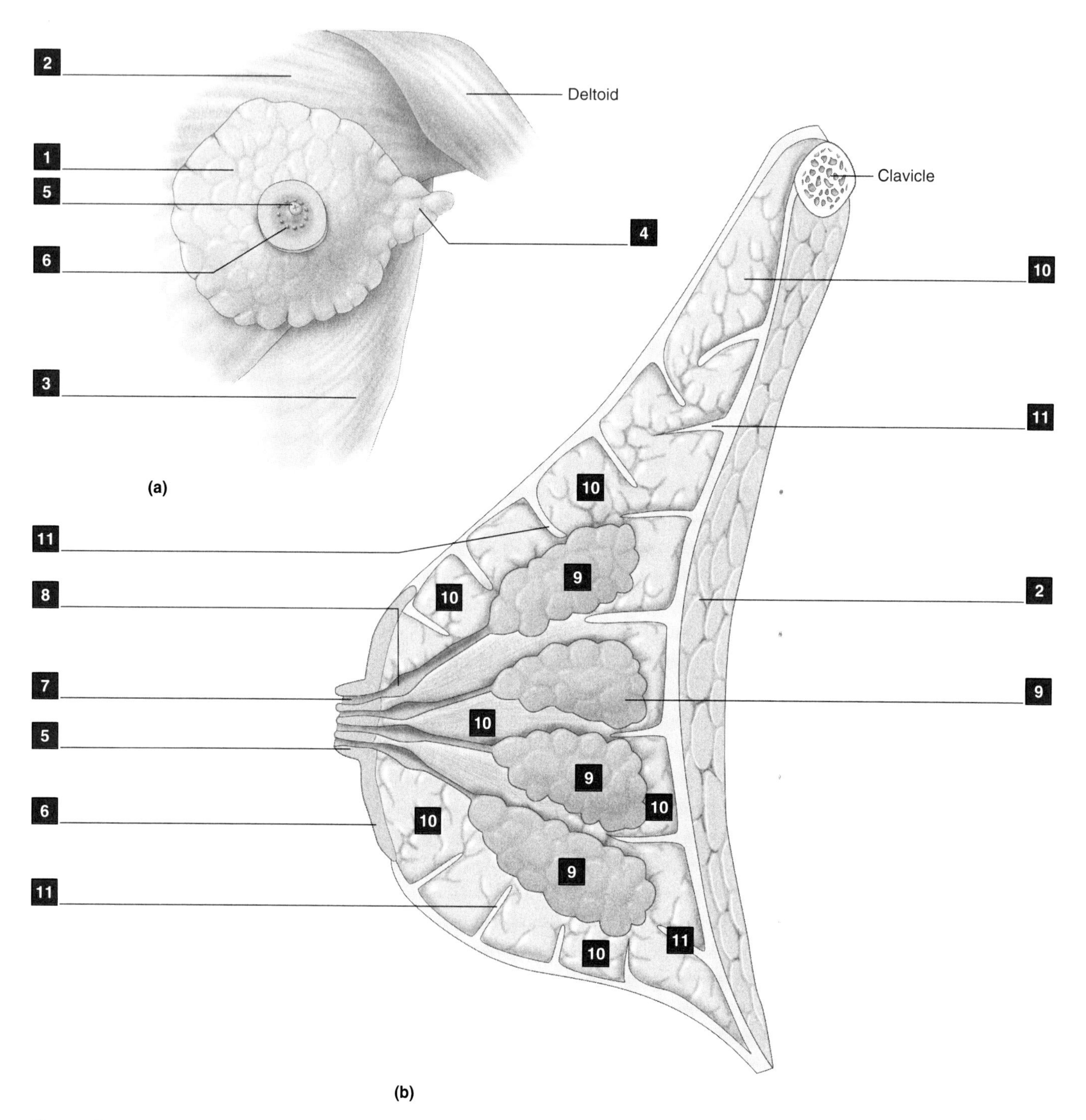

Figure 9.1 **The Breast**
(a) anterior view; (b) sagittal section.

9.2 THE THORACIC SKELETON

The bony thorax consists of 12 thoracic vertebrae, 12 pairs of ribs, the sternum, and the costal cartilages by which the ribs articulate with the sternum. This protective framework surrounds the *Thoracic Cavity* posteriorly, laterally, and anteriorly.

Vertebrae

There are 12 **Thoracic Vertebrae** **1**. Turn to laboratory 2 (p. 27) to review the structure of these bones.

Sternum

The **Sternum** **2** forms a skeletal brace in the midline anteriorly. It has three separate components that are united by fibrocartilaginous joints: manubrium, sternal body, and xiphoid process.

The **Manubrium** **3** is the uppermost element. Its superior border has a depression known as the *Jugular Notch*, which lies at the level of the intervertebral disc between T1 and T2. Its superolateral corners have concave facets that articulate with the clavicles. Immediately below each clavicular facet is a facet that articulates with the costal cartilage of the first rib. The inferior border joins the sternal body at the junction of the second rib. The manubriosternal joint has a slight anterior angle (*Sternal Angle*) that can be felt as a ridge just under the skin. The sternal angle lies at the level of vertebra T5.

The **Sternal Body** **4** is the middle element. It is made up of four segments (*Sternebrae*) that fuse between puberty and age 20. Their junctions can be palpated as slight transverse ridges on the front of the bone. The second through seventh costal cartilages articulate with the sternal body.

The **Xiphoid Process** **5** is the inferior, tongue-shaped portion of the sternum. It articulates with the body of the sternum at the level of vertebra T10.

Ribs and Costal Cartilages

There are 12 **Ribs** **6** on each side. All articulate posteriorly by synovial joints with the vertebrae.

Ribs 1–10 articulate anteriorly with the sternum by rods of hyaline cartilage known as the **Costal Cartilages** **7**.
Ribs 1–7 articulate directly with the sternum by their own costal cartilages.
Ribs 8–10 articulate with the sternum indirectly, by connecting with the seventh costal cartilage.
Ribs 11–12 are short and do not reach the sternum.

Color and label the aforementioned structures in figure 9.2. Use different shades for ribs 1–7 (the so-called "true ribs") that articulate directly with the sternum, ribs 8–10 (the so-called "false" ribs) that articulate indirectly with the sternum, and ribs 11–12 (the so-called "floating" ribs) that have free anterior ends.

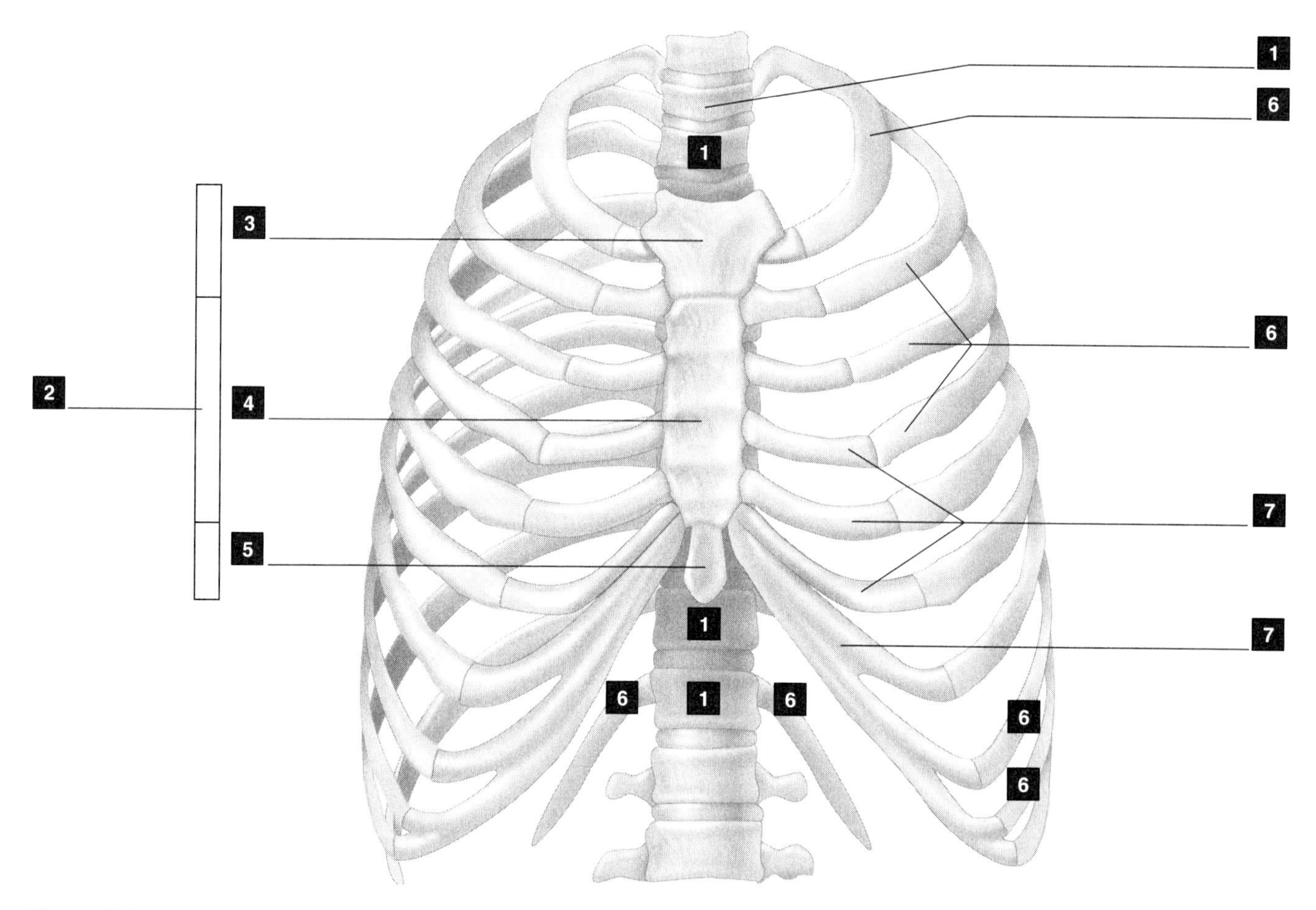

Figure 9.2 The Thoracic Skeleton

Respiratory Movement of the Thoracic Skeleton

The *Heads of Ribs 2–9* articulate by synovial joints with two adjacent *vertebral bodies* (its own and the one above) across the intervertebral disk. The *Tubercles of Ribs 1–10* articulate by synovial joints with the *transverse processes* of the vertebra.

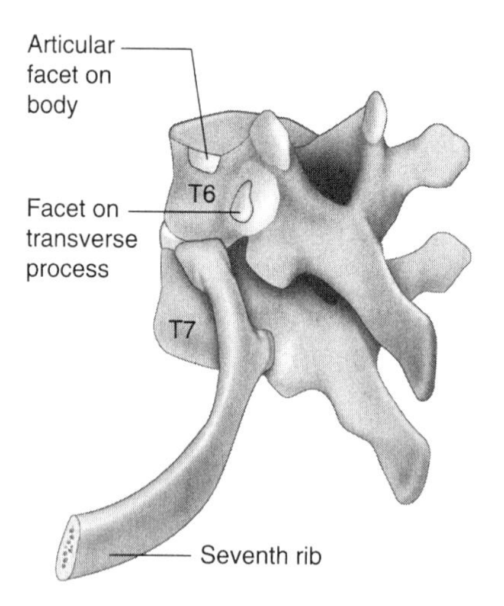

Costovertebral articulation of a typical rib (posterolateral view).

The synovial joints by which the ribs articulate with the vertebrae and sternum permit free movement of the costal cage during forced inspiration and expiration. The ribs are elevated during inspiration, which expands the anteroposterior dimension of the thoracic cavity in a movement analogous to the motion of a bucket handle. An increase in thoracic volume creates a partial vacuum, which is filled by air rushing into the lungs.

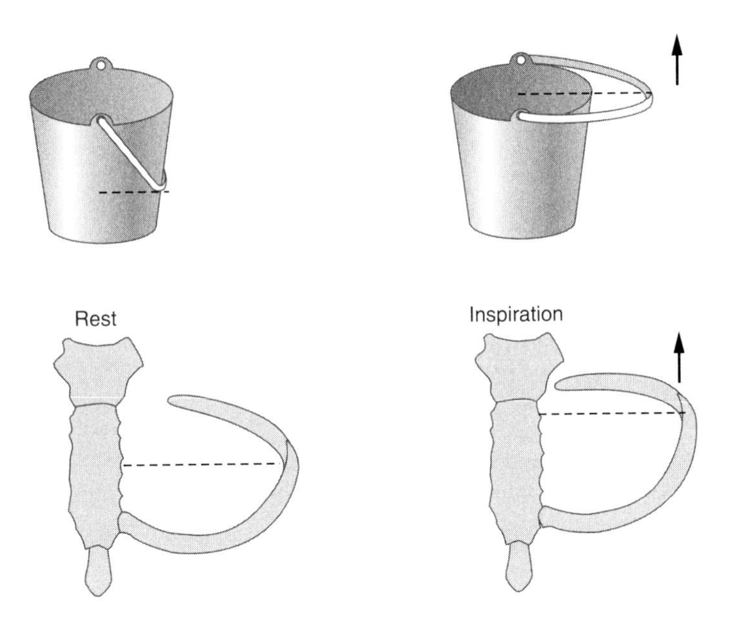

Bucket handle movement of a typical rib.

9.3 SKELETAL MUSCLES OF THE THORAX

A number of muscles attach to the bones of the thoracic skeleton. We have already examined most of them.

Muscles of the Back (laboratory 2, pp. 36–40)

Recall that there are four hypaxial muscles that attach to thoracic vertebrae and/or ribs:

Trapezius *Latissimus dorsi* *Rhomboideus* *Quadratus lumborum*

Three epaxial muscles attach to thoracic vertebrae:

Erector spinae *Semispinalis* *Multifidus*

Muscles of the Upper Limb (laboratory 3, pp. 60–77)

Recall that there is one muscle that attaches to the sternum:

Pectoralis major

In addition, there are two costoscapular muscles, three intercostal muscles, a sternocostal muscle, and the abdominal diaphragm.

Costoscapular Muscles

Two muscles run from the rib cage onto the scapula. Both protract (pull forward) the scapula. One also rotates the scapula upward; the other rotates it downward.

Serratus anterior 1 arises from the upper nine ribs by a series of fingerlike digitations, and inserts ventrally onto the medial border of the scapula. It protracts and rotates the scapula so that the glenoid fossa faces superiorly.

Innervated by *Long Thoracic Nerve* (= ventral rami C5–C7)

Pectoralis minor 2 arises from the anterior surfaces of ribs 3 to 5, and inserts onto the coracoid process of the scapula. It protracts and rotates the scapula so that the glenoid fossa faces inferiorly.

Innervated by *Medial Pectoral Nerve* (branch of Brachial Plexus)

Identify the costoscapular muscles in figure 9.3.

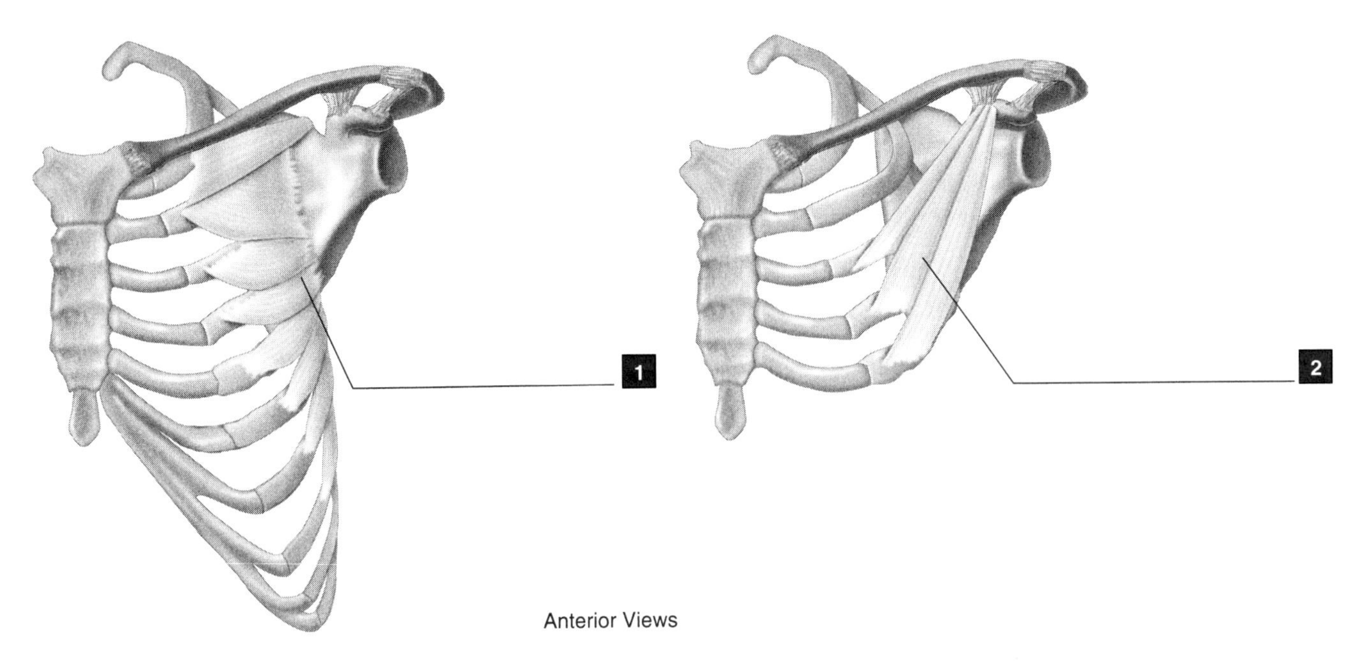

Figure 9.3 **Costoscapular Muscles**

Sternocostal Muscle

Transversus thoracis 1 arises from the posterior surface of the sternal body and xiphoid to insert onto the backs of costal cartilages 2 to 6. It draws the costae downward in forced expiration.

Innervated by *Intercostal Nerves*

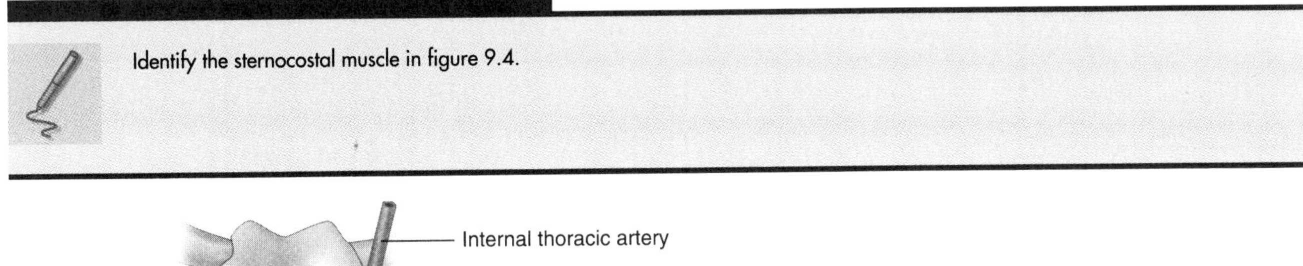

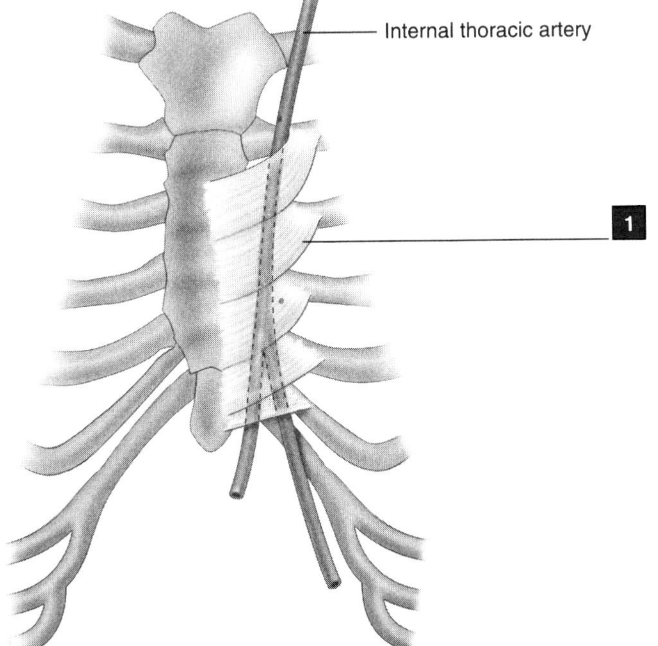

Figure 9.4 **Sternocostal Muscle**
Posterior view of sternum and costal cartilages.

Intercostal Muscles

There are three layers of thin muscles that run in the intercostal space between successive ribs. Between the middle and innermost layers is the so-called *Neurovascular Plane*. The *Intercostal Nerve*, *Artery*, and *Vein* run along this plane.

The three muscle layers differ in the extent to which they fill the intercostal space. The fibers of the external layer are orientated differently from those of the inner two, which helps strengthen the thoracic wall.

All three layers are innervated by *Intercostal Nerves*.

External Intercostal 1 occupies the space between the tubercle of the rib and the edge of the costochondral junction. Its fibers are directed downward and forward.

Internal Intercostal 2 occupies the space between the angle of the rib and the edge of the costosternal junction. Its fibers are directed downward and backward.

Innermost Intercostal 3 lies deep to the neurovascular plane. Its fibers are directed downward and backward.

Identify the three intercostal muscles in figure 9.5.

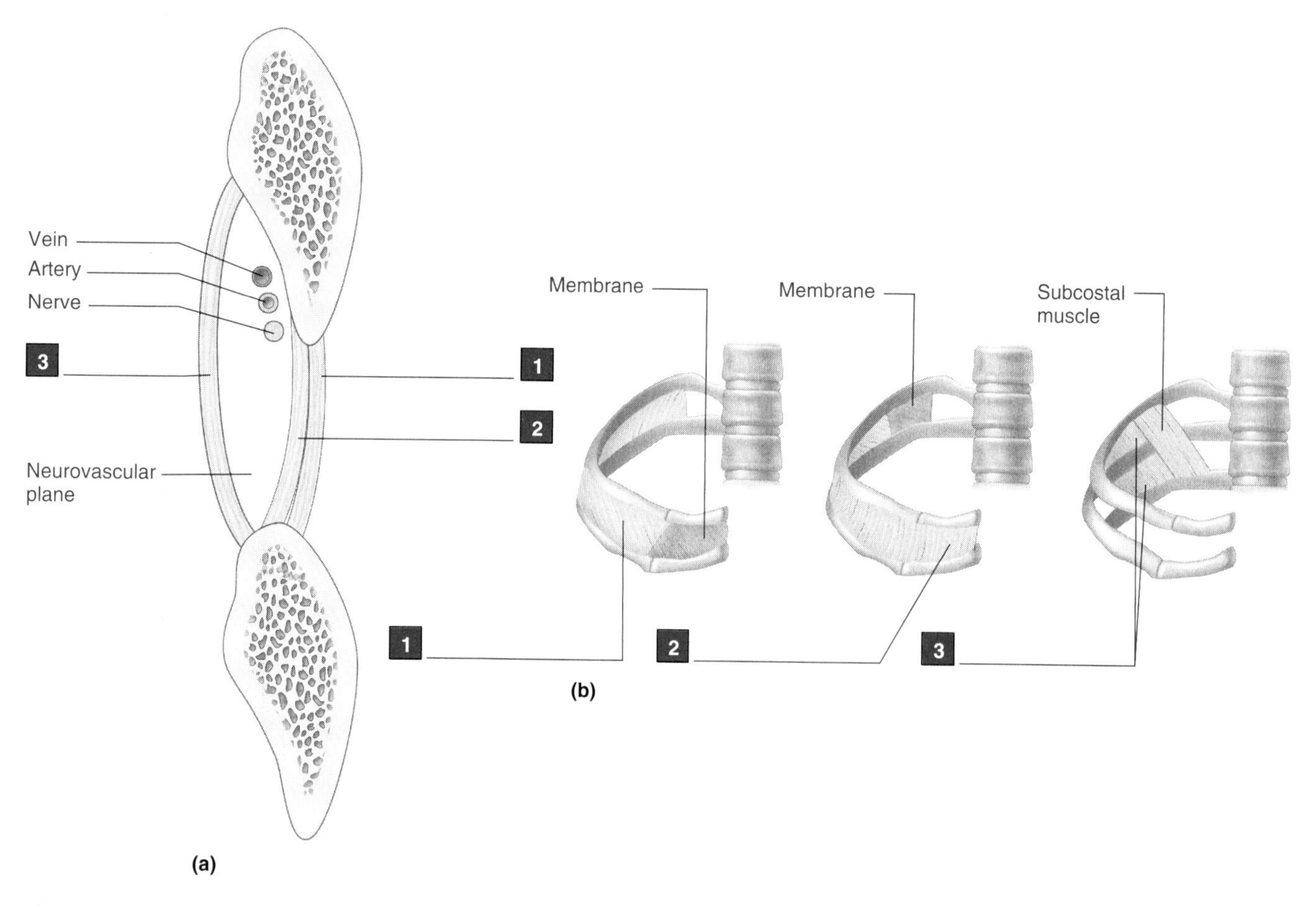

Figure 9.5 **Intercostal Muscles**
(a) coronal cross section; (b) anterior view.

Abdominal Diaphragm

The thoracic and abdominal cavities are separated by a muscular diaphragm. It is the principal force behind respiration.

Abdominal Diaphragm 1 is a dome-shaped muscle that is convex upward. Its fibers arise peripherally from the back of the xiphoid process, the internal surfaces of ribs 9–12, costal cartilages 6–8, the epimysium over the *Quadratus lumborum* and *Psoas major* muscles, and from the bodies of L1–L3.

These fibers converge upon a thin, crescent-shaped *Central Tendon*. This tendon is divisible into three portions, known as leaves. The left and right leaves lie below the lungs. The middle leaf lies below and is fused to the lower surface of the fibrous pericardium around the heart.

The abdominal diaphragm has three major orifices that permit the passage of structures between the thorax and abdomen. From front to back these are:

Vena Caval Foramen 2 through the middle leaf of the central tendon, for the inferior vena cava. It lies at the level of vertebra T8.

Esophageal Hiatus 3 through posterior muscular fibers, for the esophagus. It lies at the level of vertebra T10.

Aortic Hiatus 4 which is really behind the diaphragm, for the descending aorta. It lies at the level of vertebra T12.

The abdominal diaphragm is innervated by the *Phrenic Nerve*.

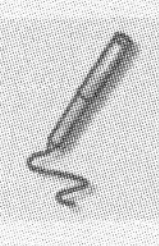

Color and label the aforementioned structures in figure 9.6.

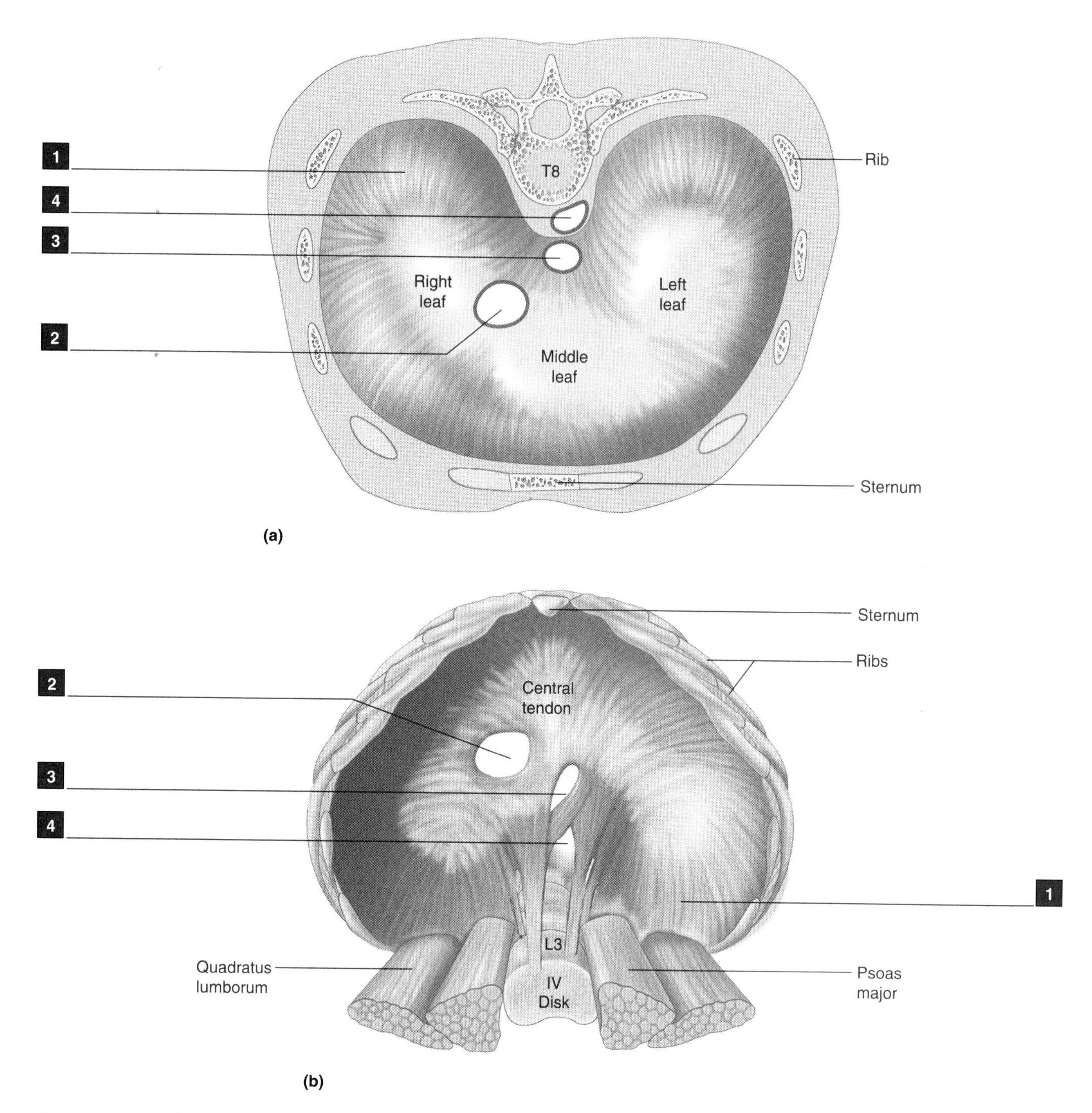

Figure 9.6 **Abdominal Diaphragm**
(a) thoracic surface; (b) abdominal surface.

9.4 THE THORACIC CAVITY

The ribs and abdominal diaphragm enclose the *Thoracic Cavity*. This contains the trachea and lungs, the heart and some very large blood vessels, the esophagus, as well as nerves and lymphatic channels.

In the embryo, a large blood vessel known as the *Heart Tube*, the *Trachea* and its *Lung Buds*, and the *Gut Tube* develop in front of the primordial vertebral column and behind a membranous sac, known as the **Coelom** **1**. The coelom extends throughout the thorax and abdomen (Figure 9.7b "Stage One").

As the heart tube grows in size into the heart, and as the lung buds increase in size into the lungs, the tissues of the coelomic sac are pushed inward, forming septa that subdivide the original single coelomic cavity into three separate cavities (Figure 9.7b "Stage Two").

The middle cavity, into which the heart grows, is known as the **Pericardial Cavity** **2**.

The two lateral cavities, into which the lungs grow, are known as the **Pleural Cavities** **3**.

The sacs of the pericardial and pleural cavities are composed of *Serous Membrane*. Part of the membrane lies against the outer thoracic wall. This forms the *Parietal Layer* of the sac. Part of the membrane is adherent to the organ (heart or lung). This forms the *Visceral Layer* of the sac.

The space between the parietal and visceral layers contains *Serous Fluid* which permits free movement of the heart and lungs, and their adherent visceral membranes against the parietal membrane and thoracic wall.

Some of the fibrous connective tissue from the septa around the pericardial cavity adheres to the outside of its parietal layer, forming a *Fibrous Pericardium*.

Identify the aforementioned structures in figure 9.7.

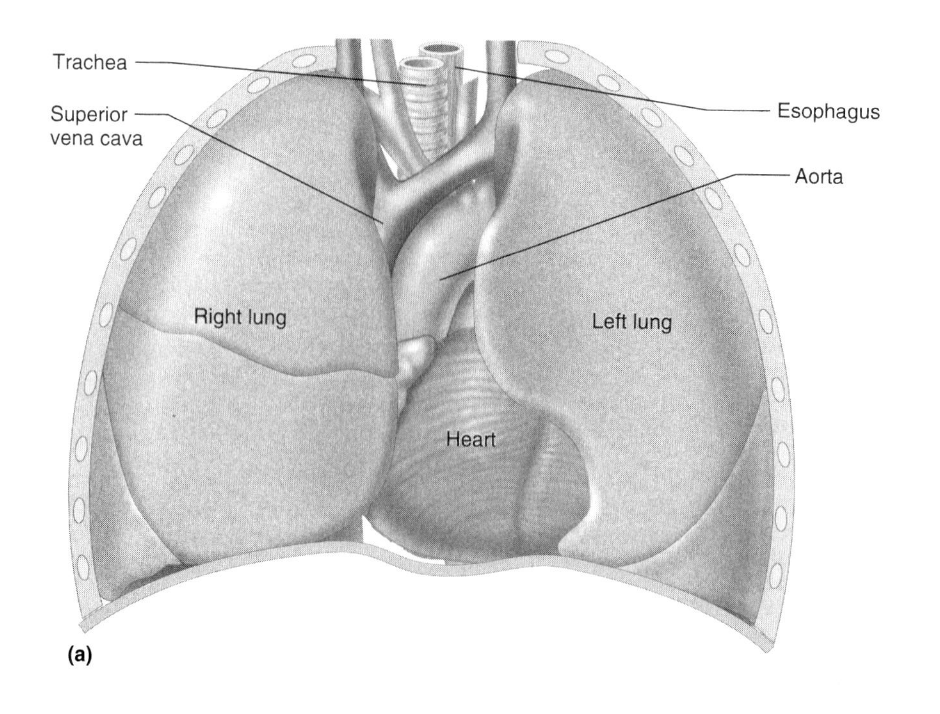

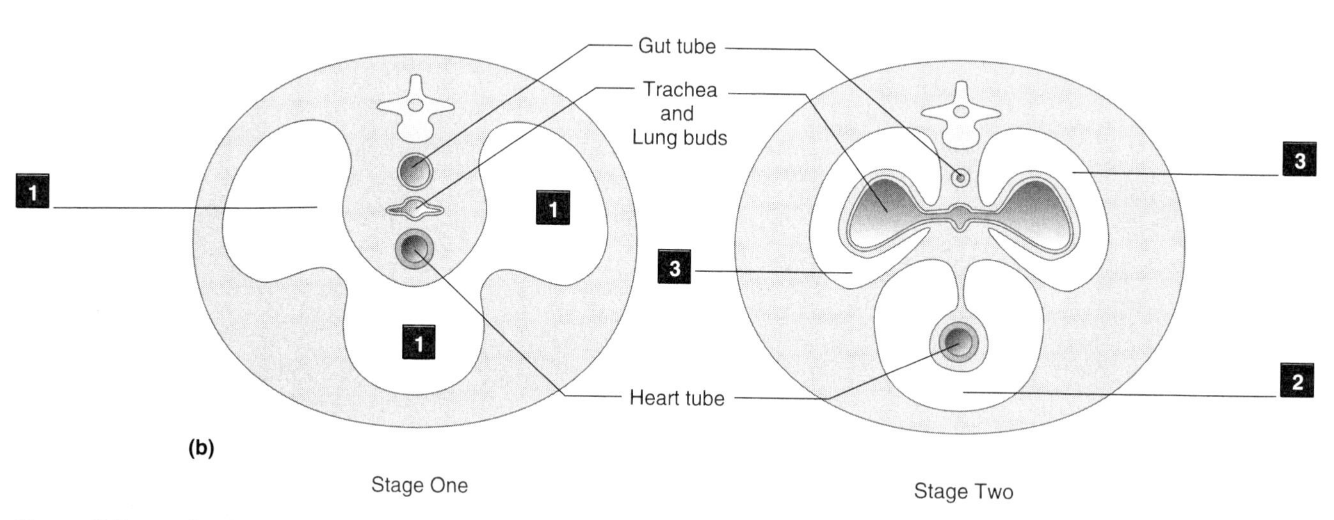

Figure 9.7 **The Thoracic Cavity**
(a) principal contents of the thorax (anterior view); (b) development of pleural and pericardial cavities (transverse sections).

9.5 THE MEDIASTINUM

The structures that occupy the space between the lungs form a partition between them known as the mediastinum. The largest structure of the mediastinum is the *Heart*.

If the heart and its pericardial sac are removed from the thoracic cavity, the other parts of the mediastinum become readily visible. Entering the top of the pericardial sac on the right is the **Superior Vena Cava 1**, into which drain the left and right **Brachiocephalic Veins 2**.

Immediately to the left of the superior vena cava is the **Aortic Arch 3**, which emerges from the pericardial sac. From this arch come the **Brachiocephalic Trunk 4**, the **Left Common Carotid Artery 5**, and the **Left Subclavian Artery 6**.

Below the superior vena cava and aortic arch are the **Pulmonary Arteries** **7**. They emerge by the pulmonary trunk from the heart and run into the lungs.

Behind the superior vena cava and aorta is the **Trachea** **8**. The left and right bronchi bifurcate from the trachea behind the pulmonary arteries, at about the level of the sternal angle.

Behind the trachea is the **Esophagus** **9**, which runs through the abdominal diaphragm.

The aortic arch runs upward and then curves back over the left bronchus of the trachea. From here it descends through the thorax as the **Descending Aorta** **10** between the vertebral column and esophagus. The descending aorta also runs through the abdominal diaphragm.

The mediastinum also contains the main lymphatic channel of the thorax, the *Thoracic Duct*, as well as the *Phrenic* and *Vagus Nerves*, and the *Azygos* and *Hemiazygos Veins* that run along either side of the vertebral column. We will examine these structures separately a bit later in this lab.

Identify the structures of the mediastinum in figure 9.8.

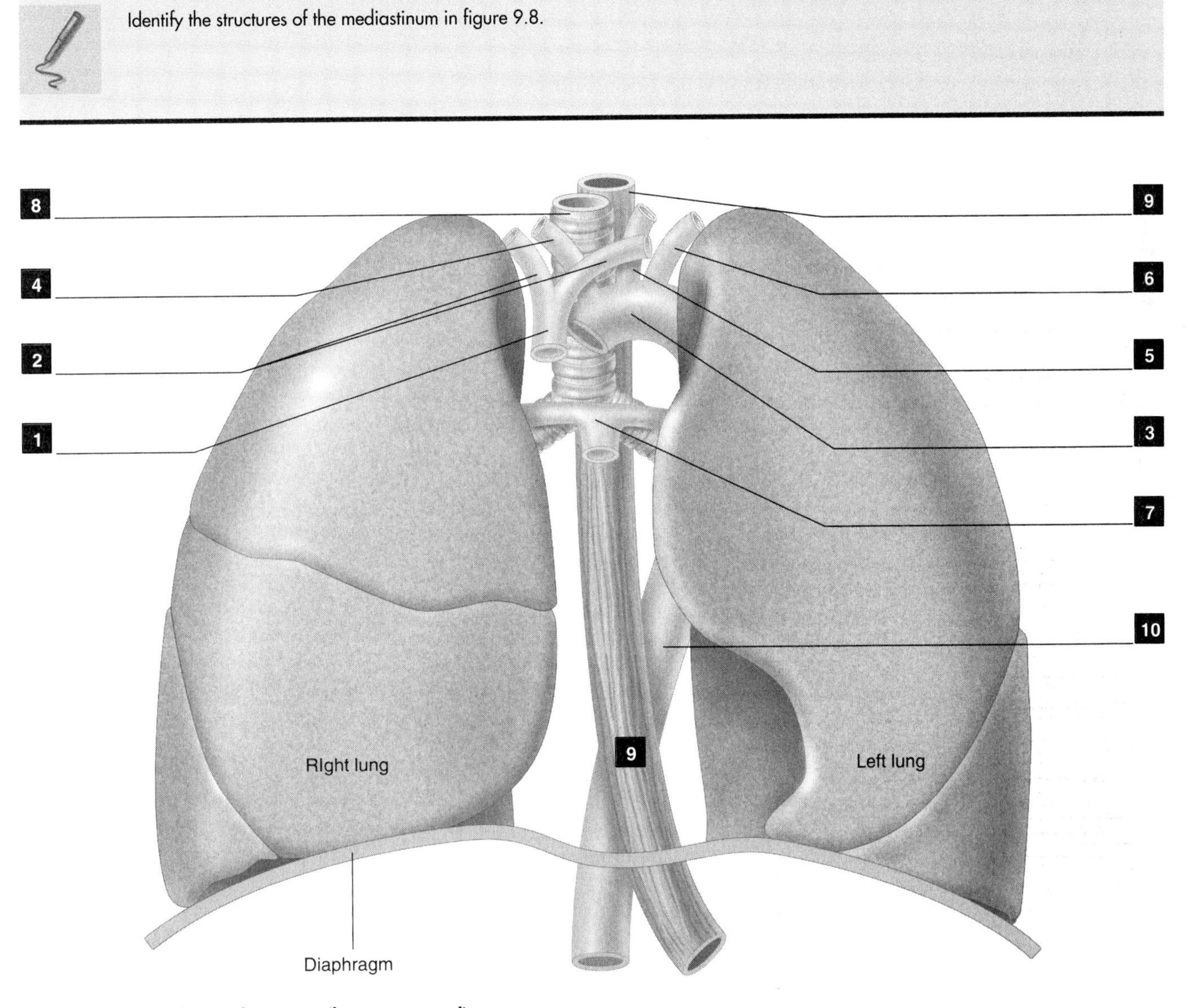

Figure 9.8 **The Mediastinum (heart removed)**

9.6 THE RESPIRATORY APPARATUS

The respiratory apparatus comprises the oral and nasal cavities in the head; the pharynx, larynx, and trachea in the neck, and by the trachea, bronchial tree, and lungs within the thorax. We have already studied those components in the head and neck. Let us now examine the thoracic structures.

Trachea and Bronchial Tree

The **Trachea** 1 is 4 to 5 inches long, and has a diameter of about 1 inch. It begins immediately below the cricoid cartilage. The tracheal lumen is kept open by a series of C-shaped **Tracheal Cartilages** 2 that partially surround it except on its posterior aspect. The cartilages are connected by fibroelastic tissue and smooth muscle fibers. At the inferior end of the trachea is the bronchial tree.

The *Bronchial Tree* has three levels of division:

Bronchus	**Primary Bronchus** **Secondary (Lobar) Bronchus** **Tertiary (Segmental) Bronchus**
Bronchiole	**Bronchiole** **Terminal (Respiratory) Bronchiole**
Alveolar Portion	**Alveolar Duct** **Alveolar Sac** **Alveolus**

The trachea bifurcates into left and right **Primary Bronchi** 3. The right primary bronchus is about 1 inch long. The left is longer by about another inch because the entrance (*Hilum*) of the left lung is pushed laterally by the heart.

Once the primary bronchus enters the lung it divides into **Secondary (Lobar) Bronchi** 4. There are *Three Lobar Bronchi* in the *Right Lung*, because it has three lobes. There are *Two Lobar Bronchi* in the *Left Lung*, because it has only two lobes.

The lobar bronchi divide into **Tertiary (Segmental) Bronchi** 5. A segmental bronchus serves one *Bronchopulmonary Segment* of the lung. Each lung has *10 Segmental Bronchi*.

The segmental bronchi subdivide into **Bronchioles** 6. They differ from bronchi because they lack cartilaginous rings. The bronchioles divide into even smaller **Respiratory Terminal Bronchioles** 7.

The respiratory bronchioles, in turn, branch into a number of **Alveolar Ducts** 8. They enter into **Alveolar Sacs** 9, from which bud numerous microscopic **Alveoli** 10. Gas exchange takes place in the alveoli.

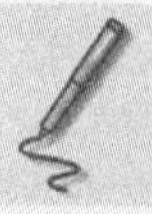

Identify the trachea and components of the bronchial tree in figure 9.9.

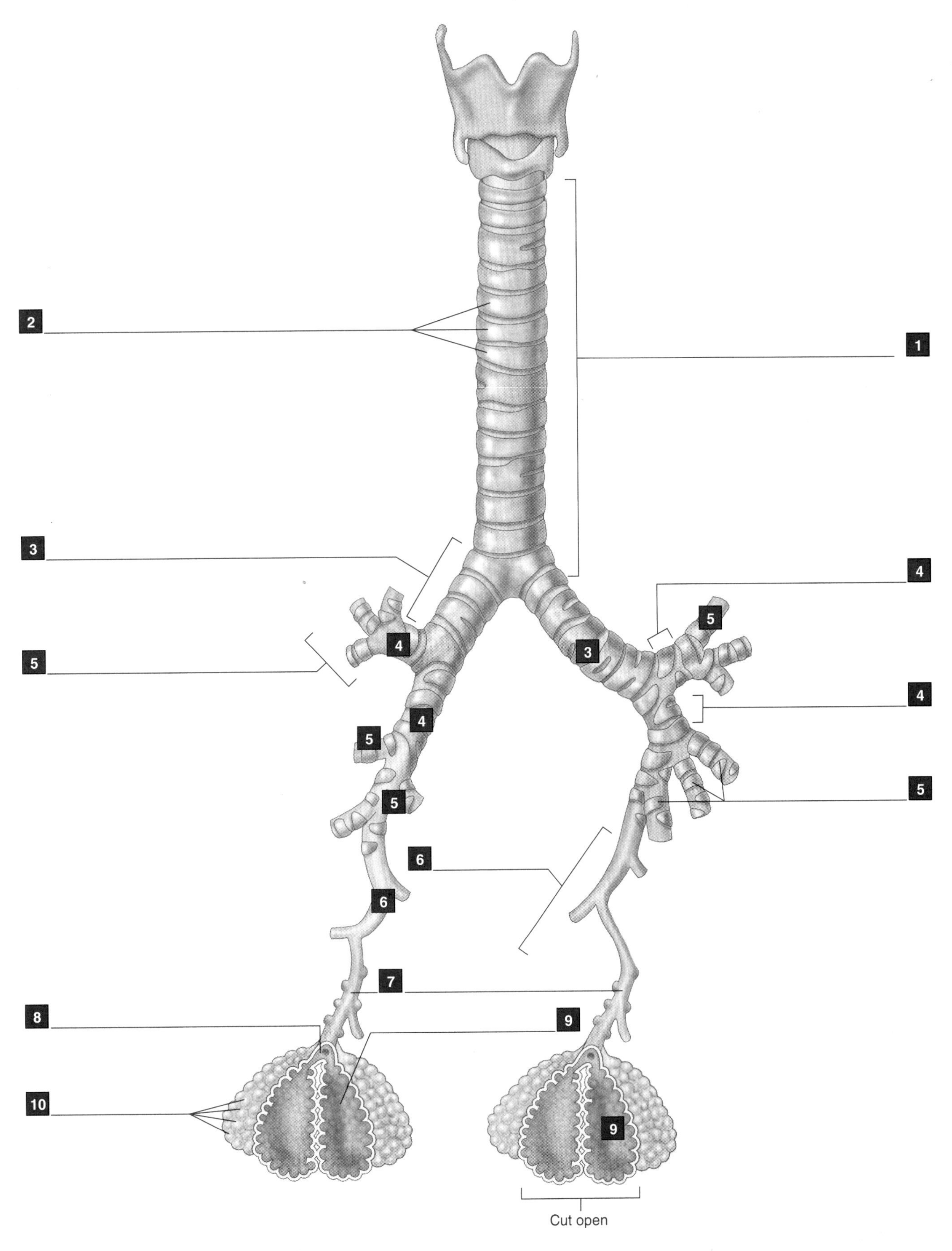

Figure 9.9 Trachea and Bronchial Tree

The Lungs

The lungs fill most of the thoracic cavity. Each is composed of a mass of spongy tissue—the alveoli—that constitutes the *Respiratory Portion* of the lung. Each also contains a branching network of tubes—bronchi and bronchioles—that comprises the *Conducting Portion* of the lung. The lung is covered by intimately adherent *Visceral Pleura*, which is separated from the *Parietal Pleura* that lines the thoracic wall by an intrapleural space with a slightly lower pressure than that of the atmosphere. This space also contains *Serous Fluid* that facilitates movement of the lungs during inspiration and expiration.

The top, or *Apex* of the lung is somewhat pointed. The bottom, or *Base*, of the lung is concave because it rests upon the dome-shaped abdominal diaphragm.

Because the heart takes up more space on the left side of the thoracic cavity, the left lung differs slightly in its gross appearance from the right lung.

Right Lung

The right lung has *three Lobes* separated by two fissures. Uppermost is the **Superior Lobe** 1. Below it is the **Middle Lobe** 2. Lowermost is the **Inferior Lobe** 3. The **Horizontal Oblique Fissure** 4 separates the superior and middle lobes. The **Oblique Fissure** 5 separates the middle and inferior lobes.

Left Lung

The left lung has *two Lobes* separated by one fissure. Uppermost is the **Superior Lobe.** It is separated from the **Inferior Lobe** by the **Oblique Fissure** 5. The left lung has no middle lobe. Instead, it has a concavity, the **Cardiac Notch** 6, into which the heart fits. Below the cardiac notch, the anterior aspect of the superior lobe has a tonguelike protrusion known as the **Lingula** 7.

Both Lungs

Despite the different numbers of lobes in the left and right lungs, both have *10 Bronchopulmonary Segments*. Each of these blocks of lung tissue receives its air supply from a separate segmental bronchus and its blood supply from a separate branch of the *Lobar Artery*.

The bronchopulmonary segments are independent enough that infection may affect one and not others. Because of their autonomy, it is possible to surgically remove a single bronchopulmonary segment without affecting the others.

Identify the aforementioned structures in figure 9.10.

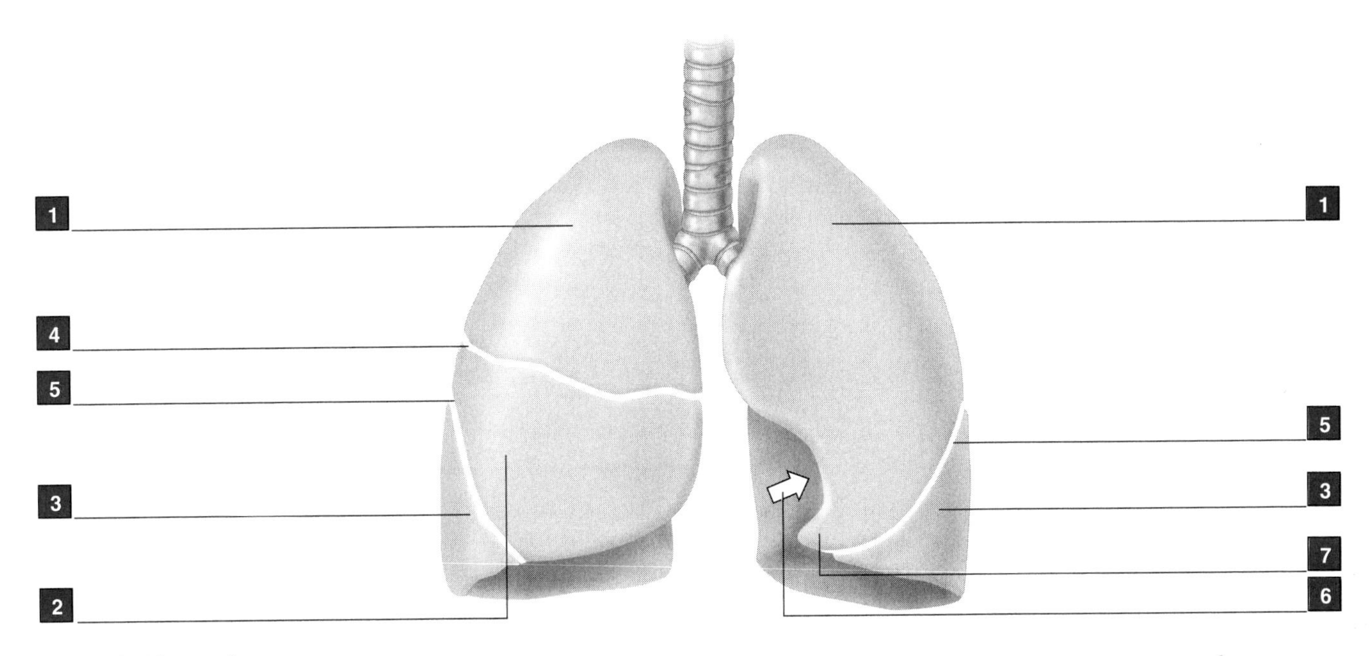

Figure 9.10 **The Lungs**

Respiration

Carbon dioxide–rich blood is pumped into the lungs from the right ventricle of the heart through the pulmonary arteries. In the lungs, the pulmonary arteries branch extensively so that each microscopic air sac (alveolus) is surrounded by a capillary network. As deoxygenated arterial blood passes through the capillaries around the alveoli, the carbon dioxide diffuses out of the blood through the wall of the alveolus into its cavity. Oxygen in the alveolar cavity diffuses into the capillary blood and is carried away by the pulmonary veins to the left atrium of the heart.

9.7 THE HEART

A large part of the mediastinum is occupied by the heart. It is a four-chambered muscular pump. It takes deoxygenated blood from the body and pumps it into the lungs, and then pumps the oxygenated blood from the lungs into the body. It also provides blood to its own *Cardiac Muscle* via a separate network of vessels.

Location of the Heart

The heart is about the size of a clenched fist. It is somewhat cone-shaped, and is located in the center of the chest. Most of its mass is to the left of the midline, and it lies closer to the front than the back of the thorax.

Let us locate your heart. We will do this by projecting and drawing its margins onto the front of your chest. For this exercise, use a washable marker, and stand in front of a large mirror (in the privacy of your own room). We will refer again to the drawing on your chest when we locate the heart valves on p. 273.

Upper Margin of the Heart

The upper margin is marked by a straight line connecting points 1 and 2.

Point 1 is located on the lower border of the *left second costal cartilage*. It is one finger breadth from the edge of the sternum. Recall that the second costal cartilage articulates with the sternum at the *Sternal Angle* (see p. 251).

Point 2 is on the upper border of the *right third costal cartilage*. It is also one finger breadth from the edge of the sternum.

Lower Margin of the Heart

The lower margin is marked by a straight line connecting points 3 and 4.

Point 3 is on the lower border of the *right sixth costal cartilage*. It is one finger breadth from the edge of the sternum.

Point 4 is in the *left fifth intercostal space*. It is three thumb breadths from the midline of the sternum.

Left and Right Margins of the Heart

The left and right margins are marked by lines that are slightly convex laterally between the left and right ends of the upper and lower margins.

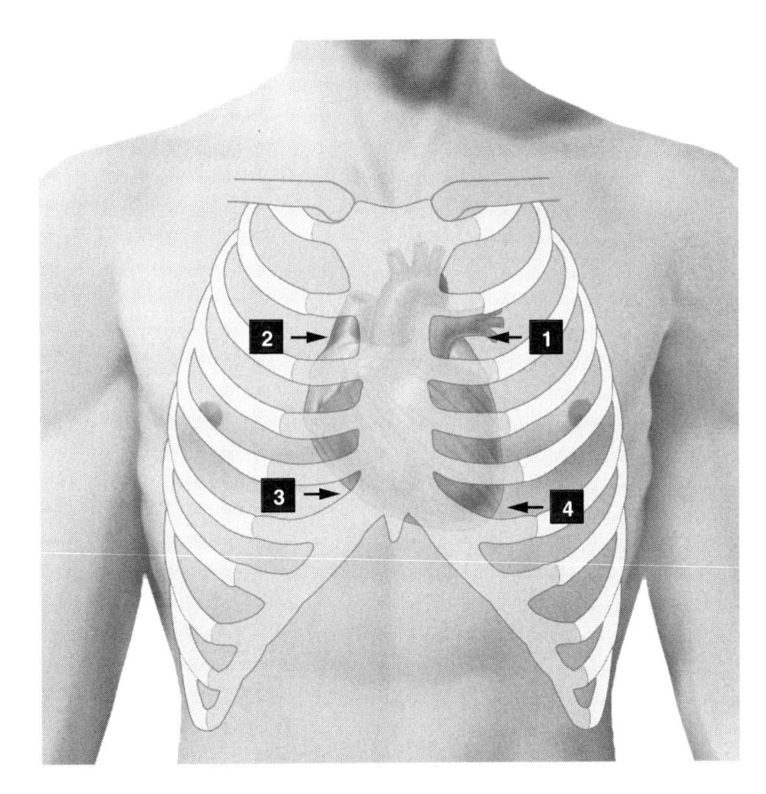

Surface Markings of the Heart

The Pericardium

The heart hangs in a membranous sac known as the ***Pericardium,*** which is suspended from the roots of the large blood vessels that exit and enter the heart. The pericardium serves to restrict excessive movement of the heart, while providing a lubricated container within which it can beat freely.

The pericardium is comprised of two sacs. The outermost sac of a thick, fibrous layer known as the **Fibrous Pericardium 1.** This sac is attached to the coverings of the great blood vessels that pass through it: aorta, superior and inferior venae cavae, pulmonary trunk, and pulmonary veins. The fibrous pericardium is firmly attached to the central tendon of the diaphragm, and to the back of the sternum.

Within the fibrous pericardium resides the membranous **Serous Pericardium 2.** Like other serous sacs, it has two layers. The **Parietal Serous Pericardium 3** adheres to the fibrous pericardium. It reflects around the roots of the great vessels to become continuous with the **Visceral Serous Pericardium 4,** or *Epicardium*, which is adherent to the outer surface of the heart.

The narrow space between the parietal and visceral layers of serous pericardium constitutes the **Pericardial Cavity 5.** It contains a small amount of lubricating fluid that facilitates the beating movements of the heart.

Identify the components of the pericardial sacs in figure 9.11.

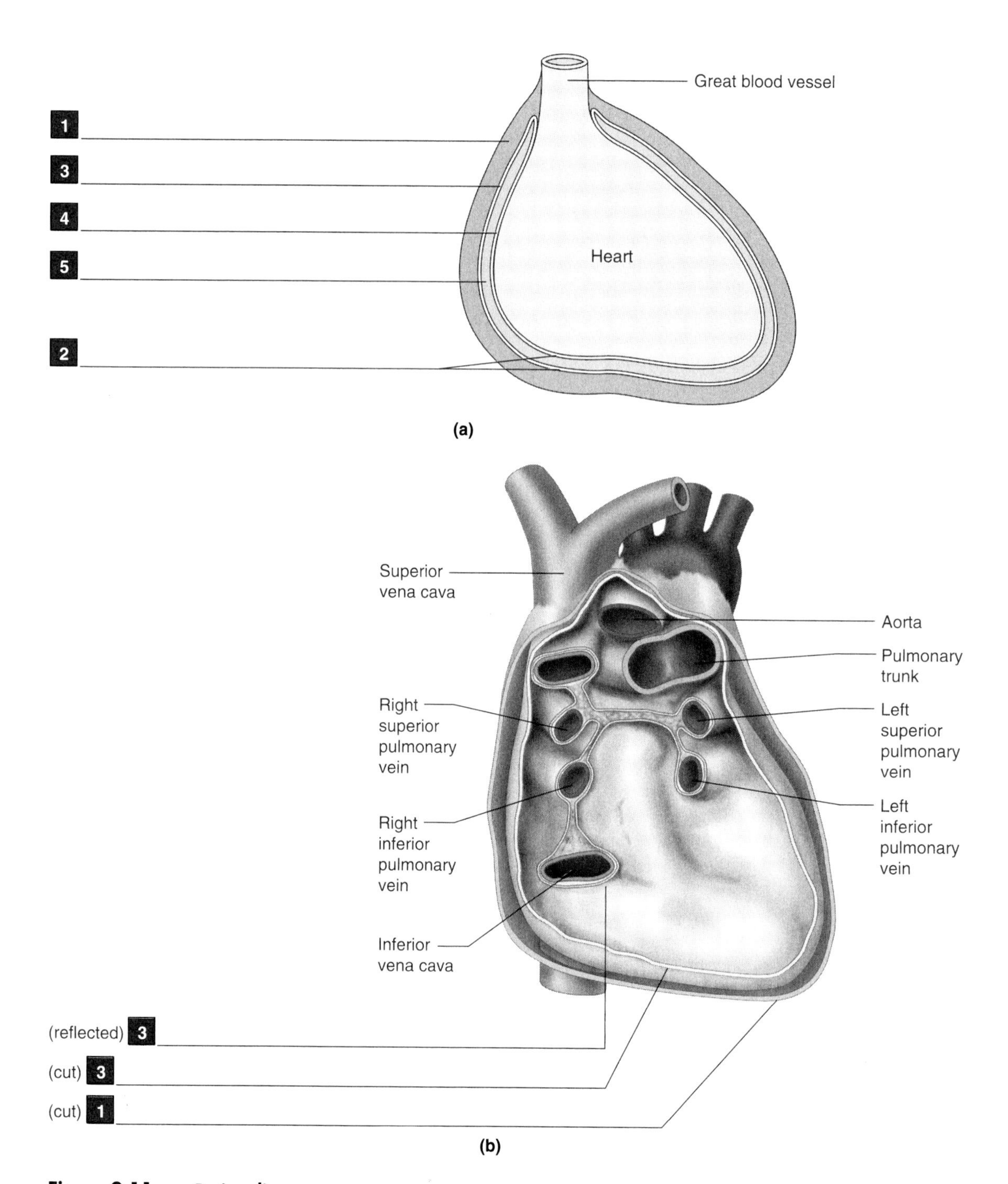

Figure 9.11 **Pericardium**
(a) schematic structure; (b) relationship to great vessels (anterior view of pericardium with heart removed).

The Heart as a Pump

The heart is a four-chambered, muscular pump. The chambers are separated by internal septa. The superior chambers are the *Atria*. The inferior chambers are the *Ventricles*. The atria receive blood from outside the heart, and because they pump it only into the ventricles, the muscular walls of the atria are comparatively thin. The ventricles are more heavily muscled because they pump blood out of the heart and, ultimately, through high-resistance capillary networks.

The heart pushes deoxygenated blood into the lungs, and receives oxygenated blood from the lungs by a series of vessels that constitute the **Pulmonary Circulatory System 1.** It pumps oxygenated blood into, and receives deoxygenated blood from the rest of the body by a series of arteries and veins that constitute the **System Circulatory System 2.**

Deoxygenated Blood 3 from the body flows from the capillary beds and through systemic veins into one of two large vessels known as the **Vena Cavae 4.** The vena cavae empty into the **Right Atrium** 5 of the heart.

The right atrium pumps this deoxygenated blood into the **Right Ventricle 6.** The right ventricle, in turn, pumps it into the **Pulmonary Arteries 7.** The pulmonary arteries carry the deoxygenated blood to the lungs, where it receives oxygen.

Oxygenated Blood 8 is returned from the lungs to the heart through the **Pulmonary Veins 9.** The pulmonary veins empty into the **Left Atrium** 10 of the heart.

The left atrium pumps this oxygenated blood into the **Left Ventricle 11.** The left ventricle, in turn, pumps it into the **Aorta 12.** The aorta branches into a series of subsidiary arteries, through which the oxygenated blood reaches the capillary networks of the body.

Identify the aforementioned structures in figure 9.12. Color the arrows designated 3 (deoxygenated blood) dark blue. Color the arrows designated 8 (oxygenated blood) dark red. Color the blood vessels and the chambers of the heart that carry these arrows lighter shades of blue and red, respectively.

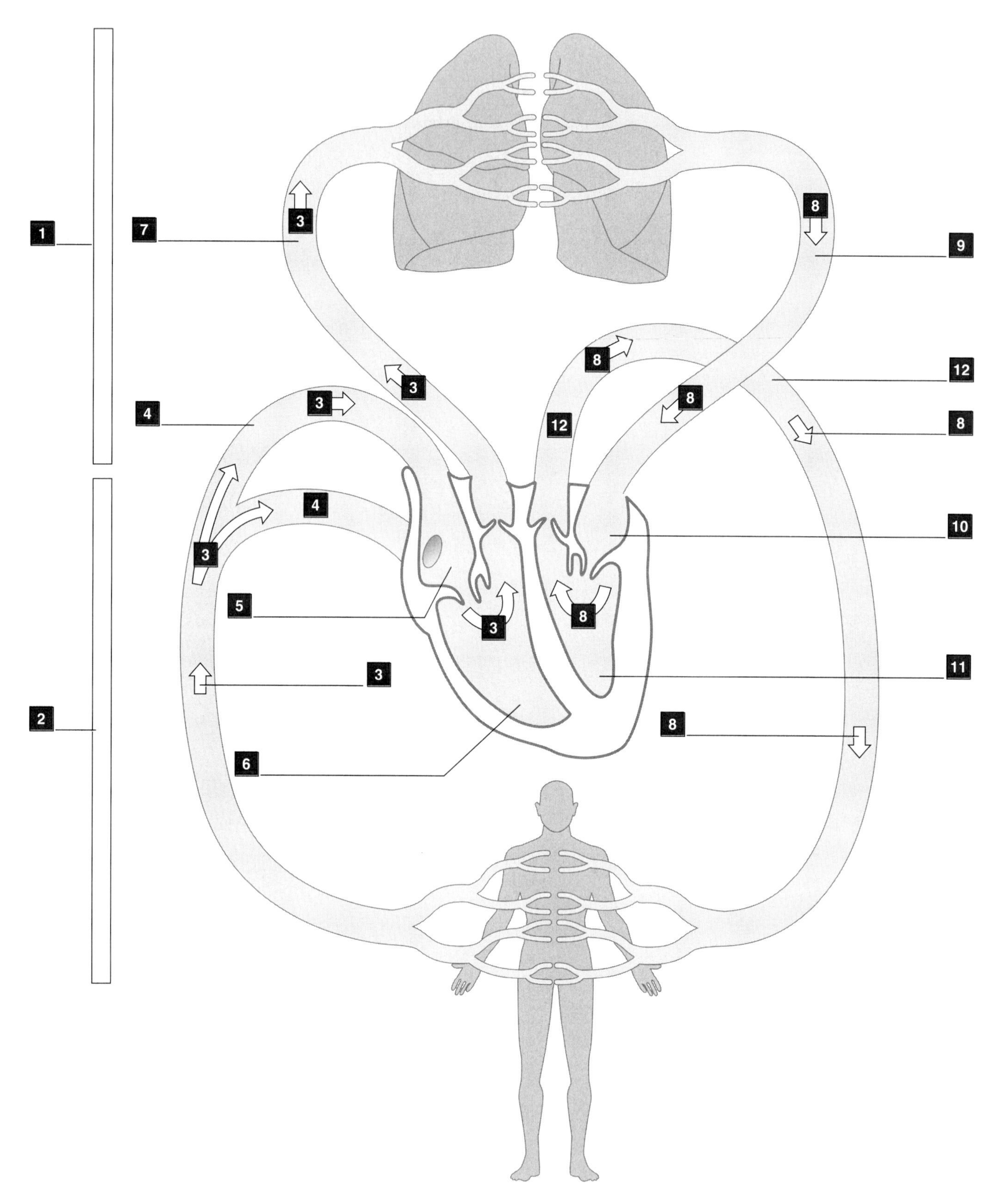

Figure 9.12 The Heart as a Pump for the Pulmonary and Systemic Circulatory Systems

The Heart and Great Vessels

The chambers of the heart can be visualized on its external surface. Entering these chambers are some large blood vessels.

Anterior Aspect of the Heart

The front of the heart is dominated by the **Right Ventricle 1.** A shallow groove separates it from the **Left Ventricle 2.** Above and to the right of the right ventricle is the **Right Atrium 3.** A fleshy flap that projects anteriorly from it is the **Auricle of the Right Atrium 4.** Above the left ventricle is the **Auricle of the Left Atrium 5.**

Posterior Aspect of the Heart

The back of the heart is dominated by the **Left Atrium 6;** its auricle was visible in anterior view. Between the left atrium and the left ventricle is a sulcus.

The Great Vessels

Entering the Right Atrium. Entering the *top* of the right atrium is the **Superior Vena Cava 7.** It receives blood from its main tributaries, the left and right **Brachiocephalic Veins 8.** These drain deoxygenated blood from the head, neck, and upper limbs. Also draining into the back of the superior vena cava is the **Azygos Vein 9.** It carries blood from the wall of the thorax.

Entering the *bottom* of the right atrium is the **Inferior Vena Cava 10.** It carries deoxygenated blood from the abdomen, pelvis, and lower limbs

Exiting the Right Ventricle. Emerging from the *top* of the right ventricle is the **Pulmonary Trunk 11.** It soon divides (below the arch of the aorta) into left and right **Pulmonary Arteries 12.** They carry deoxygenated blood into the lungs.

Entering the Left Atrium. Entering the *posterior sides* of the left atrium are four **Pulmonary Veins 13.** They carry oxygenated blood from the lungs into the heart.

Exiting the Left Ventricle. Emerging from the *top* of the left ventricle (between the pulmonary trunk and the right auricle) is the **Ascending Aorta 14.** It soon turns to the left and arches backward, as the **Aortic Arch 15,** over the pulmonary arteries. The aortic arch gives off three main branches—**Brachiocephalic Trunk 16, Left Common Carotid Artery 17** and **Left Subclavian Artery 18**—before turning downward as the descending aorta.

Identify the aforementioned structures in figure 9.13.

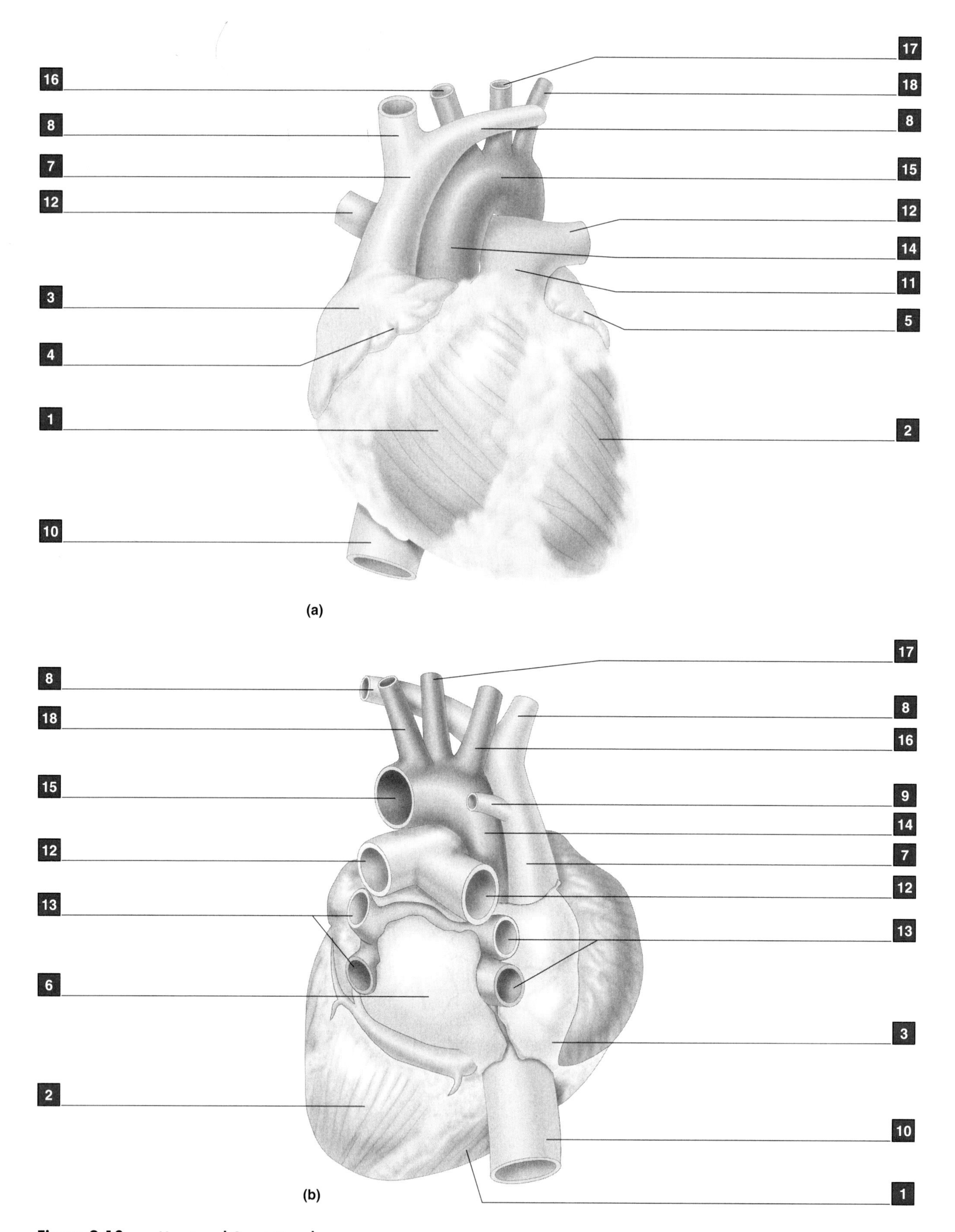

Figure 9.13 **Heart and Great Vessels**
(a) anterior view; (b) posterior view.

Internal Aspect of the Heart (Chambers and Valves)

The internal walls of the **Right Atrium** 1 are smooth at the back, but roughened in front by bundles of muscle fibers called the *Pectinate Muscles*. The medial wall of the right atrium has an oval depression just above the entrance of the inferior vena cava. This is known as the **Fossa Ovalis** 2. It is the remnant of a foramen (*Foramen Ovale*) that existed in fetal life between the right and left atria. The foramen ovale enables blood to bypass the right ventricle and the non-functional fetal lungs.

The right atrium opens into the **Right Ventricle** 3 by an orifice that has three flaps of tissue (*Cusps*) around its circumference. The cusps project into the right ventricle forming the **Tricuspid Valve** 4.

The internal walls of the right ventricle have a number of muscular ridges called *Trabeculae Carneae*. Three of them project toward the cusps of the tricuspid valve, and are attached to the cuspal margins by tendons. The muscular extensions are called **Papillary Muscles** 5, and their tendons are called **Chordae Tendineae** 6. They prevent the tricuspid valve cusps from being pushed up into the right atrium when the right ventricle contracts.

The entrance to the **Pulmonary Trunk** 7 is guarded by another valve with three cusps, each of which is shaped like a half-moon. Thus, it is called the **Pulmonary Semilunar Valve** 8. Blood tending to flow back into the right ventricle from the pulmonary trunk runs into the space between the cusps and the wall of the trunk. This causes the cusps to balloon out and contact one another, closing the orifice.

The inner walls of the **Left Atrium** 9 are smooth.

The left atrium communicates with the **Left Ventricle** 10 by an orifice with two cusps that project into the ventricle. This valve is called the **Mitral Bicuspid Valve** 11. The edges of these cusps are connected to the two **Papillary Muscles** 12 of the left ventricle by **Chordae Tendineae** 13. They prevent the mitral valve cusps from being pushed up into the left atrium when the left ventricle contracts. The walls of the left ventricle are the thickest, and they are lined with Trabeculae Carnae.

The entrance to the **Ascending Aorta** 14 is guarded by a three-cusped valve called the **Aortic Semilunar Valve** 15. It is like the pulmonary semilunar valve in its structure and mechanism of operation.

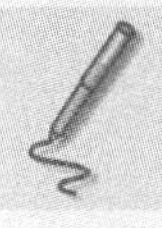

Identify the aforementioned structures in figure 9.14.

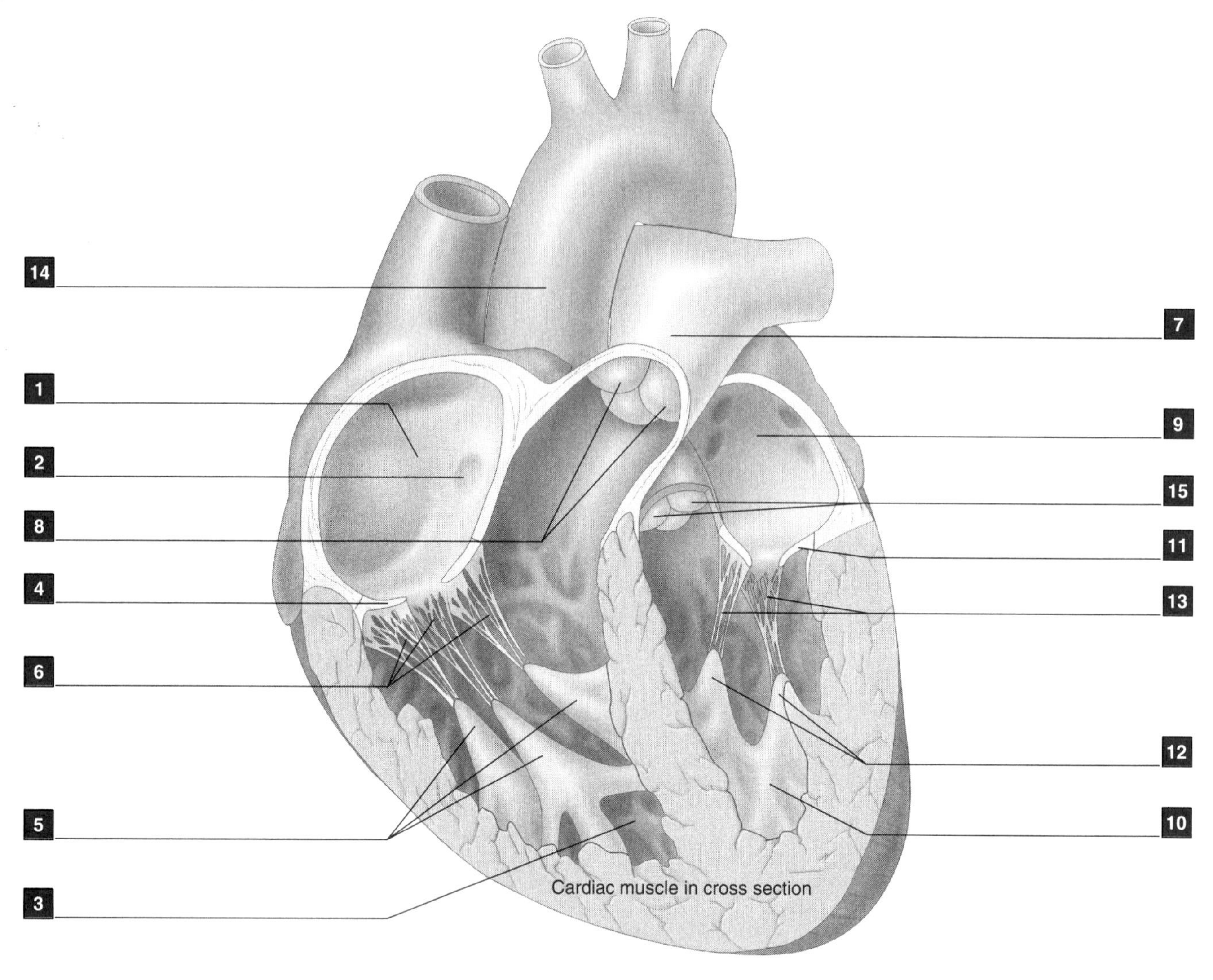

Figure 9.14 **Chambers and Valves of the Heart**
The anterior wall of the heart has been cut away.

Surface Projection and Auscultation of the Heart Valves

It is possible to hear the sound that each valve produces as it closes. If you listen to the heart with a stethoscope (surface auscultation), you will hear two beats: "*lub-dup*."

The "*lub*" is produced by the contraction of the ventricles and the concomitant closure of the *Tricuspid* and *Mitral* valves.

The "*dup*" is produced by the sharp closure of the *Pulmonary Semilunar* and *Aortic Semilunar* valves.

However, the noise that each produces upon closure is not necessarily heard best directly over the valve.

Let us project the actual positions of the valves onto the outline of your heart that you drew on the surface of your chest. We will then locate the place where each valve is best auscultated.

Anterior Projection and Auscultation of the Valves

Tricuspid Valve

- lies behind the right half of the sternum at the level of the fourth intercostal space **1**.
- is best heard over the left or right edge of the sternum at the level of the fifth intercostal space **2**.

Mitral Valve

- lies behind the medial end of the left fourth costal cartilage **3**.
- is best heard over the left fifth intercostal space about 3½ inches from the midline **4**.

Pulmonary Semilunar Valve

- lies behind the medial end of the left third costal cartilage **5**.
- is best heard over the medial end of the left second intercostal space **6**.

Aortic Semilunar Valve

- lies behind the left half of the sternum at the level of the third intercostal space **7**.
- is best heard over the medial end of the right second intercostal space **8**.

Identify the projected location and the auscultation point for each of the four valves in figure 9.15. Use light shading for the location and darker shading for the auscultation point of each valve.

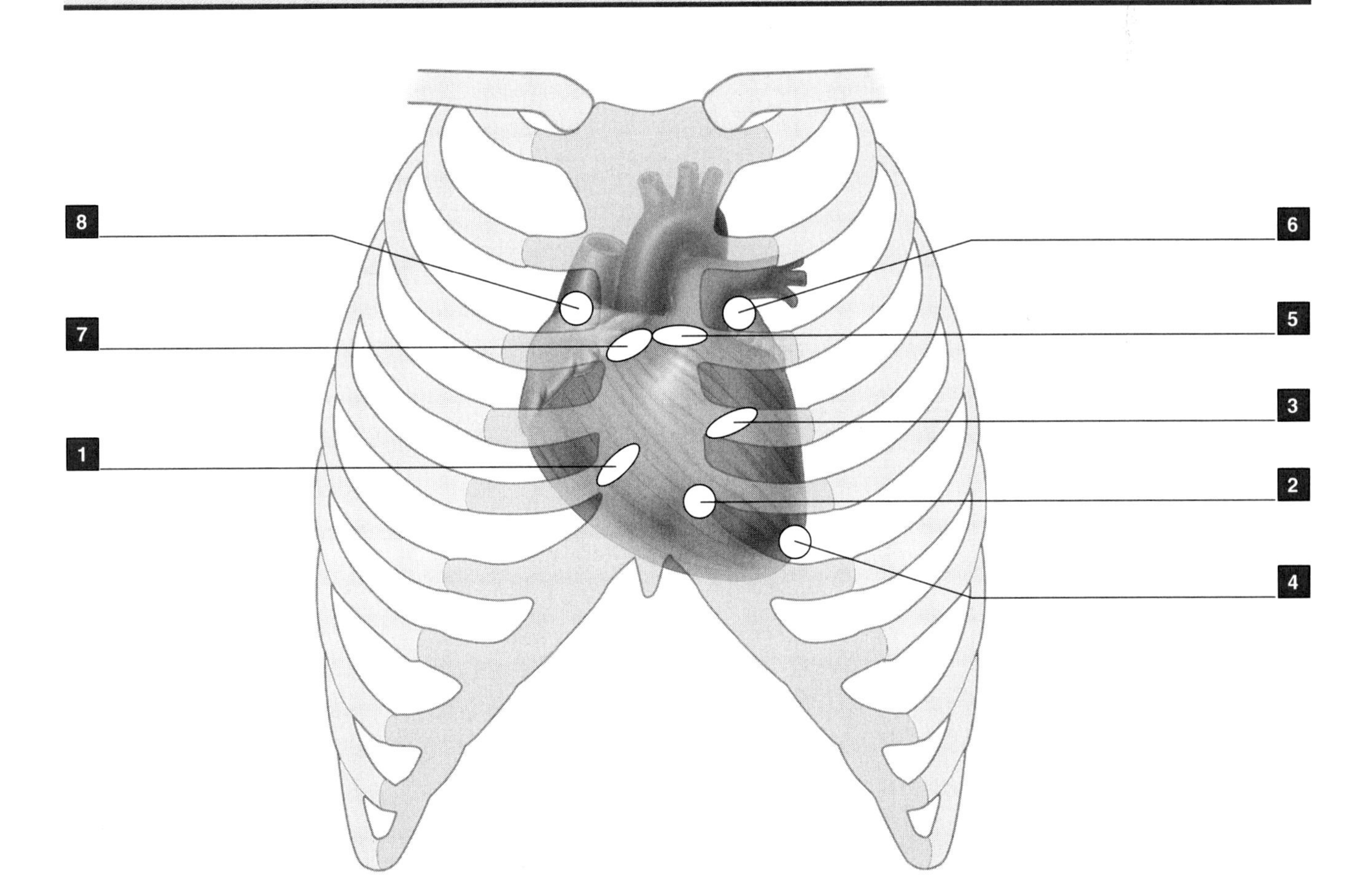

Figure 9.15 **Projected Location and Auscultation Points of the Heart Valves**

Conducting System and Innervation of the Heart

Conducting System

Cardiac muscle cells have an intrinsic ability to generate and conduct impulses. The cells contact one another at their ends, across which they communicate via gap junctions. Interconnected cardiac muscle cells form a functional mass called a ***Syncytium.*** The entire heart would be a single syncytium were it not for the fact that it has a fibrous internal "skeleton" that supports its valves. This fibrous tissue effectively divides the heart into two functional masses: an *Atrial Syncytium* and a *Ventricular Syncytium*.

In order to spread impulses across the "skeletal" boundary between the atria and ventricles, and in order to speed the rate at which impulses spread throughout the ventricular syncytium, the heart has a conducting system composed of specialized cardiac muscle cells.

The impulse that initiates each heartbeat begins at the **Sinoatrial (SA) Node 1.** This is a group of cells that lies in the wall of the *Right Atrium* near the opening of the superior vena cava. The SA node is the cardiac *Pacemaker*. Impulses from it spread throughout the atrial syncytium, causing both atria to contract simultaneously. These impulses reach a second node, by which they are transmitted to the ventricular syncytium.

The second node is the **Atrioventricular (AV) Node 2.** It is located in the lower part of the interatrial septum. The impulse travels from the AV node through a series of cellular fibers into and around the ventricles. The impulse crosses the fibrous partition into the ventricular mass along the fibers of the **Atrioventricular Bundle 3,** which runs into the interventricular septum. This fiber bundle soon divides into left and right **Bundle Branches 4** that traverse the septum. At the base of the septum, the fibers turn upward into the external walls of the ventricles. The recurrent fibers are known as **Purkinje Fibers 5.** These speed impulse transmission throughout the ventricular syncytium.

Innervation

The heart's inherent rate of contraction is set by the SA node. However, the depolarization rate of the SA node can be altered by extrinsic motor impulses carried by the *Autonomic Nervous System*.

Parasympathetic innervation, by means of the *Vagus Nerve*, *slows* the rate of heart contraction.

Sympathetic innervation, by fibers from sympathetic chain ganglia in the neck and thorax, *increases* the rate of heart contraction.

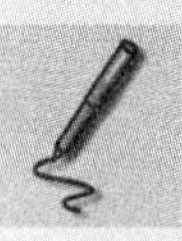

Identify the components of the cardiac conducting system in figure 9.16.

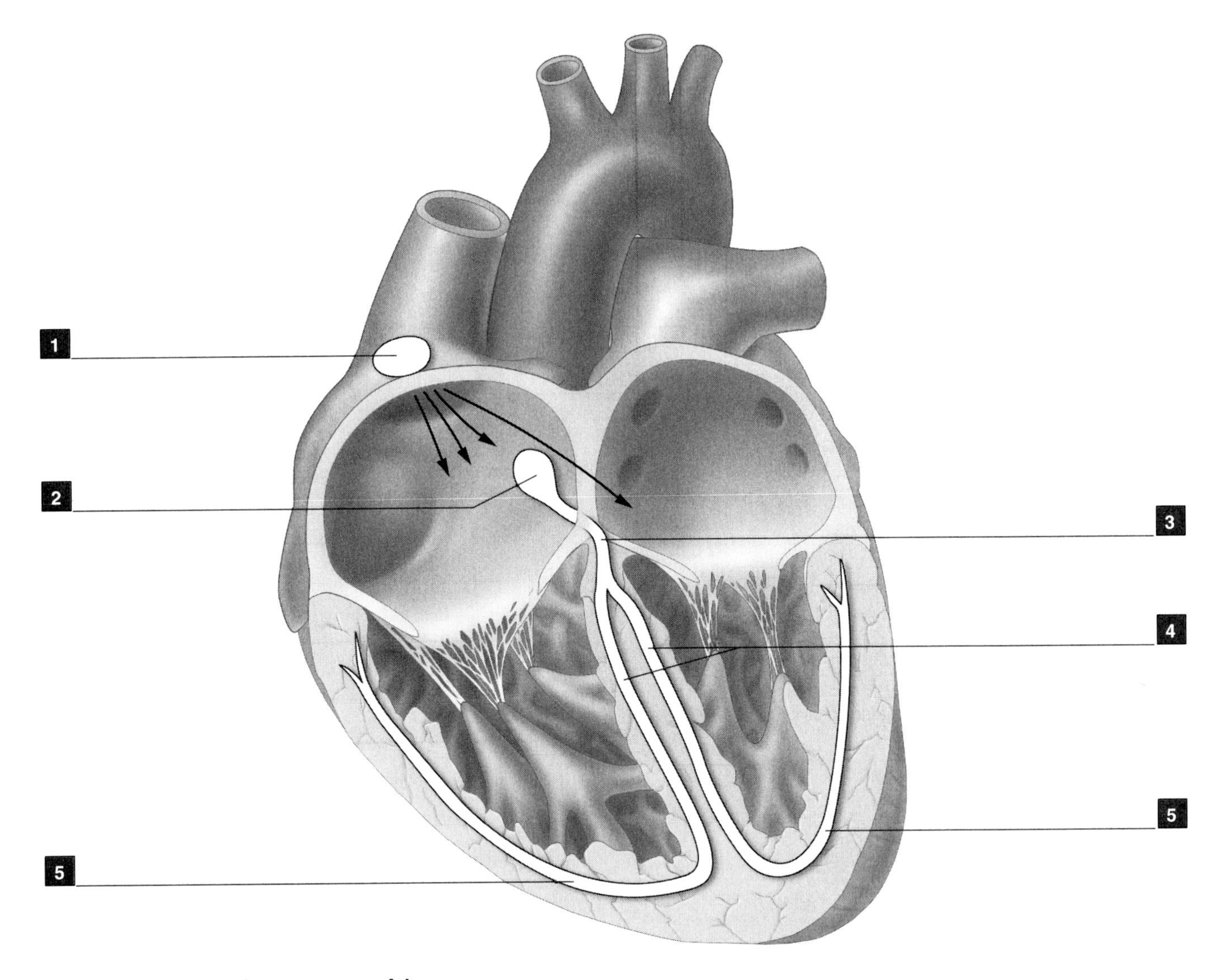

Figure 9.16 **Conducting System of the Heart**

Blood Supply of the Heart

Although nearly 3 quarts of blood move through the heart every minute, the cardiac muscle cells do not receive any nutrients from it. This is because the internal walls of the atria and ventricles are lined with a layer of tissue called *Endocardium* and because, at any rate, the blood in the right atrium and ventricle is deoxygenated! Thus, the heart is supplied by a separate system of arteries and veins that form the *Coronary Circulation*.

Arteries

Two coronary arteries branch from the *Aorta* as it emerges from the heart. Each has two principal branches. There are some anastomotic connections between the branches of the coronary arteries.

The **Left Coronary Artery** 1 emerges between the left atrium and the pulmonary trunk. One of its main branches is the **Anterior Interventricular Artery** 2. It runs to the apex of the heart in a groove on the anterior surface between the left and right ventricles. The other main branch is the **Circumflex Artery** 3. It curves around to the back of the heart in a groove between the left ventricle and the auricle of the left atrium.

The **Right Coronary Artery** 4 emerges from the aorta between the auricle of the right atrium and the pulmonary trunk. It curves around toward the back of the heart in a groove between the right auricle and right ventricle. It gives off a **Marginal Branch** 5, which runs down the right side of the heart toward the apex. The right coronary artery runs to the back of the heart, and turns toward the apex as the **Posterior Interventricular Artery** 6. It runs in a groove between the left and right ventricles.

Veins

Blood from the heart drains into the *Right Atrium* just to the left of the inferior vena cava via the **Coronary Sinus** 7. The coronary sinus receives blood from three major tributaries. The **Great Cardiac Vein** 8 accompanies the anterior interventricular and circumflex arteries. The **Small Cardiac Vein** 9 accompanies the marginal branch of the right coronary artery. It is joined by the **Middle Cardiac Vein** 10, which accompanies the posterior interventricular artery.

Identify the cardiac arteries and veins in figure 9.17.

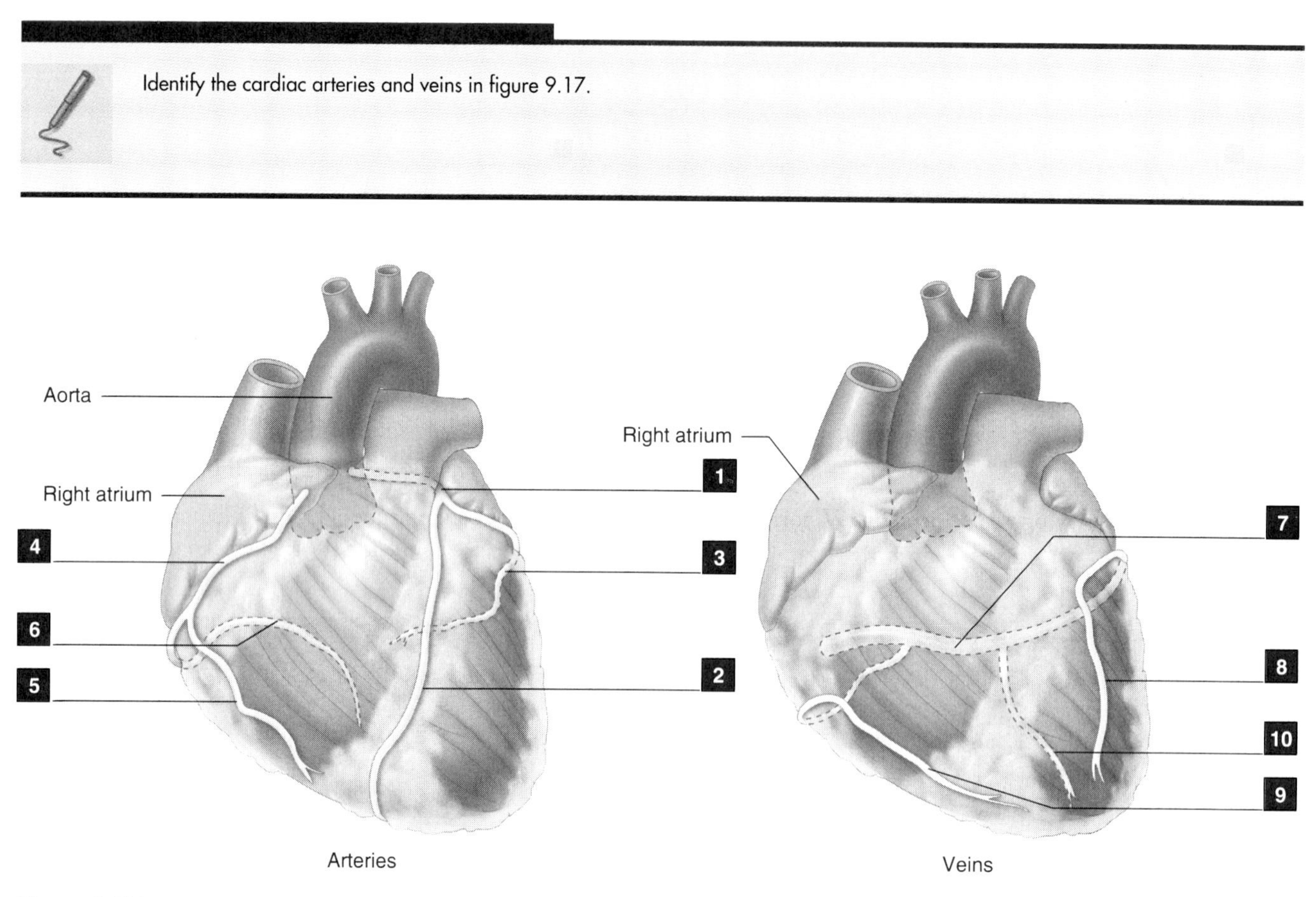

Figure 9.17 Coronary Circulation (anterior view of the heart)

9.8 LYMPHATICS IN THE THORAX

Recall that the lymphatic drainage of the breast is primarily into nodes at the end of its axillary tail (p. 250). These *Axillary Nodes* also receive lymph from the thoracic wall.

The deep structures of the thorax—the lungs, heart, and esophagus—drain their lymph into ducts on the left and right sides known as *Bronchomediastinal Trunks*. The right bronchomediastinal trunk drains into a large channel known as the **Thoracic Duct 1.**

The thoracic duct runs along the front of the vertebral column behind the esophagus, and between the azygos vein and descending aorta. It originates just below the abdominal diaphragm as a dilated midline sac, the **Cisterna Chyli 2.** The cisterna chyli receives lymph from the abdomen, pelvis, and lower limbs.

Just above the level of the first rib, the thoracic duct bends to the left to enter the **Left Brachiocephalic Vein 3.**

The thoracic duct also receives lymph from the left upper limb and left axillary nodes by a channel known as the *Left Subclavian Trunk*. This joins the thoracic duct just before it enters the left brachiocephalic vein. The thoracic duct also receives lymph from the left side of the head and neck by a channel known as the *Left Jugular Trunk*.

The right upper limb and right axillary nodes are drained of lymph by the *Right Subclavian Trunk*. Lymph is drained from the right side of the head and neck by the *Right Jugular Trunk*. These trunks merge with the right bronchomediastinal trunk to form the short **Right Lymphatic Duct 4** just above the first rib. The right lymphatic duct drains into the **Right Brachiocephalic Vein 5.**

Thus, lymph is drained from the right upper limb, right side of the head and neck, the right breast, and the right thoracic cavity and wall by the *Right Lymphatic Duct*. The rest of the body is drained by the *Thoracic Duct*.

Identify the components of the lymphatic system in figure 9.18.

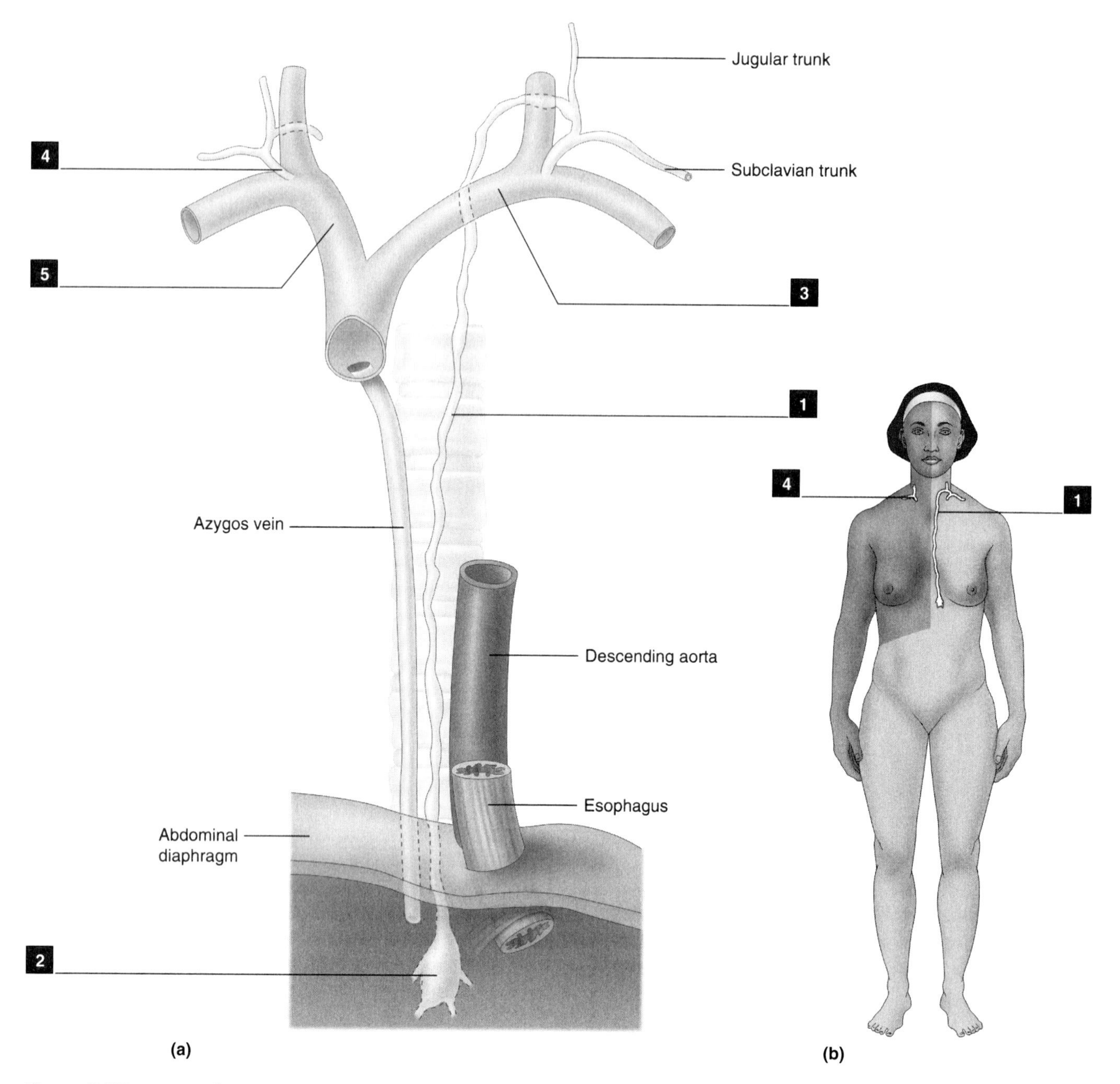

Figure 9.18 **Lymphatics**
(a) lymphatic structures; (b) regions drained by thoracic and right lymphatic ducts.

9.9 BLOOD VESSELS OF THE THORAX

We have examined the great vessels of the thorax, and the coronary vessels. The great vessels of the thorax (p. 271) are summarized in the following table.

Systemic Vessels	
Arteries	**Veins**
Aorta Brachiocephalic Trunk Left Common Carotid Artery Left Subclavian Artery	Inferior Vena Cava Superior Vena Cava Right Brachiocephalic Vein Left Brachiocephalic Vein
Pulmonary Vessels	
Arteries	**Veins**
Pulmonary Trunk Pulmonary Arteries	Pulmonary Veins

The coronary vessels (pp. 277–278) are summarized in the following table.

Coronary Vessels	
Arteries	**Veins**

Left Coronary
  Anterior Interventricular
  Circumflex
Right Coronary
  Marginal
  Posterior Interventricular
	Coronary Sinus
	  Great Cardiac Vein
	  Small Cardiac Vein
	  Middle Cardiac Vein

There are several other arteries and veins of some significance that run along the walls of the thoracic cavity.

Arteries

The descending aorta gives off 12 pairs of unnamed *Intercostal Arteries* into the neurovascular plane as it passes by the intercostal spaces. These arteries supply the posterior wall of the thorax. The aorta also gives off several pairs of unnamed *Subcostal Arteries* that supply the posterior abdominal wall. The subclavian and axillary arteries give off branches that supply the anterior wall of the thorax.

The *Subclavian Artery* gives off the **Internal Thoracic Artery** 1. It descends behind the costal cartilages, wedged between them and the *Transversus thoracis* muscle (p. 255). At the bottom of the rib cage it divides into two terminal branches, the *Superior Epigastric* and *Musculophrenic* arteries.

The internal thoracic artery gives off branches that perforate the intercostal spaces anteriorly. Some of these perforating arteries give off **Medial Mammary Branches** 2 to the breast.

The *Axillary Artery* gives off the **Lateral Thoracic Artery** 3 just lateral to the rib cage. It descends along the side of the cage, providing **Lateral Mammary Branches** 4 to the breast.

Identify the arteries of the thoracic wall in figure 9.19.

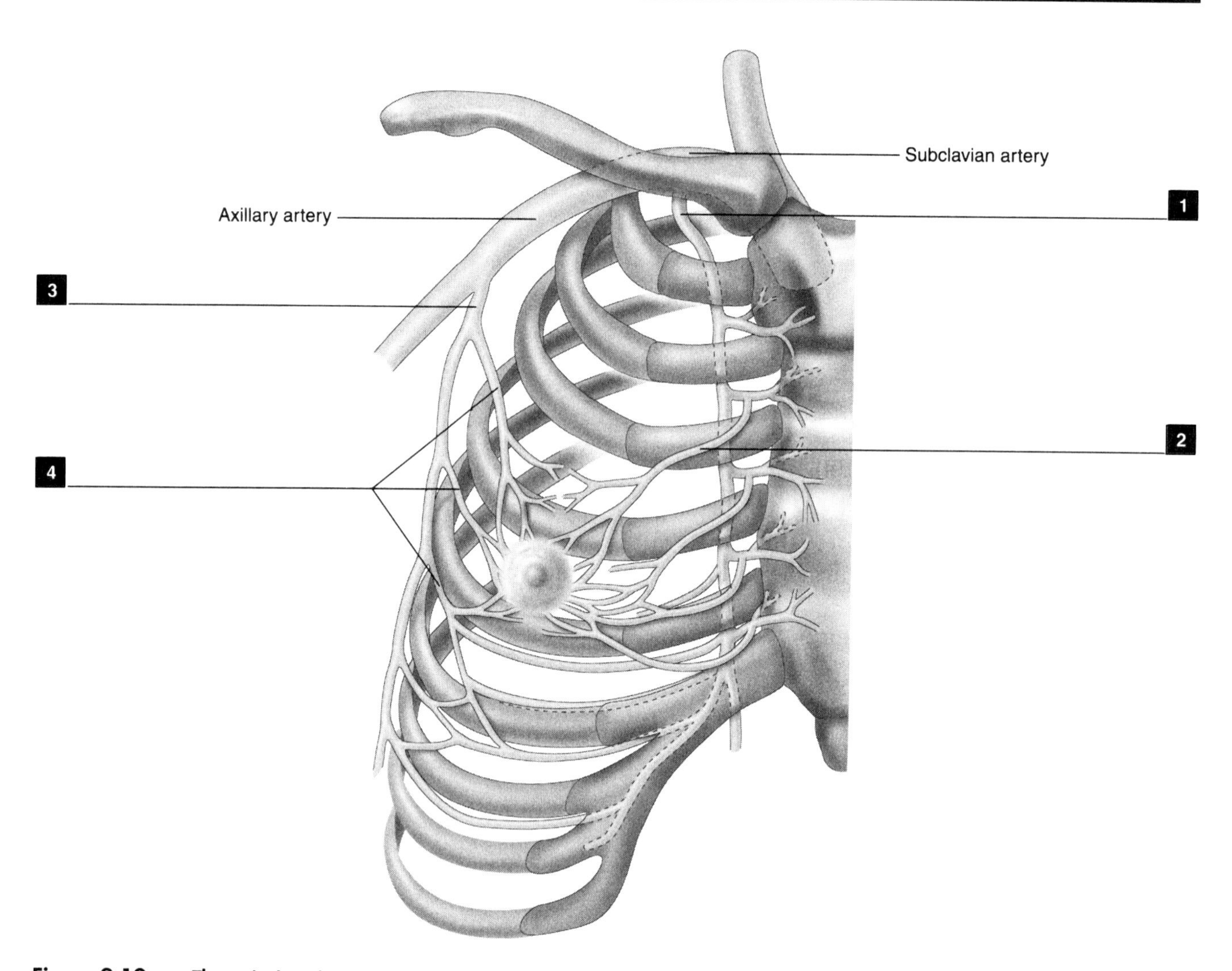

Figure 9.19 **Thoracic Arteries**

Veins

The descending aorta gives off intercostal arteries as it passes by each intercostal space, but there is no great vessel that runs along the back of the thoracic wall into which the intercostal veins can drain. Rather, the intercostal veins that drain blood from the posterior wall of the thorax form vessels that ultimately empty blood into the superior vena cava.

Each intercostal space is drained by an *Intercostal Vein*.

On *both sides of the body*, the first intercostal space is drained by a vessel known as the *Supreme Intercostal Vein*. It empties into the *Brachiocephalic Vein* on each side.

On the *right side of the body*, the second through twelfth intercostal veins empty into a vessel that ascends alongside the vertebral column. This vessel, which originates in the abdomen, is known as the **Azygos Vein 1.** The second through fourth right intercostal veins usually form a single vessel, the *Superior Intercostal Vein*, that joins the azygos vein. All other right intercostal veins empty directly into the azygos vein.

The azygos vein ascends behind the pericardium and the root of the lung. At the top of the root of the lung, it arches forward to enter the **Superior Vena Cava 2.**

On the *left side of the body*, the second through fourth left intercostal veins form a single vessel, the *Superior Intercostal Vein*, that empties into the *Left Brachiocephalic Vein*. The vein from the abdomen is joined by the lower four left intercostal veins as the **Hemiazygos Vein 3.** It runs to the right, across the front of the vertebral column at the level of T9, to join the azygos vein. The middle thoracic intercostal veins on the left join together to form the **Accessory Hemiazygos Vein 4.** It runs to the right, across the front of the vertebral column, at the level of T8, to join the azygos vein.

Identify the veins of the thoracic wall in figure 9.20.

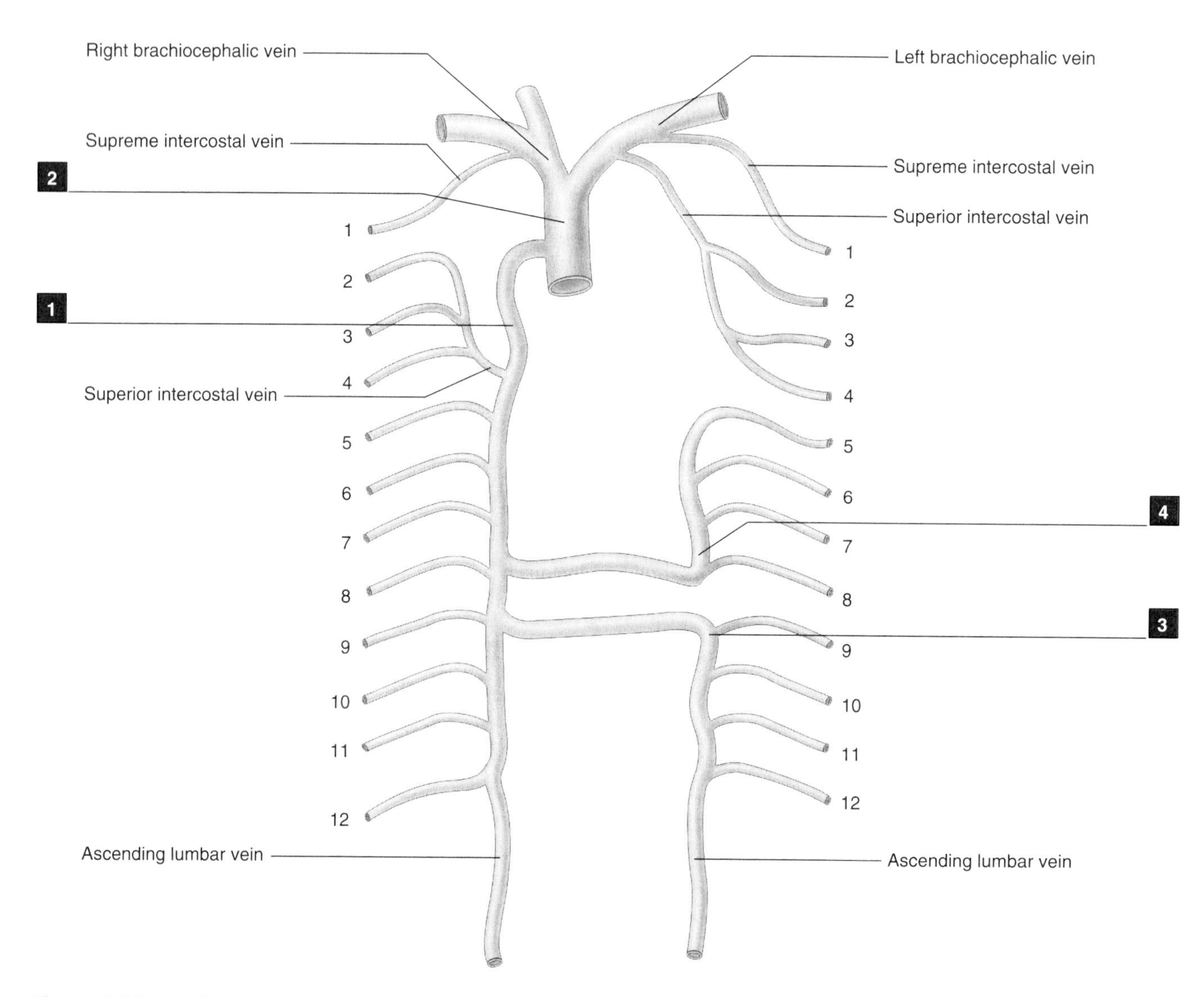

Figure 9.20 **Thoracic Veins**

9.10 NERVES OF THE THORAX

Thoracic skeletal muscles are supplied by spinal nerves. The heart, lungs, and lower part of the esophagus are supplied by autonomic fibers—parasympathetic from the vagus and sympathetic from cervical and thoracic ganglia of the sympathetic chain.

Spinal Nerves in the Thorax

We examined the structure of a typical spinal nerve in laboratory 2 (pp. 33–34). With the exception of T1, the **Thoracic Spinal Nerves** 1 are not involved in the formation of a plexus. Instead, each runs right into the neurovascular plane of an intercostal space. Recall that *T1*, contributes to the *Brachial Plexus* (laboratory 3, pp. 56–57).

The **Phrenic Nerve** 2 derives from spinal nerves in the neck, and makes its way into the thorax. We studied it in Laboratory 5 (p. 138). Recall that it derives from the ventral rami of C3–C5, and enters the thoracic cavity behind the subclavian vein. It descends through the thoracic cavity in contact with the superior vena cava (*right side*) or aorta (*left side*); and crosses the pericardium to reach the abdominal diaphragm, which it innervates.

Cranial Nerve in the Thorax

The **Vagus Nerve (CN X)** 3 is the only cranial nerve to enter the thoracic cavity. It provides parasympathetic fibers to the heart and lungs. We examine it in laboratory 7 (pp. 220–221).

In the thorax, the *right vagus* runs along the trachea, and onto the esophagus adjacent to the azygos vein. It descends into the abdomen on the esophagus as the *Posterior Vagal Trunk*. In the thorax, the *left vagus* sends off the **Recurrent Laryngeal Nerve** 4, which curves around the aortic arch to run back up into the neck. The *left vagus* runs onto the esophagus, becoming the *Anterior Vagal Trunk*. Both trunks contribute to the *Pulmonary Plexus* and *Cardiac Plexus*. Axons destined for the lungs and heart come from these plexuses.

Identify the aforementioned nerves in figure 9.21.

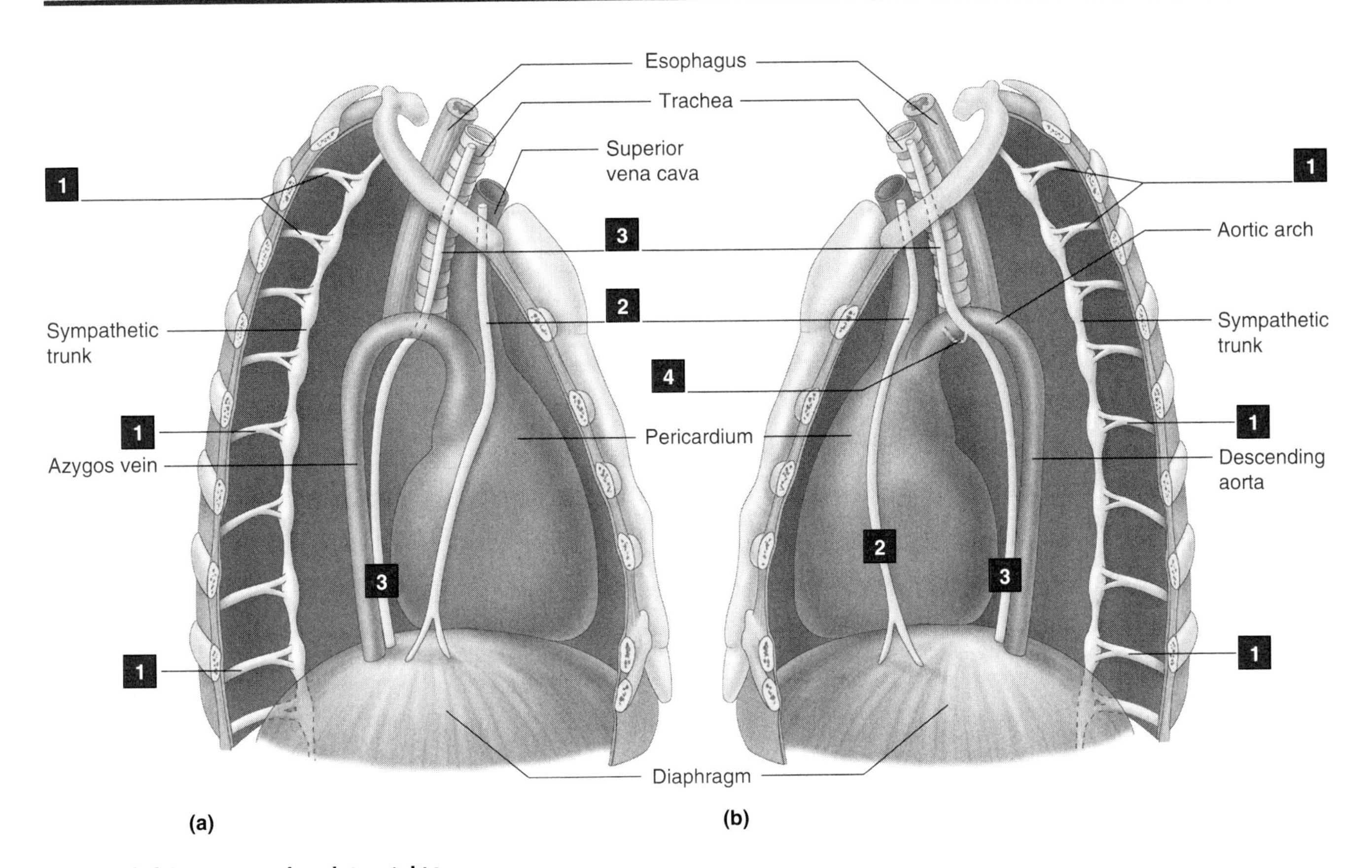

Figure 9.21 **Spinal and Cranial Nerves**
(a) right side of mediastinum; (b) left side of mediastinum.

Sympathetic Chain in the Thorax

We examined the sympathetic chain briefly in laboratory 5 (pp. 142–144) with reference to the ganglia that extend into the neck. Let us return to it now, and examine it in a bit more detail, because the chain is complete in the thorax.

Recall that the cell bodies of the *preganglionic* sympathetic neurons lie in the spinal cord. The *myelinated axons* of these neurons travel through the **Ventral Root** 1 and into the **Ventral Ramus** 2 of a *Spinal Nerve*.

These *myelinated preganglionic* axons then jump off the ventral ramus, and enter a **Paravertebral Sympathetic Ganglion** 3 through the **White Ramus Communicans** 4. Once the axon enters the ganglion it can do one of two things. It can either synapse in the paravertebral ganglion, or not synapse in the paravertebral ganglion.

- The preganglionic neuron synapses with a second neuron, known as a *postganglionic neuron*, whose cell body resides in the ganglion. Postganglionic axons are *unmyelinated*, and they exit the ganglion.

  The postganglionic may take one of two paths:

  a. Travel directly to a target organ (such as the heart). In the thorax, these axons join together to form an **"Organ" Nerve** 5 (e.g., a cardiac nerve) that synapses in the wall of the target organ.

  b. Reenter the ventral ramus of the spinal nerve through the **Gray Ramus Communicans** 6. Along the ventral ramus, the gray ramus communicans is *proximal* to the white ramus communicans. Once in the spinal nerve, these axons will travel to body wall structures (e.g., *Arrector pili* muscles) that respond to autonomic innervation.

- The axon does not synapse in the sympathetic ganglion that it enters, but instead ascends or descends to another paravertebral ganglion through the **Sympathetic Trunk** 7. Once it reaches another ganglion, it may synapse there, or it may simply pass through on its way to a ganglion that is even further away along the chain.

  Once the preganglionic axon reaches a ganglion in which it will synapse, it synapses on a postganglionic neuron, whose axon will either travel directly to a target organ or enter the ventral ramus of an adjacent spinal nerve for distribution to the body wall.

Identify the aforementioned structures in figure 9.22. Color sympathetic chain components shades of brown, and spinal nerve components shades of green.

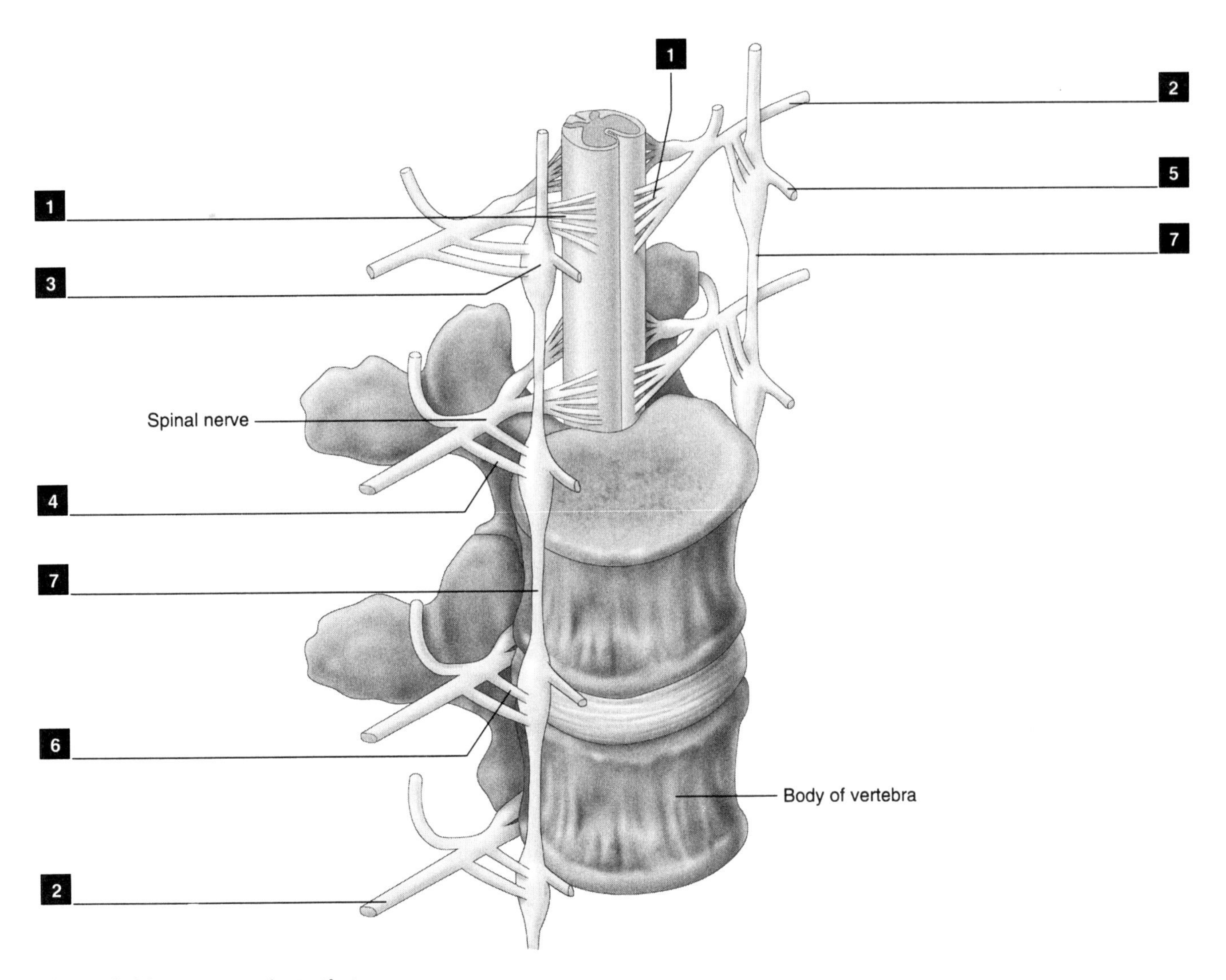

Figure 9.22 **Sympathetic Chain**

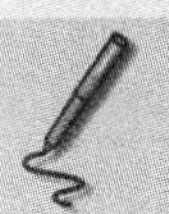

Some Sympathetic Possibilities

In figures 9.23, 9.24, 9.25, and 9.26, draw the possible courses of preganglionic and postganglionic fibers in relation to the thoracic sympathetic chain.
Use *red* for *preganglionic* (myelinated) fibers.
Use *blue* for *postganglionic* (unmyelinated) fibers.

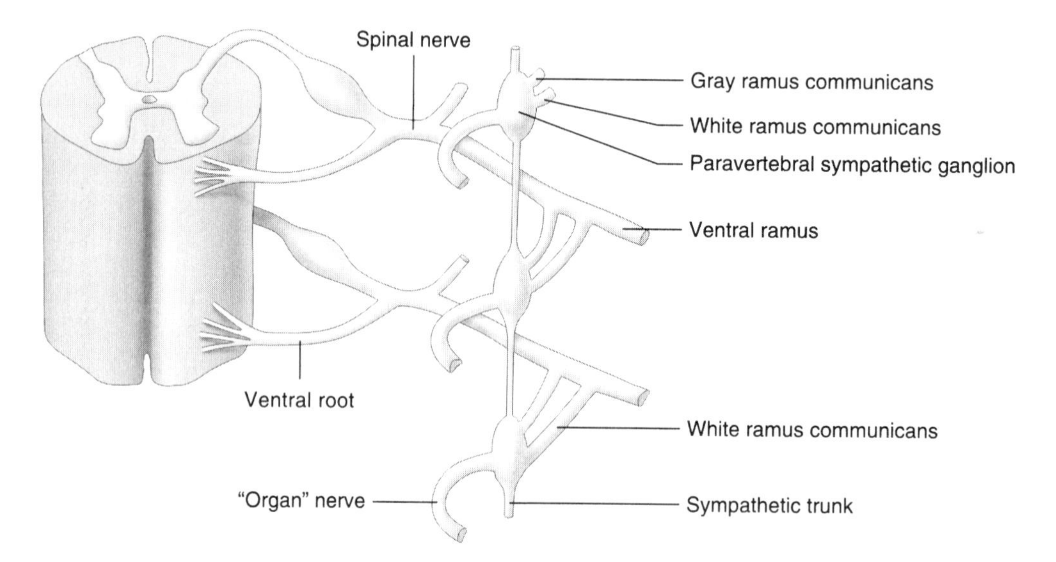

Figure 9.23 Sympathetic Chain

- Preganglionic fiber enters a paravertebral ganglion via the white ramus communicans and synapses there.
- Postganglionic fiber travels directly to target organ through an organ nerve.

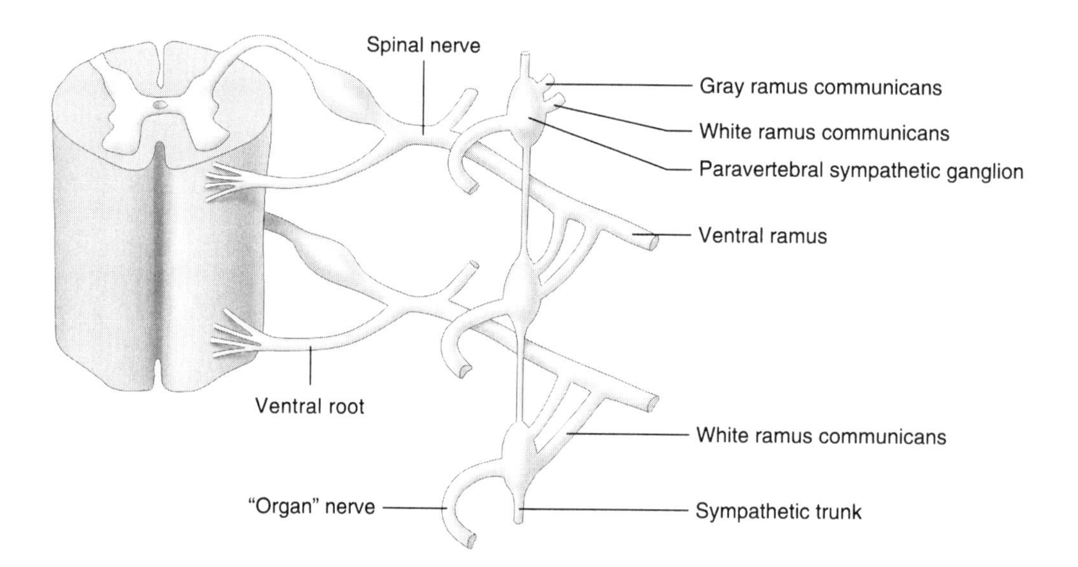

Figure 9.24 Sympathetic Chain

- Preganglionic fiber enters a paravertebral ganglion via the white ramus communicans and synapses there.
- Postganglionic fiber travels into the ventral ramus via the gray ramus communicans for distribution to body wall.

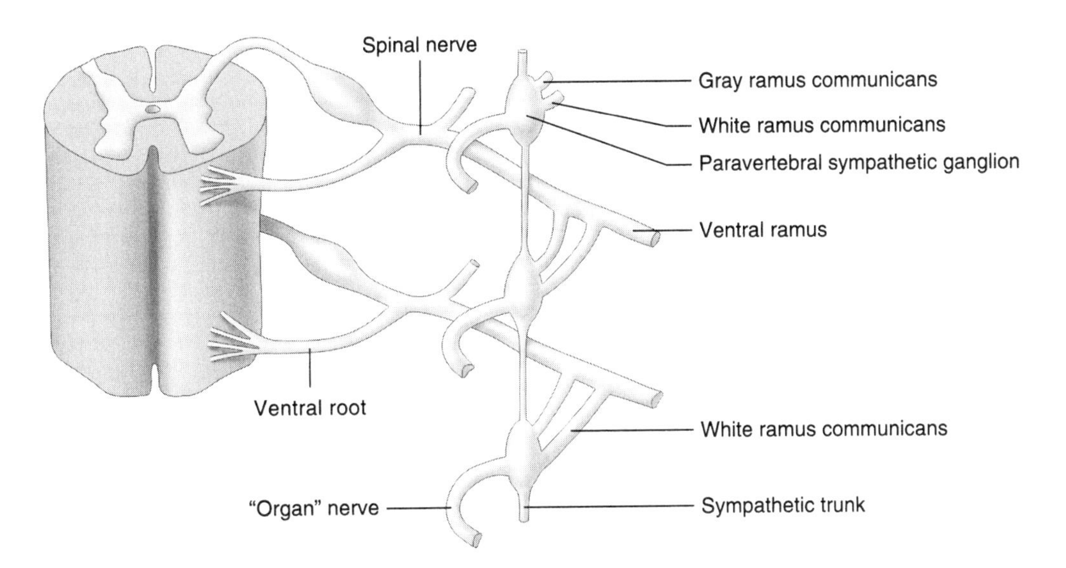

Figure 9.25 Sympathetic Chain

- Preganglionic fiber enters the paravertebral ganglion via the white ramus communicans and turns up or down through sympathetic trunk to synapse in a higher or lower paravertebral ganglion.
- Postganglionic fiber travels to a target organ through an organ nerve.

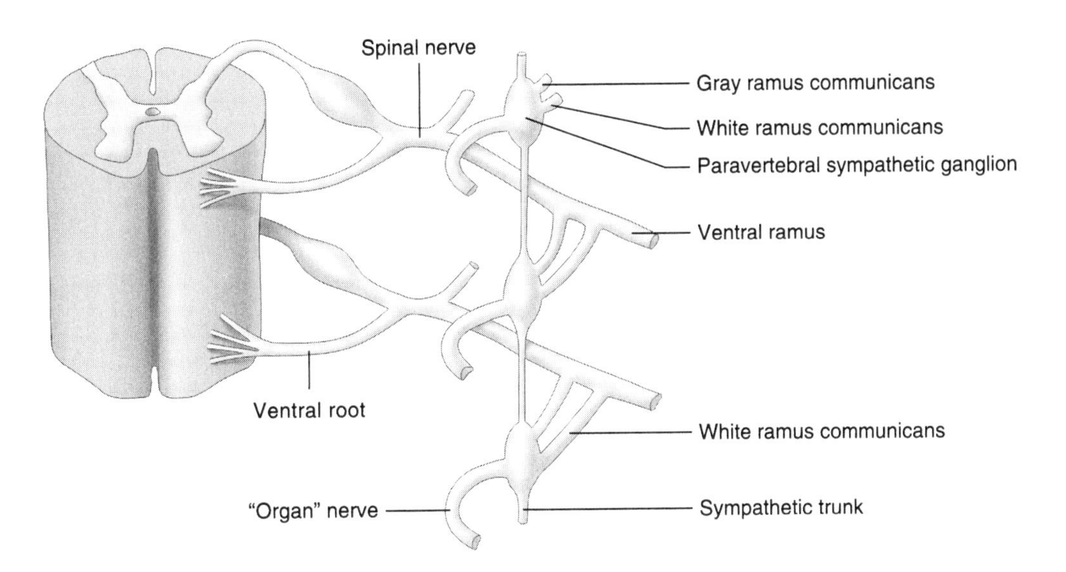

Figure 9.26 Sympathetic Chain

- Preganglionic fiber enters the paravertebral ganglion via the white ramus communicans and turns up or down through sympathetic trunk to synapse in higher or lower paravertebral ganglion.
- Postganglionic fiber travels into the ventral ramus via the gray ramus communicans for distribution to body wall.

LABORATORY

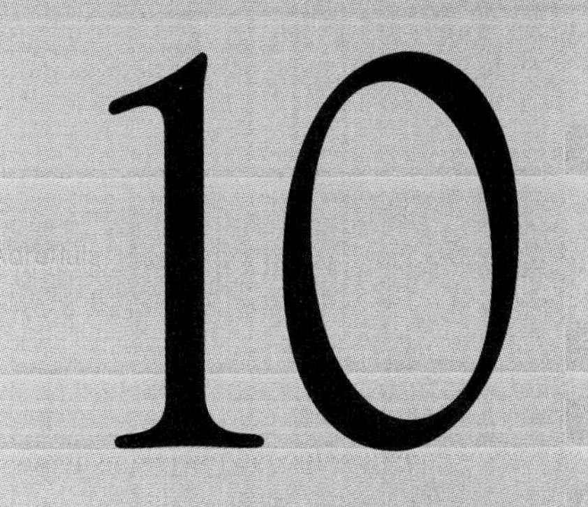

The Abdomen

10.6 LYMPHATIC ORGANS AND LYMPH DRAINAGE 320

Lymph Drainage 320
The Spleen 320

10.7 THE KIDNEYS AND ADRENAL GLANDS 322

Blood Vessels of the Kidneys and Adrenal Glands 322
- Arteries 322
- Veins 322

Structure of the Kidney 324
The Nephron 326
- Tubular Component 326
- Vascular Component 326

10.8 GONADAL BLOOD VESSELS 328

Gonadal Arteries 328
Gonadal Veins 328

10.9 NERVES IN THE ABDOMEN 329

Spinal Nerves in the Abdomen 329
Cranial Nerve in the Abdomen 329
Sympathetic Chain in the Abdomen 329

10.1 THE ABDOMINAL SKELETON

The abdomen is bounded superiorly by the abdominal diaphragm, and inferiorly by the upper margin of the bony pelvis. The abdomen consists of a large cavity that is walled primarily by muscle and serous membrane. The abdominal cavity is continuous inferiorly with the pelvic cavity, which is enveloped by the bony pelvis.

The structure of the rib cage and the tilt of the pelvis cause the *Thoracopelvic Gap* between them to be higher in front than behind.

The nature of the thoracopelvic gap might lead you to conclude that the abdominal skeleton is restricted to the five **Lumbar Vertebrae** 1. However, the dome of the abdominal diaphragm is very high—extending from the margins of the twelfth ribs up to the level of the fifth ribs. Thus, some abdominal organs are covered by the lower five or six **Ribs** 2 and their **Costal Cartilages** 3.

These osseous elements have been examined previously in laboratory 2 (p. 27) and laboratory 9 (pp. 251–252).

Identify the components of the abdominal skeleton in figure 10.1.

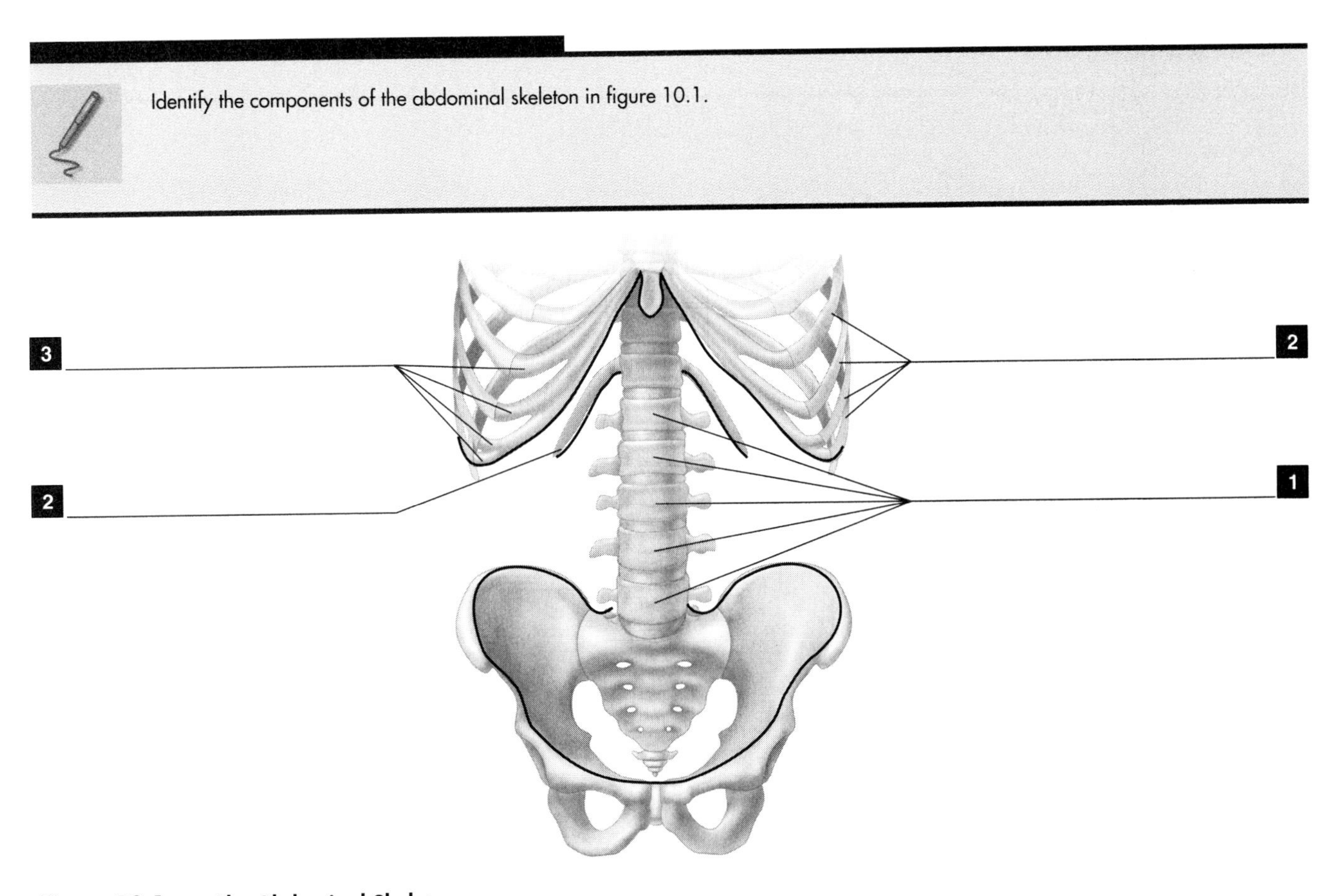

Figure 10.1 **The Abdominal Skeleton**
The thoracopelvic gap is bounded by the dark lines.

Relationship of Abdominal Viscera to the Rib Cage

Several abdominal organs are partially, largely, or even entirely covered by the rib cage.

Much of the **Liver** 1 is deep to the rib cage. It is covered anteriorly by the bottom of the sternum and some of the inferior costal cartilage to the left. On the right it is covered by the fifth to eleventh ribs and their costal cartilages. Posteriorly it is under the ninth to eleventh ribs.

The proximal third of the **Stomach** 2 is covered anteriorly on the left side by the fifth to eighth ribs.

Posteriorly on the left, the **Spleen** 3 is entirely under cover of the ninth to eleventh ribs.

Also posteriorly, portions of the **Kidneys** 4 are under the ribs. Both **Adrenal (Suprarenal) Glands** 5 are covered by the twelfth ribs.

Identify the aforementioned abdominal viscera in figure 10.2.

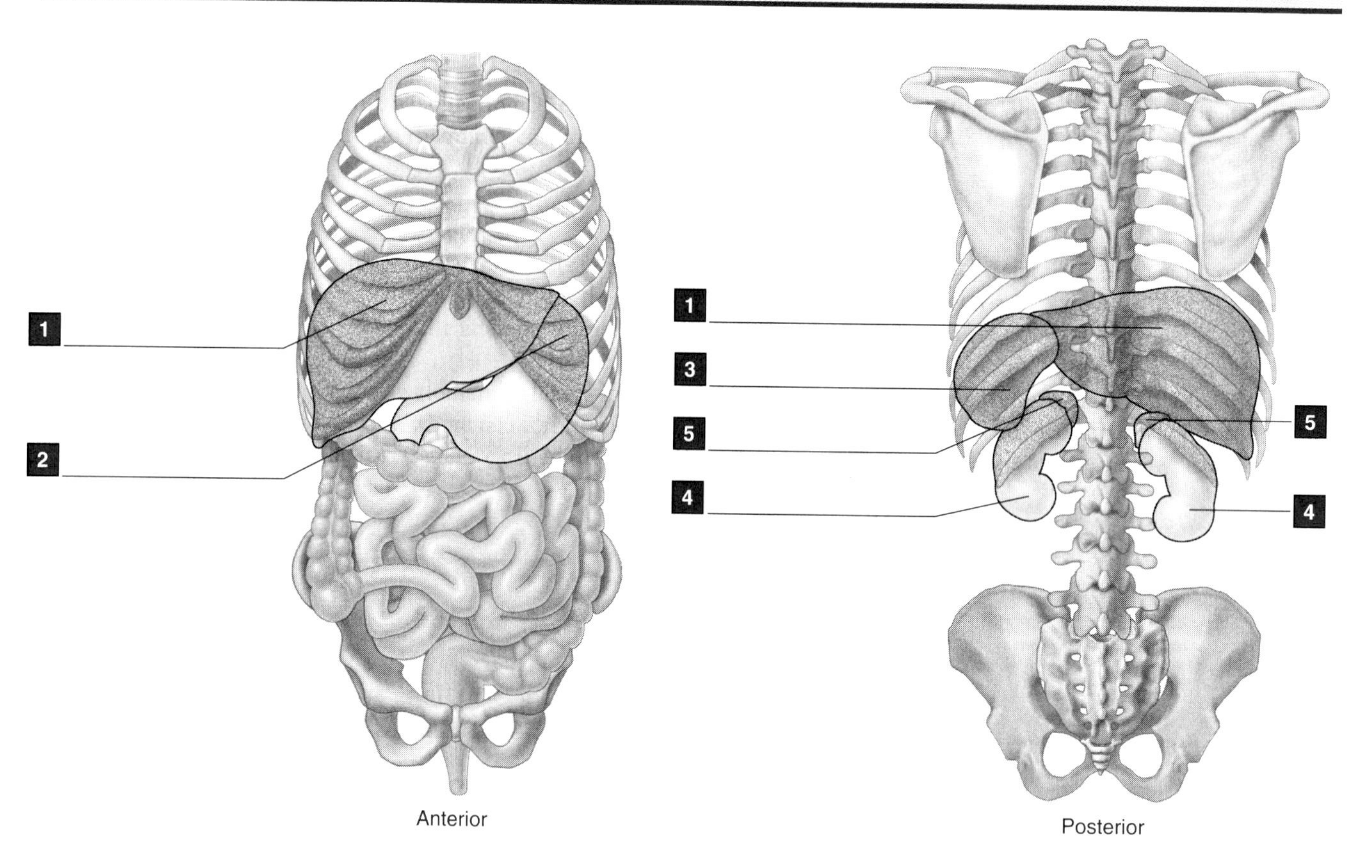

Figure 10.2 **Abdominal Organs within the Rib Cage**
The portions of the abdominal organs under cover of the rib cage are stippled. Note the difference in the height of the thoracopelvic gap anteriorly and posteriorly.

10.2 SKELETAL MUSCLES OF THE ABDOMEN

The abdominal cavity is separated from the thoracic cavity by the abdominal diaphragm. The external walls of the abdominal cavity are composed almost entirely of muscles that span the thoracopelvic gap.

Roof of the Abdominal Cavity

Abdominal Diaphragm forms the roof of the abdominal cavity. We examined this muscle in laboratory 9 (pp. 257–258).

Innervated by *Phrenic Nerve* (C3–C5)

Anterolateral Walls of the Abdominal Cavity

The anterolateral aspect of the abdominal cavity is covered by four muscles. In the front, a longitudinal muscle runs on either side of the midline from the sternum to the pubis. It is variably covered by aponeuroses of three sheetlike muscles that cover the sides of the abdomen from the ribs to the iliac crest.

Anterior Muscle

Rectus abdominis 1 is a straplike muscle that arises from the xiphoid and the fifth to seventh costal cartilages. It inserts on the superior edge of the pubis, and flexes the trunk. The muscle belly is interrupted by three (sometimes four) bands of connective tissue called *Tendinous Intersections*. The left and right muscles are separated by a longitudinal band of connective tissue called the *Linea Alba*. The tendinous intersections and linea alba create the "six-pack."

Innervated by *Spinal Nerves T7–T12*

The aponeuroses of the three sheetlike lateral wall muscles variably enclose *Rectus abdominis* in what is called the ***Rectus Sheath.*** Aponeuroses of the left and right *External oblique* cover *Rectus abdominis* anteriorly and fuse in the midline, forming the linea alba.

About 80% of people have a small muscle, *Pyramidalis*, that arises from the pubis and inserts into the linea alba.

Identify Rectus abdominis in figure 10.3.

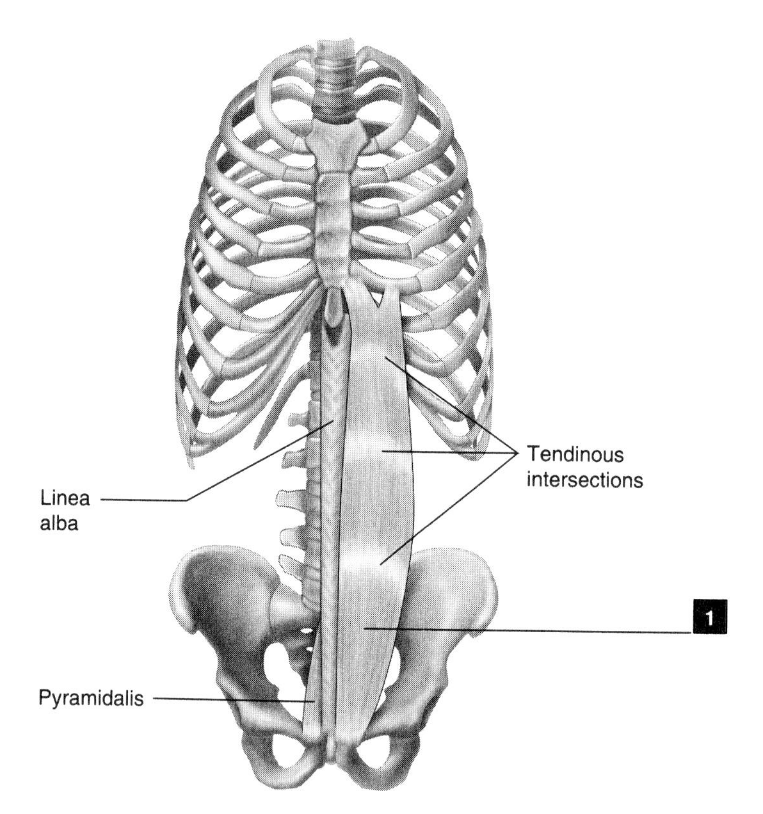

Figure 10.3 **Anterior Abdominal Muscle**

Anterolateral Muscles

The anterolateral wall of the abdomen is covered by three broad, thin sheets of muscle that are aponeurotic in front. They are trilaminar, like the intercostal muscles that ring the thoracic cavity, with which they are homologous. They flex the trunk laterally, rotate the trunk, and compress the abdominal organs. They are innervated by *Spinal Nerves T7–L1*.

External oblique 1 is the most superficial of the three. It arises from the lower eight ribs and the iliac crest; its fibers run inferiorly and medially, and become tendinous anteriorly to form the linea alba. The inferior free edge of the aponeurosis is thickened where it stretches between the *Anterior Superior Iliac Spine* and the *Pubic Tubercle*. This thickened band is known as the **Inguinal Ligament** 2. Just above the pubic tubercle, the aponeurosis has a hole known as the **Superficial Inguinal Ring** 3.

Internal oblique 4 is the intermediate of the three. It arises from the front of the iliac crest and the inguinal ligament; its fibers run superiorly and medially to insert onto ribs 10 to 12. Its inferior edge forms an arc above the inguinal ligament that is deep to and just a little lateral to the superficial inguinal ring. Above the umbilicus, its aponeurosis splits into superficial and deep layers around *Rectus abdominis*. Below the umbilicus, its aponeurosis runs superficial to *Rectus abdominis* to mesh with the linea alba.

Transversus abdominis 5 is the deepest of the three. It arises posteriorly from an aponeurosis that passes behind *Quadratus lumborum* to attach to the transverse processes of L1–L5. It is also attached to the lower six costal cartilages, the iliac crest, and the inguinal ligament. Its fibers run anteriorly to become aponeurotic and mesh with the linea alba. Above the umbilicus, its aponeurosis passes behind rectus abdominis. Below the umbilicus, its aponeurosis passes in front of rectus abdominis.

The internal aspect of the abdominal wall is lined with a layer of tissue known as the ***Transversalis Fascia.***

Identify the anterolateral abdominal muscles in figure 10.4.

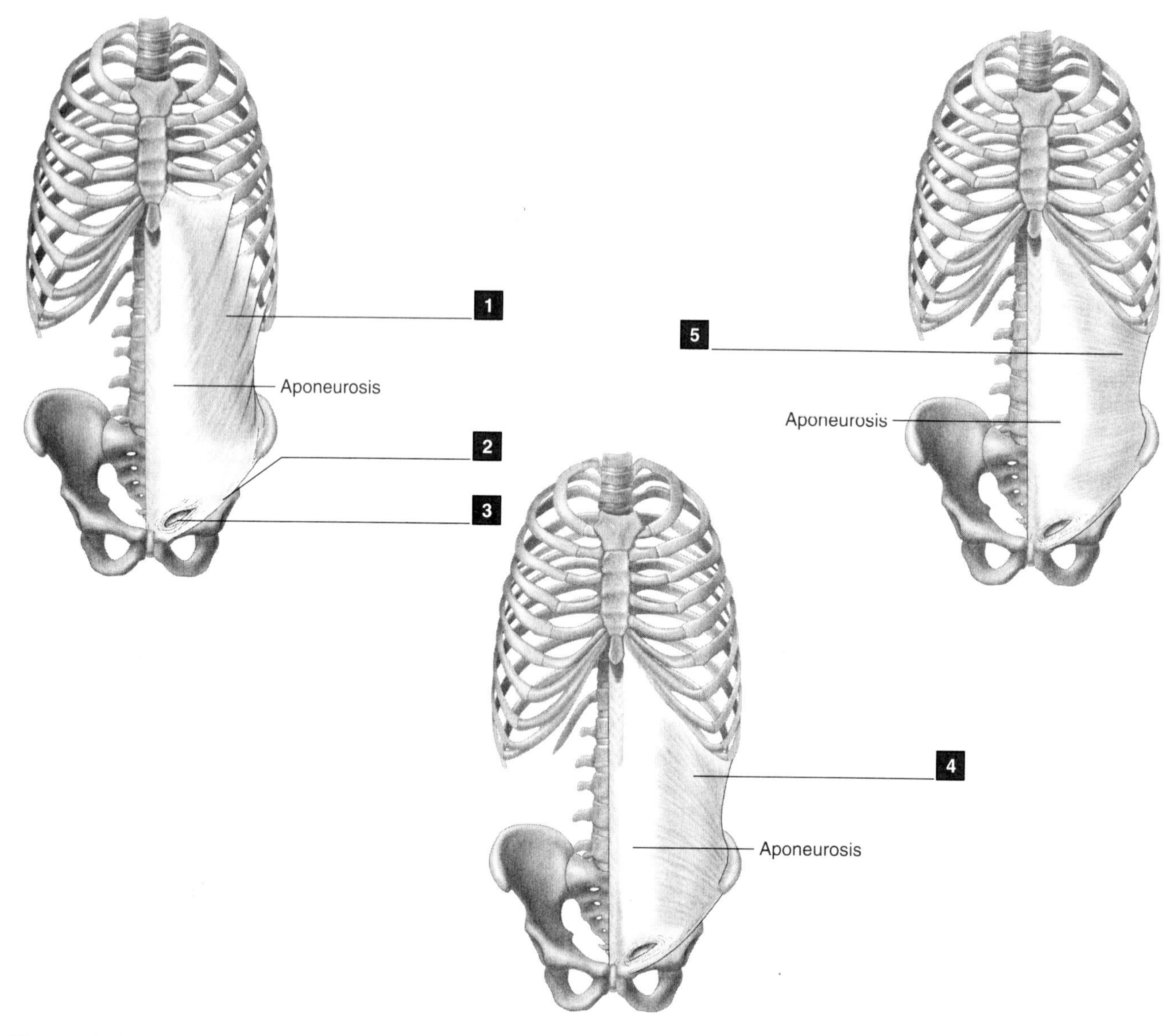

Figure 10.4 **Anterolateral Abdominal Muscles**

Posterior Wall of the Abdominal Cavity

The thoracopelvic gap is bridged posteriorly by three or four muscles: *Quadratus Lumborum*, *Psoas major*, *Psoas minor*, and the *Abdominal Diaphragm*. All except Psoas minor have been studied previously. Nevertheless, let us examine them here briefly as a group.

Abdominal Diaphragm 1 forms the superoposterior wall of the abdominal cavity. It was discussed in laboratory 9 (pp. 257–258), and again earlier in this chapter.

Quadratus lumborum 2 forms the bulk of the posterior abdominal wall. It was discussed in laboratory 2 (pp. 36–37). Recall that this back muscle runs along the transverse processes of L1–L4 from the iliolumbar ligament and iliac crest inferiorly to the twelfth rib.

Psoas major 3 forms the medial part of the posterior abdominal wall adjacent to the vertebral column. It was discussed in laboratory 4 (pp. 107–108). Recall that it arises from the bodies and intervertebral discs of T12 to L5. Its fibers merge with those of *Iliacus* to form a conjoint *Iliopsoas* tendon that inserts onto the lesser trochanter of the femur.

Psoas minor **4** is a thin muscular strip that is present in only about 50% to 60% of people. When present, it arises from the bodies and disc of T12–L1, and runs along the front of *Psoas major* to insert onto the iliopubic eminence of the pelvis.

Identify the posterior abdominal muscles in figure 10.5.

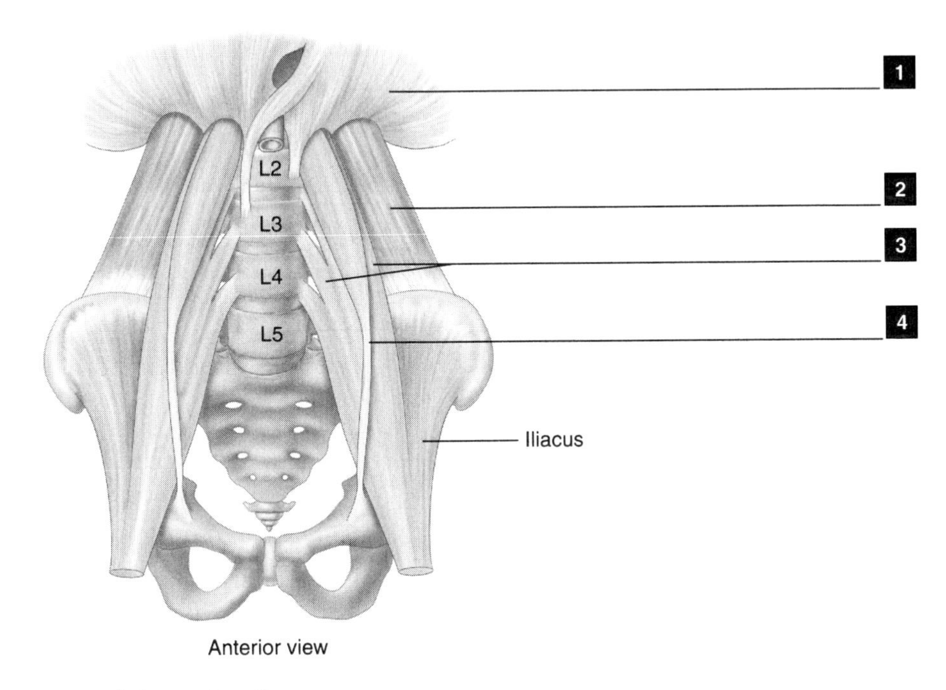

Figure 10.5 **Posterior Abdominal Walls**

10.3 ABDOMINAL CAVITY AND PERITONEUM

The cavity that is enclosed by the abdomen and pelvis is lined with a serous membrane called the ***Peritoneum.*** The membranous sac into which many, but not all of the abdominal organs protrude is called the *Peritoneal Cavity*.

The peritoneal sac has an outer parietal component, as well as a visceral component that lies against the walls of organs that have pushed into it. The visceral component is known as *Visceral Peritoneum*. The parietal component is known as *Parietal Peritoneum*.

Organs that are *Retroperitoneal* sit between the parietal peritoneum and the posterior wall of the abdomen. The kidneys and adrenal glands, pancreas, spleen, ascending colon, descending colon and rectum are retroperitoneal.

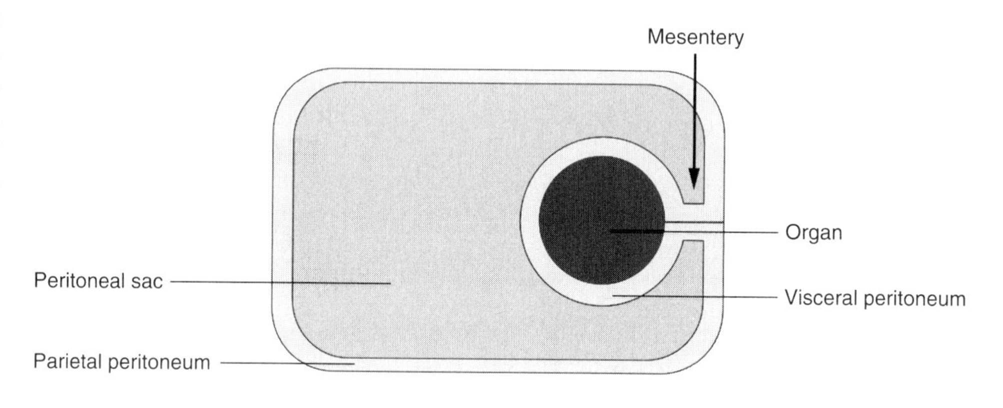

Organs that are not retroperitoneal push into the peritoneal sac from behind. Some push into the sac so deeply that the peritoneum stretches out behind them, becoming bilaminar between the organ and the posterior abdominal wall. This reflected peritoneum is known as a ***Mesentery.*** It permits the organ a degree of movement, and it is the conduit for blood vessels and nerves to the organ.

The *small intestine* is suspended by a mesentery that is simply called the **Mesentery** 1. The *transverse colon* is suspended by a mesentery that is called the **Transverse Mesocolon** 2.

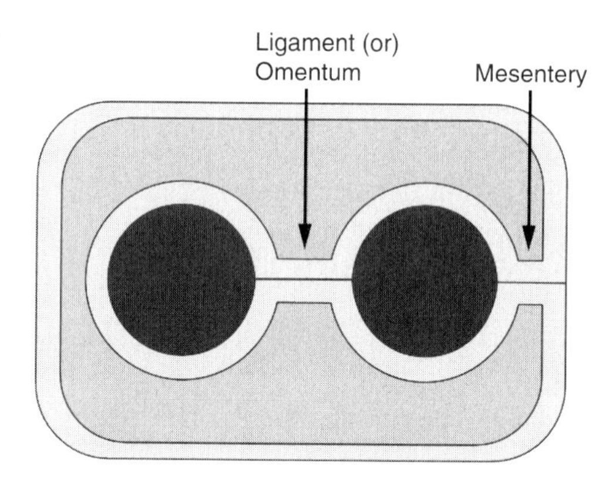

Some organs push into the peritoneal sac even more deeply. In such a case, one organ will push into the sac in front of another, so the first becomes suspended by a bilaminar sheet of peritoneum that is reflected off of the second. A suspensory fold such as this is referred to as either a ***Ligament*** or an ***Omentum.***

The *stomach* is suspended from the *liver* by the **Lesser Omentum** 3. The *stomach* and *transverse colon* are connected by a large flap of peritoneum known as the **Greater Omentum** 4. It hangs down like an apron to cover the front of most of the intestinal tract. It will adhere to the surface of the gut around a site of infection or perforation, which helps seal off and localize such conditions.

Label the peritoneal structures in figure 10.6.

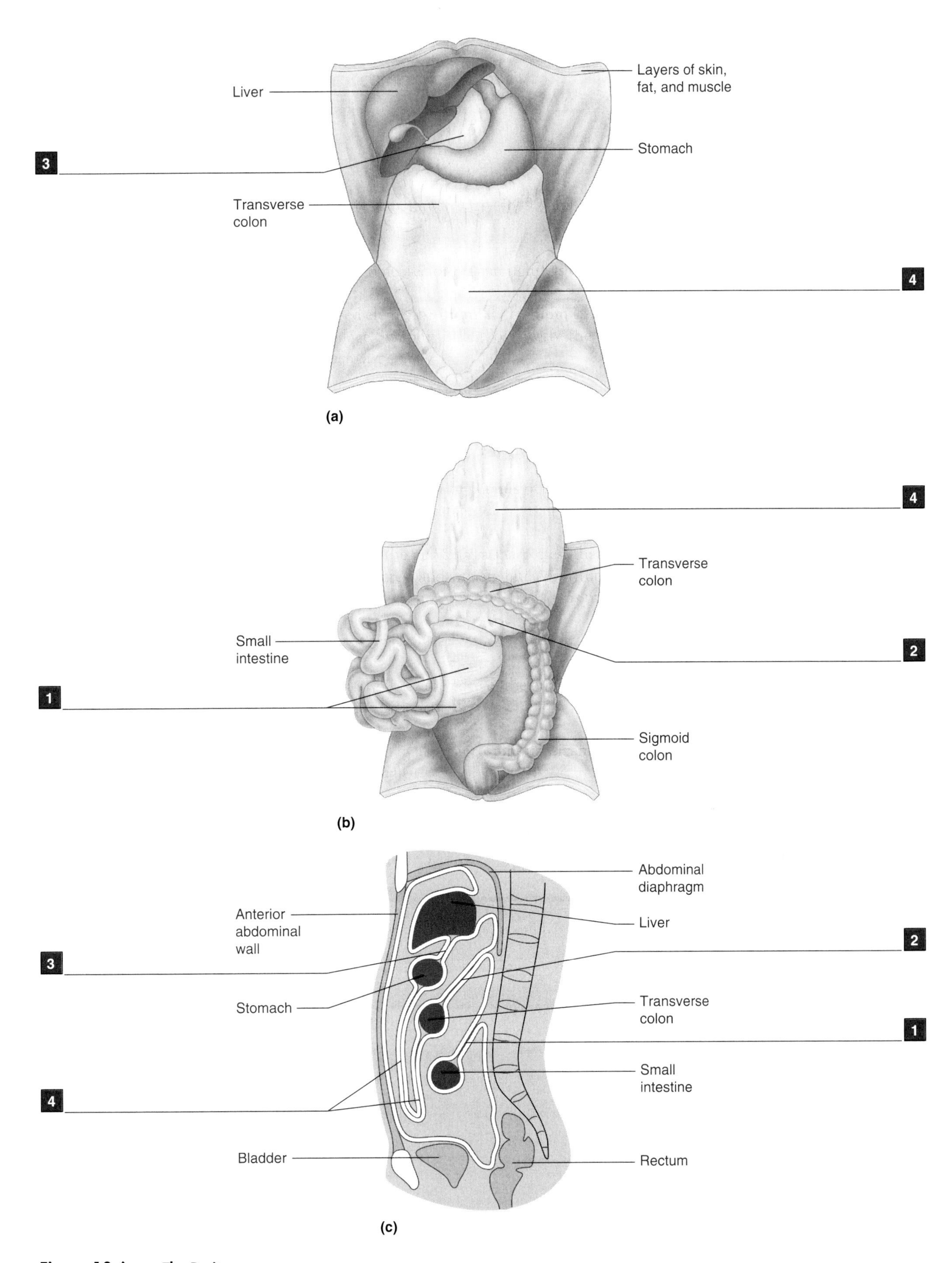

Figure 10.6 **The Peritoneum**
(a) the anterior abdominal wall has been opened and the liver and gallbladder have been retracted upward; (b) the stomach and greater omentum have been lifted up and the small intestine has been pulled forward; (c) schematic median sagittal section through the abdominopelvic cavity of a male.

10.4 DIGESTIVE CANAL AND ORGANS

The digestive system consists of an *Alimentary Canal* and three (or four) abdominal organs of digestion that are associated with it. The alimentary canal begins with the oral cavity, includes the oropharynx and pharynx in the head and neck, and the esophagus through the neck and thorax. The esophagus passes through the abdominal diaphragm to enter the *Gut*.

The gut is contained within the abdominopelvic cavity. About 1/2 inch below the diaphragm, the **Esophagus** 1 opens into the **Stomach** 2**.** The stomach leads to the **Small Intestine** 3**,** which is about 21 feet in length. The small intestine opens into the **Large Intestine** 4**,** which is just over 4 feet long.

The *small intestine* is divided into three parts:

Duodenum, Jejunum, Ileum

The *large intestine* is divided into eight parts:

Cecum, Appendix, Ascending Colon, Transverse Colon,
Descending Colon, Sigmoid Colon, Rectum, Anal Canal

The digestive organs associated with the alimentary canal are the **Liver** 5**, Gallbladder** 6 and **Pancreas** 7**.** The *Spleen* is primarily a lymphatic organ, but it produces a substance that is used by the liver to make bile, which is used in the digestive process.

The liver and gallbladder respectively produce and store bile, which is delivered into the small intestine by a network of tubes called the **Biliary Tree** 8**.** It opens into the duodenum. The liver also filters blood from the intestines, and metabolizes the carbohydrates, fats, and proteins that are carried to it by that blood.

The pancreas produces the hormone insulin, which plays a key role in carbohydrate metabolism, but most of its substance manufactures digestive enzymes. These are delivered to the small intestine by tubes that are linked to the biliary tree.

The gut is developmentally divided into three parts:

Foregut, Midgut, Hindgut

Each of these three parts receives its blood supply from a separate vessel that arises from the abdominal aorta. The three arteries are:

Celiac, Superior Mesenteric, Inferior Mesenteric

Blood from the gastrointestinal tract has to be filtered by the liver before it reaches the heart. This is accomplished by a *Portal Venous System* that carries blood from the gut to the liver.

Identify the components of the alimentary canal and its associated organs in figure 10.7.

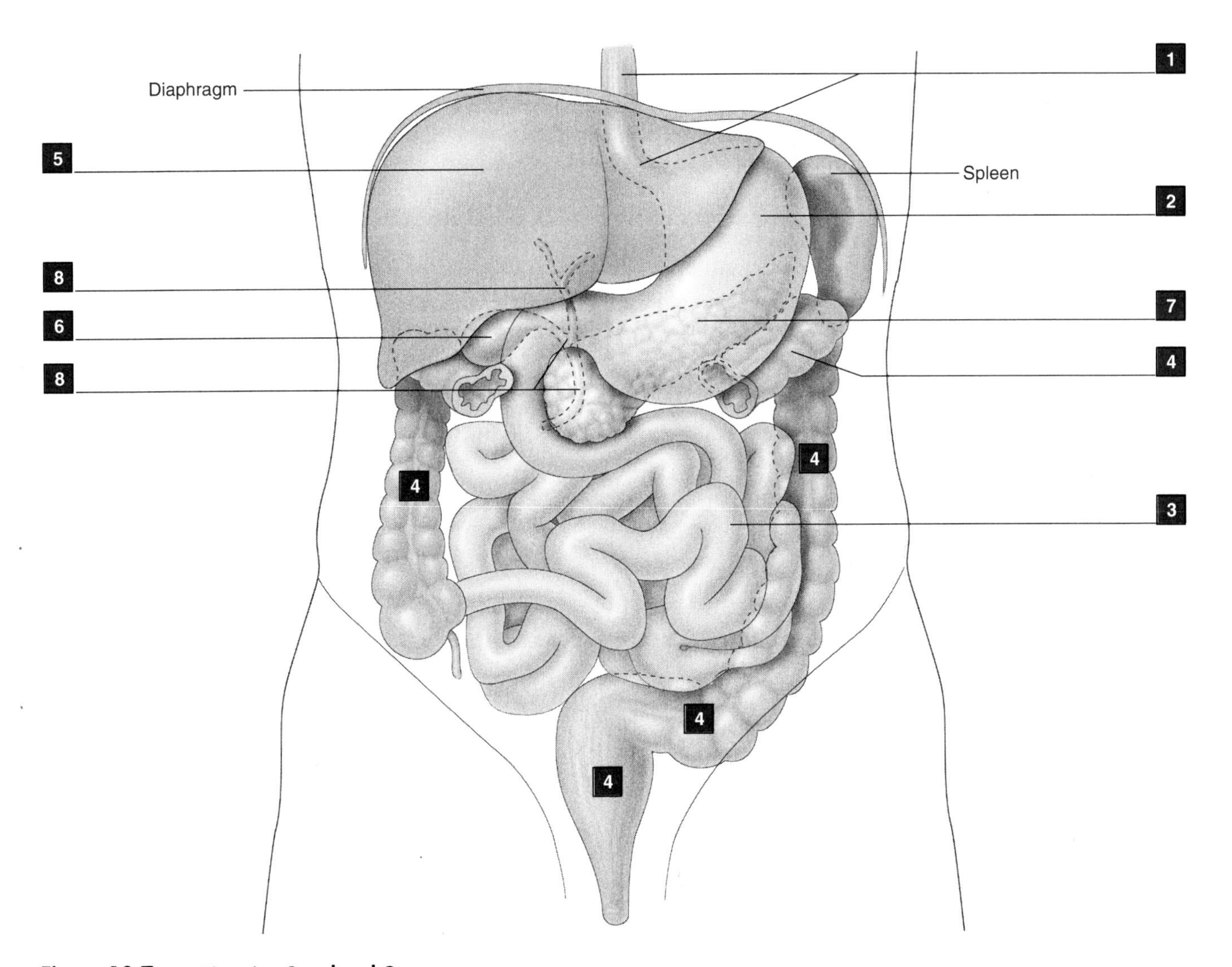

Figure 10.7 **Digestive Canal and Organs**
Note that part of the stomach and most of the gallbladder lie behind the liver. Most of the pancreas lies behind the stomach. The middle of the transverse colon has been cut away.

The Stomach

The stomach is a large, dilated portion of the alimentary canal. It stores food for processing, and mixes the food with acidic secretions from glands in its walls to form a semifluid known as *chyme*. It delivers chyme to the small intestine periodically.

The **Esophagus** 1 is continuous with the stomach just below the abdominal diaphragm. Its opening into the stomach is known as the **Cardiac Orifice** 2 because of its proximity to the heart. A circular band of smooth muscle that surrounds this orifice prevents the regurgitation of food and drink. This is referred to as the **Cardiac Sphincter** 3.

The stomach is shaped somewhat like the letter "J". Its longer left margin forms the **Greater Curvature** 4. Its shorter right margin forms the **Lesser Curvature** 5.

The stomach is divided into three parts. The **Fundus** 6 is the dome-shaped region that extends upward from the cardiac orifice. The **Body** 7 extends from the level of the cardiac orifice to the end of the lesser curvature. The **Pylorus** 8 extends to the point at which the stomach becomes continuous with the small intestine.

The stomach opens into the small intestine at the **Pyloric Orifice** 9. The walls of this orifice are thickly muscled, forming the **Pyloric Sphincter** 10. This regulates the discharge of chyme into the small intestine.

The most proximal part of the small intestine is the **Duodenum** 11.

Stomach Wall

The mucous membrane that lines the inside of the stomach wall is thick with numerous folds, known as **Rugae** 12. These are mainly longitudinally oriented, and they flatten out as the stomach becomes distended with food. On average, the stomach can expand to accommodate a bit more than a quart of ingested material, although there is considerable individual variation in this capacity.

The wall of the stomach comprises three layers of smooth muscle. They are named according to their fiber orientation. From superficial to deep they are: **Longitudinal** 13, **Circular** 14, and **Oblique** 15.

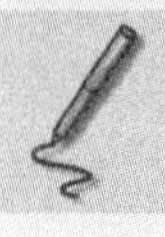

Identify the aforementioned structures in figure 10.8.

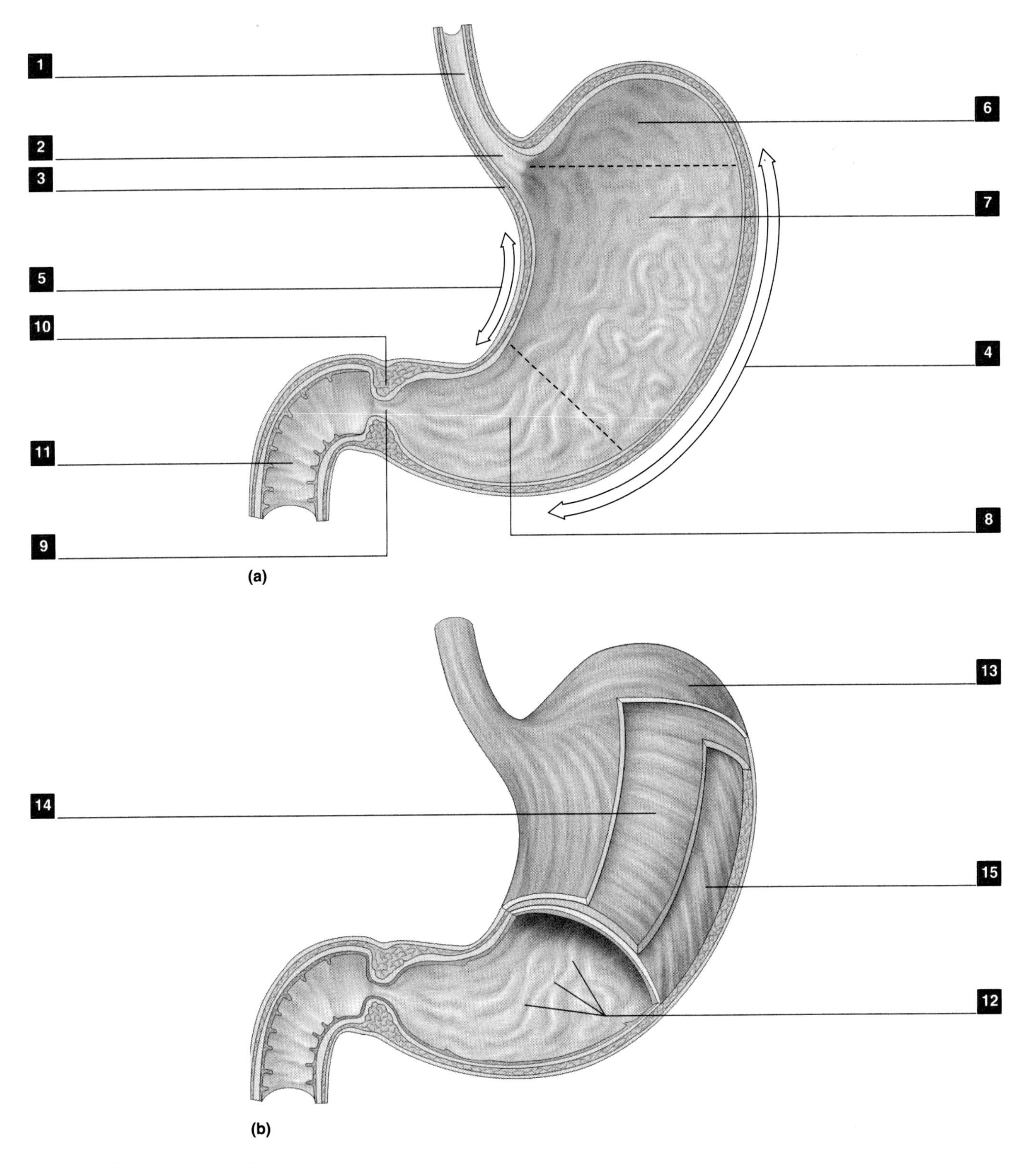

Figure 10.8. **The stomach**
(a) parts of the stomach; (b) structure of the stomach wall.

The Small Intestine

The small intestine is responsible for most of the digestion and absorption of food. Its wall is composed of two layers of smooth muscle. The fibers of these layers from superficial to deep are longitudinal (peristaltic contractions), and circular (segmenting contractions). The small intestine is divided into three anatomical regions from beginning to end:

Duodenum *Jejunum* *Ileum*

The duodenum is fixed in place. The long coils of the jejunum and ileum are freely mobile, being attached to the posterior abdominal wall by a fold of peritoneum, the *Mesentery* of the small intestine.

The **Duodenum** 1 is shaped like the letter "C", which curves around the head of the pancreas. It runs from the *pyloric orifice* for about 2 inches to the right, then makes a sharp turn downward and runs vertically for about 3 inches, when it turns to the left and upward for about 5 inches along the margin of the pancreas. It is continuous with the second part of the small intestine, the *jejunum*.

The vertical portion of the duodenum is pierced by the conjoined **Main Pancreatic Duct** 2 and **Bile Duct** 3**,** which enters through an orifice guarded by the *Sphincter of Oddi*. The **Accessory Pancreatic Duct** 4**,** if present, opens a little proximal to that sphincter.

The mucous membrane lining the inside of the duodenum has numerous transverse, permanent folds, the **Plicae circulares** 5**.** They serve to increase its surface area for absorption.

The **Jejunum** 6 lies primarily in the upper left part of the peritoneal cavity, whereas the **Ileum** 7 lies mostly in the lower right part of the abdomen. The ileum opens into the *cecum of the large intestine*.

The jejunum and ileum differ in several ways:

1. *Size of lumen and thickness of wall:* The jejunum has a wide lumen, and a thick wall. The ileum has a narrow lumen, and a thin wall.

2. *Arrangement of arterial branches:* The jejunum and ileum are both supplied by the **Superior Mesenteric Artery** 8**,** the branches of which form anastomosing arcades. However, branches to the jejunum form only one or two arcades, and their terminal portions are long. Branches to the ileum form several arcades, and their terminal portions are short.

3. *Lining of lumen:* The jejunum is like the duodenum. It has numerous, closely set plicae circulares. In the proximal part of the ileum, the plicae are set further apart, and they are absent from its distal part.

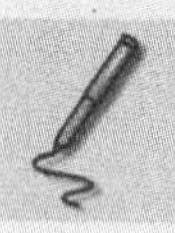

Identify the components of the small intestine in figure 10.9.

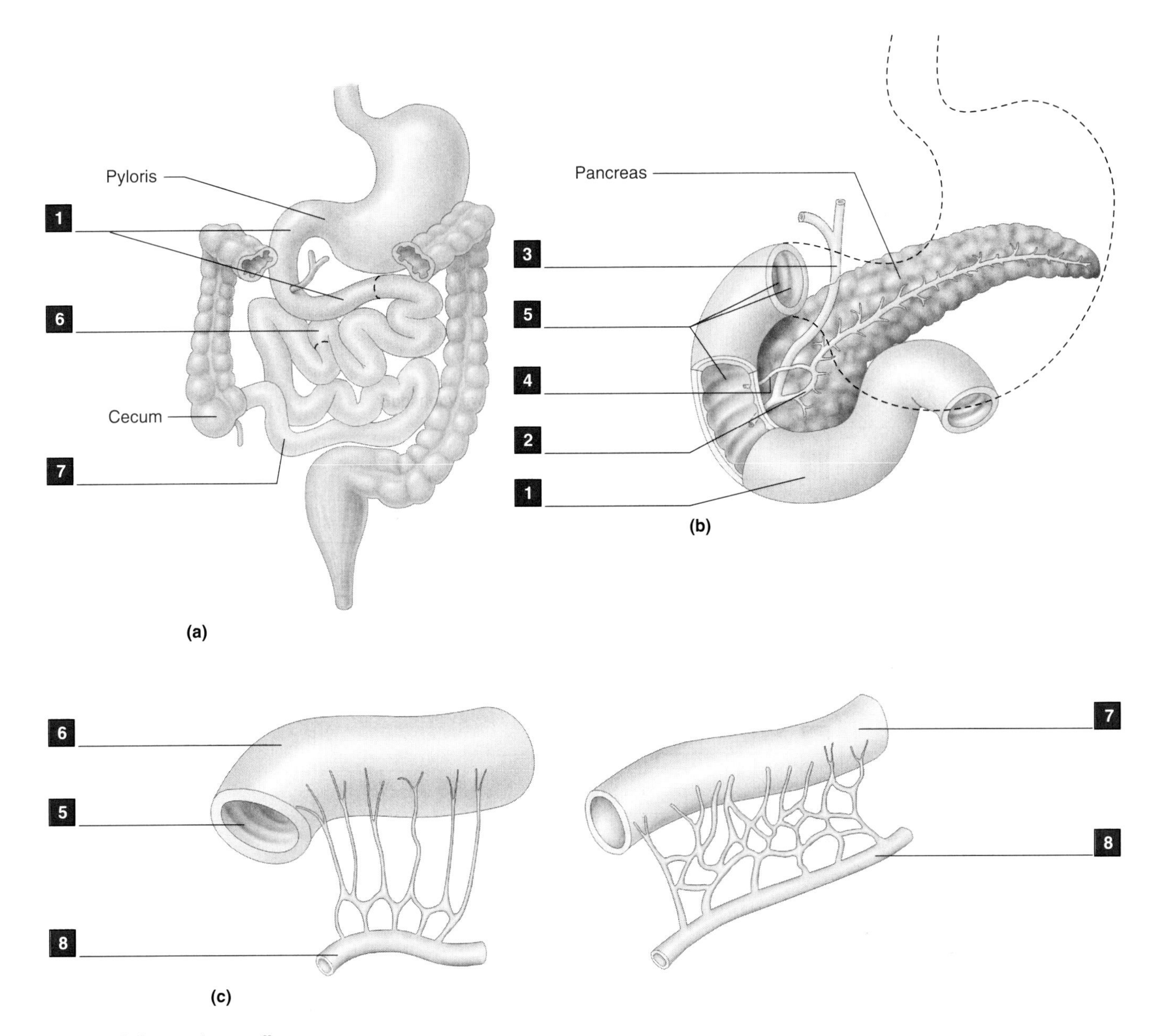

Figure 10.9 **The Small Intestine**
(a) the parts of the small intestine; (b) the duodenum (position of stomach indicated by dotted line); (c) note the differences between the jejunum and ileum.

The Large Intestine

The large intestine is responsible for the absorption of water, and the storage of undigested material until it can be expelled as feces. At about 5 feet, it is shorter than the small intestine, but it has a larger diameter. The large intestine forms a frame around the small intestine.

The large intestine has eight anatomical regions:

Cecum, Vermiform Appendix, Ascending Colon, Transverse Colon
Descending Colon, Sigmoid Colon, Rectum, Anal Canal

The **Cecum** 1 forms a pouch below its junction with the **ileum** at the *Ileocecal Orifice*. This orifice is guarded by the *Ileocecal Valve*. Hanging from the end of the cecum is the 2 to 6 inch long **Vermiform Appendix** 2**,** which opens into the cecum a little below the ileocecal orifice. The base of the appendix lies directly behind a point on the anterior abdominal wall (known as *McBurney's Point*) that is one-third of the way up a line drawn between the right anterior superior iliac spine and the umbilicus.

The **Ascending Colon** 3 extends up the right side of the abdomen from the cecum to the liver, where it turns to the left at the **Right Colic (= Hepatic) Flexure** 4**.**

At this flexure, the ascending colon becomes the **Transverse Colon** 5**.** This hangs down a bit as it crosses the abdomen to a point just below the spleen. Here it turns downward at the **Left Colic (= Splenic) Flexure** 6**.**

At this flexure, the transverse colon becomes the **Descending Colon** 7**.** It runs downward along the left side of the abdomen to become the **Sigmoid Colon** 8 at the brim of the pelvis. The sigmoid colon leads to the rectum.

The **Rectum** 9 begins at the level of the third sacral vertebra (S3), and runs downward along the sacrum to the level of the tip of the coccyx. Here it pierces the *Pelvic Diaphragm* to become the **Anal Canal** 10**.**

The *Longitudinal* layer of muscle that completely surrounds the outside of the small intestine is restricted to three separate bands in the large intestine. These bands are called **Taeniae coli** 11**.** They cause regular bulges and constrictions along most of the large intestine. These bulges are called **Haustra** 12**,** and they can be visualized by x-rays when the colon is coated internally with radiopaque material such as barium.

The taeniae coli and haustra extend the entire length of the colon up to the rectum. The rectum and anal canal lack these.

The circular muscle fibers of the anal canal form an *Anal Sphincter,* which is aided by a ring of skeletal muscle around it. The anal canal opens at the *Anus*.

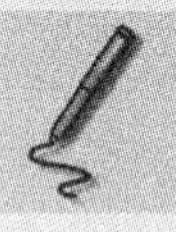

Identify the features of the large intestine in figure 10.10.

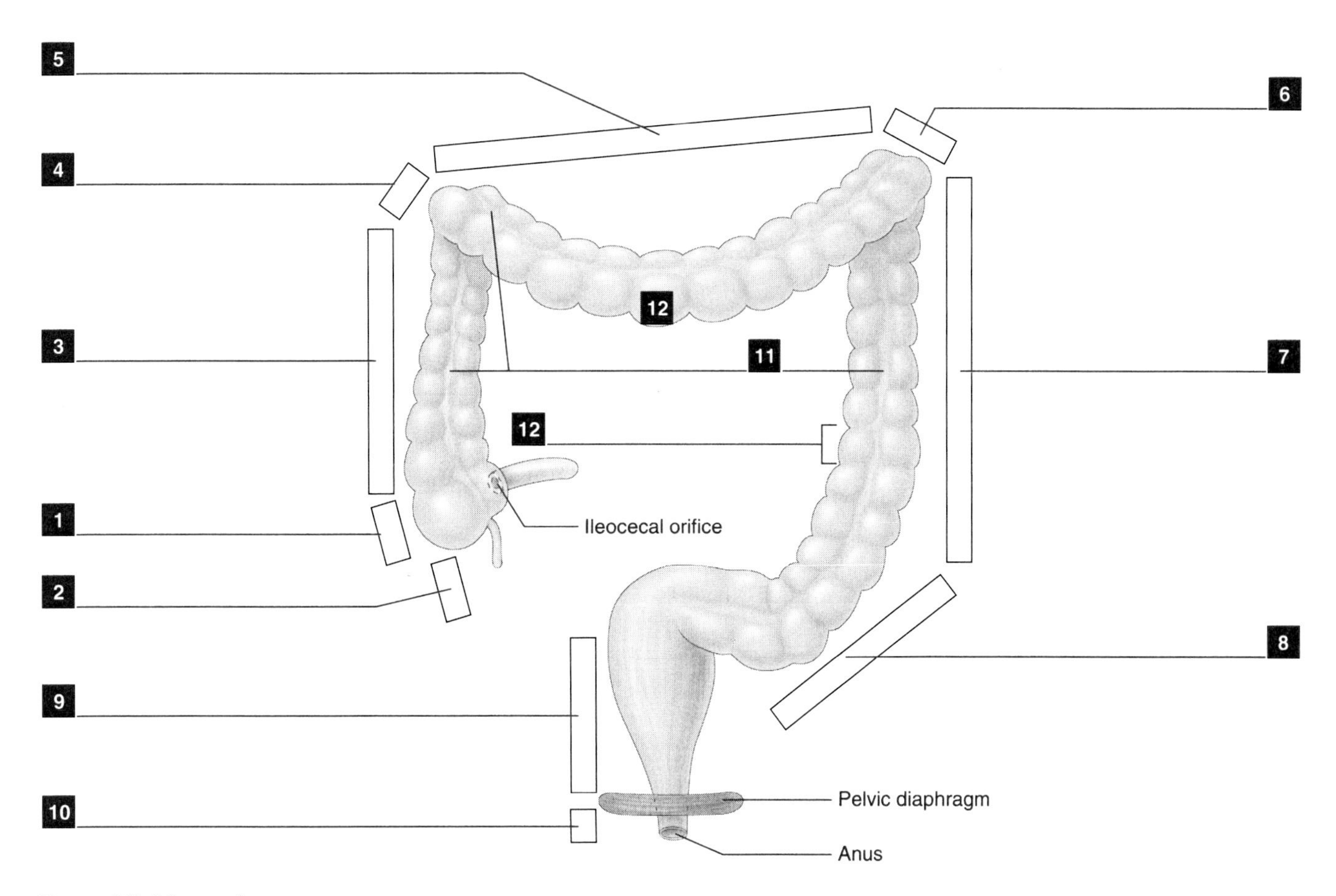

Figure 10.10 **The Large Intestine**
Anterior view.

The Liver

The liver has numerous functions. In relation to the digestive system, it filters blood from the gastrointestinal tract, metabolizes carbohydrates, fats, and proteins carried by that blood, and produces bile.

The liver has a roughly pyramidal shape, with its base on the right side. Much of it lies under cover of the rib cage, and its convex upper (*diaphragmatic*) surface is molded to the inferior surface of the abdominal diaphragm. Its inferior (*visceral*) surface is slightly concave. These surfaces are separated in front by the sharp *inferior border*.

The **Gallbladder** 1 peeks out from the visceral surface under the inferior border at a point just behind the tip of the ninth right costal cartilage.

The liver is commonly divided into *Four Lobes*. There is a large **Right Lobe** 2, and a smaller **Left Lobe** 3. The so-called **Quadrate Lobe** 4 and **Caudate Lobe** 5 are centrally placed on the visceral surface, and they are identified anatomically by their surface relationships. Functionally, they are part of the left side of the liver.

Most of the liver is covered with peritoneum, except for a *Bare Area* on the diaphragmatic surface around the inferior vena cava. The left and right lobes are separated by a mesentery that attaches the liver to the middle of the anterior abdominal wall. This is the **Falciform Ligament** 6, and its thickened inferior, free edge is the **Ligamentum Teres** 7. The ligamentum teres extends from the liver to the umbilicus; it is the remnant of the fetal umbilical vein. The peritoneum reflects off the liver's visceral surface between the quadrate and caudate lobes as the *Lesser Omentum*. As noted previously, this extends to the stomach.

The **Porta Hepatis** 8 of the liver is the place through which the blood vessels and ducts enter and exit the organ. These are:

Hepatic Artery 9, which supplies the liver with oxygenated blood

Portal Vein 10, which drains blood from the entire gastrointestinal tract into the liver

Common Hepatic Duct 11, which drains bile from the liver

Deoxygenated, filtered blood is drained from the liver into the **Inferior Vena Cava** 12 by two very short *Hepatic Veins* that emerge from its posterior surface.

Identify the features of the liver in figure 10.11.

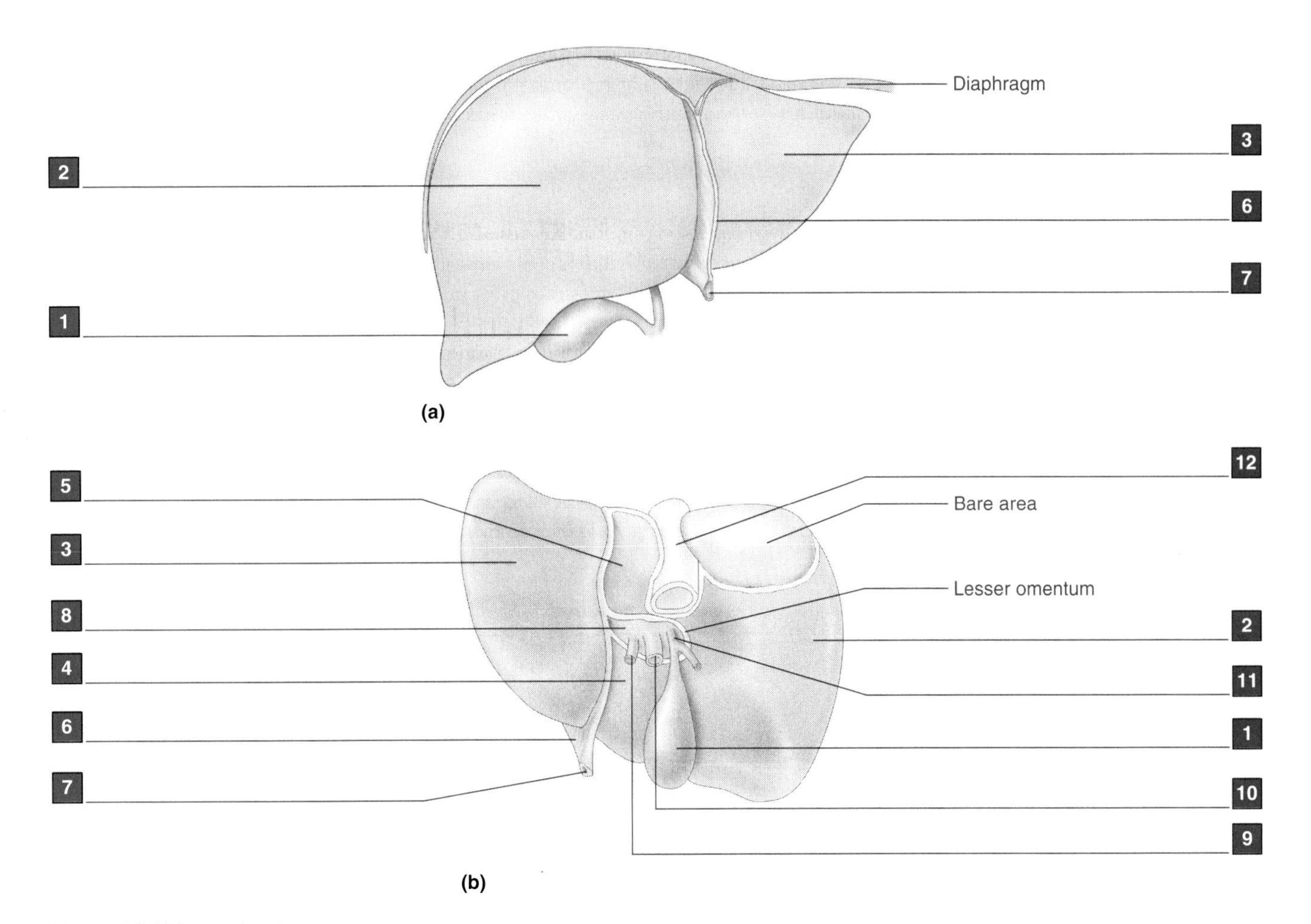

Figure 10.11 **The Liver**
(a) anterior (diaphragmatic) aspect; (b) inferior (visceral) aspect.

The Gallbladder and Biliary Tree

The **Gallbladder** 1 is a pear-shaped sac that lies on the visceral surface of the liver. Recall that part of it peeks out anteriorly from under the right inferior border of the liver.

The gallbladder has three parts. The *Fundus* is the rounded end that projects below the inferior margin of the liver. The *Body* lies against the visceral surface of the liver. The *Neck* extends from the body to a tube known as the cystic duct.

Tubes that carry bile connect the liver to the gallbladder, and connect the liver and gallbladder to the duodenum. These tubes comprise the ***Biliary Tree.*** It is arranged as follows:

The **Left Hepatic Duct** 2 and **Right Hepatic Duct** 3 drain bile from the left and right lobes of the liver.

The left and right hepatic ducts unite to form the **Common Hepatic Duct** 4. It exits the liver through the porta hepatis.

The common hepatic duct is joined by the **Cystic Duct** 5 from the gallbladder. The union of the common hepatic and cystic ducts forms the **Bile Duct** 6.

The bile duct runs inferiorly and joins the **Main Pancreatic Duct** 7 from the pancreas just before opening into the duodenum of the small intestine. These two ducts enter the duodenum through a common opening, the *Hepatopancreatic Orifice*, that is guarded by the *Sphincter of Oddi*.

The Pancreas

The **Pancreas** 8 traverses the posterior abdominal wall behind the stomach at the level of L2–L3. It is about 6 inches long, and extends from the vertical part of the duodenum to the spleen. It is retroperitoneal.

The pancreas has four parts. The *Head* fits into the C-shaped bend of the duodenum. The *Neck* connects the head to the *Body*. The body has a *Tail* that contacts the spleen.

Digestive enzymes are drained from the pancreas by the main pancreatic duct, which runs the length of the organ. It joins with the bile duct to open into the duodenum at the hepatopancreatic orifice. The main pancreatic duct usually has a branch in the head of the pancreas known as the **Accessory Pancreatic Duct** 9. It opens into the duodenum a little proximal to the hepatopancreatic orifice.

Identify the gallbladder, biliary tree, and pancreas in figure 10.12.

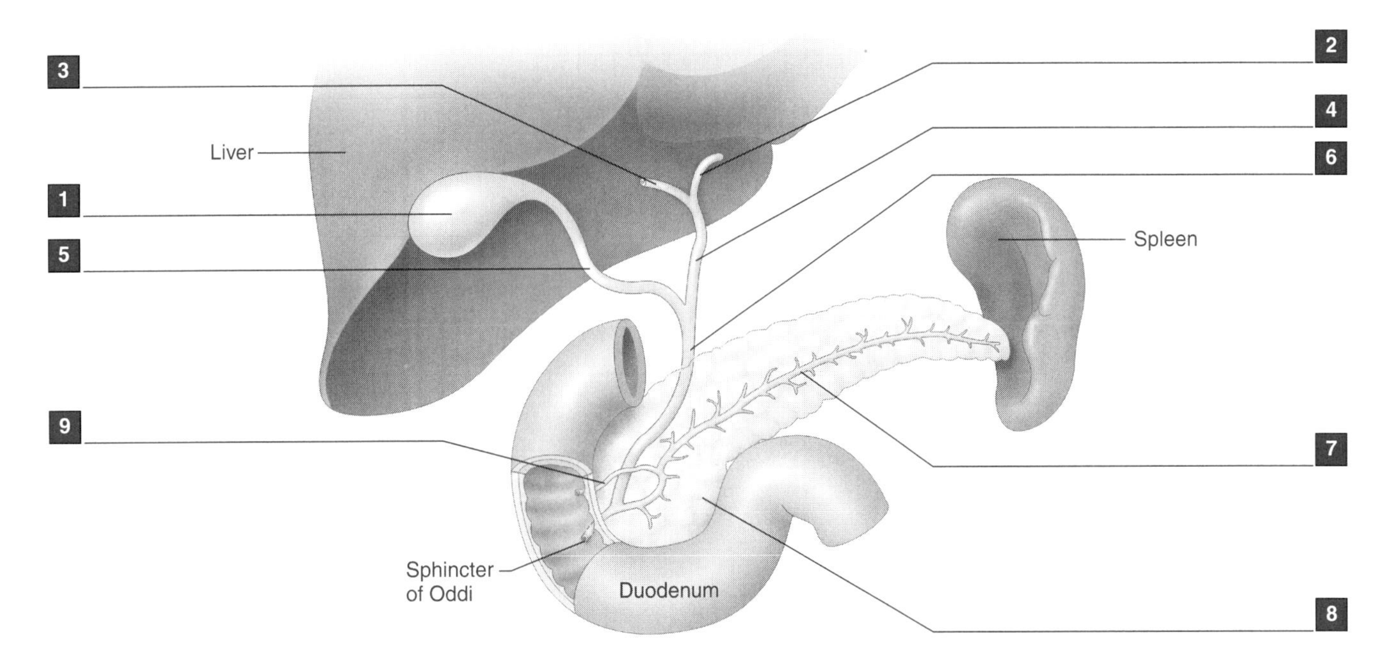

Figure 10.12 **Gallbladder, Biliary Tree, and Pancreas**

10.5 BLOOD VESSELS OF THE GUT

The alimentary canal and associated digestive organs are supplied by three arteries that arise from the front of the abdominal aorta. Venous blood from the gut contains products absorbed from digested food, so it must be processed by the liver before it can be added to the general circulatory system. In order to accomplish this, the blood is taken to the liver by the *Portal Venous System*, which bypasses the inferior vena cava. After filtering and processing it, the liver then returns the blood to the inferior vena cava for general circulation.

Arterial Supply

The blood supply to the gut is related to its developmental division into three parts. These three parts are the foregut, midgut, and hindgut.

The *Foregut* gives rise to the esophagus, stomach, and duodenum. It also contains the liver, gallbladder, pancreas, and spleen.

The *Midgut* gives rise to the rest of the small intestine, the cecum, and vermiform appendix, as well as the ascending colon, and most of the transverse colon.

The *Hindgut* gives rise to the terminal portion of the transverse colon, together with the descending colon, sigmoid colon, rectum, and anal canal.

Each of these three parts gets its blood from an unpaired artery that arises from the front of the abdominal aorta.

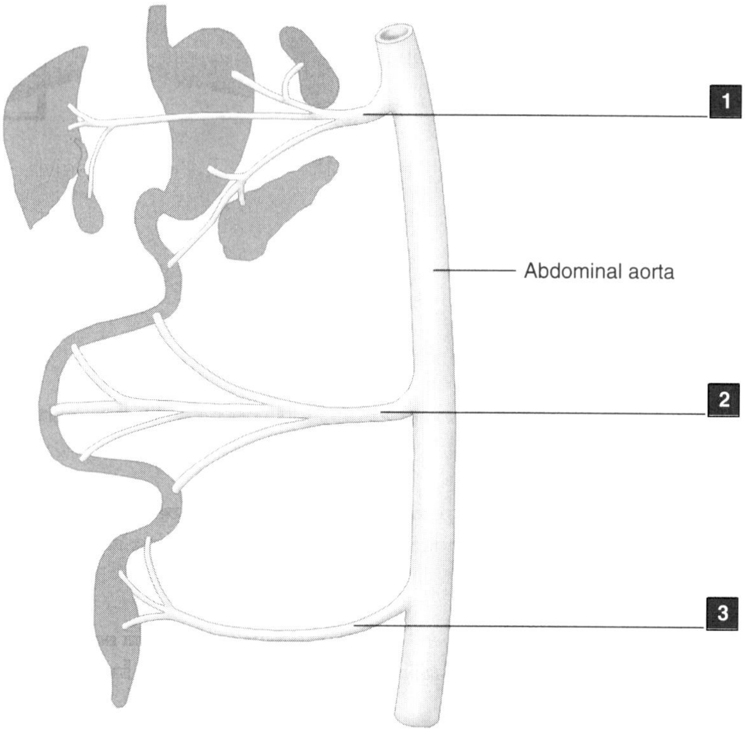

Figure 10.13 Blood Vessels of the Gut

These three vessels are:

Celiac Artery (Trunk) 1 serving the foregut structures
Superior Mesenteric Artery 2 serving the midgut structures
Inferior Mesenteric Artery 3 serving the hindgut structures

Identify these three arteries in figure 10.13.

Celiac Artery

The very short **Celiac Artery** 1 (= **Celiac Trunk**) arises from the aorta just below the abdominal diaphragm, and behind the head of the pancreas. It divides into three branches: the left gastric artery, the splenic artery, and the hepatic artery.

The **Left Gastric Artery** 2 ascends toward the junction of the stomach and esophagus, supplying each.

The **Splenic Artery** 3 runs to the left along the top of the pancreas into the spleen, supplying both. Just before entering the spleen, it gives off the **Left Gastroepiploic (= Gastro-omental) Artery** 4**,** which runs back to the right along the greater curvature of the stomach.

The **Hepatic Artery** 5 runs to the right into the liver. It has three branches: gastroduodenal artery, right gastric artery, and cystic artery.

First, the **Gastroduodenal Artery** 6 comes off behind the duodenum, and supplies it. It gives off a branch, the **Right Gastroepiploic Artery** 7**,** which runs to the left along the greater curvature of the stomach to anastomose with the left gastroepiploic artery.

Second to come off is the **Right Gastric Artery** 8**.** It runs to the left and upward along the lesser curvature of the stomach to anastomose with the left gastric artery.

Finally, the **Cystic Artery** 9 arises just before the porta hepatis of the liver to supply the gallbladder.

Identify the branches of the celiac artery in figure 10.14.

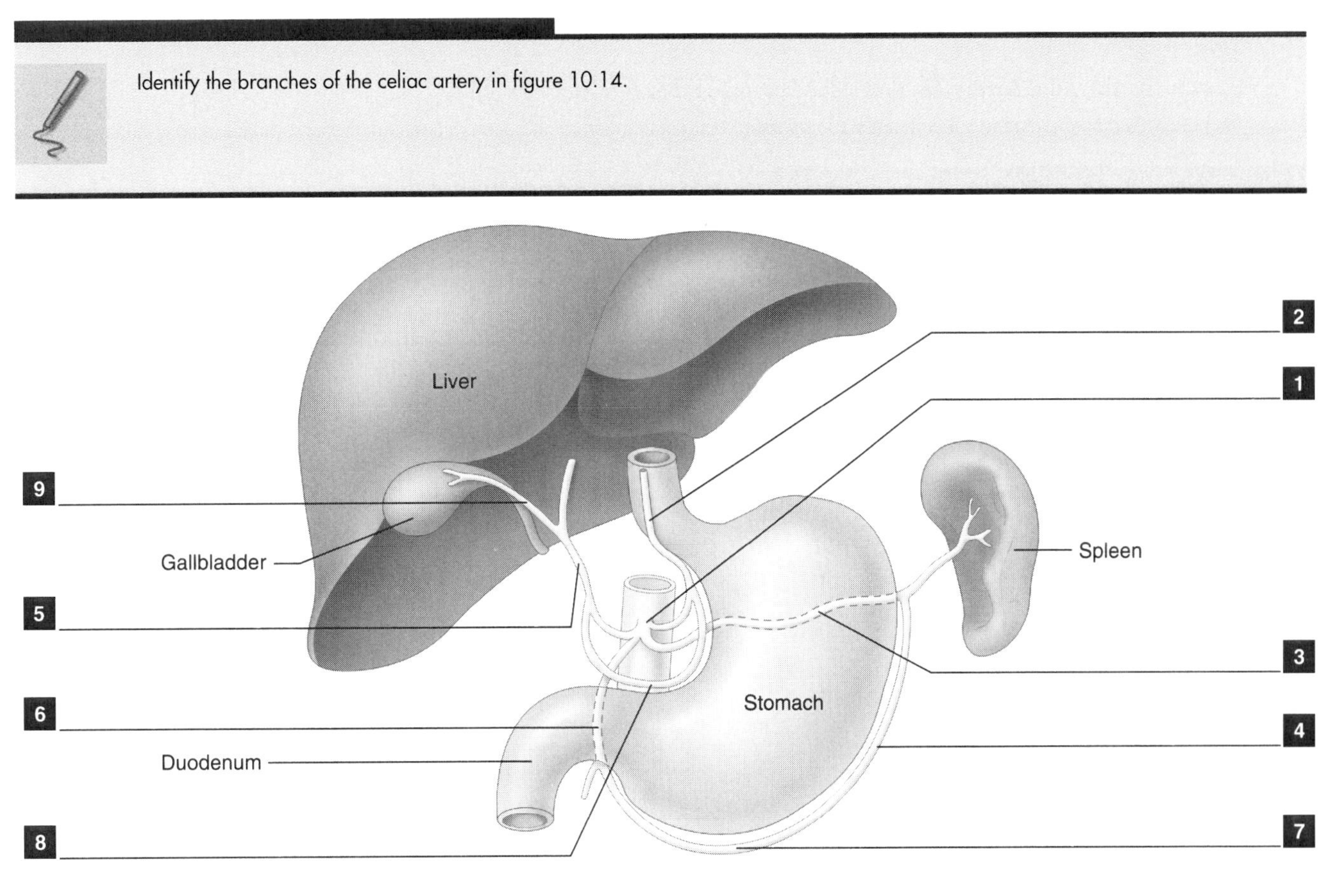

Figure 10.14 **Celiac Artery and Its Branches**

Superior Mesenteric Artery

The **Superior Mesenteric Artery** **1** arises from the aorta just below the celiac artery. It runs between the pancreas and the top of the horizontal part of the duodenum to enter the *Mesentery* of the small intestine. It supplies the distal third of the duodenum, the entire small intestine, and the proximal part of the large intestine.

As it runs through the mesentery it gives off 12 to 15 branches to the jejunum and ileum. Recall that the **Jejunal Branches** **2** form only one or two anastomosing arcades, and their terminal portions are long. The **Ileal Branches** **3**, by comparison, have several arcades, and their terminal portions are short.

The superior mesenteric artery supplies the *proximal half* of the large intestine by three large branches: middle colic, right colic, and ileocolic.

First to arise is the **Middle Colic Artery** **4**. It supplies the terminal portion of the ascending colon, and most of the transverse colon.

Second is the **Right Colic Artery** **5**. It supplies most of the ascending colon.

Finally, the **Ileocolic Artery** **6** supplies the cecum and appendix.

Identify the branches of the superior mesenteric artery in figure 10.15.

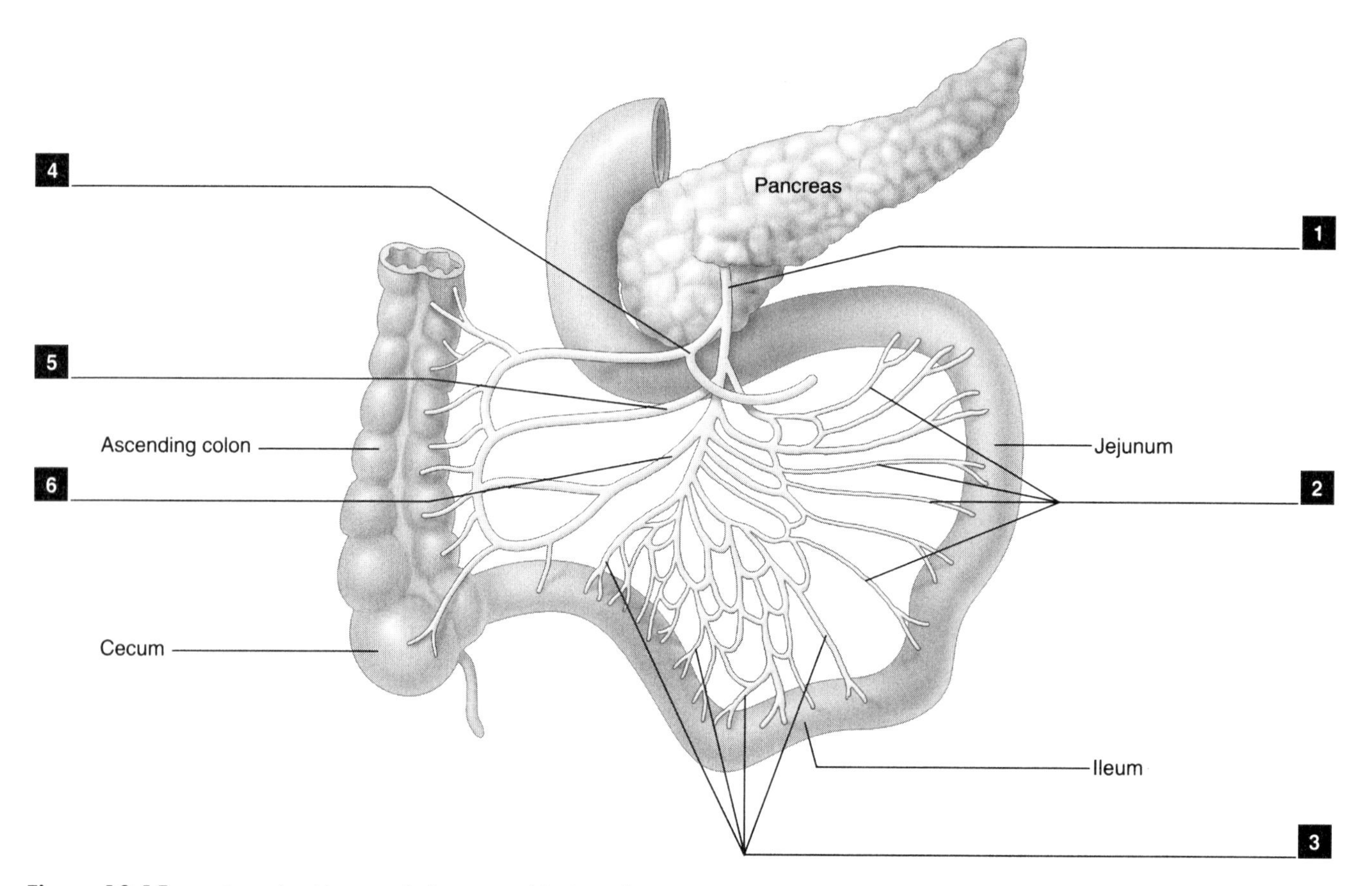

Figure 10.15 Superior Mesenteric Artery and Its Branches

Inferior Mesenteric Artery

The **Inferior Mesenteric Artery** 1 arises at the level of L3, about 1½ inches above the bifurcation of the aorta into the left and right common iliac arteries. The inferior mesenteric artery has a smaller diameter than either the celiac artery or the superior mesenteric artery. It supplies the distal part of the large intestine and the proximal part of the rectum.

The inferior mesenteric artery gives off the **Left Colic Artery** 2**.** It supplies the end of the transverse colon and beginning of the descending colon.

The inferior mesenteric artery gives off several **Sigmoid Branches** 3**.** These supply the descending colon and sigmoid colon.

The terminal portion of the inferior mesenteric artery changes its name to the **Superior Rectal Artery** 4 as it crosses over the left common iliac artery. The superior rectal artery supplies the proximal third of the rectum.

Identify the branches of the inferior mesenteric artery in figure 10.16.

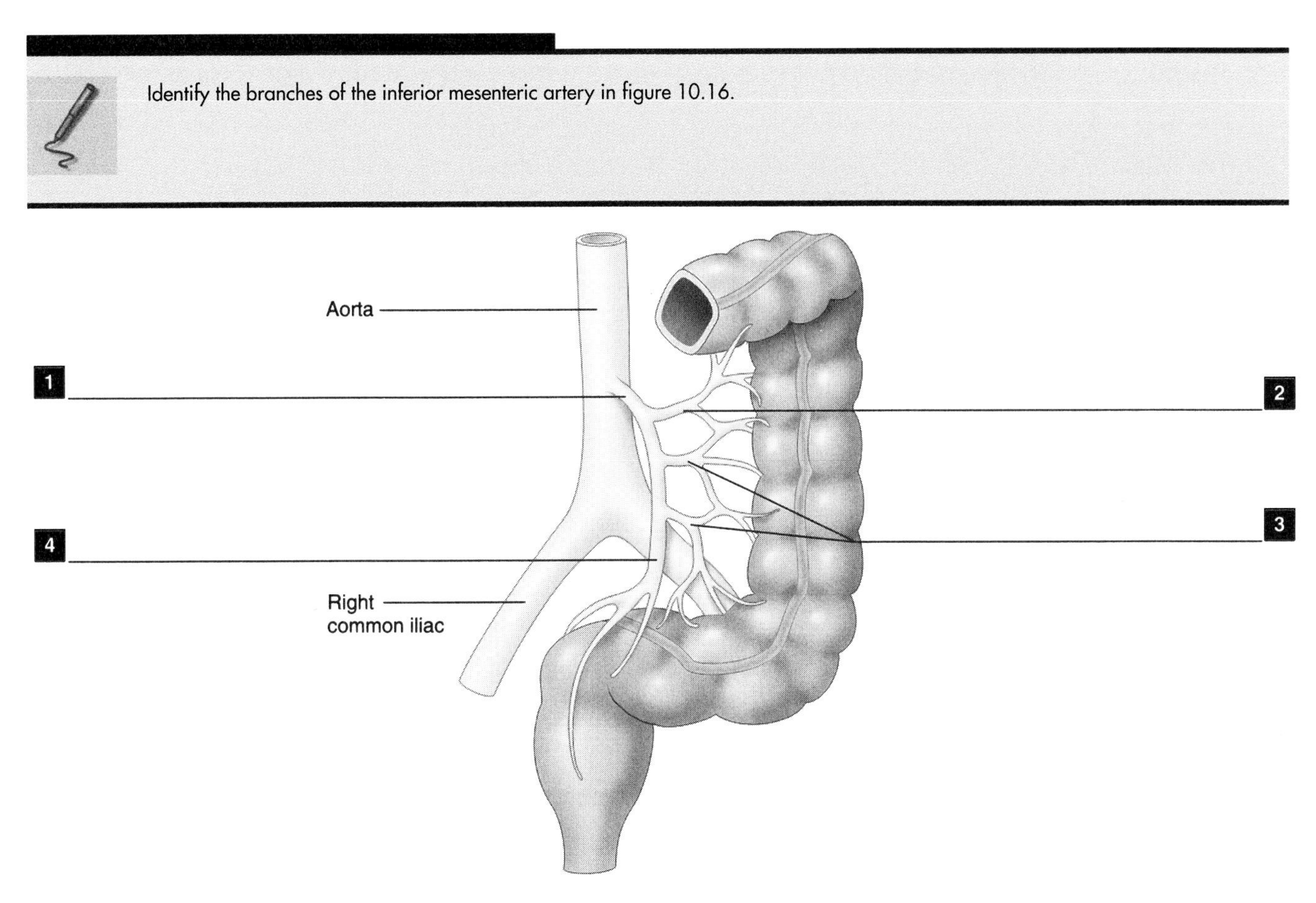

Figure 10.16 Inferior Mesenteric Artery and Its Branches

Venous Drainage

Nutrient-rich blood from the gastrointestinal tract is transported to the liver for filtration and processing by means of a network of veins that comprise the *Hepatic Portal Venous System*. The veins that drain blood from the digestive organs merge to form the short (Hepatic) *Portal Vein*, which enters the liver.

The veins that drain blood from the digestive organs have the same names as the arteries that feed them. Thus, for example, the spleen is supplied by the splenic artery and drained by the splenic vein. The distal parts of the veins also follow the paths taken by the arteries of the same name. The more proximal portions of the veins have somewhat different courses.

The (Hepatic) **Portal Vein** 1 is about 2 inches long. It is formed behind the pancreas by the union of the **Splenic Vein** 2 and the **Superior Mesenteric Vein** 3.

The **Superior Mesenteric Vein** drains the small intestine, cecum, ascending colon, and transverse colon. It is fed by about a dozen jejunal and ileal veins, and by three large tributaries: the **Middle Colic Vein** 4, the **Right Colic Vein** 5, and the **Ileocolic Vein** 6. The superior mesenteric vein also receives blood from the *Pancreaticoduodenal Vein* and *Gastroepiploic Vein*, which drain the duodenum and greater curvature of the stomach.

The **Splenic Vein** drains the spleen and has a large tributary: the inferior mesenteric vein.

The **Inferior Mesenteric Vein** 7 drains the descending colon, sigmoid colon, and the proximal third of the rectum. It is fed by three principal tributaries: the **Left Colic Vein** 8, the **Sigmoid Veins** 9, and the **Superior Rectal Vein** 10.

The (hepatic) portal vein is formed by the union of the superior mesenteric and splenic veins. Three veins drain into it: the **Left Gastric Vein** 11 and **Right Gastric Vein** 12 drain the lesser curvature of the stomach, and the **Cystic Vein** 13 drains the gallbladder.

The portal vein enters the liver via the *Porta Hepatis* and then splits into left and right branches that take blood to the liver's lobules. The blood that is filtered through the lobules is collected by numerous *Central Veins* that merge to form the two **Hepatic Veins** 14. The hepatic veins drain directly into the *Inferior Vena Cava*.

Blood Supply to the Liver

Although the hepatic artery carries oxygenated blood, it provides only about 20% of the liver's supply. The other 80% is supplied by the portal vein. Even though it carries deoxygenated blood, the portal vein supplies the majority of the liver's oxygen because of the sheer volume of blood that it carries!

Identify the components of the hepatic portal venous system in figure 10.17.

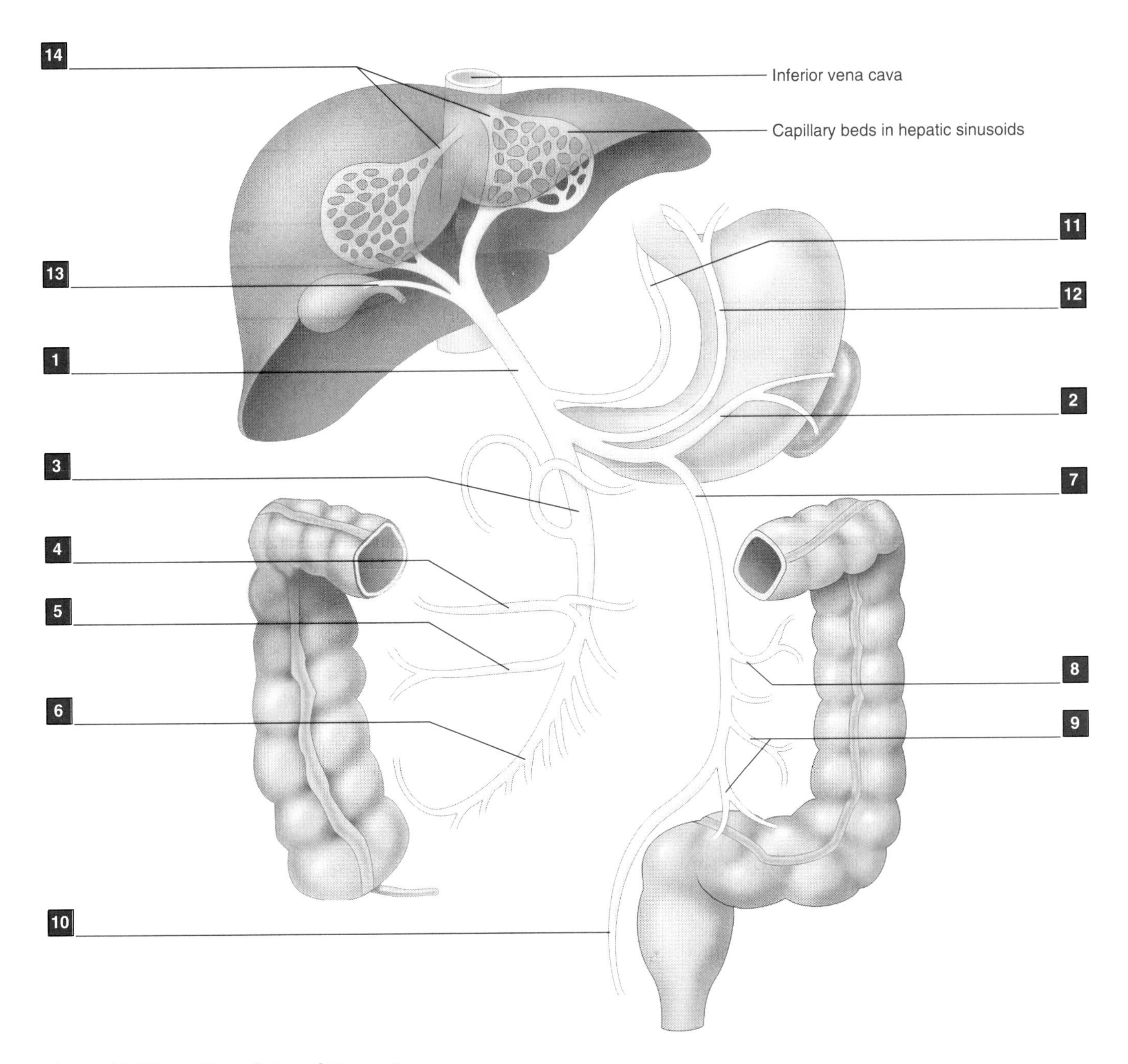

Figure 10.17 **Hepatic Portal Venous System**

10.6 LYMPHATIC ORGANS AND LYMPH DRAINAGE

The abdomen, pelvis, and lower limbs are drained of lymph by two sets of trunks that run along the posterior abdominal wall. The trunks merge to form the thoracic duct. The largest lymphatic structure in the body is the spleen.

Lymph Drainage

Lymph from the anterior abdominal wall and lower limbs drains into very large *Inguinal Nodes* that lie in the *Femoral Triangle* just below the inguinal ligament (laboratory 4, pp. 107–108). These nodes are easily palpated, even in the absence of infection.

Lymph ducts from retroperitoneal abdominal organs (e.g., the kidneys) drain into nodes that lie along either side of the abdominal aorta. These *Lumbar Nodes* are connected by large ducts known as *Lumbar Trunks*.

The gut is drained by ducts that follow one of the three arteries that arise from the front of the abdominal aorta to supply it. These ducts lead to *Preaortic Nodes* that lie along the front of the aorta. They are connected by a large duct known as the *Intestinal Trunk*.

The lumbar and intestinal trunks merge at the level of L1–L2 to form the *Thoracic Duct*, which is dilated here as the *Cisterna Chyli*. Recall that the thoracic duct ascends through the thorax to empty into the left brachiocephalic vein (laboratory 9, pp. 279–280).

The *Spleen* is the principal lymphatic organ. It is composed mainly of lymphatic tissue that filters blood passing through it. It is also involved in the production of bilirubin, a constituent of bile. This is transported to the liver by the splenic and hepatic portal veins.

The Spleen

The **Spleen** 1 lies behind and to the left of the stomach between the ninth and eleventh ribs. It resides above the left colic flexure and anterior to the left kidney. Blood and lymph vessels enter its concave visceral surface via the **Hilum** 2.

Recall that the spleen receives blood from the **Splenic Artery** 3, which is a primary branch of the **Celiac Artery** 4.

Recall also that blood from the spleen is drained by the **Splenic Vein** 5. It is joined by the inferior mesenteric vein to drain into the **Hepatic Portal Vein** 6.

Identify the aforementioned structures in figure 10.18.

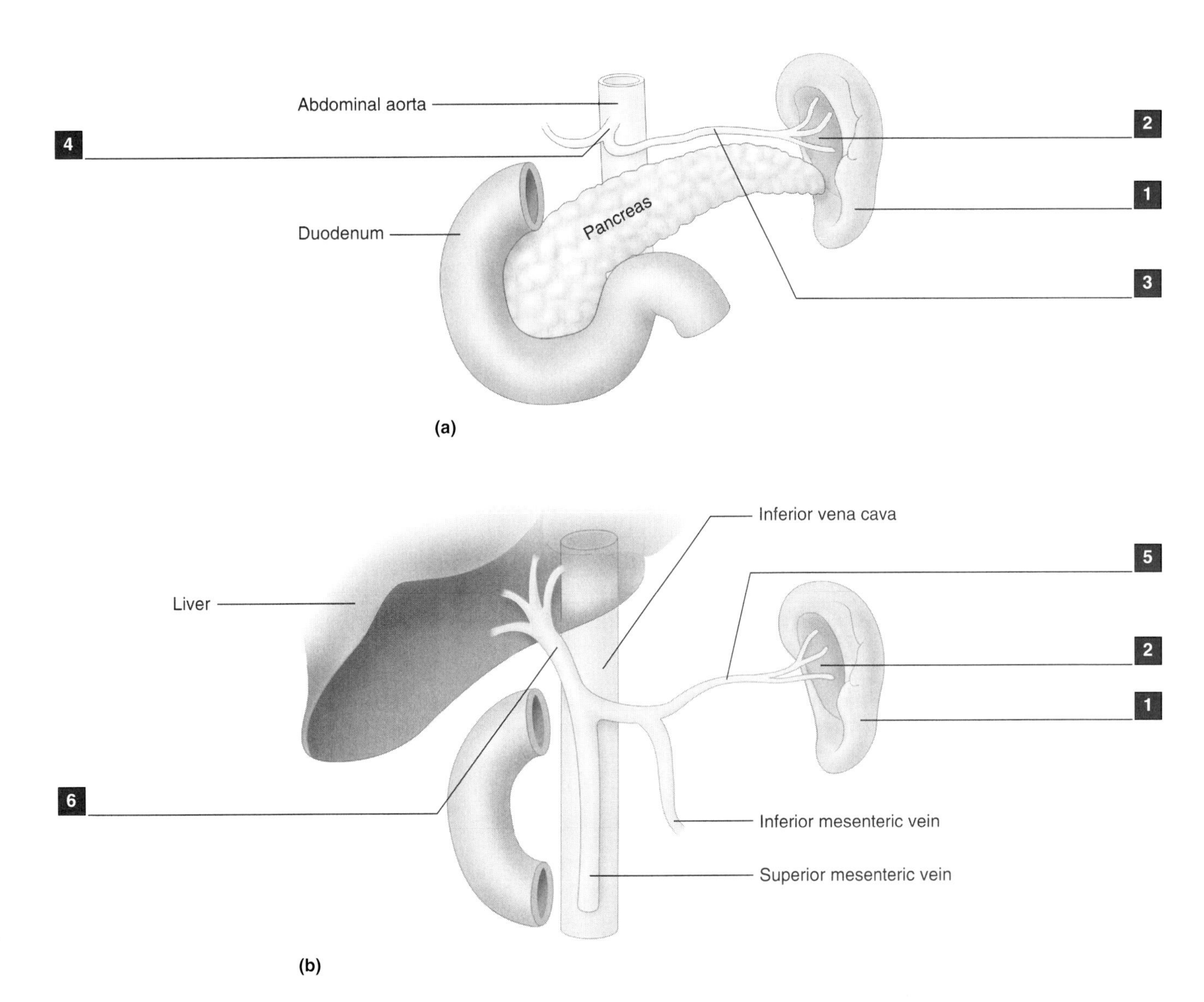

Figure 10.18 **The Spleen**
(a) arterial supply; (b) venous drainage.

10.7 THE KIDNEYS AND ADRENAL GLANDS

The kidneys excrete most of the body's metabolic waste products, and help maintain water and electrolyte balance. Waste products leave the kidney as urine, which passes down a muscular tube, the ureter, to the bladder. Lying on the superior pole of each kidney is an adrenal (= suprarenal) gland. Each gland consists of two separate masses of tissue: the cortex secretes over 30 steroid hormones; the medulla secretes amine hormones.

The kidneys and adrenal glands are retroperitoneal. Each kidney is covered with four protective layers of tissue. From deep to superficial these are: a *Fibrous Capsule*, a layer of *Perirenal Fat*, the *Renal Fascia*, and a layer of *Pararenal Fat*.

The medial, concave aspect of the kidney has a vertical slit called the *Hilum*. It transmits the renal artery, the renal vein, the renal pelvis, and autonomic nerve fibers.

Blood Vessels of the Kidneys and Adrenal Glands

Arteries

Each kidney is supplied by a **Renal Artery 1.** They arise from the left and right sides of the *Abdominal Aorta* a bit below the origin of the superior mesenteric artery. The right renal artery passes behind the inferior vena cava.

The adrenal glands are richly supplied with blood by three arteries. The first is the **Superior Suprarenal Artery 2.** It is a branch of the *Inferior Phrenic Artery*, which arises from either side of the aorta immediately below the diaphragm. The second is the **Middle Suprarenal Artery 3.** It arises from either side of the *Aorta* just below the level of the celiac artery. The third is the **Inferior Suprarenal Artery 4.** It runs upward to the gland from the *Renal Artery*.

Veins

Each kidney is drained by a **Renal Vein 5.** These empty into the *Inferior Vena Cava*. The left renal vein crosses anterior to the aorta just below the root of the superior mesenteric artery.

Each adrenal gland is drained by the **Suprarenal Vein 6.** On the *right* it enters the *Inferior Vena Cava* directly. On the *left* it enters the *Left Renal Vein*.

Identify the blood vessels of the kidneys and adrenal glands in figure 10.19.

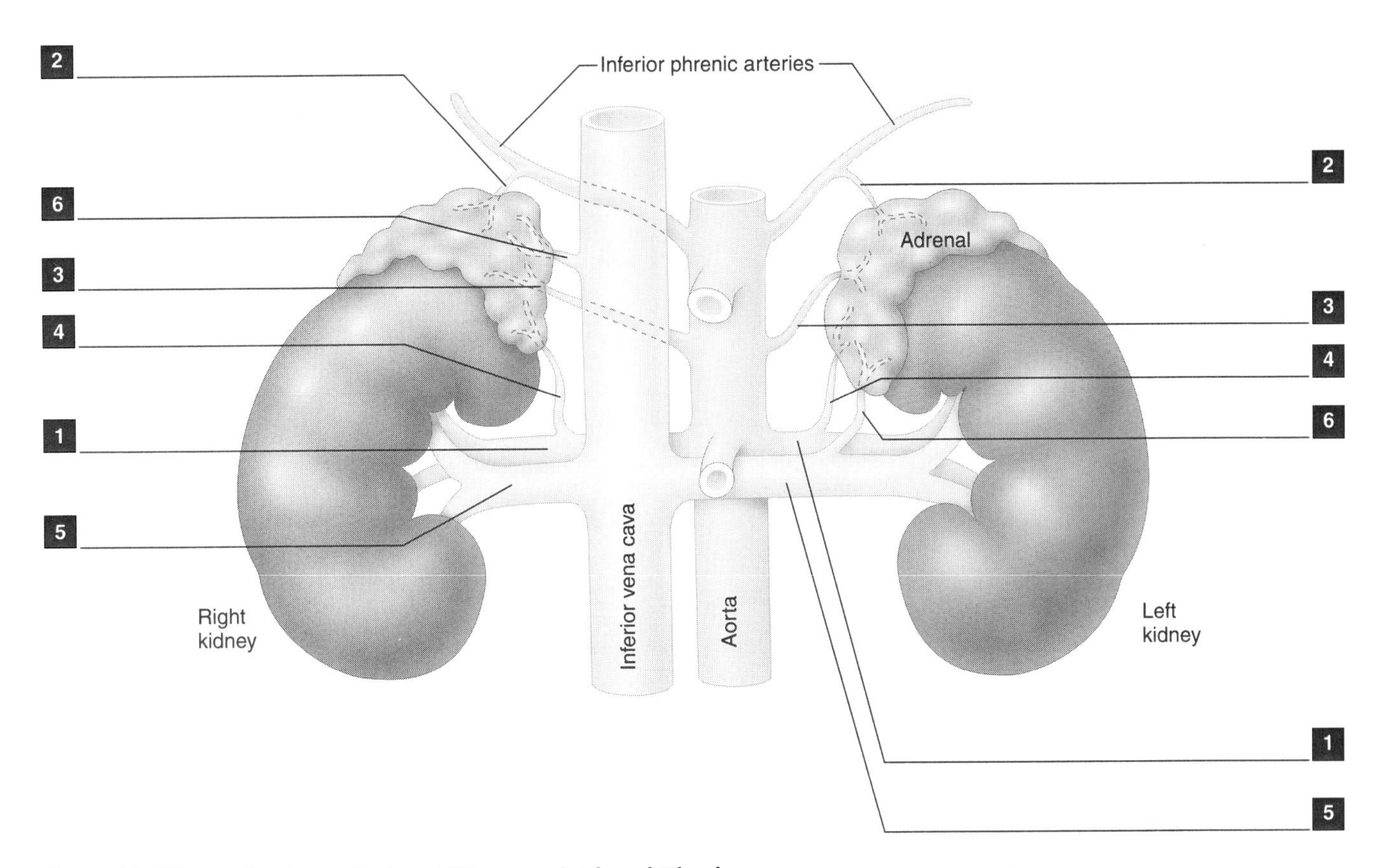

Figure 10.19 Blood Vessels of the Kidneys and Adrenal Glands

Structure of the Kidney

Recall that the medial, concave aspect of the kidney has a vertical slit called the *Hilum*. It transmits the **Renal Artery 1**, **Renal Vein 2**, and the **Renal Pelvis 3**. The renal pelvis is the enlarged beginning of the **Ureter 4**.

The hilum extends into the substance of the kidney for a short distance, creating the *Renal Sinus*.

A longitudinal section through the kidney reveals that it is composed of three distinct regions around the renal sinus.

The first region comprises the collection chambers for urine. The renal pelvis branches into several chambers called **Major Calyces 5** (singular = *Calyx*). Each major calyx branches into several smaller, funnel-shaped chambers known as **Minor Calyces 6**. There are about a dozen minor calyces. Urine drains from the renal pyramids into the minor calyces.

The **Renal Medulla 7** is composed of about a dozen **Renal Pyramids 8**. Each pyramid is a cone-shaped structure that contains numerous straight urine *Collecting Ducts*. The ducts open at the apex, or **Papilla 9**, of the renal pyramid into the minor calyx.

The **Renal Cortex 10** contains numerous twisted (convoluted) tubes and capillaries and the *Renal Corpuscles* (described later, p. 326). Between adjacent renal pyramids is a pillar of tissue that descends from the renal cortex. This pillar is called the **Renal Column 11**. It contains arteries and veins.

A *Renal Pyramid* together with the *Renal Cortex* capping it constitute a **Renal Lobe 12**. The *Nephrons* are contained within the renal lobes.

The nephrons are the functional units of the kidney. They are responsible for the production of urine by means of blood filtration, and the selective reabsorption by the blood of previously filtered substances that are useful to the body. Part of each nephron lies within the renal medulla, and part lies within the adjacent renal cortex. We will examine the structure and function of a nephron next.

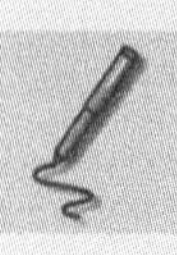

Identify the components of the kidney in figure 10.20.

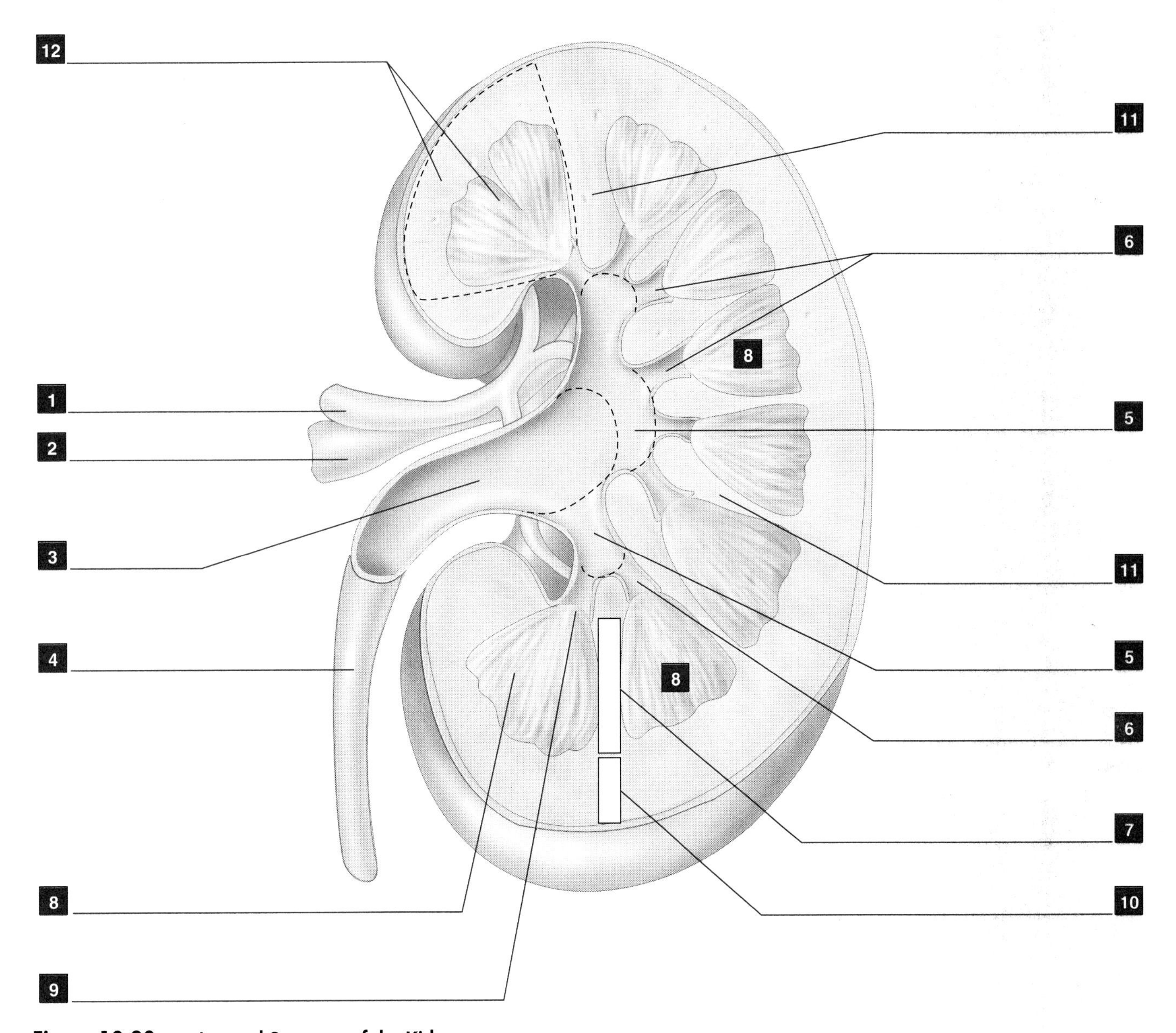

Figure 10.20 Internal Structure of the Kidney

The Nephron

The nephron is the functional unit of the kidney. Most of the nephron is located within the renal cortex above a renal pyramid. It produces urine by a two-stage filtration process. In the first instance, blood plasma filters from a capillary bed into the end of a tube that ultimately drains into a minor calyx. Because it is desirable to conserve water, salt, and sugar, these are reabsorbed back into blood vessels from the tube. In this way, excreted urine is hyperosmotic, and largely devoid of salt and sugar.

Each nephron has a tubular component and a vascular component.

Tubular Component

This is comprised of a *Renal Tubule* that is closed at one end, and open at the other. The closed beginning of the tubule has a cup-shaped expansion known as the **Glomerular (Bowman's) Capsule** 1. It surrounds a tightly twisted capillary bed, known as the *Glomerulus*, like a hand grasping a ball of string. Together, the glomerulus and glomerular capsule constitute the ***Renal Corpuscle.***

A short distance from the glomerular capsule, the renal tubule becomes twisted as the **Proximal Convoluted Tubule** 2. It then straightens out and loops into and out of the renal pyramid as the **Loop of Henle** 3. The limbs of the Loop of Henle descend into and ascend from the renal pyramid. After ascending, the tubule becomes twisted again as the **Distal Convoluted Tubule** 4. A short distance from the glomerular capsule, the renal tubule becomes surrounded by a second capillary network.

The renal tubule opens into a **Collecting Duct** 5, which drains several nephrons. The collecting duct opens at the papillae of the renal pyramid into the minor calyx.

Vascular Component

This consists of two capillary networks connected by an arteriole. Thus, it is a *Portal System*. The first capillary bed is the **Glomerulus** 6. Blood reaches the glomerulus through a small vessel called an **Afferent Arteriole** 7. About 20% of the blood plasma moving through the afferent arteriole filters through the glomerulus.

Blood that passes through the glomerulus leaves it by another small vessel called an **Efferent Arteriole** 8. This leads to the second capillary network, where reabsorption of water, salt and sugar takes place. The second network is intertwined around the renal tubule; it is therefore called the **Peritubular Capillary** 9. Small vessels lead from the peritubular capillary to the renal vein.

Identify the tubular and vascular components of the nephron in figure 10.21.

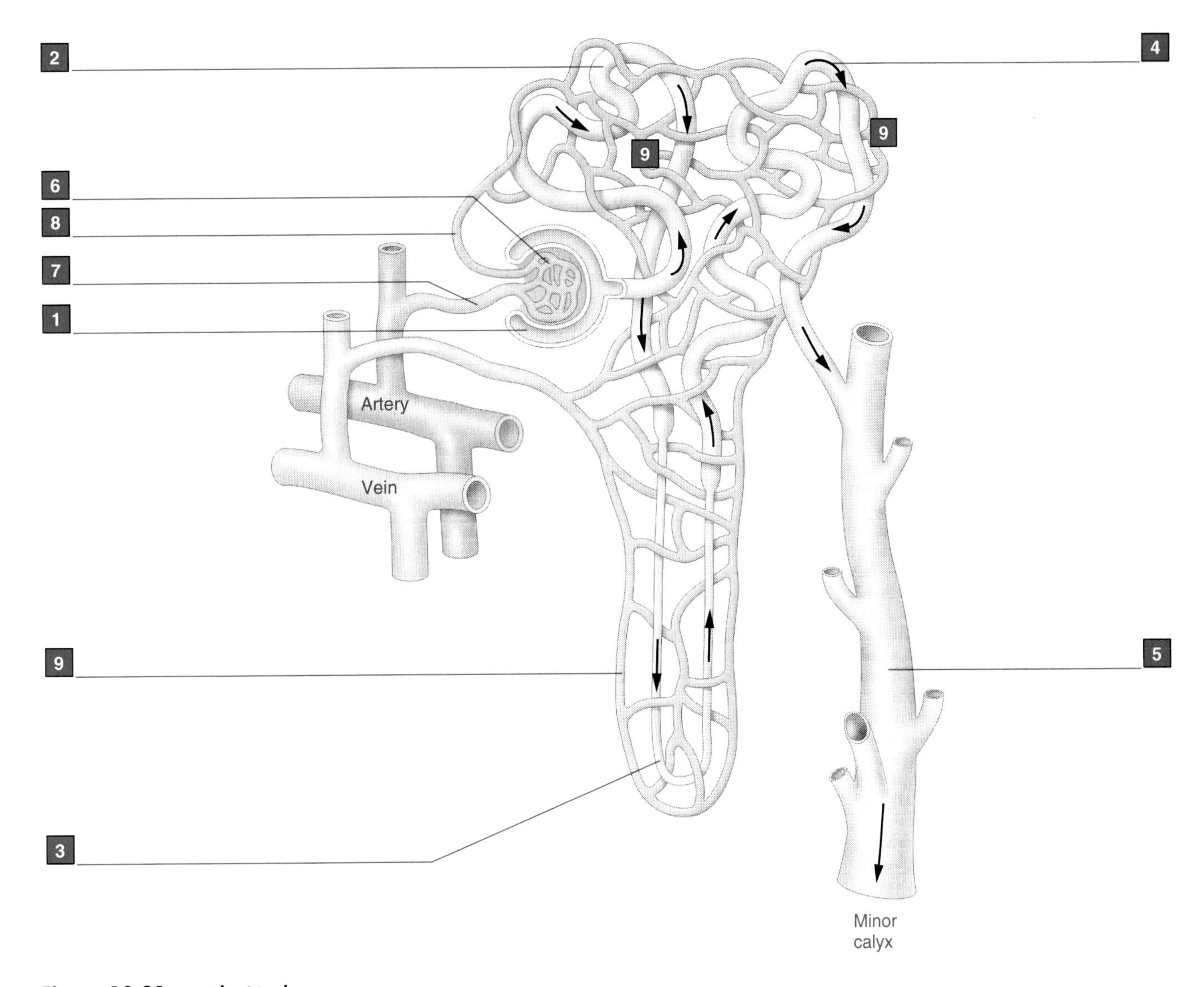

Figure 10.21 **The Nephron**

10.8 GONADAL BLOOD VESSELS

Although the gonads (ovaries and testicles) are within or below the pelvis, their vasculature originates high up in the abdomen. The vessels that supply and drain their blood are the same, whether the gonads are ovaries or testicles.

Gonadal Arteries

The **Left** 1 and **Right** 2 **Gonadal Arteries** arise from the front of the **Abdominal Aorta** 3 just below the level at which the **Renal Arteries** 4 come off. The gonadal arteries descend to the left and right to supply the respective gonad.

Gonadal Veins

The **Right Gonadal Vein** 5 follows the course of the right gonadal artery for much of its length. It enters the front of the **Inferior Vena Cava** 6 just below the level at which the **Renal Veins** 7 come off.

The **Left Gonadal Vein** 8 follows the course of the left gonadal artery for much of its length, but as it approaches the kidney, it departs company from the artery. It runs straight upward to drain into the **Left Renal Vein.**

Identify the aforementioned blood vessels in figure 10.22. Color arteries shades of red and veins shades of blue.

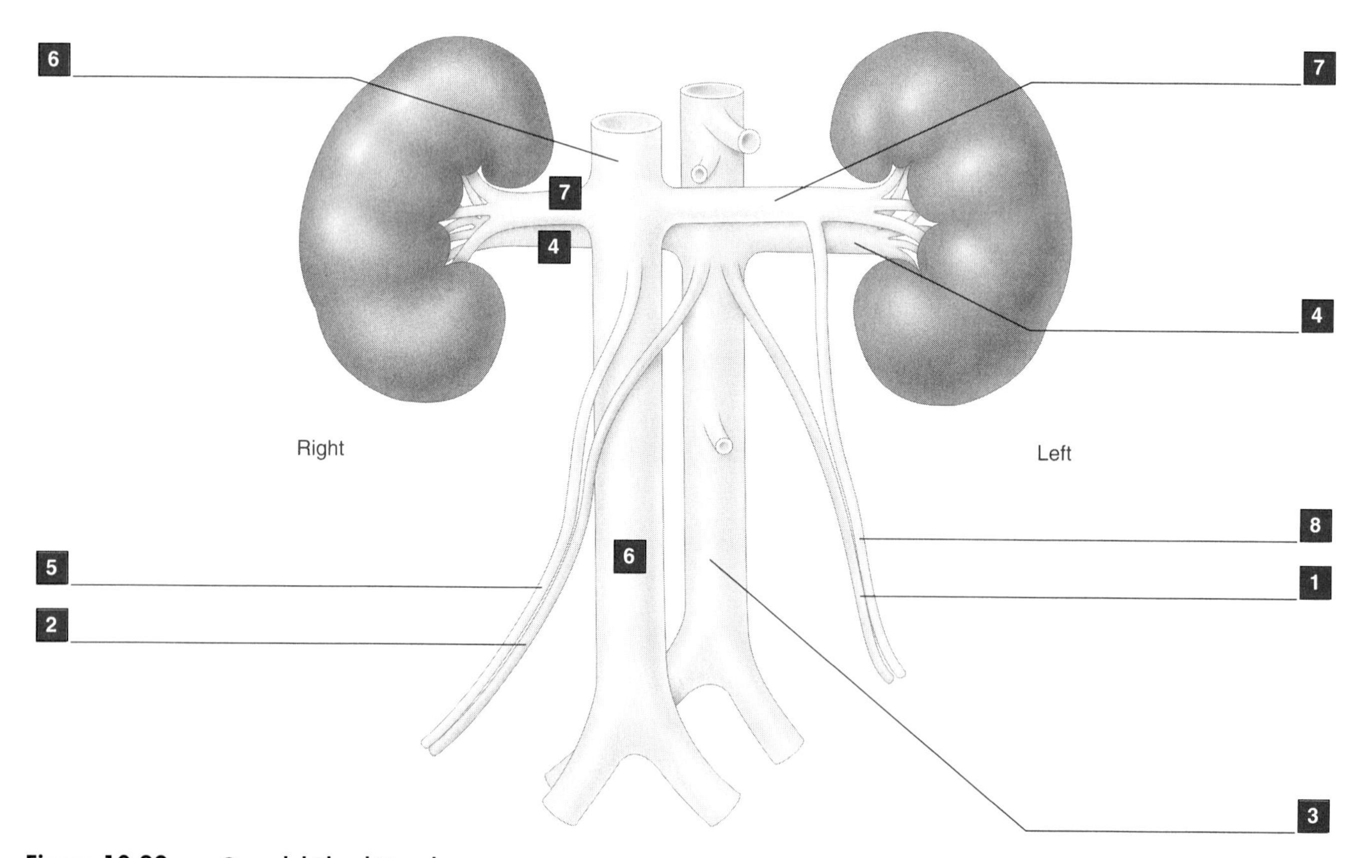

Figure 10.22 **Gonadal Blood Vessels**

10.9 NERVES IN THE ABDOMEN

The skeletal muscles of the abdomen are supplied by spinal nerves. Spinal nerves that emanate in the abdomen also innervate muscles of the lower limb. The gastrointestinal tract and organs of digestion, as well as the spleen, kidneys, and adrenal glands are supplied by autonomic nerves.

Spinal Nerves in the Abdomen

The skeletal muscles of the abdominal wall are supplied by spinal nerves T7–L1. The structure of a typical spinal nerve was studied in laboratory 2 (pp. 33–34).

The muscles of the lower limb are innervated by the ventral rami of L1–L4, which form the *Lumbar Plexus*. The lowermost lumbar spinal nerve (L5) contributes to the *Sacral Plexus*. These plexuses are connected by the *Lumbosacral Trunk* between L4 and L5. These structures were studied in laboratory 4 (pp. 102–103).

Cranial Nerve in the Abdomen

The *Vagus Nerve* (*CN X*) is the only cranial nerve to enter the abdominal cavity. It was studied in laboratory 7 (pp. 220–221).

Recall that the *left* vagus nerve enters the abdomen in front of the esophagus as the *Anterior Vagal Trunk*. It gives branches to the liver, gallbladder, stomach and duodenum, and to the kidneys and adrenal glands.

The *right* vagus enters the abdomen behind the esophagus as the *Posterior Vagal Trunk*. Its fibers are distributed with the branches of the celiac and superior mesenteric arteries. Thus, it innervates the pancreas, spleen, stomach, small intestine, and the large intestine as far as the left colic (splenic) flexure.

Sympathetic Chain in the Abdomen

We examined the anatomy of the sympathetic chain in the neck and thorax in laboratories 5 (pp. 142–144) and 9 (pp. 286–287). The sympathetic chain also runs through the abdomen into the pelvis. Only spinal nerves *L1–L2* contribute *preganglionic* fibers to the chain. Thus, sympathetic ganglia below L2 have gray but no white rami communicantes. In this regard, the sympathetic chain below L2 is similar to the sympathetic chain in the neck.

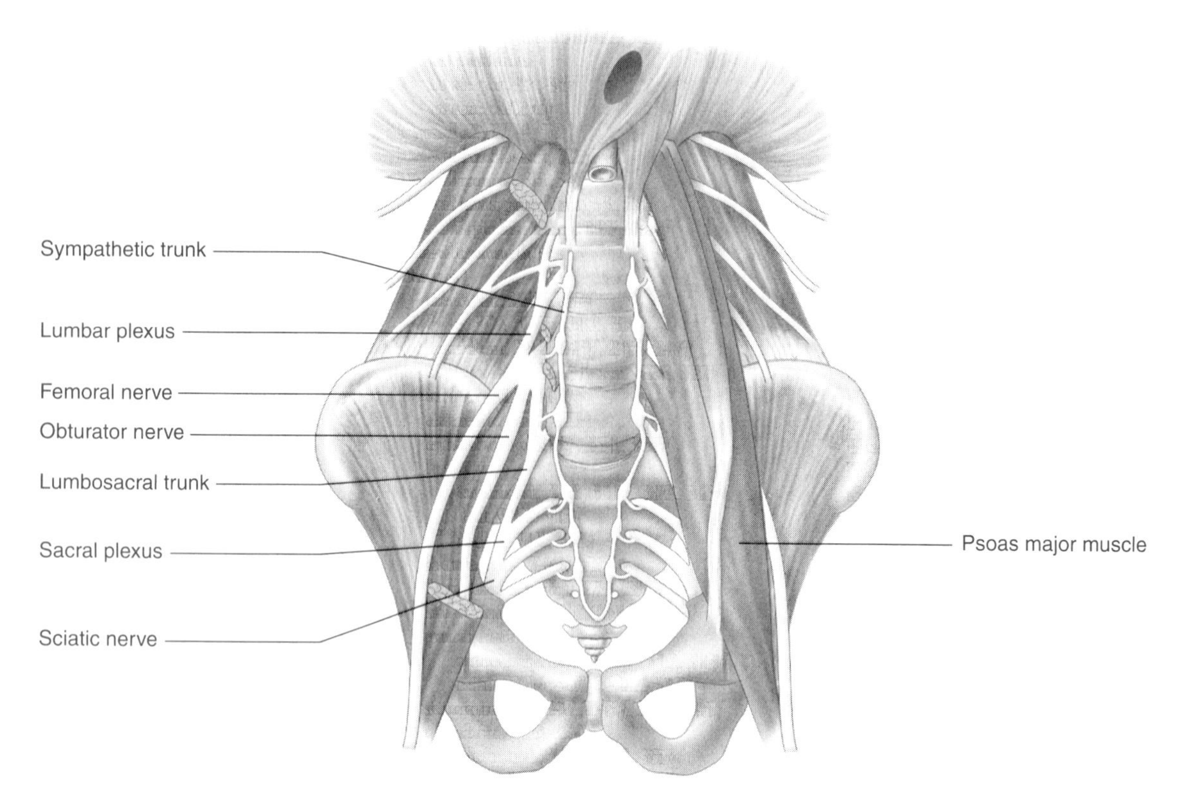

Spinal Nerves and the Sympathetic Chain.

LABORATORY

The Pelvis

11.1 THE PELVIC SKELETON

Bony Pelvis

We examined the bony pelvis previously (laboratory 2, pp. 38–39 and laboratory 4, pp. 85–86). Nevertheless, a brief review of its principal features is warranted.

Recall that the bony pelvis of an adult is composed of left and right **Os Coxae 1**, and the **Sacrum 2** and **Coccyx 3**.

The os coxae is composed of three fused elements: **Ilium 4**, **Ischium 5**, and **Pubis 6**. The pubic bones articulate with one another anteriorly by a fibrocartilage **Symphysis 7**. Posteriorly, the ilium articulates with the sacrum at the **Auricular Surface 8**, which forms the *Sacroiliac Joint*.

The os coxae is "waisted" below the auricular surface by the **Greater Sciatic Notch 9**, which is delimited inferiorly by the **Ischial Spine 10**. Below the spine, the ischium is indented by the **Lesser Sciatic Notch 11**. This notch is bounded inferiorly by the **Ischial Tuberosity 12**.

The ischial and pubic bodies are joined by the **Superior Ramus of the Pubis 13** superiorly, and the **Ischiopubic Ramus 14** inferiorly. These rami surround the **Obturator Foramen 15**.

The upper margin of the superior ramus of the pubis has a sharp crest that continues posteriorly as a ridge, known as the *Arcuate Line*, to the auricular surface. The crest and arcuate line constitute the **Pelvic Brim 16**. It separates the so-called *False Pelvis* above from the *Pelvic Cavity* (also known as the *True Pelvis* or *Obstetrical Pelvis*) below.

Pelvic Ligaments

The ilium and sacrum are joined anteriorly along the edge of the auricular surface by the **Anterior Sacroiliac Ligament 17**, and posteriorly by the **Posterior Sacroiliac Ligament 18**.

The ischial spine and sacrum are connected by the **Sacrospinous Ligament 19**. It closes the open end of the greater sciatic notch, creating the *Greater Sciatic Foramen*.

The ischial tuberosity and sacrum are connected by the **Sacrotuberous Ligament 20**. Together with the sacrospinous ligament, it closes the open end of the lesser sciatic notch, creating the *Lesser Sciatic Foramen*.

Identify the aforementioned bony and ligamentous features in figure 11.1.

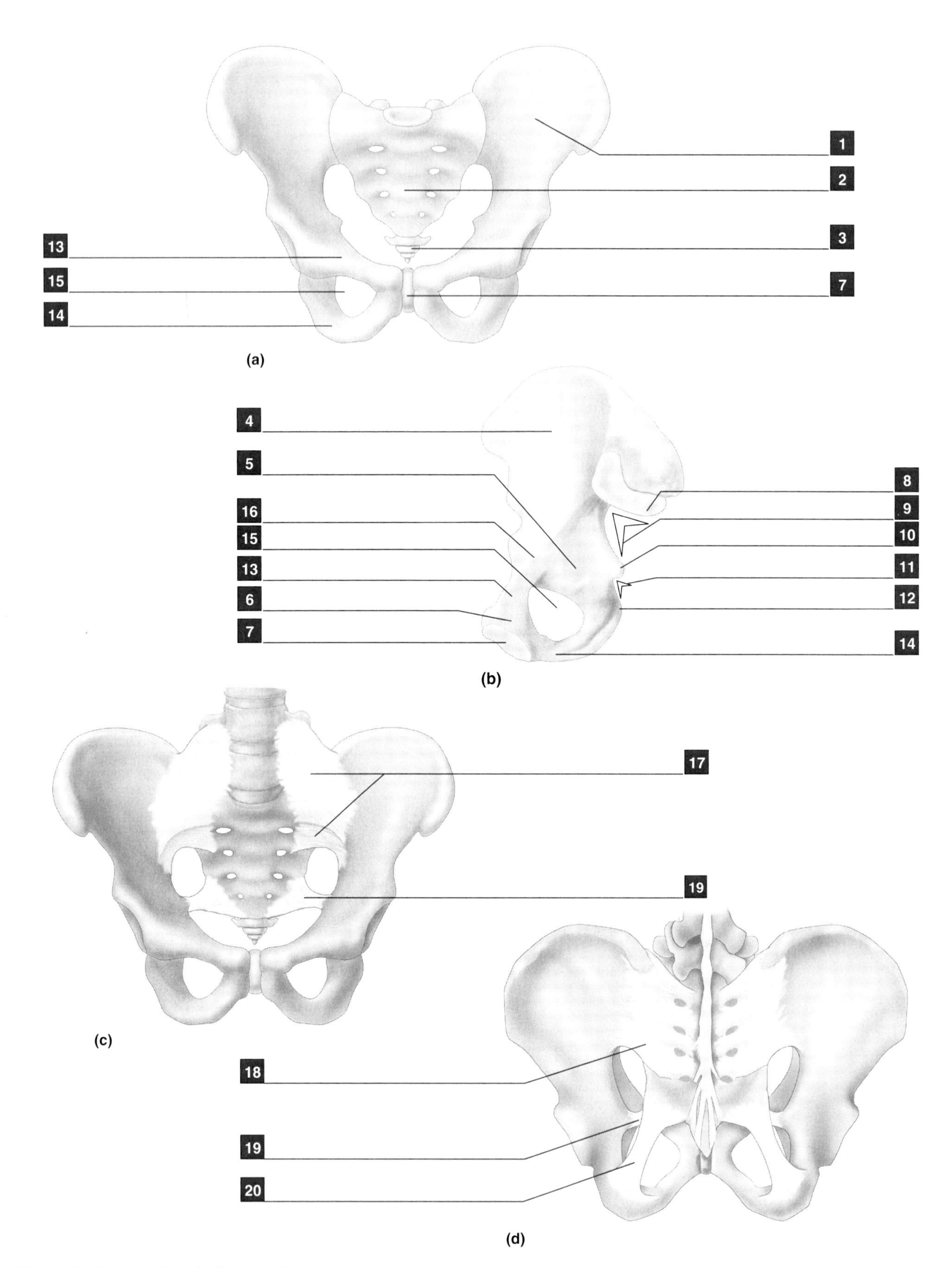

Figure 11.1 **Pelvic Skeleton and Ligaments**
(a) pelvis in anterior view; (b) medial view of right os coxae; (c) ligaments in anterior view; (d) ligaments in posterior view.

Sex Differences in the Bony Pelvis

There are several morphological differences between males and females in the bony pelvis. These differences make their appearance during puberty and are fully developed by the time sexual maturity is attained. The differences relate to obstetrical demands, and act to increase the diameter of the outlet of the pelvic cavity in females.

Thus, the separation of the sacrum and ischial spines, and the distance between the two ischial tuberosities are relatively greater in females than in males.

These morphological differences are of considerable importance to forensic anthropologists, because they permit the correct sexual identification of adult skeletal remains in well over 90% of the cases.

Some of the principal morphological differences between male and female pelves are:

Subpubic Angle 1

In males, the angle formed by the left and right ischiopubic rami below the symphysis is narrow (about 50 to 70 degrees). In females, the subpubic angle is more obtuse (about 80 to 90 degrees).

Ischiopubic Ramus 2

In males, the inferior margin of the ischiopubic ramus is straight or even slightly convex. In females, it is slightly concave.

Body of Pubis 3

In males, the body of the pubis is tall and narrow. In females, it is short and broad.

Greater Sciatic Notch 4

In males, the greater sciatic notch is narrow, forming an angle of approximately 30 degrees. In females, the notch has a broader curve, attaining an angle of about 60 degrees.

Auricular Surface 5

In males, the auricular surface is large, extending well over the first three sacral elements. In females, this joint surface is smaller, usually extending over the first two sacral elements.

Identify the sexually dimorphic features of the pelvis skeleton in figure 11.2.

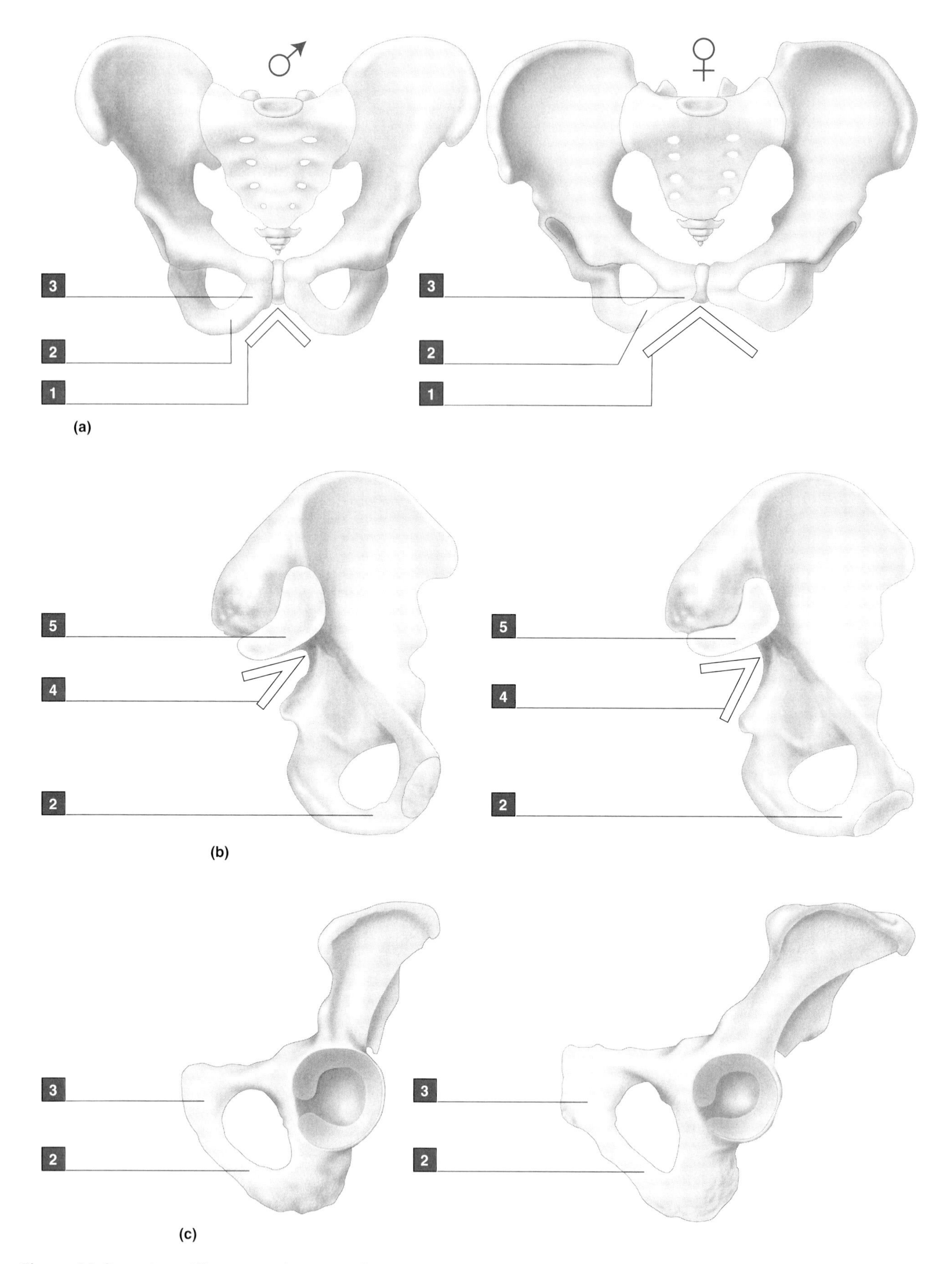

Figure 11.2 **Sex Differences in the Bony Pelvis**
(a) anterior view; (b) medial view; (c) lateral view.

11.2 MUSCLES OF THE PELVIS

Numerous muscles that move the trunk and lower limbs arise from the bony pelvis. We have already examined these muscles of the back (laboratory 2, pp. 36–40), lower limb (laboratory 4, pp. 107–113), and abdomen (laboratory 10, pp. 295–299).

These 30-odd muscles need not concern us here.

The pelvis also provides attachment to a group of muscles that make up the pelvic diaphragm. This is like the abdominal diaphragm because it is dome-shaped, and has openings that permit the passage of certain structures. However, its dome is convex downward, and it is composed of more than one muscle.

The Pelvic Diaphragm

The **Pelvic Diaphragm** 1 is composed of several muscles, whose fibers arise along a line that extends around the inside of the pelvis from the body of the pubis, thickened fascia that covers *Obturator internus*, the body of the ischium, and the sacrospinous ligament.

It divides the pelvis into two parts. The part above the diaphragm is the **Pelvic Cavity** 2**;** the part below it is known as the **Perineum** 3**.**

The pelvic diaphragm has two components: *Levator Ani* and *Coccygeus*. Levator Ani is composed of three muscles. Coccygeus is a single muscle. All four muscles are innervated by *Ventral Rami of S2–4*.

Identify the aforementioned structures in figure 11.3.

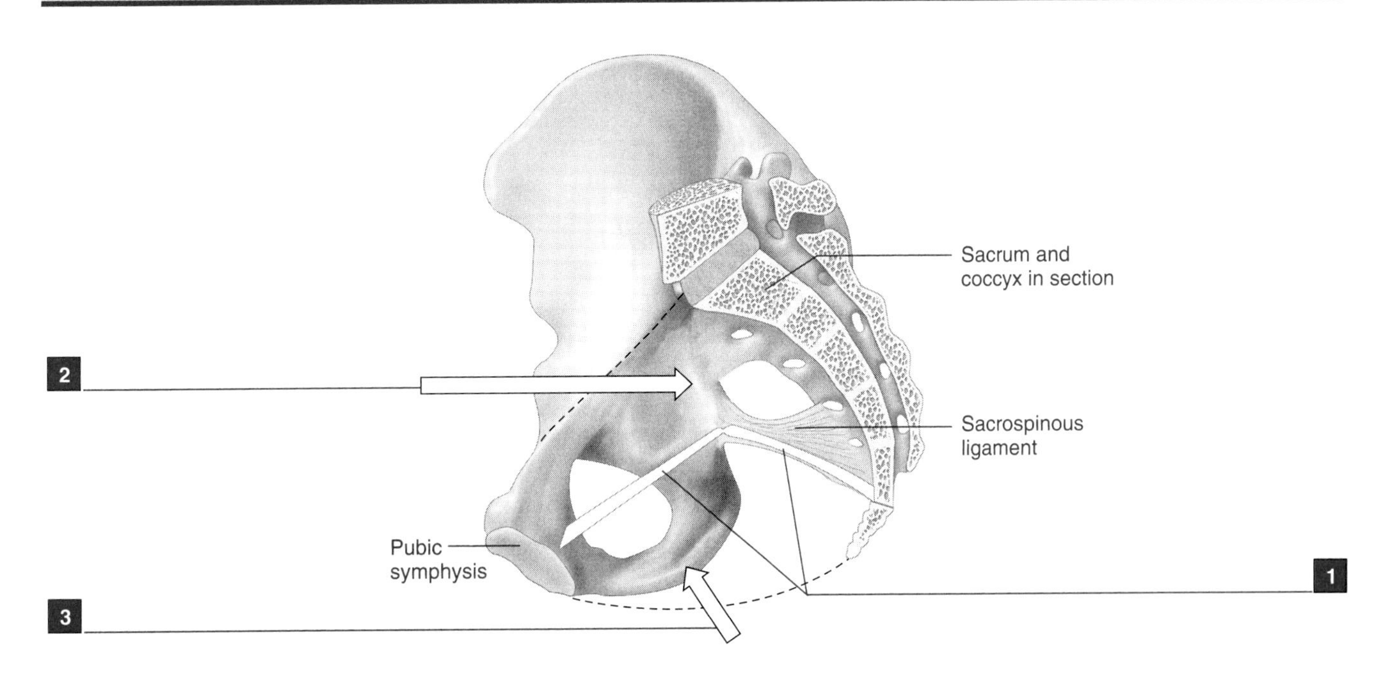

Figure 11.3 **Pelvic Diaphragm**
Medial view showing separation of the pelvic cavity and perineum by the diaphragm.

Levator Ani

The fibers of **Levator Ani** 1 run posteriorly and inferiorly to converge in the midline on a linear "tendon" that begins about 1 inch behind the pubic symphysis and stretches to the tip of the coccyx. This "tendon" is pierced by the *Rectum*, which divides it into anterior and posterior portions. The anterior part is known as the **Perineal Body** 2. The posterior part is called the **Anococcygeal Raphé** 3.

The gap between the pubic symphysis and the perineal body is known as the **Urogenital Hiatus** 4. It permits passage of the *Urethra* in both sexes, and the *Vagina* in females.

The three muscles that constitute the Levator Ani are: Pubococcygeus, Puborectalis, and Iliococcygeus.

Pubococcygeus 5 is the most anterior. Its fibers arise from the body of the pubis and pass around the urogenital hiatus to insert into the perineal body.

Puborectalis 6 arises from the pubis and the fascia over *Obturator internus*. It inserts into the anococcygeal raphé, forming a sling that pulls the rectum forward. This is known as the "*Puborectal Sling*," and it must be relaxed in order for fecal matter to pass from the rectum into the anal canal during defecation.

Iliococcygeus 7 arises from the fascia over *Obturator internus* and the ischium, and inserts into the anococcygeal raphé.

Coccygeus

Coccygeus 8 is sometimes referred to as ***Ischiococcygeus.*** Its fibers arise from the ischial spine and run along the sacrospinous ligament to the sacrum and coccyx. Coccygeus is only variably present (it is a much more profound muscle in our primate relatives with tails).

Identify the muscles that comprise the pelvic diaphragm in figure 11.4.

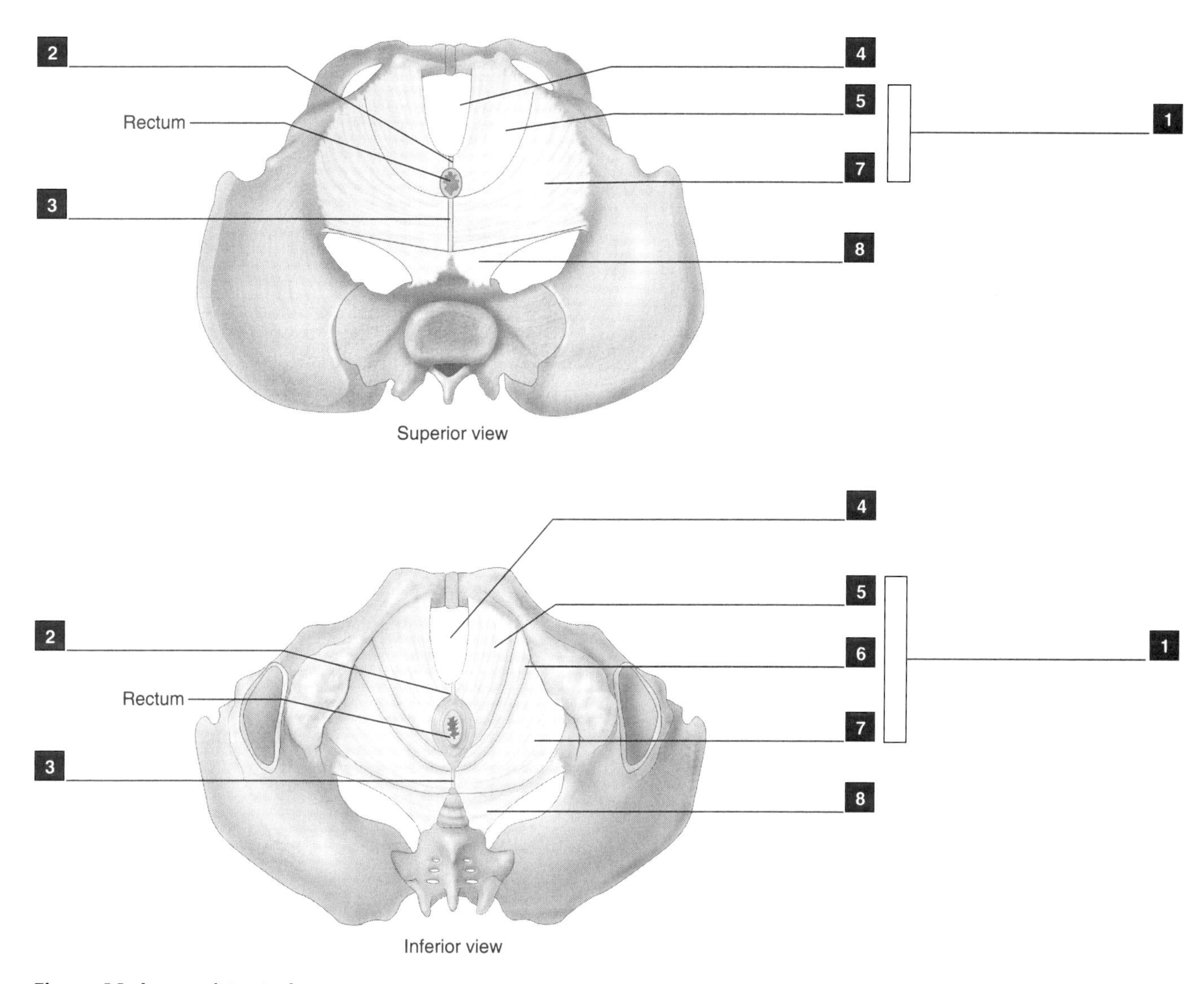

Figure 11.4 **Pelvic Diaphragm**

Muscles of the Perineum

Recall that the *Perineum* is the region of the pelvis below the pelvic diaphragm. The perineum has a diamond-shaped outline when viewed from below. It is bounded anteriorly by the *pubic symphysis*, posteriorly by the tip of the *coccyx*, and laterally by the two *ischial tuberosities*.

The roof of the perineum is formed by the pelvic diaphragm. The floor of the perineum is formed by a sheet of fibrous connective tissue (*Colle's Fascia*) that is simply an extra (deep) layer of the body's superficial fascia. There are several muscles situated within the perineum. They occupy two regions of this space, known as the *Urogenital and Anal Triangles*.

The diamond-shaped outline of the perineum can be thought of as comprising two triangles joined at their bases. The anterior is the urogenital triangle, and the posterior is the anal triangle.

Urogenital Triangle

The **Urogenital Triangle** 1 is divided into superior and inferior compartments by the **Perineal Membrane** 2. This is a flat, fibrous sheet that stretches between the left and right ischiopubic rami. It has a free anterior margin with a small gap between it and the back of the pubic symphysis. Its posterior margin is free except in the midline, where it is attached to the perineal body.

The perineal membrane has an opening that permits passage of the *Urethra* and *Vagina*, which also traverse the urogenital hiatus above it. Five muscles occupy the urogenital triangle. Two muscles lie above the perineal membrane, and three muscles lie below it within the urogenital triangle.

The Anal Triangle

The **Anal Triangle** 3 differs from the urogenital triangle because it does not possess a flat membranous sheet that divides it into upper and lower parts.

The anal triangle is traversed by the **Anal Canal** 4. This is the continuation of the **Rectum** 5 below the pelvic diaphragm, and it extends to the anus. The anal canal is situated behind the perineal body and in front of the anococcygeal raphé. On either side it is surrounded by fat that occupies a wedge-shaped space called the **Ischiorectal (= Ischioanal) Fossa** 6. These fat-filled fossae permit the anal canal to distend during defecation.

Only one muscle (the *External anal sphincter*) occupies the anal triangle.

Identify the components of the urogenital and anal triangles in figure 11.5.

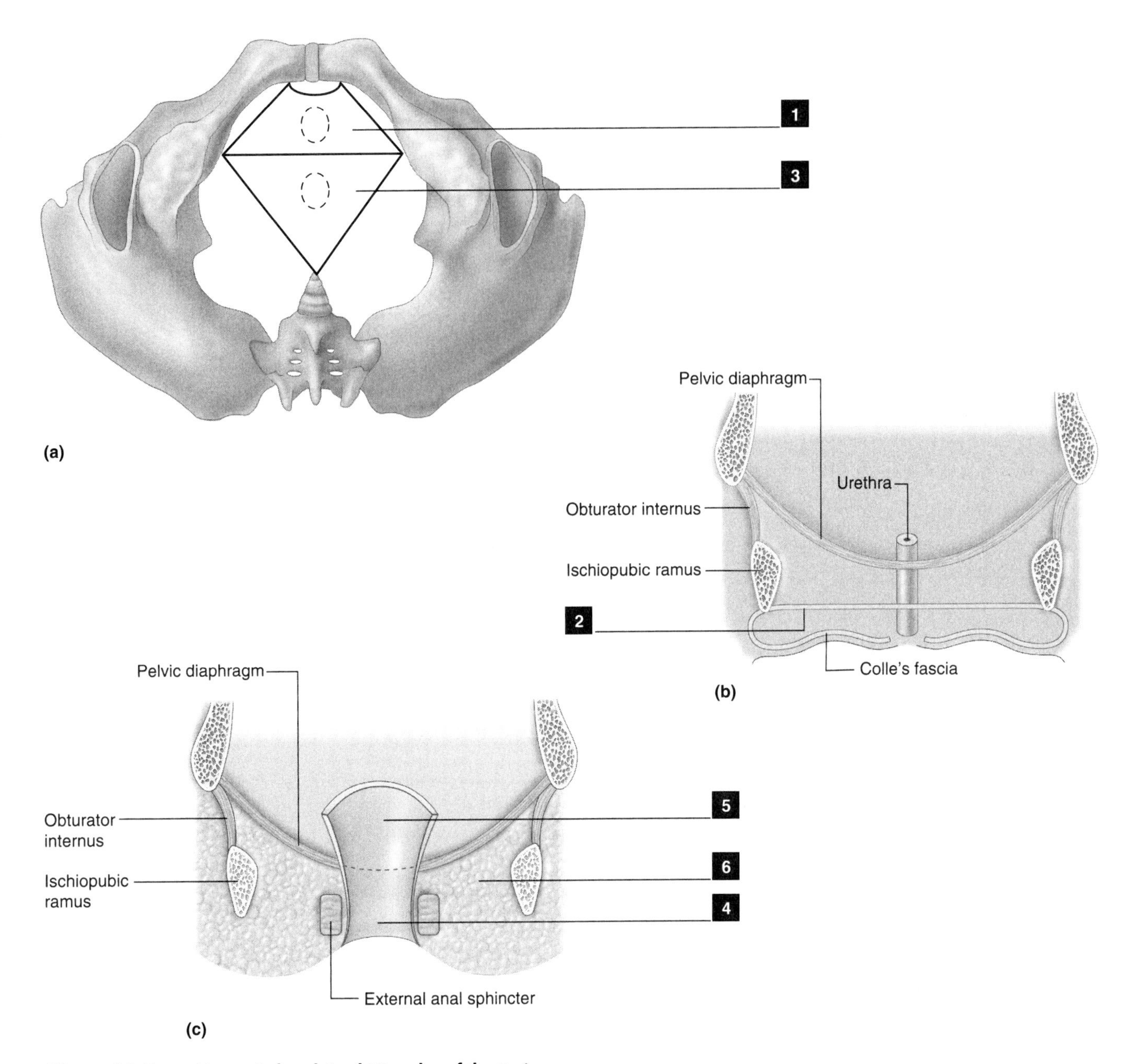

Figure 11.5 **Urogenital and Anal Triangles of the Perineum**
(a) inferior view; (b) coronal section through the urogenital triangle; (c) coronal section through the anal triangle.

Muscles of the Urogenital Triangle

Two muscles lie above the perineal membrane, and three lie below it.

Muscles Superior to the Perineal Membrane

Sphincter Urethrae 1 is a circular muscle that surrounds the urethra in both sexes. It is situated between the pelvic diaphragm and perineal membrane.

In *Males*, the paired *Bulbourethral Glands*, which open into the urethra, are embedded in this muscle.

In *Females*, muscle fibers spin off the sphincter urethrae to encircle the inferior portion of the vagina. These fibers form the *Sphincter urethrovaginalis*.

The second muscle is ***Deep transverse perineus.*** It is more trivial than the sphincter urethrae. Deep transverse perineus arises from the ischial tuberosity and runs medially to insert into the perineal body in both sexes. Its function is unknown.

Muscles Inferior to the Perineal Membrane

Ischiocavernosus 2 arises from the ischiopubic ramus and covers the crus of either the penis or clitoris. The crus is a highly vascular tube of erectile tissue that is attached to the inner surface of the ischiopubic ramus just below the perineal membrane.

Bulbospongiosus 3 arises from the perineal membrane medial to ischiocavernosus.

In *Males*, this single, midline muscle covers the bulb of the penis and the root of the corpus cavernosum of the penis.

In *Females*, this bilateral muscle covers the bulb of the vestibule (homologous with the bulb of the penis). It also covers the *Greater Vestibular Gland* (*Bartholin's Gland*), which is located at the posterior end of the bulb of the vestibule, on either side of the vaginal orifice.

The third muscle is ***Superficial Transverse Perineus.*** It is more trivial than ischiocavernosus and bulbospongiosus. Superficial transverse perineus arises from the edge of the ischial tuberosity and passes medially to insert into the perineal body in both sexes. Its function is unknown.

Identify the muscles of the urogenital triangle in figures 11.6 and 11.7.

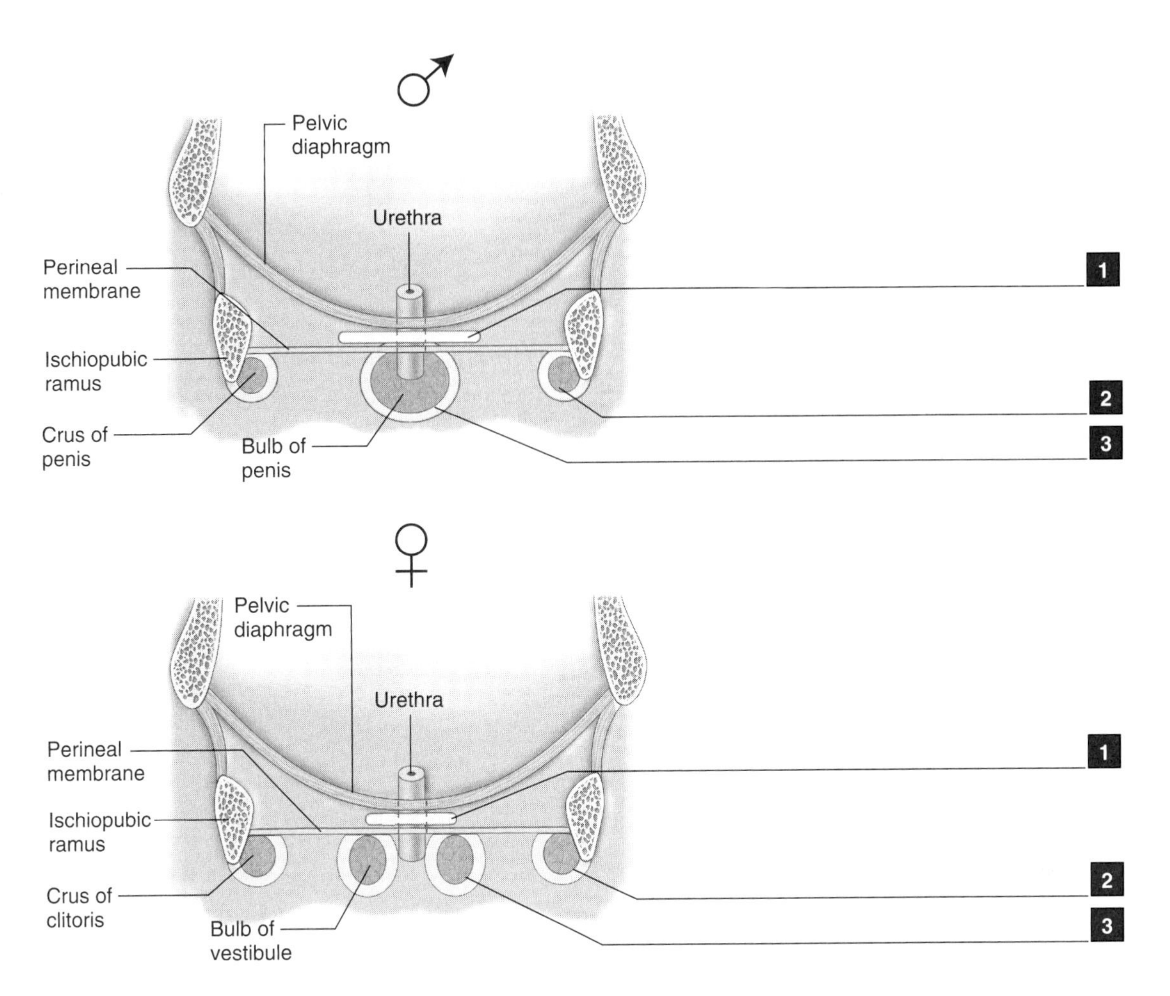

Figure 11.6 **Muscles of the Urogenital Triangle**
Coronal sections through male and female pelves showing muscles above and below the perineal membrane.

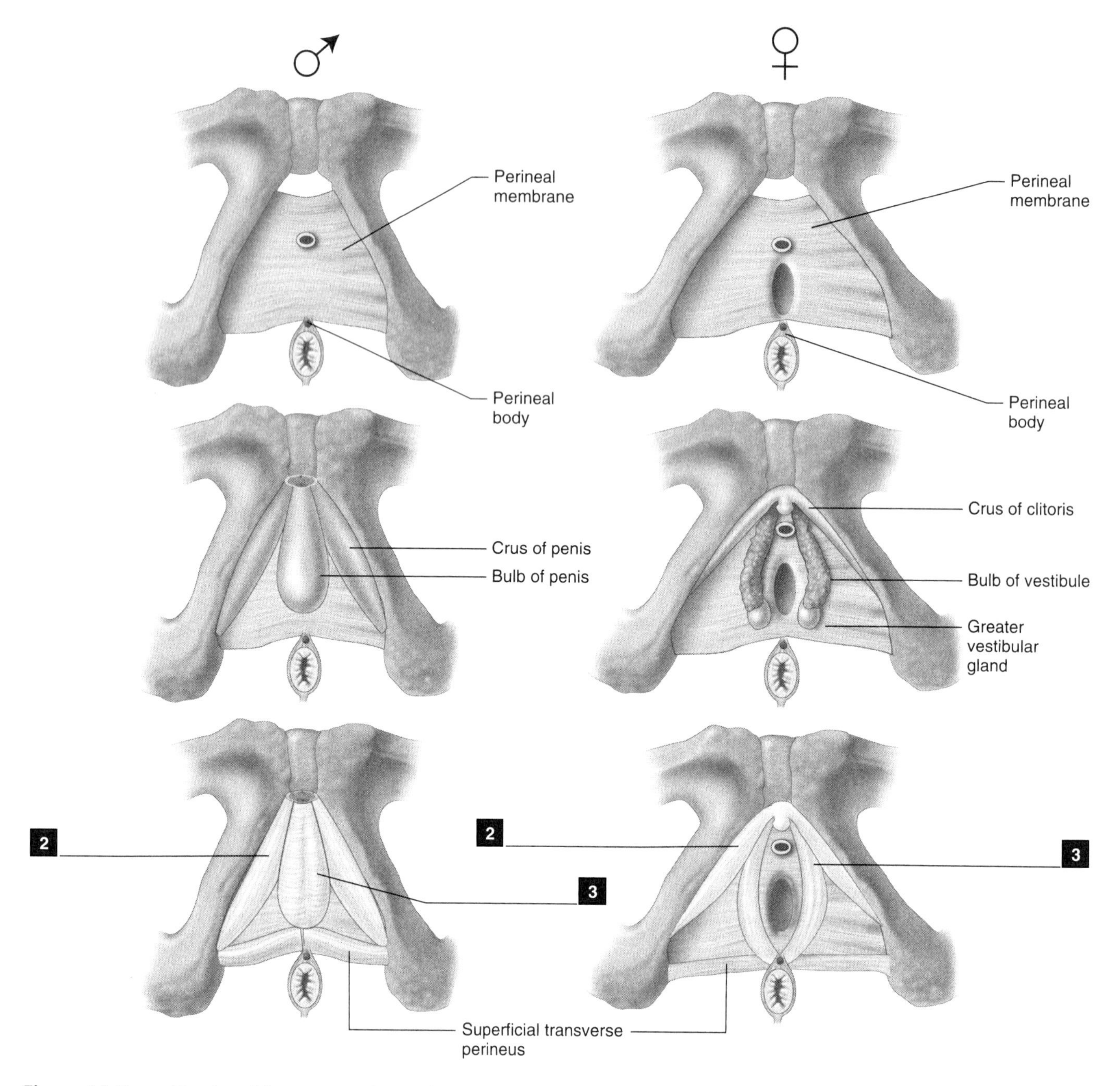

Figure 11.7 **Muscles of the Urogenital Triangle**
Inferior views of male and female pelves showing muscles below the perineal membrane, and their relationship to the penis and clitoris.

Muscle of the Anal Triangle

Recall from laboratory 10 (pp. 302–303) that the wall of the *Anal Canal* is composed of outer longitudinal and inner circular smooth muscle fibers. A thickening of the circular smooth muscle fibers in the upper half of the canal forms an *involuntary internal sphincter.*

There is also a voluntary external anal sphincter that encircles the canal. This is the only skeletal muscle of the anal triangle.

External Anal Sphincter 1 arises from the perineal body and sends fibers around either side of the anal canal to insert into the anococcygeal raphé. It constricts the anal canal, enabling one to be continent.

Identify the external anal sphincter in figure 11.8.

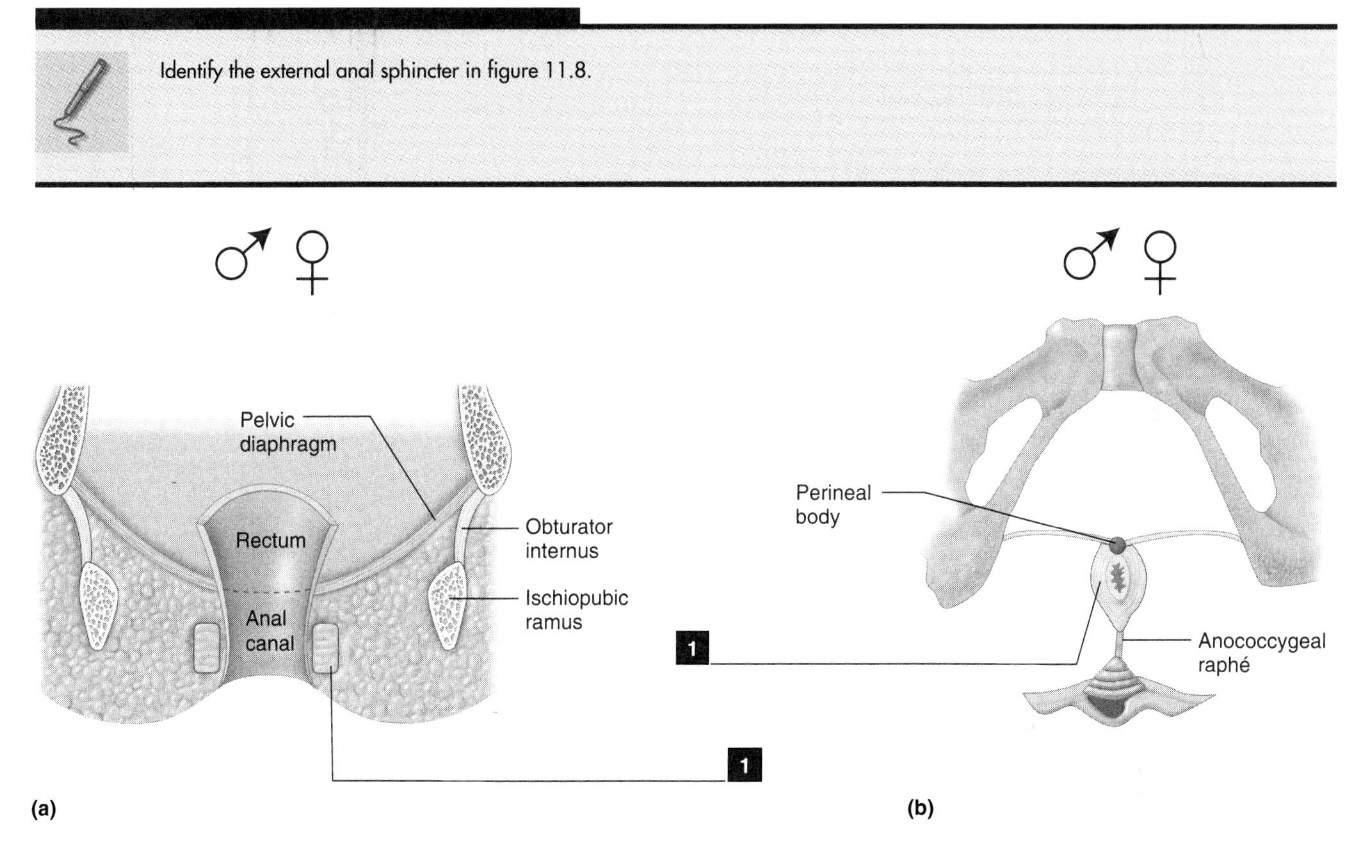

Figure 11.8 **Muscle of the Anal Triangle**
(a) coronal section through pelvis; (b) inferior view of pelvis.

11.3 PERITONEUM IN THE PELVIC CAVITY

Recall that the pelvic cavity is continuous superiorly with the abdominal cavity. The peritoneum that lines the walls of the abdominal cavity extends into the pelvic cavity. Peritoneum covers the small and large intestines within the "false pelvis." The organs that reside within the "true pelvis" are retroperitoneal.

In *Males*, the peritoneum reflects away from the anterior abdominal wall onto the upper surface of the urinary bladder. It runs across the bladder and then reflects onto the anterior surface of the rectum. The gap between the bladder and rectum forms a peritoneal recess known as the **Rectovesical Pouch 1.**

In *Females*, the peritoneum reflects away from the anterior abdominal wall onto the upper surface of the urinary bladder. It runs across the bladder and then reflects onto the anterior surface of the uterus. The gap between the bladder and uterus forms a peritoneal recess known as the **Uterovesical Pouch 2.** The sheet of peritoneum runs over the uterus, where it jumps off onto the anterior surface of the rectum. The gap between the uterus and rectum forms a second, larger peritoneal recess known as the **Rectouterine Pouch 3.**

Identify the peritoneal reflections within the pelvis in figure 11.9.

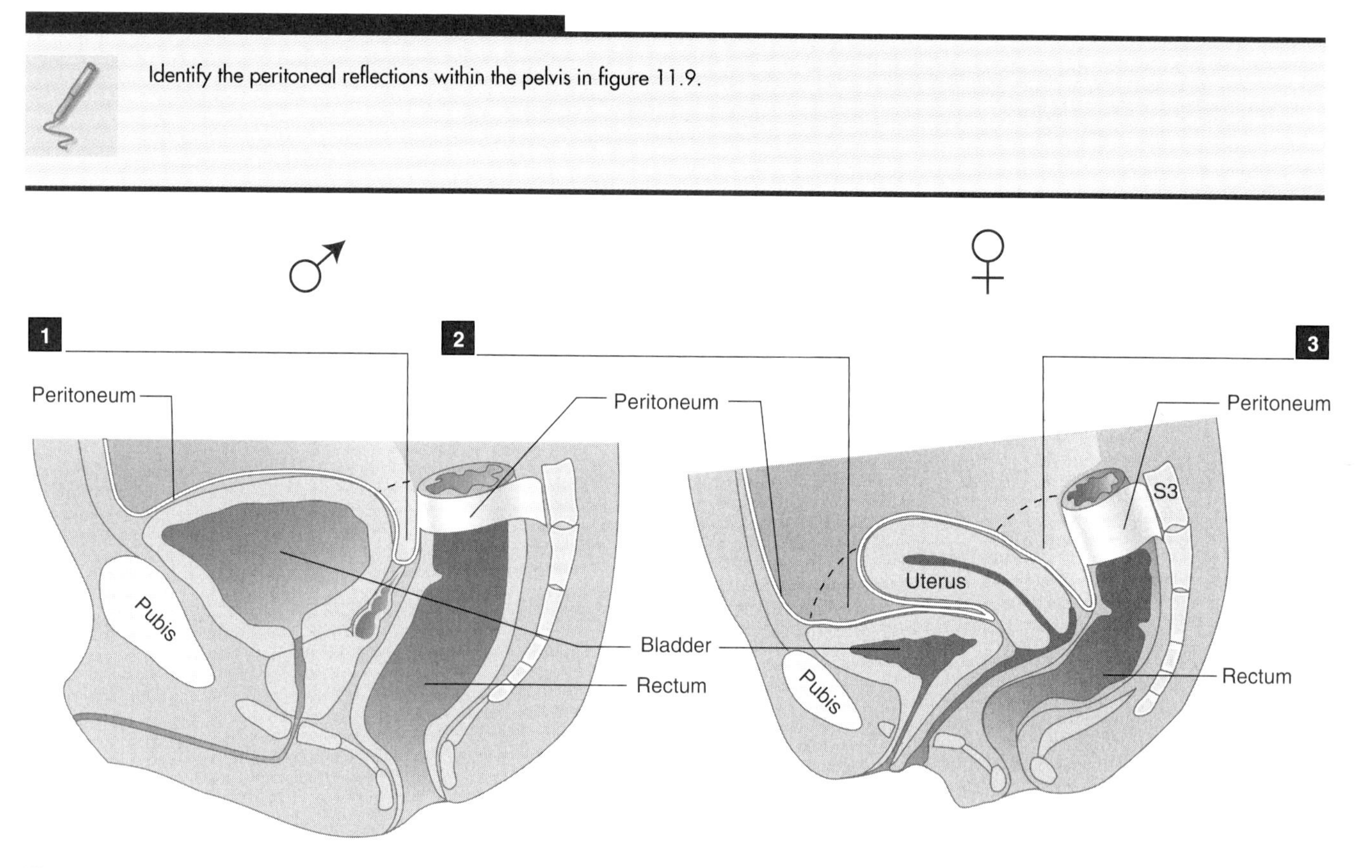

Figure 11.9 **Peritoneal Reflections in the Pelvis**
Sagittal sections.

11.4 COMMON PELVIC VISCERA

The pelvic components of the gastrointestinal and urinary systems occur in both sexes. That is, there is no difference in the anatomy of the rectum and anal canal, and there is no difference in the anatomy of the urinary bladder and ureters. The length and relationships of the urethra differ between the sexes.

We have already studied the gastrointestinal tract (laboratory 10, pp. 302–303). Let us turn our attention to the pelvic components of the urinary system.

Urinary Bladder and Ureters

The urinary bladder is a storage receptacle for urine from the kidneys. Urine passes to the bladder from the kidneys through the two ureters. In adults, the bladder has a capacity of about 1 1/2 pints.

The **Bladder** 1 is situated directly behind the pubic bones, and is composed of three layers of tissue. It sits a little higher in the pelvis in males because of the prostate gland below it.

The innermost layer of the bladder consists of a highly distendable epithelium (urothelium) that is thrown into folds called **Rugae** 2 when the bladder is empty. These flatten out as the bladder fills. The thick middle layer is composed of smooth muscle known as the **Detrusor Muscle** 3. The outermost layer is the **Adventitia** 4. The adventitia is peritoneum, and therefore covers only the upper surface of the bladder.

The **Ureters** 5 are muscular tubes that run along the posterior wall of the abdomen on the front of *Psoas major*. They enter the pelvic cavity at the sacroiliac joint, crossing the bifurcation of the common iliac artery into the external and internal iliac arteries. They then follow the internal iliac artery along the lateral wall of the pelvis to the ischial spine, where they turn medially to enter the posterolateral walls of the bladder.

The ureters enter the bladder obliquely. Thus, each **Ureteric Orifice** 6 can be compressed shut by internal pressure, which prevents a reverse flow of urine as the bladder fills.

Urine passes out of the bladder through the **Urethral Orifice** 7. Here, the circular fibers of the detrusor muscle are thickened to form an involuntary sphincter, known as the **Internal Sphincter Urethrae** 8, or *Sphincter Vesicae*.

The triangular area between the ureteric and urethral orifices is called the **Trigone** 9. The mucous membrane lining the trigone is always smooth, even when the bladder is empty.

Identify the ureters and the components of the bladder in figure 11.10.

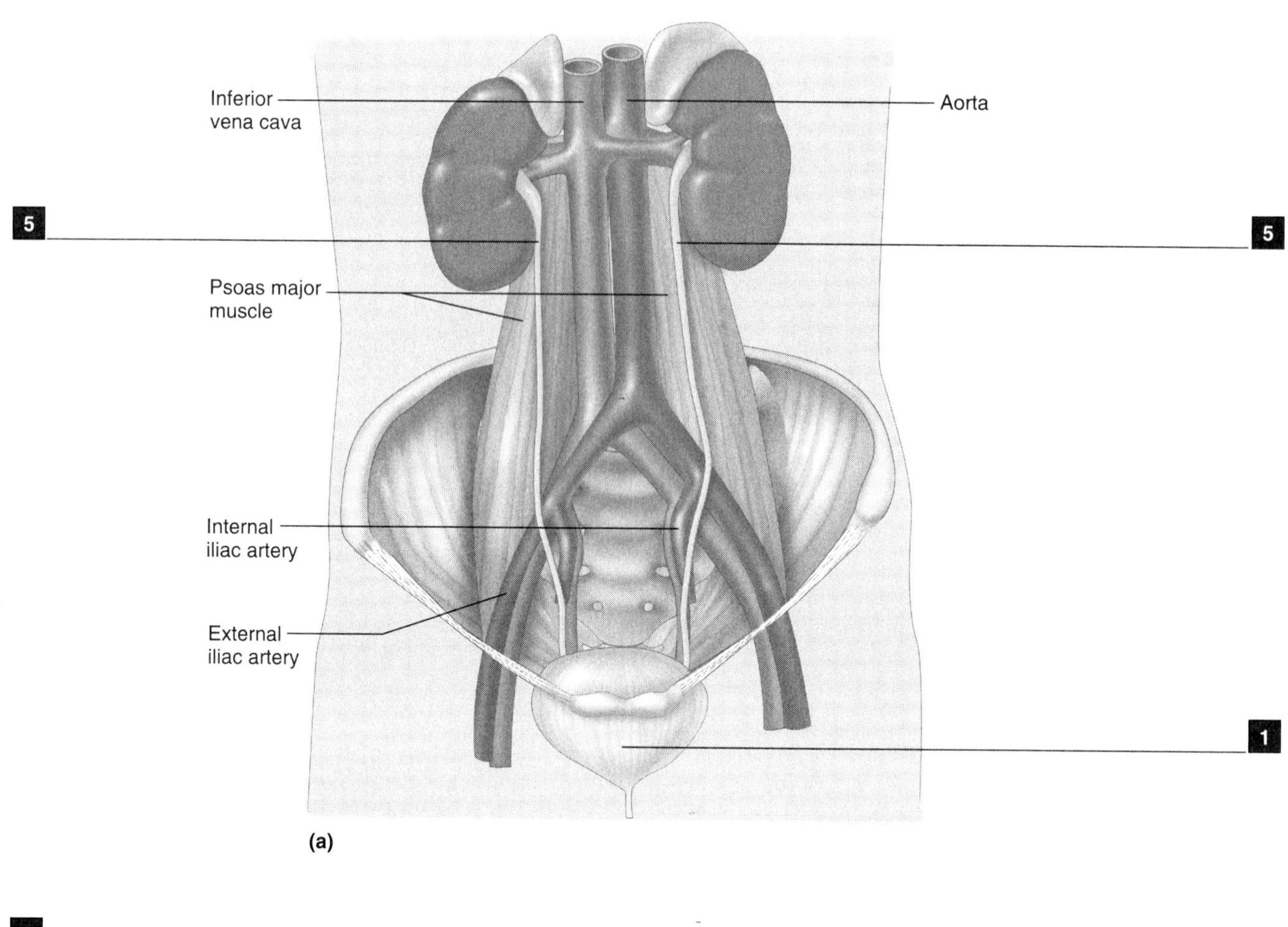

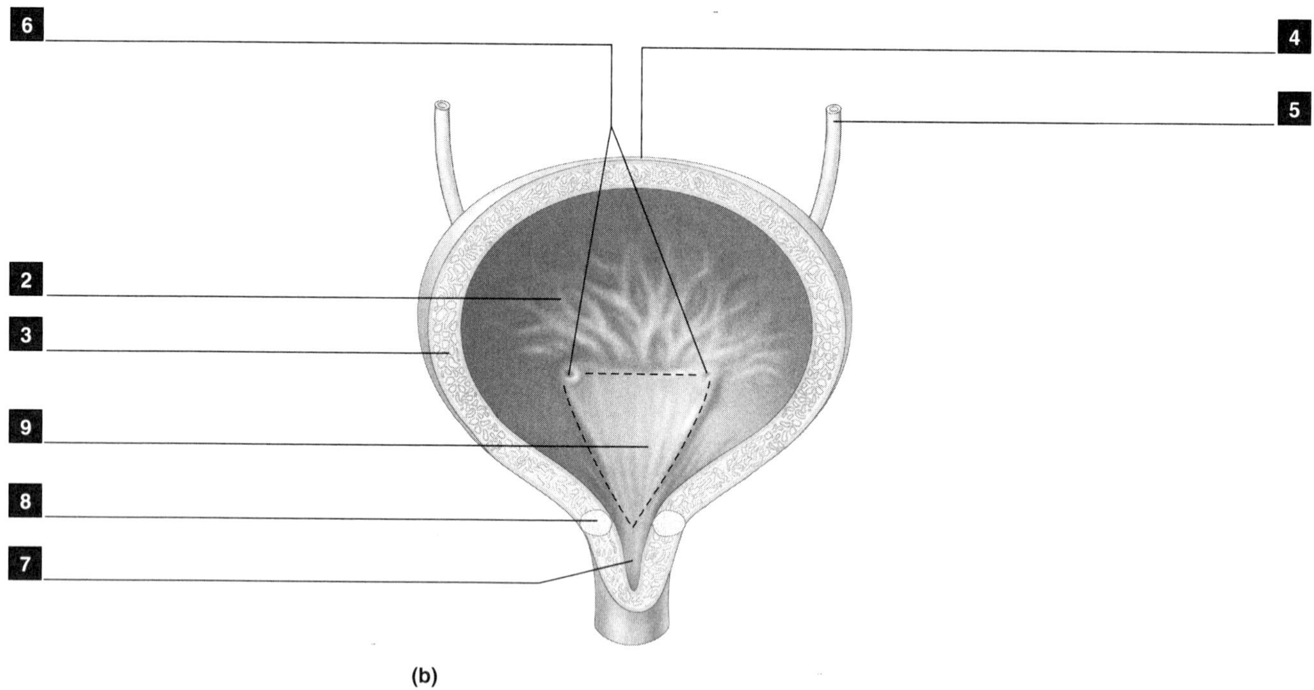

Figure 11.10 **Urinary Bladder and Ureters**
(a) anterior view showing the anatomical relationships of the ureters; (b) bladder cut open.

11.5 MALE GENITALIA

The role of the male reproductive system is to produce sperm, and deliver them into the vagina of the female. The following structures comprise the male genitalia:

Scrotum	1	Ejaculatory Duct	6
Testis	2	Prostate Gland	7
Epididymis	3	Bulbourethral Gland	8
Ductus Deferens	4	Urethra	9
Seminal Vesicle	5	Penis	10

Identify each of these in figure 11.11. We will examine each in more detail in turn.

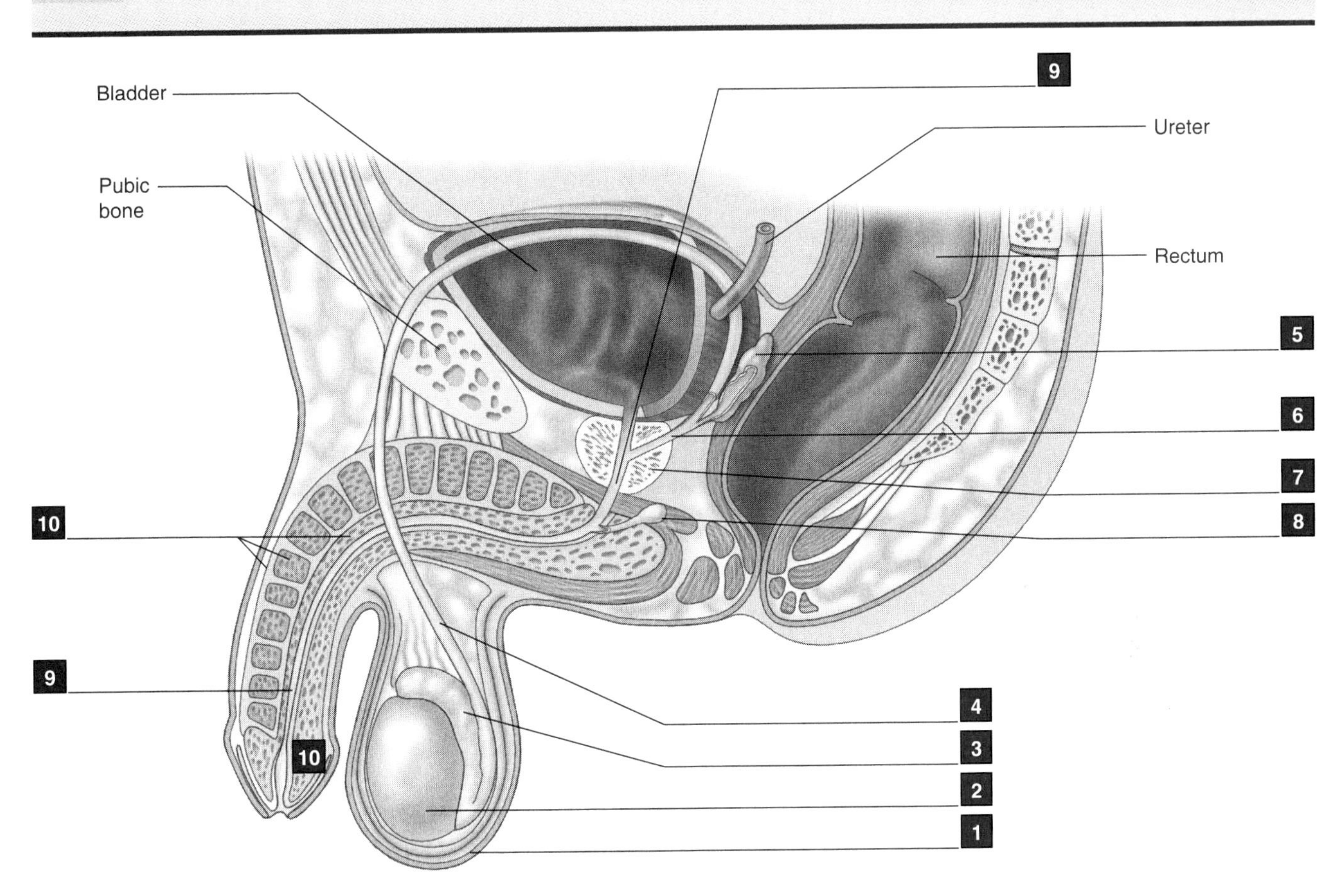

Figure 11.11 **Male Genitalia**
Schematic sagittal section through a male pelvis.

Scrotum

The scrotum is an outpouching of the lower part of the anterior abdominal wall. It is partly divided into two compartments by a fibrous median septum, which is expressed externally by a midline ridge of skin, the *Scrotal Raphé*. Each compartment contains a testis, epididymis, and spermatic cord.

The scrotum is composed of a number of tissue layers that cover each testis.

- The outermost layer is skin.
- Beneath the skin is a layer of smooth muscle known as the **Dartos Muscle** 1. It is responsible for the wrinkled appearance of the scrotal skin.
- Deep to the dartos muscle there are three layers of tissue that constitute the **Spermatic Fascia** 2. They derive from the muscle and fascial layers of the anterior abdominal wall (*External oblique* muscle, *Internal oblique* muscle, *Transversalis fascia*). The middle layer of the spermatic fascia is the **Cremaster Muscle** 3. Its contraction brings the testis closer to the body when scrotal temperature falls.
- The spermatic fascia surrounds the ductus deferens, as well as the lymphatic ducts, autonomic nerves, and blood vessels that supply the testis. This "package" of ducts, nerves, and blood vessels constitutes the **Spermatic Cord** 4.
- The deepest layer, the *Tunica Vaginalis*, adheres to the testis.

The structures within the spermatic cord reach the testis from the abdominopelvic cavity by way of the *Inguinal Canal*.

The inguinal canal runs along the **Inguinal Ligament** 5. Its entrance from the abdominopelvic cavity is known as the **Deep Inguinal Ring** 6, and its exit into the scrotum is known as the **Superficial Inguinal Ring** 7.

In fetal development, the testis descends from the abdominopelvic cavity into the scrotum through the inguinal canal. In its descent, the testis encounters the muscle and facial layers of the anterior abdominal wall, and drags these layers with it into the scrotum through the canal.

Excessive intraabdominal pressure can force part of the small intestine into the inguinal canal. This constitutes an ***Inguinal Hernia.*** An inguinal hernia is diagnosed by feeling for a bulge through the superficial inguinal ring. Inguinal hernias are more common in males than females. Why do you think this should be so? (The last paragraph provides the answer.)

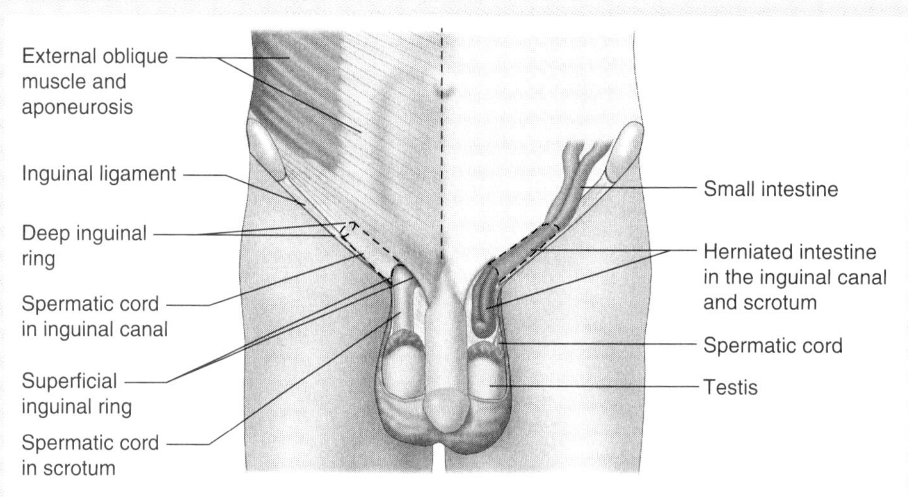

Inguinal Canal and Inguinal Hernia

Identify the components of the scrotum in figure 11.12.

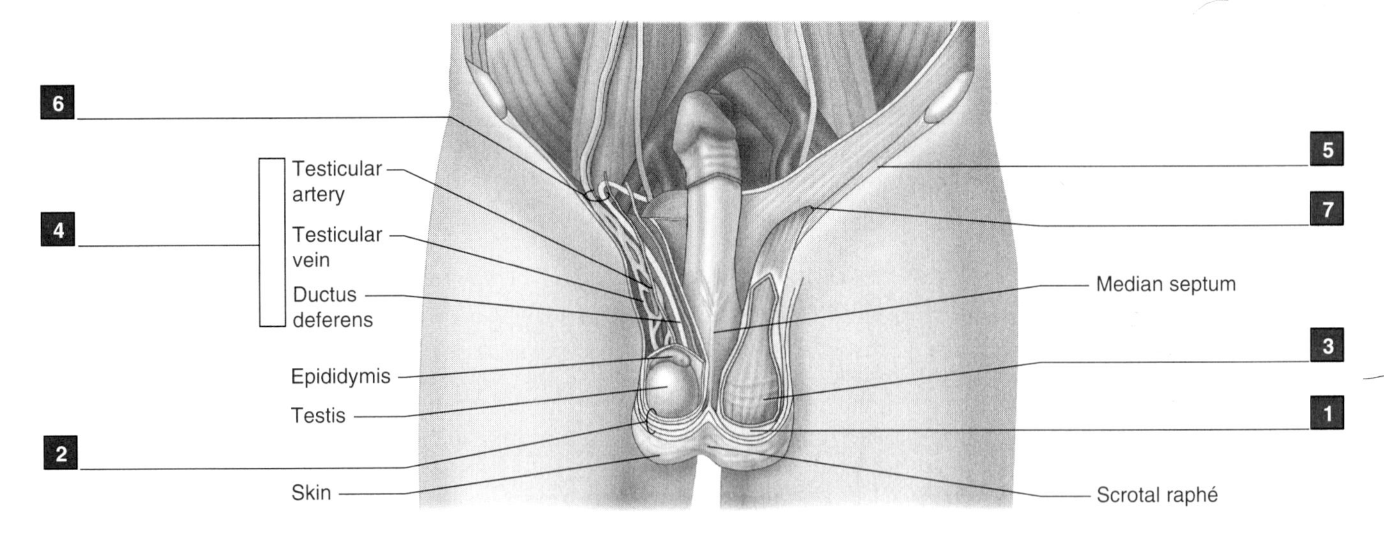

Figure 11.12 **Scrotum and Its Contents**

Testis

The egg-shaped **Testis** 1 is responsible for the production of sperm and androgens (testosterone). It is covered by a fibrous capsule, the **Tunica Albuginea** 2, from which extend a number of fibrous septa that divide the interior of the testis into **Lobules** 3. Each lobule contains several **Seminiferous Tubules** 4, which are responsible for the production of sperm. The seminiferous tubules open into a network of channels called the **Rete Testis** 5. These drain into a series of **Efferent Ducts** 6 that pierce the tunica albuginea and open into the epididymis.

The testicular artery and vein enter and leave the testis at its *Hilum*. The vein forms a dense network, known as the *Pampiniform Plexus*, around the ductus deferens within the spermatic cord.

Epididymis

The **Epididymis** 7 is a highly coiled tube that is responsible for the storage of sperm until they are ejaculated. The epididymis has a **Head** 8, into which open the efferent ducts from the testis. Its **Body** 9 extends down the posterolateral side of the testis, at the bottom of which the **Tail** 10 of the epididymis begins to straighten out and enlarge to turn upward as the *Ductus Deferens*.

Ductus Deferens

The **Ductus Deferens (= Vas Deferens)** 11 is a continuation of the tube that comprises the epididymis. It is about 18 inches long. It ascends through the scrotum covered by the fascial layers of the *Spermatic Cord*. It traverses the inguinal canal and, leaving the cover of the spermatic cord, enters the abdominopelvic cavity over body of the pubic bone.

Upon entering the abdominal cavity, the ductus deferens passes onto the urinary bladder. It runs posteriorly over the bladder, crossing the ureter, and then turns downward to run along the medial border of an accessory gland, the *Seminal Vesicle*, which is located on the back of the bladder.

The terminal portion of the ductus deferens is expanded. This is known as the *Ampulla*. The seminal vesicle opens into the ampulla, at which point it becomes known as the *Ejaculatory Duct*. This then pierces the *Prostate Gland* to open into the lumen of the *Urethra*.

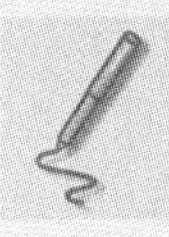

Identify the aforementioned structures in figure 11.13.

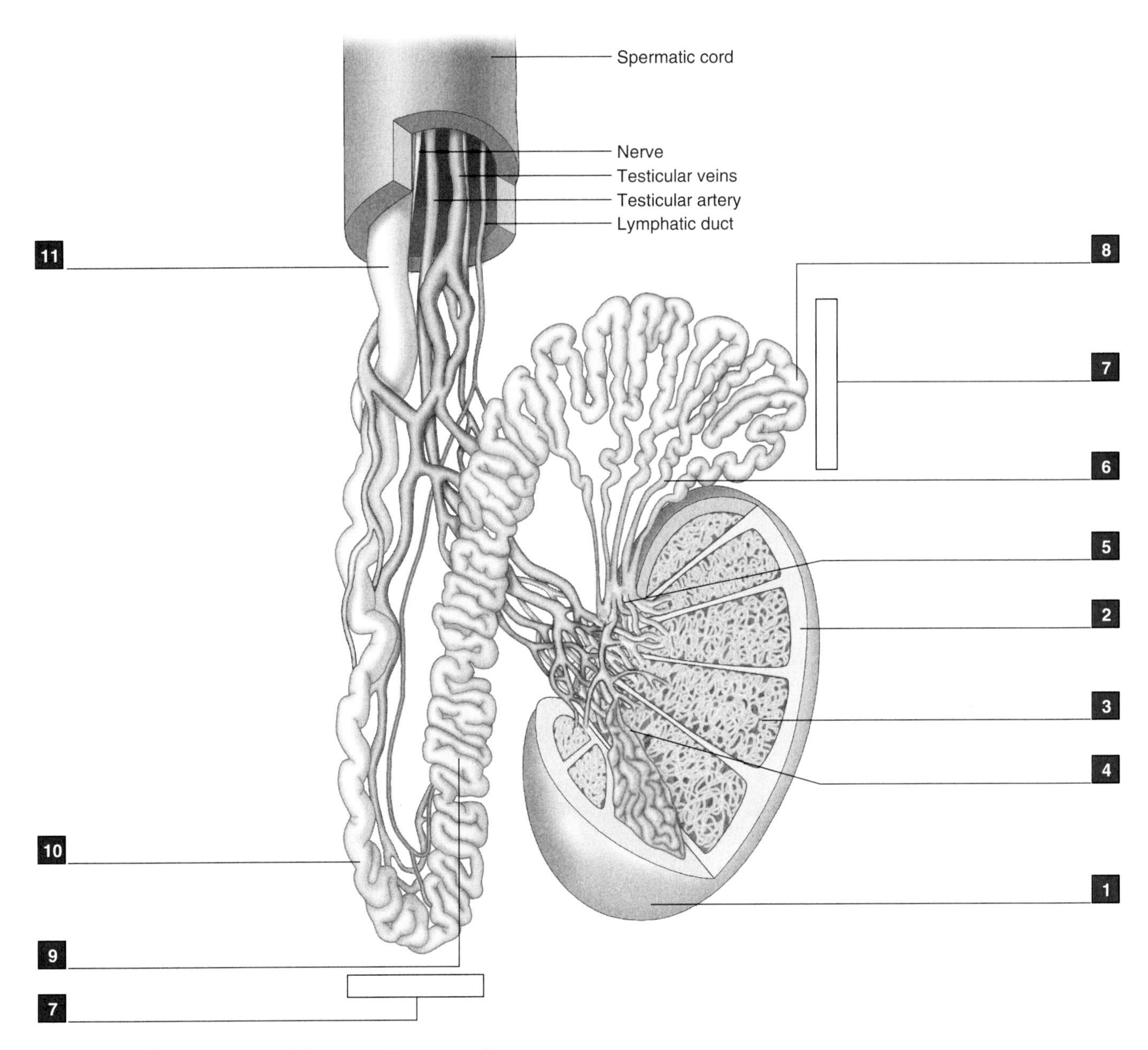

Figure 11.13 **Testis, Epididymis, and Ductus Deferens**

Seminal Vesicle and Ejaculatory Duct

The **Seminal Vesicles** 1 are lobulated organs, about 2 inches in length, that lie on the posterior surface of the bladder. They secrete a fructose-rich alkaline fluid that constitutes a good portion of the *Seminal Fluid* that is expelled upon ejaculation. This fluid nourishes the sperm, and helps neutralize the natural acidity of the vagina.

The duct that leads from the seminal vesicle opens into the distal end of the ampulla of the ductus deferens. The tube that is formed by the union of the ductus deferens and the duct from the seminal vesicle is known as the **Ejaculatory Duct** 2.

The ejaculatory duct pierces the posterior surface of the prostate gland, and opens into the urethra.

Prostate Gland

The **Prostate Gland** 3 is a cone-shaped mass of glandular tissue about the size of a chestnut. It surrounds the first part of the urethra, known as the *Prostatic Urethra*. The prostate gland is surrounded by a thin fibrous capsule and a layer of smooth muscle. The glandular secretions are squeezed through tiny openings in the prostatic urethra by this smooth muscle.

Prostatic secretions are released during ejaculation. They mix with the seminal fluid and sperm to produce *Semen*.

Enlargement of the prostate gland may partially occlude the prostatic urethra, causing difficulty in completely emptying the bladder. This condition is common in men over age 70.

Bulbourethral Glands

Immediately below the prostate gland is the *Sphincter urethrae* muscle. Embedded within the substance of this muscle are two, pea-sized **Bulbourethral Glands** 4, also known as ***Cowper's Glands***. They secrete a clear, alkaline fluid that serves as a lubricant, and to neutralize urethral acidity. Each gland has a duct that opens into the urethra below the prostate. The 1/2 inch long portion of the urethra between the prostate and the bottom of the sphincter urethrae is known as the *Membranous Urethra*.

Identify the aforementioned structures in figure 11.14.

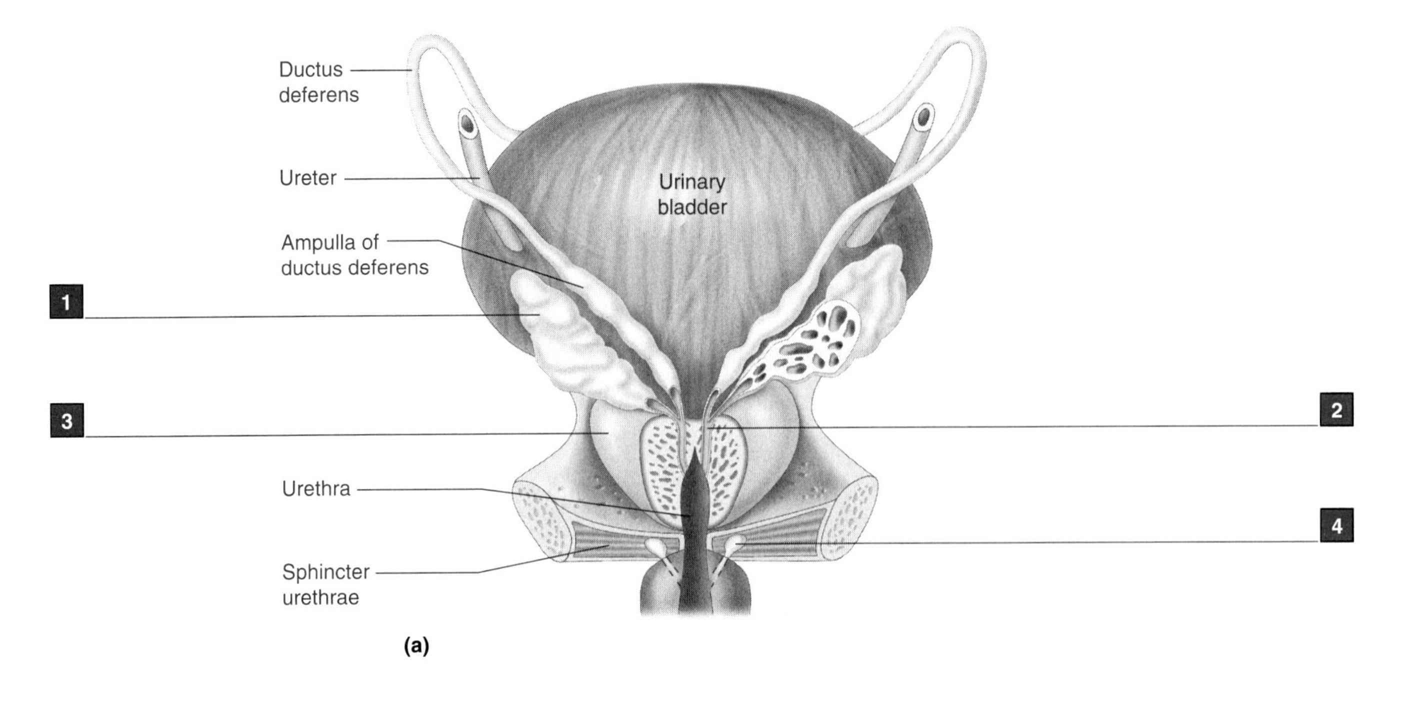

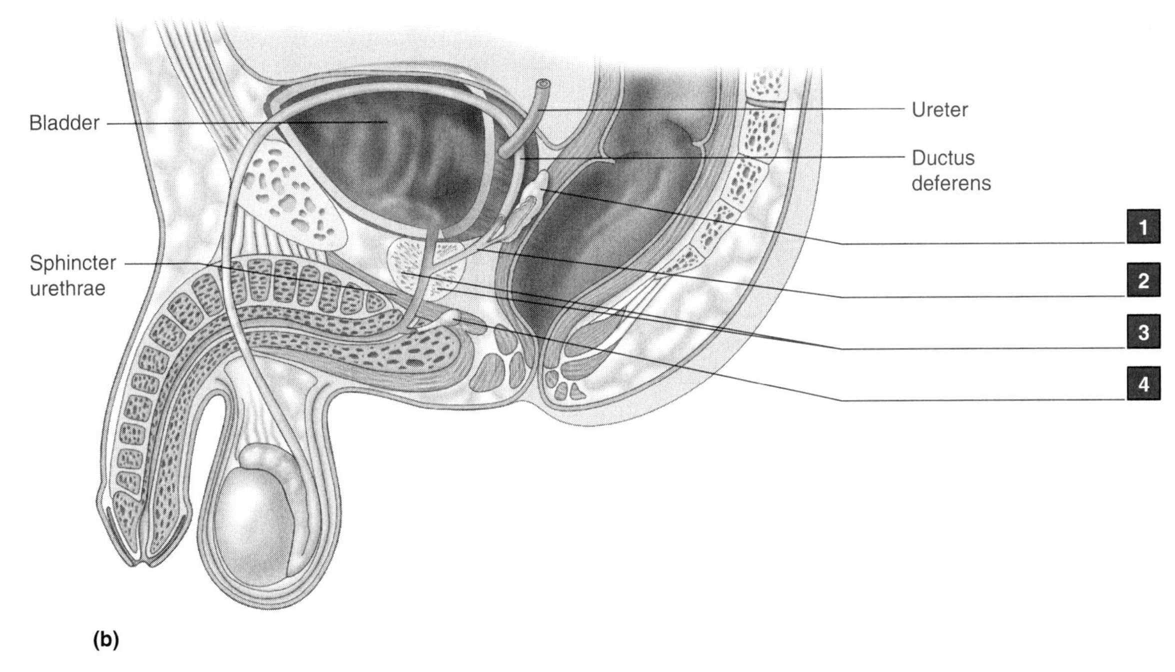

Figure 11.14 **Seminal Vesicle, Prostate, and Bulbourethral Glands**
(a) posterior view with prostate cut open; (b) schematic sagittal section.

Penis

The penis comprises three bodies of erectile tissue surrounded by tubular sheaths of fascia and skin. The erectile tissue bodies are the midline corpus spongiosum and the left and right corpora cavernosa. The urethra runs through the corpus spongiosum.

The midline **Corpus Spongiosum** **1** has an expanded root, known as the **Bulb of the Penis** **2**. This lies below the perineal membrane, and is covered by the *Bulbospongiosus* muscle. The **Urethra** **3** enters the bulb, and continues forward through the substance of the corpus spongiosum. Here it is known as the *Spongy Urethra*. The distal end of the corpus spongiosum is expanded as the **Glans Penis** **4**. The urethra opens at the tip of the glans penis by the *External Urethral Orifice*.

The root of each **Corpus Cavernosum** **5** is known as the **Crus of the Penis** **6**. They run along the underside of the perineal membrane adjacent to the ischiopubic ramus, and are covered by the *Ischiocavernosus* muscles. The corpora cavernosa extend distally along the shaft of the penis dorsal to the corpus spongiosum. A prominent artery, known as the **Deep Artery** **7**, runs the length of each corpus cavernosum. The corpora are surrounded by a layer of deep fascia, which creates a porous midline septum between them. Their distal ends are covered by the glans penis.

Along the dorsal surface of the fascia sheath covering the corpora cavernosa run the large *Dorsal Vein*, the *Dorsal Nerves* and a pair of *Dorsal Arteries* of the penis.

The skin that covers the body of the penis is loosely attached, and extends over the glans penis as a retractable sheath called the **Prepuce (= Foreskin)** **8**. It is attached to the ventral surface of the glans penis by a vertical fold known as the *Frenulum*.

Erection and Ejaculation

Erection of the penis occurs when efferent nerve impulses pass down the spinal cord to the ***Parasympathetic*** outflow from S2–S4. The parasympathetic fibers are distributed with the deep arteries in the corpus cavernosa. Here they produce vasodilation, which results in an increased blood flow into the erectile tissue. The expanded corpora cavernosa further compress the veins of the penis against the surrounding fascia. The resultant retardation of blood outflow from the erectile tissues further accentuates and helps maintain internal pressure within the erectile bodies.

Friction of glans penis results in a discharge along the ***Sympathetic*** nerve fibers to the smooth muscle of the epididymis, seminal vesicles, and prostate gland. The sperm and glandular secretions that form the semen are propelled from the penis upon ejaculation. Ejaculation results from the rhythmic contractions of the bulbospongiosus muscle, which compresses the urethra.

Identify the components of the penis in figure 11.15.

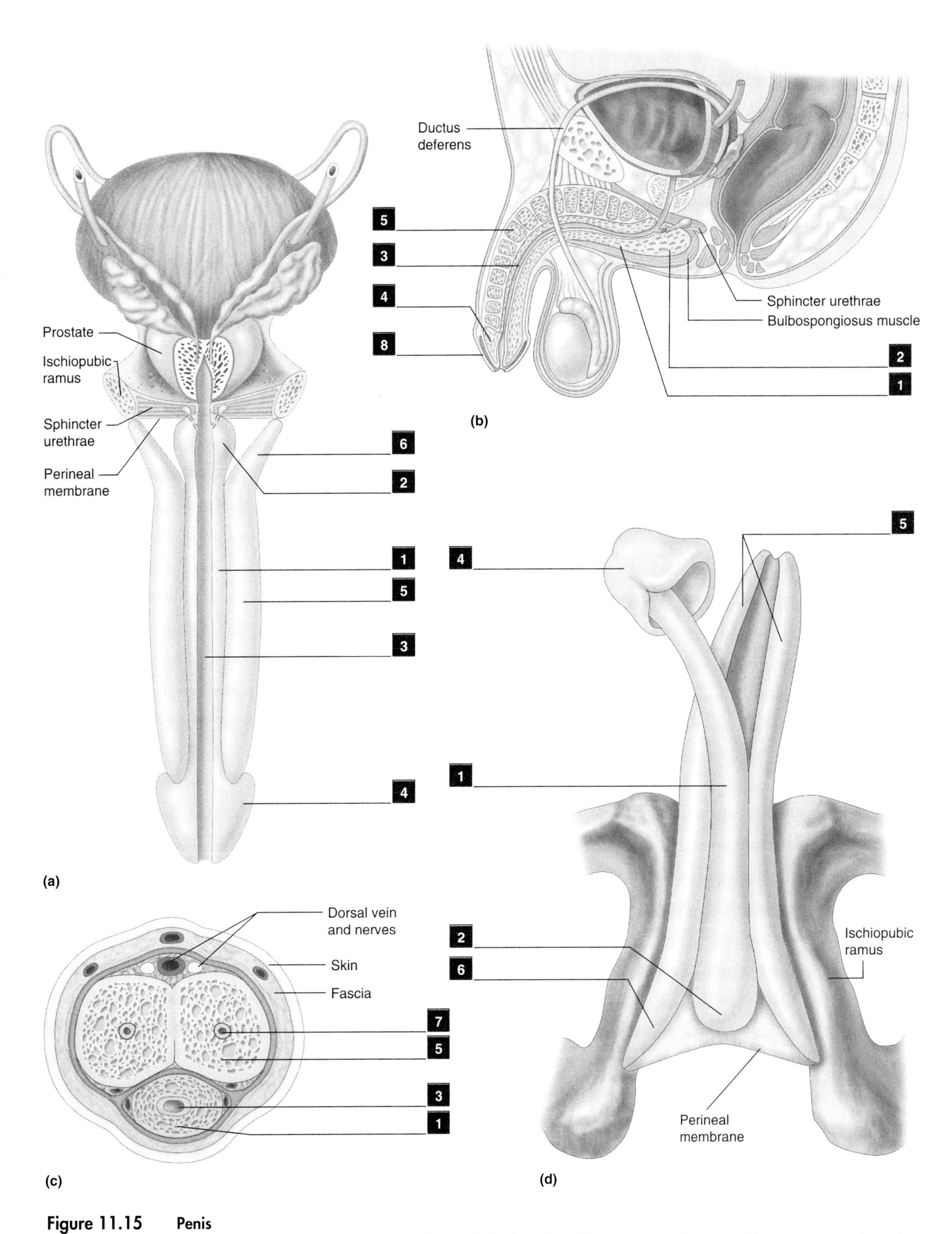

Figure 11.15 **Penis**
(a) schematic posterior-ventral view of erectile bodies; (b) schematic sagittal section; (c) cross section through shaft; (d) inferior view of erectile bodies (partially separated).

11.6 FEMALE GENITALIA

The role of the female reproductive system is to produce ova and to provide an environment for their fertilization and for embryonic and fetal development. The following structures are the major components of the female genitalia:

Labia Majora	1	Uterus	5
Labia Minora	2	Uterine Tube	6
Clitoris	3	Ovary	7
Vagina	4		

Identify each of these in figure 11.16. We will examine each in more detail in turn.

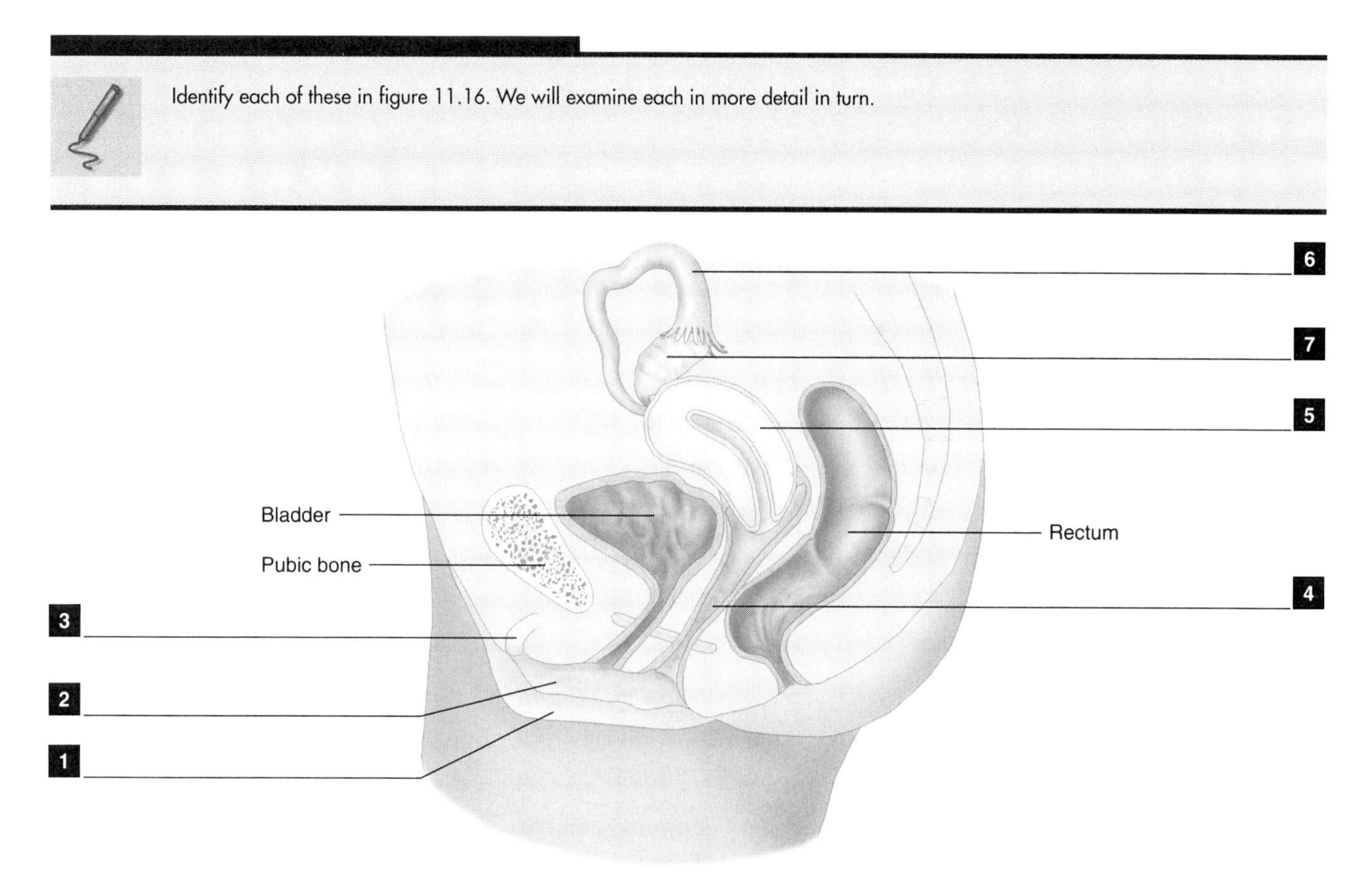

Figure 11.16 **Female Genitalia**
Schematic sagittal section through a female pelvis.

Labia Majora and Labia Minora (Vulva and Vestibule)

The labia majora and minora, and the clitoris are the principal features of the external genitalia, or *Vulva*. The Vulva consists of those structures that occupy the perineum below the urogenital triangle. These also include the mons pubis, vestibule, and vestibular glands.

The *Mons Pubis* is a mound of adipose and areolar tissue that covers the ventral aspect of the pubic bones and symphysis. It is continuous posteriorly with the two labia majora.

The **Labia Majora** 1 form the outer borders of the vulva, and contain adipose and areolar tissue from the mons pubis. Smooth muscle fibers (homologous to the *dartos* muscle of the male scrotum) lie deep to the skin. Posteriorly, the labia majora lose some of their adipose tissue where they join at the *Posterior Labial Commissure* just anterior to the anus.

The **Labia Minora** 2 are thin folds of skin medial to the labia majora. They contain connective tissue rather than fat. Anteriorly, each labium splits into two folds that join with the folds from the opposite side anterior and posterior to the clitoris. The conjoint anterior folds form the *Prepuce*, and the posterior folds form the *Frenulum* of the clitoris.

The ***Vestibule*** is the space between the labia minora.

The vestibule contains the **Urethral Meatus** 3 and, just posterior to it, the **Vaginal Orifice** 4**.** Around the circumference of the vaginal orifice there is commonly a thin fold of mucous membrane known as the **Hymen.** It may partially or even completely occlude the orifice until it is ruptured.

The **Clitoris** 5 is situated at the anterior end of the vestibule. Its free distal end is the *Glans*. It is partially shrouded by the prepuce.

Opening into the vestibule are two pairs of mucous-secreting glands. The anterior pair are the **Lesser Vestibular (*Skene's*) Glands** 6**,** which open on either side of the urethral meatus. The posterior pair are the **Greater Vestibular (*Bartholin's*) Glands** 7**.** They lie at the posterior ends of the bulb of the vestibule, and are covered by the bulbospongiosus muscle. Their ducts open on either side of the vaginal orifice. Their secretions aid in vaginal lubrication.

Identify the aforementioned structures in figure 11.17.

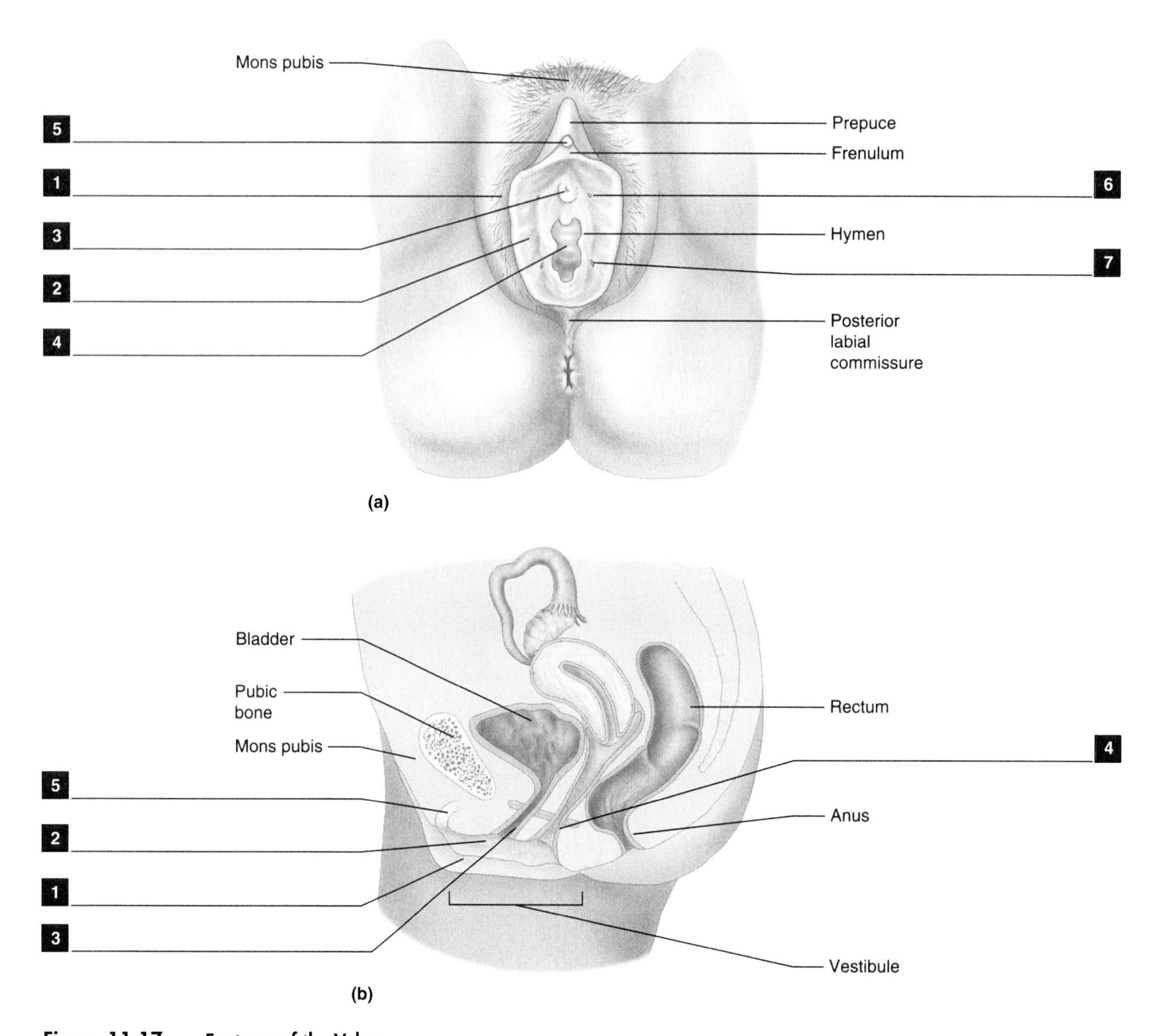

Figure 11.17 **Features of the Vulva**
(a) inferior view of vulva and vestibule; (b) schematic sagittal section through a female pelvis.

Clitoris

The clitoris is located at the anterior end of the vestibule. Its structure is homologous to that of the penis. Thus, it is composed of bodies of erectile tissue (four rather than three), and it has a crus, a bulb, and a glans.

The *Root* of the clitoris is made up of four masses of erectile tissue that lay against the inferior surface of the perineal membrane. Two of the bodies are medial and two are lateral.

The left and right **Bulb of the Vestibule** 1 constitute the medial masses of erectile tissue. They lie along either side of the urethral meatus and vaginal orifice. Recall that each is covered inferiorly by the *Bulbospongiosus* muscle. Its posterior end partially covers the *Greater Vestibular Gland*. Anteriorly, the left and right bulbs of the vestibule unite to form a short **Corpus Spongiosum** 2. The anterior (distal) extremity of the corpus spongiosum is expanded to form the **Glans Clitoris** 3. It is partially covered by the prepuce of the labia minora.

The left and right **Crus of the Clitoris** 4 constitute the lateral masses of erectile tissue. They lie along the ischiopubic ramus. Recall that each is partially covered by the *Ischiocavernosus* muscle. Their anterior ends form the **Corpora Cavernosa** 5 of the clitoris. They merge with the corpus spongiosum to form the *Body* of the clitoris. The distal ends of the corpora cavernosa are covered by the glans clitoris.

Erection and Orgasm

Erection of the clitoris occurs when efferent nerve impulses pass down the spinal cord to the ***Parasympathetic*** outflow from S2–S4. The parasympathetic fibers are distributed to the deep arteries within the corpora cavernosa. Here they produce vasodilation, which results in an increased blood flow into the erectile tissue. The expanded corpora cavernosa and the bulbospongiosus muscles that cover the bulbs of the vestibule compress the deep dorsal vein of the clitoris. This retards blood outflow from the erectile tissues. Contractions of the ischiocavernosus muscle assist in the erectile process by forcing more blood into the corpora cavernosa.

Friction of glans clitoris and/or vaginal orifice results in a climax of sensory impulses that reach the brain, resulting in the sensation of orgasm. Simultaneously, there is discharge along the ***Sympathetic*** nerve fibers to the smooth muscle of the vagina, and somatic discharge via the pudendal nerve to the bulbospongiosus and ischiocavernosus muscles, which undergo rhythmic contractions.

Identify the components of the clitoris in figure 11.18.

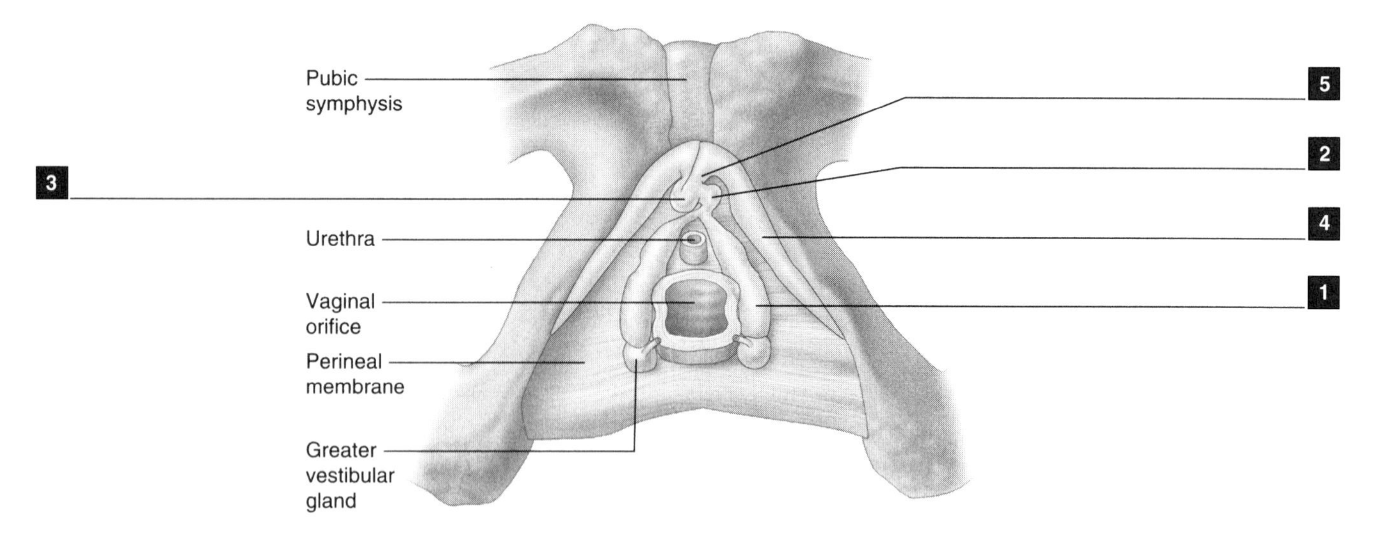

Figure 11.18 **Clitoris**

Vagina

The **Vagina** 1 is one of the structures that comprise the internal genitalia. The other structures include the uterus, uterine tubes, and ovaries.

The vagina is a tube of smooth muscle and elastic tissue that extends upward and posteriorly for about 3 inches from its orifice. It is situated between the urinary bladder and urethra, and the rectum. It traverses the *Perineal Membrane* and *Levator Ani*.

The vagina is lined with mucous membrane that is thrown into folds known as *Rugae*. Its upper end is pierced by the *Uterine Cervix*. The cervix projects downward and backward into the vaginal lumen, which forms a recess known as a **Fornix** 2, around it.

Uterus

The uterus is a hollow, pear-shaped organ with thick muscular walls. In a nulliparous woman, it is about 3 inches long, 2 inches wide, and 1 inch thick from front to back. It is divisible into three parts: fundus, body, and cervix. The **Fundus** 3 lies superior to the uterine cavity. The **Body** 4 surrounds the uterine cavity below the entrance of the uterine tubes, and it tapers inferiorly to become continuous with the cervix. The **Cervix** 5 extends into the vagina.

The uterine cavity is triangular in "coronal" projection, but merely a narrow slit in sagittal section. It extends through the cervix, where it is known as the **Cervical Canal** 6, and opens into the vagina at the **External Os** 7.

The long axis of the uterus is usually bent forward to the long axis of the vagina, such that the cervical canal and vagina form an angle of approximately 90 degrees. This is referred to as uterine ***Anteversion.*** In addition, the body of the uterus is bent forward to the long axis of the cervix by about 10 degrees. This is referred to as uterine ***Anteflexion.*** As a result of anteversion and anteflexion, the body and fundus lie nearly horizontally.

The uterus is composed of two layers of tissue, and is largely covered by peritoneum. Most of the uterine wall is composed of a layer of smooth muscle called the *Myometrium*. Internal to the myometrium is a highly vascular layer called the **Endometrium** 8. Most of it is shed each month during *Menstruation* if an ovum has not been implanted in it.

Uterine Tubes

The two **Uterine (Fallopian) Tubes** 9 connect the peritoneal cavity adjacent to each ovary with the uterine cavity. Each tube is about 4 inches long, and its lateral end is greatly expanded to form the **Infundibulum** 10. The opening of the infundibulum next to the ovary is known as the *Ostium*. It is surrounded by numerous fingerlike projections, the **Fimbriae** 11, which partly "grasp" the ovary. As a *Secondary Oocyte* (*Ovum*) is released from the ovary, it is effectively swept across the small gap to the ostium by the motion of cilia lining the fimbriae.

The uterine tube is composed of smooth muscle lined with ciliated mucous membrane. Muscular peristalsis and ciliary action carry the secondary oocyte from the ostium into the uterine cavity, a journey that takes about 3 days. Fertilization of the oocyte occurs within the uterine tube.

Identify the features of the vagina, uterus, and uterine tubes in figures 11.19 and 11.20.

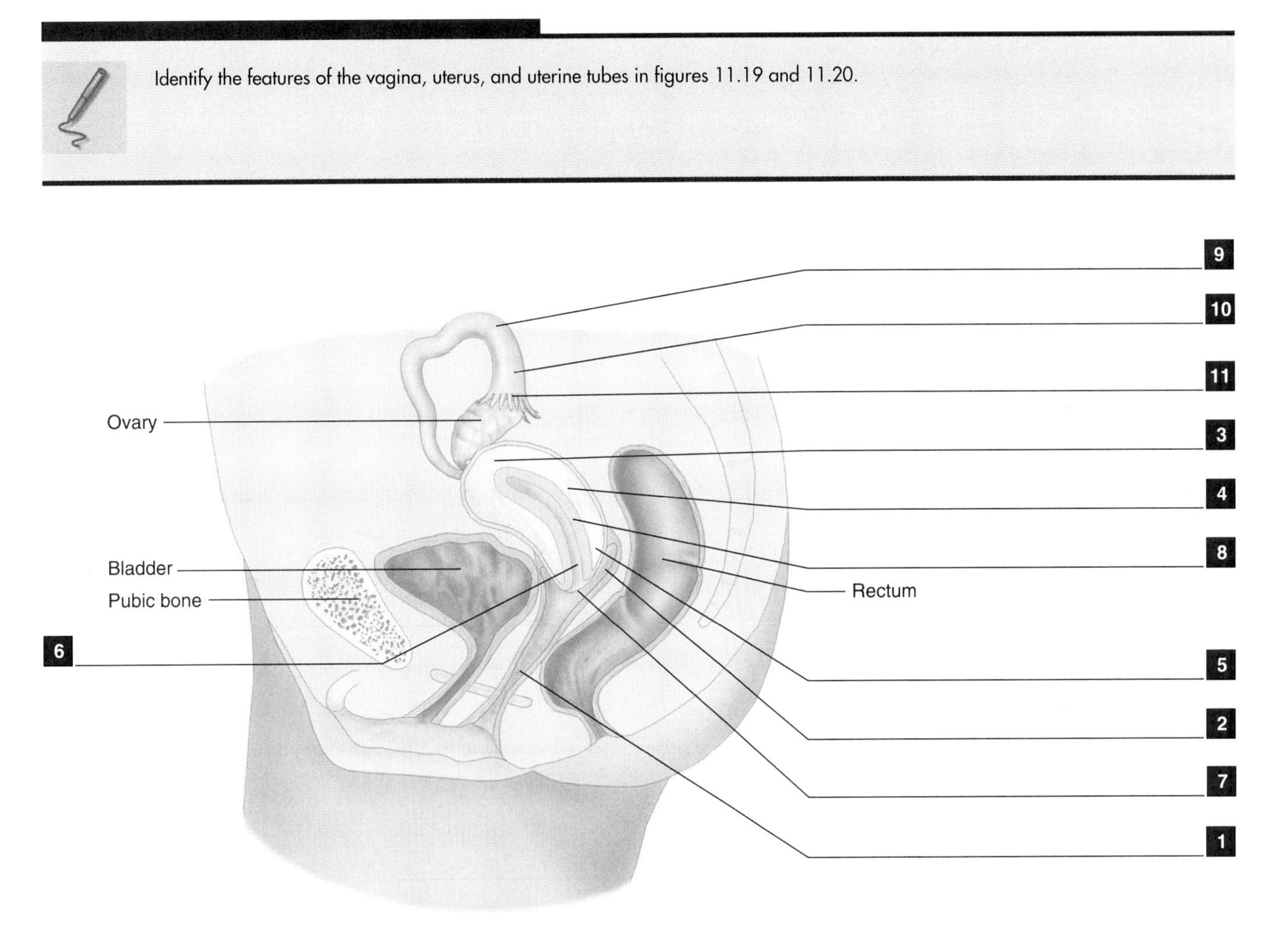

Figure 11.19 **Vagina, Uterus, and Uterine Tubes**
Schematic sagittal section.

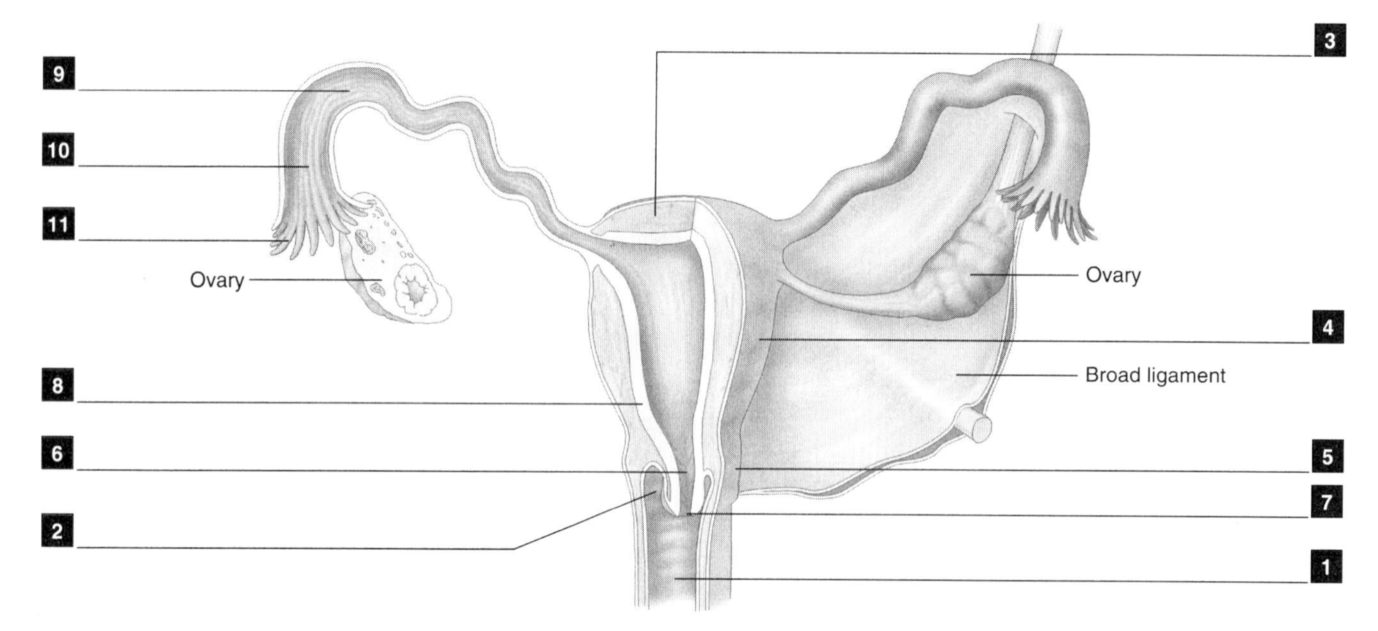

Figure 11.20 **Vagina, Uterus, and Uterine Tubes**
Posterior view of internal genitalia. The left side of the illustration has been cut away to reveal the internal structure of the uterus and uterine tube.

Ovary

The ovary is almond-shaped, and about 1 1/2 inches long by 3/4 inch broad. It is attached to a peritoneal fold that drapes over the uterine tube known as the *Broad Ligament* by a mesentery called the *Mesovarium*. The ovary is also held in place by a fibrous cord known as the *Ovarian Ligament*. It stretches between the ovary and uterus. The nerves, blood vessels, and lymphatics to the ovary run through the ovarian ligament.

The substance of the ovary is surrounded by a thin fibrous capsule, the **Tunica Albuginea 1**. External to it is a single layer of cuboidal cells that constitute what is called the **Germinal Epithelium 2**. This term, however, is a misnomer, because the germinal epithelium does not give rise to anything! The substance of the ovary that is encapsulated by the tunica albuginea is called the **Stroma 3**.

The stroma has a medulla and a cortex. The *Medulla* is highly vascular. The *Cortex* contains round epithelial vesicles that are the centers of ovum production. These vesicles are known as *Follicles*. Each follicle contains a cell that is capable of developing into an egg.

At birth there are about 200,000 follicles in each ovary, but only about 400 will actually produce and release an ovum. Follicle Stimulating Hormone (FSH), secreted by the pituitary gland, initiates egg development in several follicles each month, although only one usually reaches maturity.

Follicle Development

A follicle that has not been stimulated by FSH is known as a **Primordial Follicle 4**. Once stimulated to begin growing by FSH, it becomes known as a **Primary Follicle 5**. A primary follicle is characterized by the development of a *Zona Pellucida* around the egg. The egg at this stage of development is known as a **Primary Oocyte. 6** As development progresses, the primary follicle transforms into a **Secondary Follicle 7**. The cells of a secondary follicle secrete a fluid that fills it and transforms it into a **Mature (Graafian) Follicle 8**. This contains a fluid-filled antrum that surrounds the egg, which is now called a **Secondary Oocyte 9**. The follicular cells are primarily responsible for the secretion of *Estrogen*.

The mature follicle ruptures, releasing its secondary oocyte into the ostium of the uterine tube.

The cells of the ruptured mature follicle develop into a **Corpus Luteum 10** after release of the secondary oocyte. The corpus luteum primarily secretes *Progesterone*. If the oocyte that has been released is not fertilized, the corpus luteum degenerates into what is known as a **Corpus Albicans 11** in about 2 weeks time.

Identify the aforementioned features of the ovary in figure 11.21.

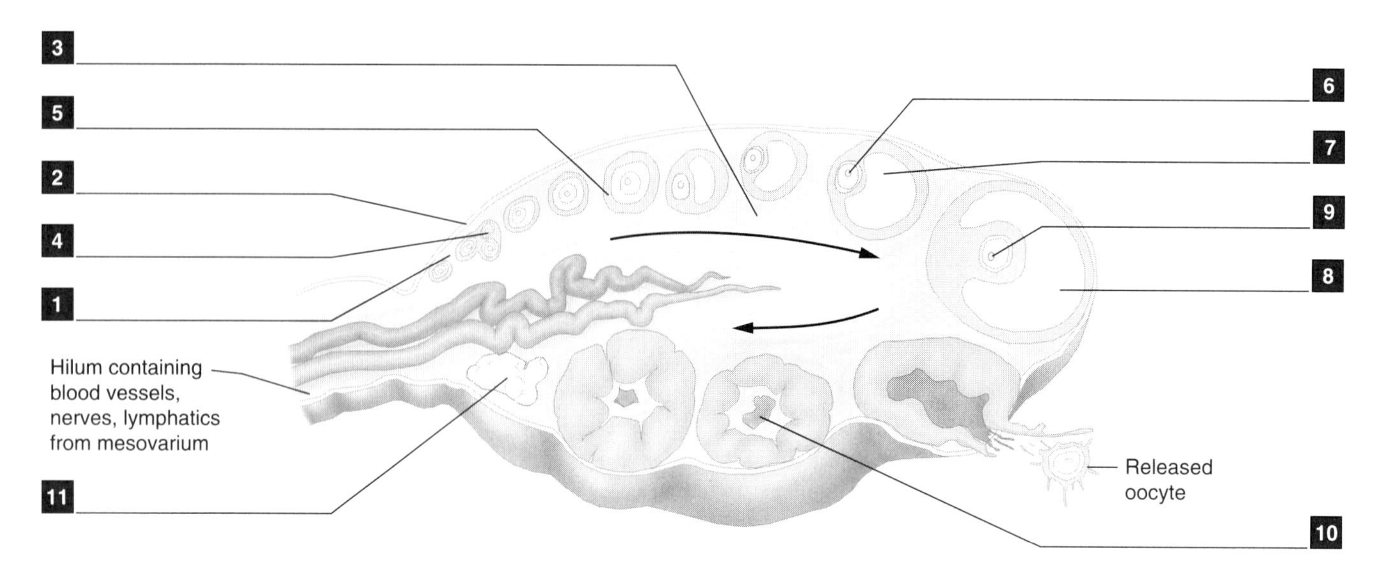

Figure 11.21 **The Ovary**
This cross section traces follicular development clockwise from its primordial stage to the formation of the corpus albicans.

Ligaments of the Uterus and Ovaries

The peritoneal covering of the uterus and uterine tubes can be envisioned easily if you imagine that they develop in the subperitoneal space between the bladder and rectum, and then push upward to become invested with the peritoneal sheet ahead of them.

In this way, the uterus becomes covered anteriorly, superiorly, and posteriorly, and the uterine tubes become invested with a double layer of peritoneum. This double layer is called the **Broad Ligament 1.** Its thickened superior margin is elongated laterally to the pelvic wall as the **Suspensory Ligament 2.** It invests the blood vessels and nerves that run to the ovary.

In the embryo, the ovary is located high up on the posterior abdominal wall. It is connected to the skin of the anterior abdominal wall (skin that will become the labia majora) by a fibrous cord known as the *Gubernaculum*. It runs past the uterus below the uterine tube, and pulls the ovary into the broad ligament.

Once it is nestled in the broad ligament, the ovary begins to enlarge and pushes out a fold of peritoneum from the posterior side of the broad ligament. This fold is pushed out so far that it reflects back onto itself, creating a "mesentery" of the ovary called the **Mesovarium 3.**

The remnant of the gubernaculum that extends between the ovary and the uterus becomes the **Ovarian Ligament 4.** The remnant of the gubernaculum that runs between the uterus and the skin of the anterior abdominal wall becomes the **Round Ligament of the Uterus 5.** The round ligament traverses the *Inguinal Canal*.

Thus, some of the ligaments that support the uterus, uterine tubes and ovaries are peritoneal reflections, while others are remnants of the gubernaculum.

Identify the uterine and ovarian ligaments in figure 11.22.

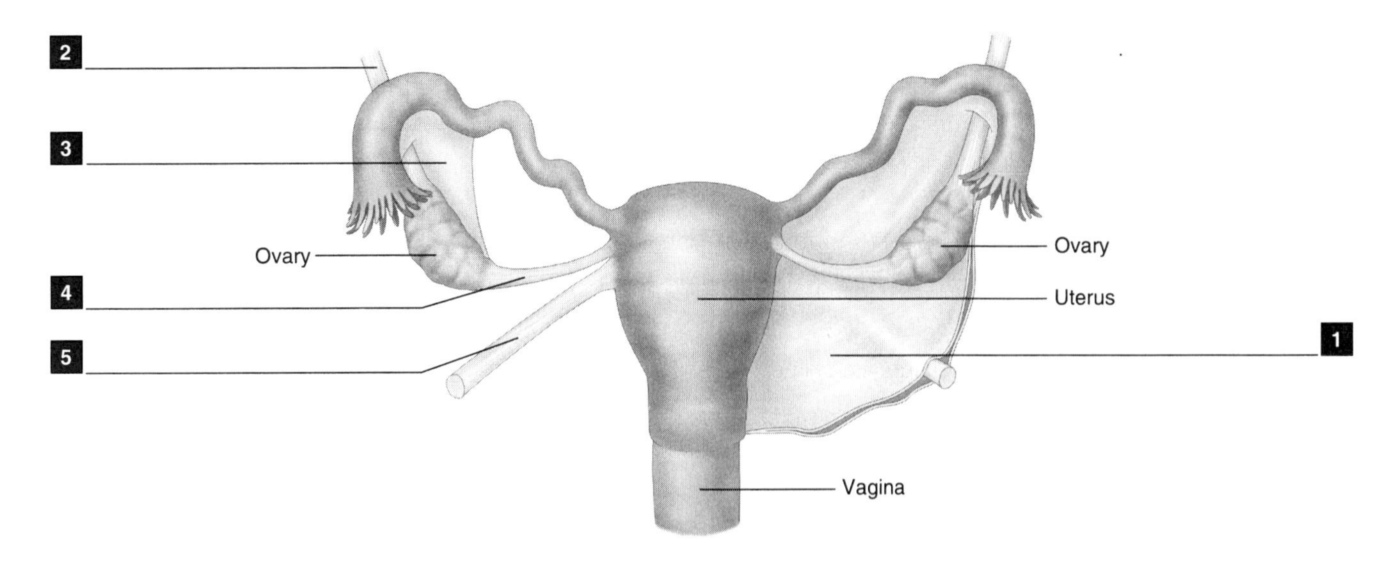

Figure 11.22 **Ligaments of the Uterus and Ovary**
This is a posterior view of the uterus, uterine tube, and ovaries to illustrate the connections of the various ligaments that support these structures. The broad ligament has been cut away from the left side.

11.7 BLOOD VESSELS OF THE PELVIS

Arteries

The pelvis and perineum receive blood from vessels that branch from the abdominal aorta, the inferior mesenteric artery, the external iliac artery, and the internal iliac artery.

Branches from the Abdominal Aorta

Recall that the *Gonadal Arteries* branch from the front of the abdominal aorta just below the level of the renal arteries (laboratory 10, p. 328).

The **Ovarian Artery** 1 descends into the pelvic cavity behind the peritoneum. It enters the suspensory ligament and passes through the broad ligament and mesovarium to reach the ovary.

The **Testicular Artery** 2 does not enter the pelvis. It descends along *Psoas major,* turns into the inguinal canal, and runs with the spermatic cord to reach the testis.

Branch from the Inferior Mesenteric Artery

The **Superior Rectal Artery** 3 is the terminal branch of the inferior mesenteric artery. It supplies the upper part of the rectum.

Branch from the External Iliac Artery

Recall that the abdominal aorta bifurcates into two **Common Iliac Arteries** 4 (laboratory 4, pp. 127–128). Each common iliac artery divides at the level of L5–S1 into an **External Iliac Artery** 5 and an **Internal Iliac Artery** 6**.**

The external iliac artery sends off a branch called the **Inferior Epigastric Artery** 7**.** This, in turn, sends a small twig to supply the cremaster muscle.

Identify these arterial branches in figure 11.23.

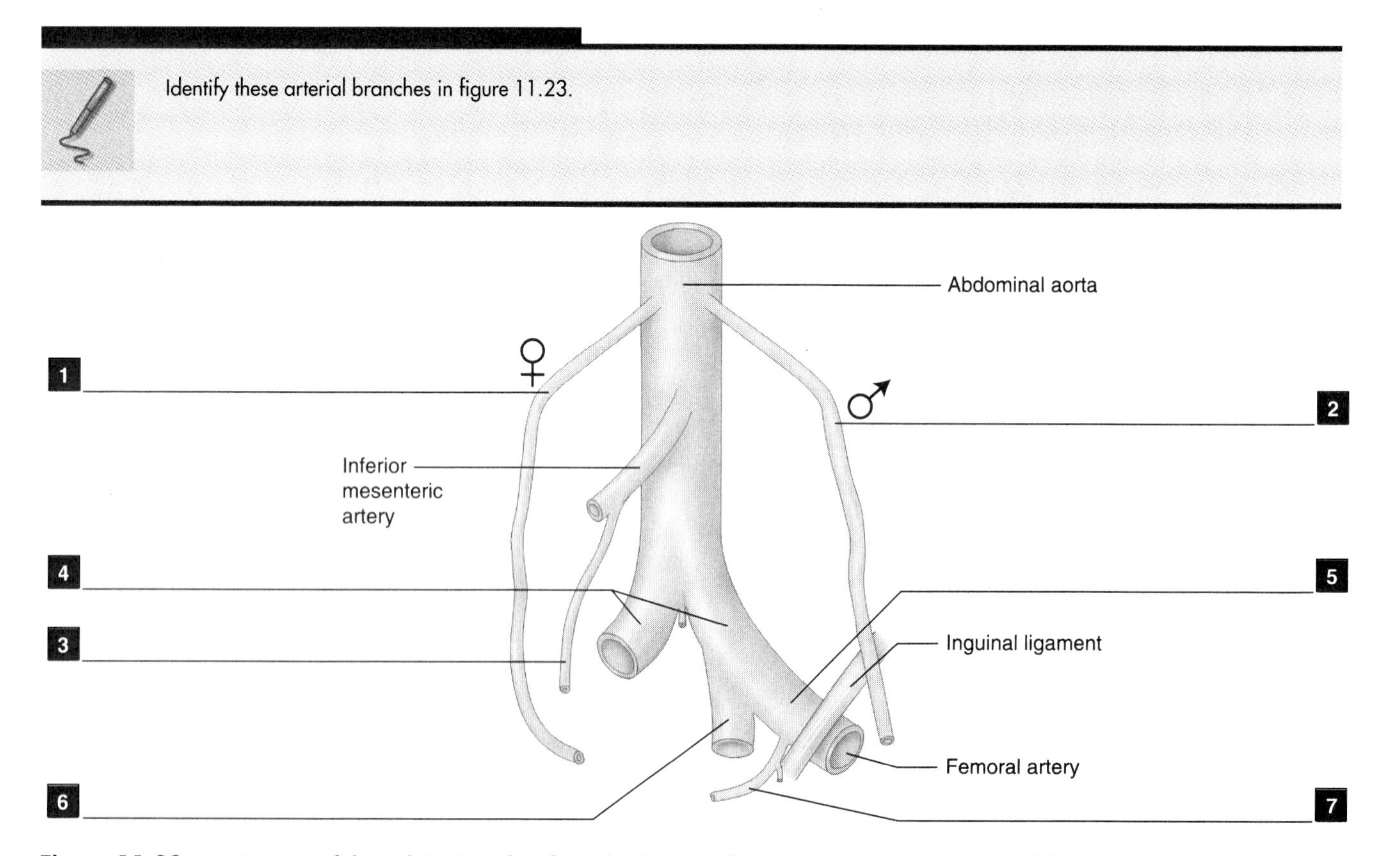

Figure 11.23 **Arteries of the Pelvis: Branches from the Aorta, Inferior Mesenteric, and External Iliac**

Branches from the Internal Iliac Artery

The internal iliac artery bifurcates after a short distance into anterior and posterior divisions. The *Posterior Division* has one branch that supplies pelvic viscera. The *Anterior Division* has several branches that supply pelvic viscera and the perineum.

From the *Posterior Division*, the **Lateral Sacral Artery** 1 descends along the sacrum, providing blood to viscera adjacent to it. The *Superior Gluteal Artery* branches from the posterior division and exits the pelvis through the *Greater Sciatic Foramen* to supply the buttocks.

The *Anterior Division* has two branches that exit the pelvis to supply adjacent muscles. These are the *Obturator Artery*, which traverses the *Obturator Foramen*, and the *Inferior Gluteal Artery*, which leaves via the *Greater Sciatic Foramen*.

From the *Anterior Division*, the **Umbilical Artery** 2 runs toward the anterior abdominal wall across the top of the bladder and gives off the **Superior Vesical Artery** 3 to it. The umbilical artery then loses its lumen to become the medial umbilical ligament.

The **Middle Rectal Artery** 4 supplies the middle and lower part of the rectum.

The next branch of the *Anterior Division* is sex-dependent. In males, it is known as the **Inferior Vesical Artery** 5**.** This supplies the prostate gland, seminal vesicles and ductus deferens. In females, it is known as the **Uterine Artery** 6**.** This runs across the pelvic floor and ascends along the vagina and the uterus between the layers of the broad ligament.

The **Internal Pudendal Artery** 7 exits the pelvis through the *Greater Sciatic Foramen*, and then enters the perineum through the *Lesser Sciatic Foramen*. It gives off three branches here. The **Inferior Rectal Artery** 8 supplies the anal canal. The **Perineal Artery** 9 supplies the muscles of the perineum. The third branch is sex-dependent. In females, it is the **Artery of the Clitoris** 10**.** In males, it is the **Artery of the Penis** 11**.**

Veins

The veins that drain the pelvic cavity accompany its arteries. ***The veins carry the same names as the arteries that they accompany.***

Identify the branches of the internal iliac artery in figure 11.24.

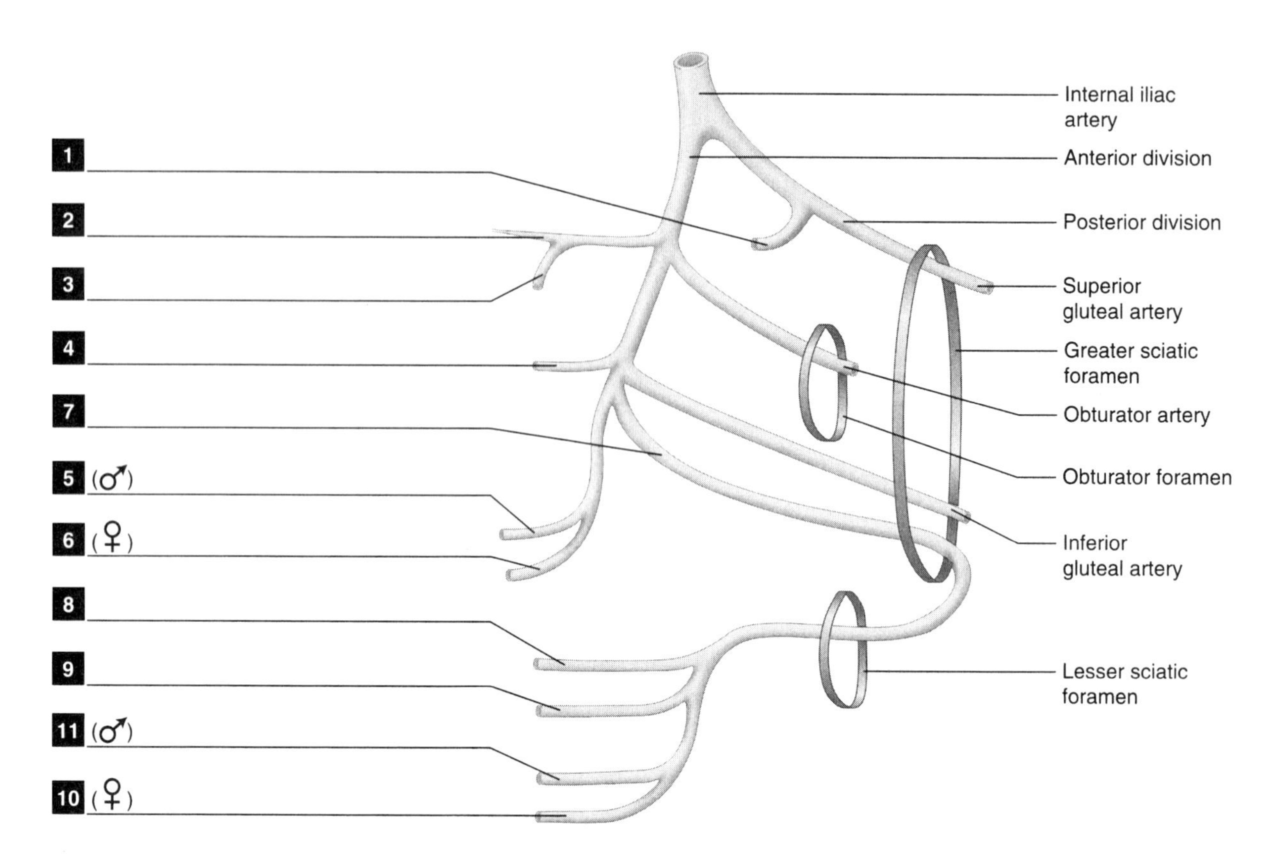

Figure 11.24 **Arteries of the Pelvis: Branches from the Internal Iliac**

11.8 NERVES OF THE PELVIS

Skeletal muscles of the pelvis are supplied by spinal nerves. The pelvic viscera, including the reproductive organs and gastrointestinal tract, are supplied by autonomic nerves.

Spinal Nerves in the Pelvis

Recall that spinal nerves from L2 to L4 form the *Lumbar Plexus*, those from L5 to S3 form the *Sacral Plexus*, and that the two are connected by the *Lumbosacral Trunk* between L4 and L5 (laboratory 4, pp. 102–103).

The ***Lumbar Plexus*** provides nerves to the thigh muscles. The *Femoral Nerve* is one branch of the lumbar plexus. The **Obturator Nerve 1** is another branch of this plexus. It barely enters the pelvic cavity below the arcuate line, and leaves it through the *Obturator Foramen*.

The ***Sacral Plexus*** also provides nerves to the muscles of the lower limb. Its largest branch is the **Sciatic Nerve 2.** It runs along the backside of the pelvic cavity for a short distance, exiting it through the *Greater Sciatic Foramen*.

The ventral rami of S2–S4 provide branches to the muscles of the *Pelvic Diaphragm*.

The ventral rami of S2–S4 also combine to form the **Pudendal Nerve 3.** It exits the pelvic cavity through the *Greater Sciatic Foramen*, curves around the sacrospinous ligament, and re-enters the pelvic cavity via the *Lesser Sciatic Foramen* in company with the internal pudendal artery.

The *Pudendal* nerve provides a branch, the **Inferior Rectal Nerve 4,** to the external anal sphincter.

The *Pudendal* nerve divides into two terminal branches at the posterior margin of the perineal membrane. One branch is the **Perineal Nerve 5.** It splays out below the membrane to supply *Bulbospongiosus* and *Ischiocavernosus* and the skin of the scrotum or labia majora. The other branch is the **Dorsal Nerve of the Penis (Clitoris) 6.** It runs across the top of the perineal membrane to supply *Sphincter urethae*, and it then pierces the membrane to run onto the clitoris or penis.

Identify the spinal nerves mentioned (above) in figure 11.25.

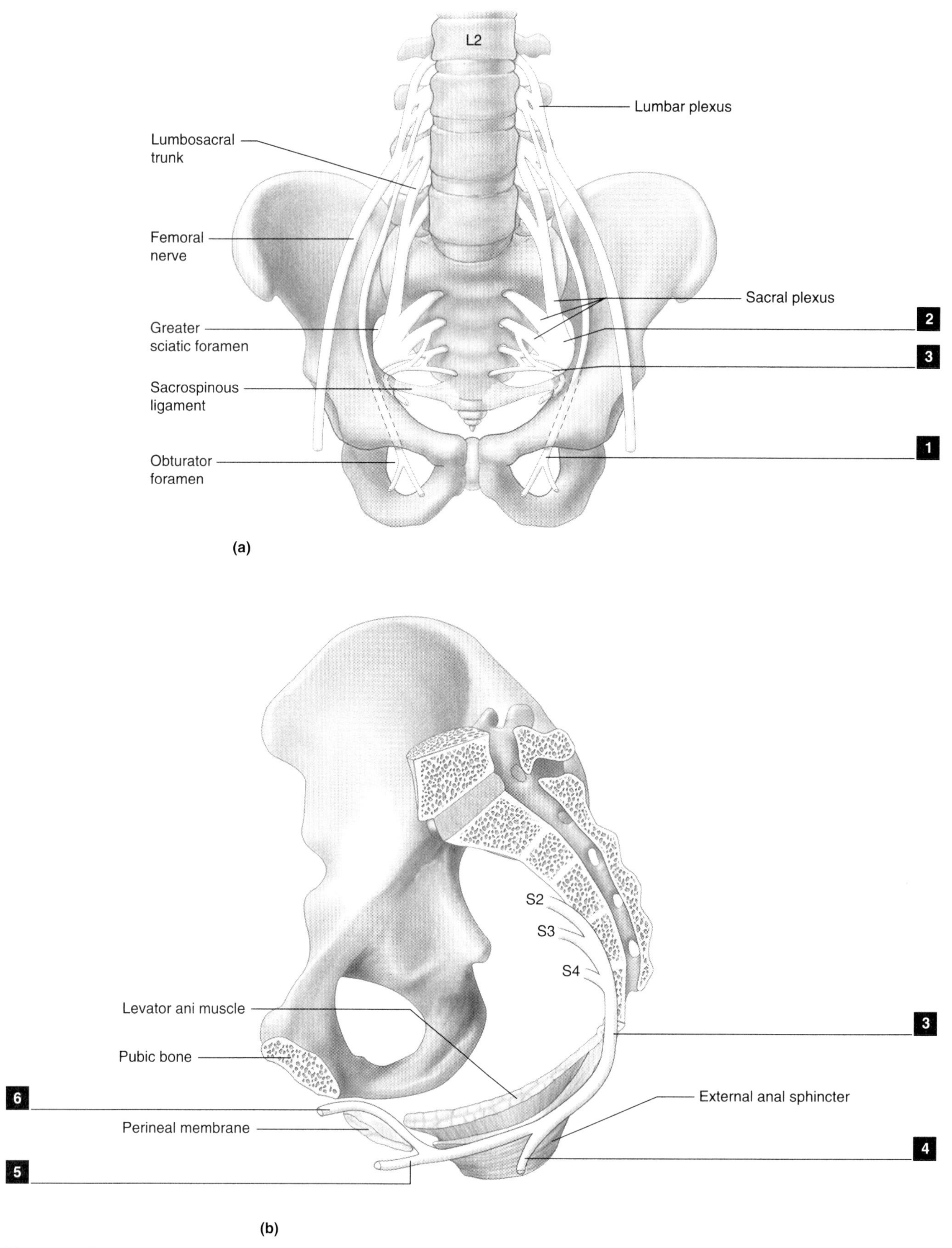

Figure 11.25 **Spinal Nerves in the Pelvis**
(a) anterior view; (b) schematic sagittal section.

Sympathetic Nerves in the Pelvis

Sympathetic fibers to pelvic viscera come from the **Sympathetic Trunks 1.** The left and right trunks run along the spinal column, cross over the sacrum, and travel medial to the anterior sacral foramina. The two trunks merge in front of the coccyx. Located along the trunks are a series of **Paravertebral Ganglia 2.**

Above L2 there are both gray and white rami communicantes between the paravertebral ganglia and associated spinal nerves. Below L2 there are gray rami communicantes, but no white rami communicantes between the paravertebral ganglia and associated spinal nerves. Why is this so?

Sympathetic fibers from the upper lumbar paravertebral ganglia may run to and synapse in a series of **Preaortic Ganglia 3** that lie on the front of the abdominal aorta around the roots of the celiac, superior and inferior mesenteric arteries. Axons descend into the pelvis from these preaortic ganglia by **Hypogastric Nerves 4.**

The hypogastric nerves intertwine to form a **Superior Hypogastric Plexus 5** where the abdominal aorta divides into the common iliac arteries. The hypogastric nerves leave these plexuses and run into the pelvis alongside the common iliac arteries.

In the pelvis, the hypogastric nerves form a second plexus, the **Inferior Hypogastric Plexus 6,** by interweaving with parasympathetic fibers (= *Pelvic Splanchnic nerves*) that emanate from S2–S4.

Nerves from the inferior hypogastric plexuses thus carry both sympathetic and parasympathetic fibers to the pelvic viscera.

Parasympathetic Nerves in the Pelvis

Recall that parasympathetic outflow is restricted to the cranial nerves and to spinal cord levels S2–S4. Because the vagus nerve does not extend into the pelvis, all of the pelvic organs receive parasympathetic stimulation from nerves that emanate from S2– S4. These are known as **Pelvic Splanchnic Nerves 7.**

Pelvic splanchnic nerve fibers mingle with sympathetic fibers in the *Inferior Hypogastric Plexuses*. From these, the parasympathetic fibers run to various target organs, where they synapse in ganglia adjacent to or within the organ wall.

Identify the aforementioned autonomic nerves in figure 11.26.

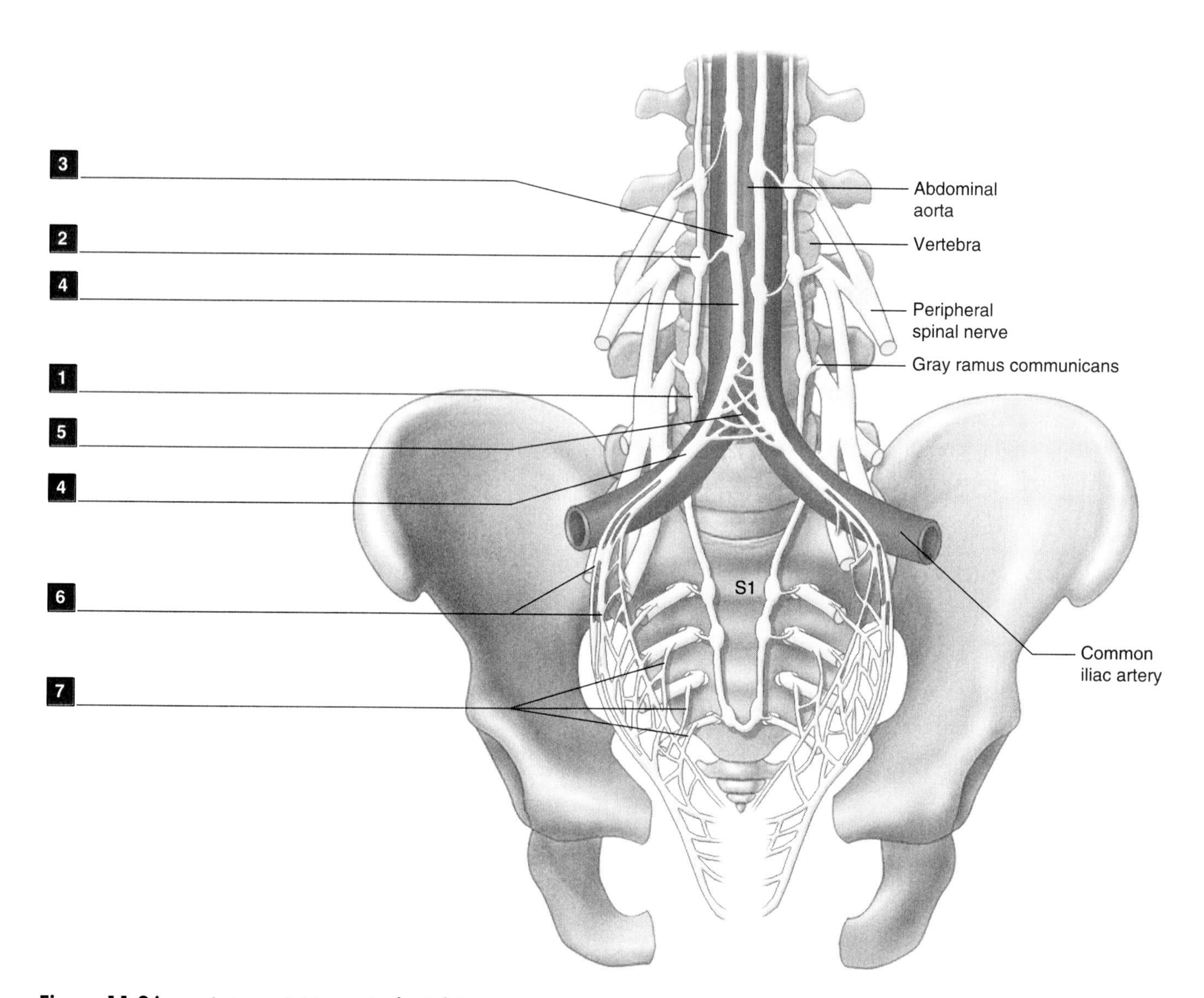

Figure 11.26 Autonomic Nerves in the Pelvis

INDEX

Note: Page numbers followed by *f* refer to figures.

Abdomen
  anterolateral walls of, 295–298, 296*f*, 298*f*
  lymphatic vessels of, 320, 321*f*
  muscles of, 295–299, 296*f*, 298*f*, 299*f*
  nerves of, 329, 330*f*
  posterior wall of, 298–299, 299*f*
Abdominal cavity, 293, 293*f*, 300, 301*f*
Abdominal diaphragm, 257, 258*f*, 298, 299*f*
Abducens nerve, 211, 212*f*
Abduction, 6, 7*f*
Abductor digiti minimi, 72, 73*f*
Abductor hallucis, 92, 125, 126*f*
Abductor pollicis brevis, 76, 77*f*
Abductor pollicis longus, 74, 75*f*
Accessory hemiazygos vein, 283, 284*f*
Accessory nerve, 140, 141*f*, 222, 223*f*
Accessory pancreatic duct, 306, 307*f*, 312, 313*f*
Acetabular labrum, 93, 94*f*
Acetabulum, 85, 86*f*, 93, 94*f*
  lunate surface of, 93, 94*f*
Achilles tendon, 90, 120
Acoustic meatus
  external, 168, 239, 240*f*
  internal, 168, 172*f*, 173*f*
Acromioclavicular joint, 43, 44*f*
Acromion, 43, 44*f*
Adduction, 6, 7*f*
Adductor brevis, 112, 113*f*
Adductor longus, 112, 113*f*
Adductor magnus, 112, 113*f*
Adductor pollicis, 76, 77*f*
Adipose tissue, of breast, 249, 250*f*
Adrenal glands, 294, 294*f*, 322, 323*f*
Adventitia, of detrusor muscle, 347, 348*f*
Alveolar bone, 177, 177*f*
Alveolar ducts, 262, 263*f*
Alveolar nerve, 177, 177*f*, 213, 214*f*
Alveolar process, 169
Alveolar sacs, 262, 263*f*
Alveoli, 262, 263*f*
Ampulla, 244, 245*f*
Anal canal, 308, 309*f*, 340, 341*f*
Anal sphincter, 345, 345*f*
Anal triangle, 340, 341*f*, 345, 345*f*
Anatomical position, 2, 2*f*, 3*f*
Anatomical snuff box, 74
Anconeus, 65, 66*f*
Ankle, 90, 91*f*, 97, 98*f*
Annular ligament, 51, 52*f*
Anococcygeal raphé, 338, 339*f*
Ansa cervicalis, 138, 139*f*
Ansa subclavia, 143, 144*f*
Antebrachial vein, median, 80, 81*f*
Anterior, 2, 3*f*
Anterior cerebral artery, 188, 189*f*
Anterior chamber, 235, 236*f*
Anterior communicating artery, 188, 189*f*
Anterior communicating vein, 160, 161*f*
Anterior cruciate ligament, 95, 96*f*
Anterior inferior cerebellar artery, 188, 189*f*
Anterior jugular vein, 160, 161*f*
Anterior longitudinal ligament, 29, 30*f*
Anterior talofibular ligament, 97, 98*f*
Anterior tibial artery, 127, 128*f*
Anterior tibial vein, 129, 130*f*
Anterior vagal trunk, 220, 221*f*
Antihelix, 239, 240*f*
Aorta, 127, 128*f*, 269, 270*f*
  abdominal, 328, 328*f*
  ascending, 271, 272*f*, 273, 274*f*
  descending, 261, 261*f*
Aortic arch, 78, 79*f*, 260, 261*f*, 271, 272*f*
Aortic hiatus, 257, 258*f*
Aortic semilunar valve, 273, 274*f*, 275, 275*f*
Apertures
  lateral, 207, 208*f*
  median, 207, 208*f*
Aponeurosis
  palmar, 68
  plantar, 92, 92*f*
Aqueous humor, 235
Arachnoid mater, 31, 32*f*, 190, 190*f*, 194
Arachnoid villi, 207, 208
Arbor vitae, 196, 197*f*
Arcuate line, 85, 333, 334*f*
Areola, 249, 250*f*
Arm
  bones of, 45, 46*f*
  muscles of, 60–64, 63*f*, 64*f*
  nerves of, 55–60, 58*f*, 60*f*
Arrector pili, 22, 23*f*
Arteriole, of glomerulus, 326, 327*f*
Artery (arteries)
  auricular, posterior, 158, 159*f*
  axillary, 78, 79*f*
  basilar, 188, 189*f*
  brachial, 78, 79*f*
  carotid
    common, 158, 159*f*
      left, 162, 163*f*, 260, 261*f*, 271, 272*f*
      right, 162, 163*f*
    external, 158, 159*f*, 164, 165*f*
    internal, 158, 159*f*, 188, 189*f*
  celiac, 320, 321*f*
  cerebellar
    inferior
      anterior, 188, 189*f*
      posterior, 188, 189*f*
    superior, 188, 189*f*
  cerebral
    anterior, 188, 189*f*
    middle, 188, 189*f*
    posterior, 188, 189*f*
  circumflex, 277, 278*f*
  of clitoris, 369, 370*f*
  colic
    left, 317, 317*f*
    middle, 316, 316*f*
    right, 316, 316*f*
  communicating
    anterior, 188, 189*f*
    posterior, 188, 189*f*
  coronary
    left, 277, 278*f*
    right, 278, 278*f*
  cystic, 315, 315*f*
  deep, 356, 357*f*
  dorsalis pedis, 127, 128*f*
  facial, 158, 159*f*
  femoral, 116, 127, 128*f*
  fibular, 127, 128*f*
  gastric
    left, 314, 315*f*
    right, 315, 315*f*
  gastroduodenal, 315, 315*f*
  gastroepiploic
    left, 315, 315*f*
    right, 315, 315*f*
  gastro-omental, 315, 315*f*
  gonadal, 328, 328*f*
  of head, 188, 189*f*
  hepatic, 310, 311*f*, 315, 315*f*
  ileocolic, 316, 316*f*
  iliac
    common, 368, 368*f*
    external, 127, 128*f*, 368, 368*f*
    internal, 127, 128*f*, 368, 368*f*
  interventricular, posterior, 278, 278*f*
  lingual, 158, 159*f*
  of lower limb, 127, 128*f*
  maxillary, 158, 159*f*
  mesenteric
    inferior, 314, 314*f*, 317, 317*f*
    superior, 306, 307*f*, 314, 314*f*, 316, 316*f*
  of neck, 158, 159*f*
  nutrient, 10, 11*f*
  occipital, 158, 159*f*
  ophthalmic, 188, 189*f*
  ovarian, 368, 368*f*
  of penis, 369, 370*f*
  perineal, 369, 370*f*
  pharyngeal, ascending, 158, 159*f*
  plantar, 127, 128*f*
  popliteal, 127, 128*f*
  profunda (deep) femoris, 127, 128*f*
  pudendal, internal, 369, 370*f*
  pulmonary, 261, 261*f*, 269, 270*f*, 271, 272*f*
  radial, 78, 79*f*
  rectal
    inferior, 369, 370*f*
    middle, 369, 370*f*
    superior, 317, 317*f*, 368, 368*f*

Artery (arteries)—*Cont.*
renal, 322, 323*f*, 324, 325*f*, 328, 328*f*
retinal, central, 235
sacral, lateral, 369, 370*f*
splenic, 315, 315*f*, 320, 321*f*
subclavian, 78, 79*f*, 158, 159*f*, 164, 165*f*
left, 162, 163*f*, 260, 261*f*, 271, 272*f*
right, 162, 163*f*
suprarenal
inferior, 322, 323*f*
middle, 322, 323*f*
superior, 322, 323*f*
suprascapular, 158, 159*f*
temporal, superficial, 158, 159*f*
testicular, 368, 368*f*
thoracic
internal, 158, 159*f*, 281, 282*f*
lateral, 282, 282*f*
thyroid
inferior, 158, 159*f*, 164, 165*f*
superior, 158, 159*f*, 164, 165*f*
tibial
anterior, 127, 128*f*
posterior, 127, 128*f*
ulnar, 78, 79*f*
umbilical, 369, 370*f*
of upper limb, 78, 79*f*
uterine, 369, 370*f*
vertebral, 158, 159*f*
vesical
inferior, 369, 370*f*
superior, 369, 370*f*
Articular cartilage, 18, 19*f*
of glenoid, 49, 50*f*
of humerus, 49, 50*f*
Articular disc, 18, 19*f*
Articular facets
of atlas, 25, 26*f*
of lumbar vertebrae, 27, 28*f*
of thoracic vertebrae, 27, 27*f*
Articular genu, 118
Articular processes, 24, 25*f*
Articular surface, 14
Aryepiglottic fold, 136, 137*f*
Arytenoideus, 154, 155*f*
Arytenoids, 134, 135*f*
Ascending aorta, 271, 272*f*, 273, 274*f*
Ascending colon, 308, 309*f*
Ascending pharyngeal artery, 158, 159*f*
Association fibers, 204, 204*f*
Atlas (C1), 25, 26*f*
Atrioventricular bundle, 276, 277*f*
Atrioventricular node, 276, 277*f*
Atrium
left, 271, 272*f*, 273, 274*f*
right, 269, 270*f*, 271, 272*f*, 273, 274*f*
Auditory (acoustic) meatus
external, 168, 239, 240*f*
internal, 168, 172*f*, 173*f*
Auditory canal, 168
external, 237, 238*f*, 239, 240*f*
Auditory (eustachian) tube, 237, 238*f*, 241, 242*f*
Auditory hiatus, 186, 187*f*
Auditory ossicles, 237, 238*f*
Auricle, 237, 238*f*, 239, 240*f*
Auricular artery, posterior, 158, 159*f*
Auricular nerve, great, 138, 139*f*
Auricular vein, posterior, 160, 161*f*
Auriculotemporal nerve, 213, 214*f*
Auscultation, of heart valves, 274
Autonomic nerves, in neck, 142–143, 144*f*
Axillary artery, 78, 79*f*
Axillary nerve, 56, 57, 59, 60*f*
Axillary tail, of breast, 249, 250*f*
Axillary vein, 80, 81*f*
Axis (C2), 25, 26*f*
Azygos vein, 271, 272*f*, 283, 284*f*

Back, 21–40. *See also* Spinal cord; Vertebral column
muscles of, 36–40
intrinsic, 38, 39*f*, 40*f*
superficial, 36, 37*f*
Bartholin's glands, 359, 360*f*
Basal nuclei, 194, 202, 205, 206*f*
Basilar artery, 188, 189*f*
Basilar membrane, 245, 246*f*
Basilic vein, 80, 81*f*
Biceps brachii, 65, 66*f*, 67
long head of, 65, 66*f*
tendon of, 49, 50*f*
short head of, 65, 66*f*
Biceps femoris, 114, 115*f*
Bicuspid valve, mitral, 273, 274*f*
Bile duct, 306, 307*f*, 312, 313*f*
Biliary tree, 302, 303*f*, 312, 313*f*
Bladder, 347, 348*f*
Blind spot, 235, 236*f*
Blood vessels. *See* Artery (arteries); Vein(s)
Body
of epididymis, 352, 353*f*
of stomach, 304, 305*f*
of uterus, 362, 363*f*
of vertebra, 24, 25*f*
Bone(s)
acetabulum, 85, 86*f*, 93, 94*f*
acromion, 43, 44*f*
alveolar, 177, 177*f*
of arm, 45, 46*f*
calcaneus, 90, 91, 91*f*, 92*f*, 97, 98*f*
cancellous (spongy), 10, 11*f*
capitate, 48, 48*f*
capitulum, 45, 46*f*, 51, 52*f*
carpal, 48, 48*f*
cervical vertebrae, 25, 26*f*, 133, 134*f*
clavicle, 43, 44*f*
coccygeal vertebrae, 28, 29*f*
coccyx, 28, 29*f*, 85, 86*f*
compact (ivory), 10, 11*f*
cranial, 168–170, 171f, 172f
cuboid, 90, 91*f*, 97, 98*f*
cuneiforms, 90, 91*f*, 134
dens (odontoid process), 25, 26*f*
distal phalanx
of finger, 48, 48*f*
of toe, 90, 91*f*
ethmoid, 169, 171*f*, 172*f*, 173*f*, 186, 187*f*, 228, 228*f*
facial, 169–170, 173f, 174*f*
femur, 87, 88*f*
fibula, 88, 89*f*
finger, 48, 48*f*
foot, 90–92, 91*f*
forearm, 45, 47*f*
frontal, 168, 171*f*, 172*f*, 173*f*, 228, 228*f*
hamate, 48, 48*f*
hand, 48, 48*f*
head of radius, 51, 52*f*
humerus, 45, 46*f*
hyoid, 133, 134*f*, 135*f*
ilium, 85, 86*f*
ischium, 85, 86*f*
lacrimal, 171*f*, 173*f*, 228, 228*f*
leg, 88, 89*f*
of lower limb, 85–92, 86*f*, 88*f*, 89*f*, 91*f*, 92*f*
lumbar vertebrae, 27, 28*f*
lunate, 48, 48*f*, 53, 54*f*
malleolus
of ankle, 97, 98*f*
of knee, 88, 89*f*
mandibular, 170, 174*f*
manubrium, 43, 44*f*
maxillary, 169, 171*f*, 172*f*, 173*f*, 228, 228*f*
metacarpal, 48, 48*f*
metatarsal, 90, 91*f*
middle phalanx
of finger, 48, 48*f*
of toe, 90, 91*f*
mylohyoid, 150, 151*f*
nasal, 169, 171*f*, 173*f*
navicular, 90, 91, 91*f*, 92*f*, 97, 98*f*
of neck, 133, 134*f*
occipital, 168, 171*f*, 172*f*, 173*f*
omohyoid, 152, 153*f*
os coxae, 85, 86*f*
palatine, 170, 172*f*, 173*f*, 186, 187*f*
palm, 48, 48*f*
parietal, 168, 171*f*, 172*f*, 173*f*
patella, 87, 95, 96*f*
of pelvis, 85, 86*f*, 333, 334*f*
phalanges
of foot, 90, 91*f*
of hand, 48, 48*f*
pisiform, 48, 48*f*
proximal phalanx
of finger, 48, 48*f*
of toe, 90, 91*f*
pubic, 85, 86*f*
radius, 45, 47*f*
head of, 51, 52*f*
sacral vertebrae, 28, 29*f*
sacrum, 85, 86*f*
scaphoid, 48, 48*f*, 53, 54*f*
scapula, 43, 44*f*
of skull, 168–178, 171*f*, 172*f*, 173*f*, 174*f*
sphenoid, 169, 171*f*, 172*f*, 173*f*, 186, 187*f*, 228, 228*f*
talus, 90, 91, 91*f*, 92*f*, 97, 98*f*
tarsal, 90, 91*f*
temporal, 168, 171*f*, 172*f*, 239, 240*f*
thigh, 87, 88*f*
thoracic vertebrae, 27, 27*f*
of thorax, 251–259, 252*f*
tibia, 88, 89*f*
trapezium, 48, 48*f*
trapezoid, 48, 48*f*
triquetral, 48, 48*f*, 53, 54*f*
tympanic, 239, 240*f*
ulna, 45, 47*f*
of upper limb, 43–48, 44*f*, 46*f*, 47*f*, 48*f*
vertebral, 24–30, 25*f*
vomer, 170, 171*f*, 172*f*, 173*f*, 186, 187*f*
wrist, 48, 48*f*
zygomatic, 169, 171*f*, 172*f*, 228, 228*f*
Bowman's capsule, 326, 327*f*
Brachial artery, 78, 79*f*
Brachial plexus, 33, 34*f*, 55, 56, 57*f*
Brachial vein, 80, 81*f*
Brachialis, 65, 66*f*
Brachiocephalic trunk, 78, 79*f*, 80, 81*f*, 158, 159*f*, 162, 163*f*, 260, 261*f*, 271, 272*f*
Brachiocephalic vein, 160, 161*f*, 162, 163*f*, 260, 261*f*, 271, 272*f*
left, 164, 165*f*, 279, 280*f*
right, 279, 280*f*
Brachioradialis, 65, 66*f*
Brain, 194–208, 195*f*, 197*f*, 199*f*, 201*f*, 203*f*, 204*f*, 206*f*

Breast, 249, 250f
Bregmatic fontanelle, coronal, 168
Broad ligament, 366, 367f
Bronchi
primary, 262, 263f
secondary (lobar), 262, 263f
tertiary (segmental), 262, 263f
Bronchial tree, 262, 263f
Bronchioles, 262, 263f
Buccal, 4, 5f
Buccinator, 178, 179f
Bulb
of clitoris, 361, 362f
of penis, 356, 357f
Bulbar conjunctiva, 229, 230f
Bulbospongiosus, 342, 343f
Bulbourethral glands, 349f, 354, 355f
Bundle branches, 276, 277f
Bursa (bursae), 18, 19f
olecranon, 51, 52f
subacromial, 49
subscapular, 49
subtendinous olecranon, 51, 52f
suprapatellar, 95, 96f

Calcaneofibular ligament, 97, 98f
Calcaneus, 90, 91, 91f, 92f, 97, 98f
Calyces, of renal pelvis, 324, 325f
Canal
anal, 308, 309f
auditory, 168, 237, 238f
cervical, 362, 363f, 364f
hypoglossal, 168, 172f, 173f
optic, 169, 171f, 172f, 228, 228f
semicircular, 242, 243f, 244, 245f
Canine, 175, 175f
Canthus
lateral, 226, 227f
medial, 226, 227f
Capitate, 48, 48f
Capitulum, humeral, 45, 46f, 51, 52f
Capsular ligament, 18, 19f
Cardiac notch, 264, 265f
Cardiac orifice, 304, 305f
Cardiac sphincter, 304, 305f
Cardiac vein
great, 278, 278f
middle, 278, 278f
small, 278, 278f
Carotid artery
common, 158, 159f
left, 162, 163f, 260, 261f, 271, 272f
right, 162, 163f
external, 158, 159f, 164, 165f, 188
internal, 158, 159f, 188, 189f
Carotid branch, of glossopharyngeal nerve, 218, 219f
Carotid canal, 168
Carotid foramen, 168, 172f
Carpal bones, 48, 48f
Carpal tunnel syndrome, 53
Cartilage(s)
articular, 18, 19f
of glenoid, 49, 50f
of humerus, 49, 50f
costal, 251, 252f, 293, 293f
cricoid, 133, 135f
laryngeal, 133, 135f
of neck, 133–137, 134f, 135f, 137f
thyroid, 133, 135f
tracheal, 133, 134f, 262, 263f
triradiate, 85
Cartilaginous joints, 16, 17f
Caruncle, lacrimal, 226, 227f
Cauda equina, 31, 32f
Caudal, 2, 3f
Caudate lobe, of liver, 310, 311f
Caudate nucleus, 205, 206f
Cavernous sinus, 191, 192f
Cecum, 308, 309f
Celiac artery, 314, 314f, 315f, 320, 321f
Celiac ganglion, 220
Cementum, 177, 177f
Central canal, 207, 208f
Central retinal artery, 235, 236f
Central sulcus, 202
Cephalic vein, 80, 81f
Cerebellar artery
inferior
anterior, 188, 189f
posterior, 188, 189f
superior, 188, 189f
Cerebellar cortex, 196, 197f
Cerebellar peduncle
inferior, 196, 197f
middle, 196, 197f
superior, 196, 197f
Cerebellum, 194, 196, 197f
Cerebral aqueduct, 194, 198, 199f, 207, 208f
Cerebral artery
anterior, 188, 189f
middle, 188, 189f
posterior, 188, 189f
Cerebral cortex, 202, 203f
Cerebral peduncles, 198
Cerebral vein, great, 191, 192f
Cerebrospinal fluid, 207, 208f
Cerebrum, 194, 202–206, 203f, 204f, 206f
Cervical canal, 362, 363f, 364f
Cervical ganglion, 142, 144f
Cervical lordosis, 29, 30f
Cervical nerves, 33, 34f
transverse, 138, 139f
Cervical plexus, 33, 34f, 138, 139f
Cervical sympathetic cardiac nerves, 143, 144f
Cervical vertebrae, 25, 26f, 133, 134f
Cervix, 362, 363f, 364f
Chorda tympani, 215, 216f
Chordae tendineae, 273, 274f
Choroid, 235, 236f
Choroid plexus, 207
Ciliary body, 235, 236f
Ciliary muscle, 235, 236f
Cingulate gyrus, 202, 203f
Circle of Willis, 188
Circular layer, of gastric smooth muscle, 304, 305f
Circulatory system, 269, 270f
Circumduction, 6, 9f
Circumflex artery, 277, 278f
Cisterna chyli, 279, 280f
Clavicle, 43, 44f
Clitoris, 358, 358f, 359, 360f, 361, 362f
artery of, 369, 370f
dorsal nerve of, 371, 372f
Coccygeal kyphosis, 29, 30f
Coccygeal nerve, 33, 34f
Coccygeal vertebrae, 28, 29f
Coccygeus, 338, 339f
Coccyx, 28, 29f, 85, 86f, 333, 334f
Cochlea, 242, 243f
Cochlear branch, of vestibulocochlear nerve, 217, 217f
Cochlear duct, 242, 243f, 245, 246f
Coelom, 259, 260f
Colic artery
left, 317, 317f
middle, 316, 316f
right, 316, 316f
Colic (hepatic) flexure
left, 308, 309f
right, 308, 309f
Colic vein
left, 318, 319f
middle, 318, 319f
right, 318, 319f
Collateral ligament
medial (tibial), 95, 96f
radial, 51, 52f, 53, 54f
ulnar, 51, 52f, 53, 54f
Collecting ducts, of kidney, 324, 325f, 326, 327f
Colliculus
inferior, 198, 199f
superior, 198, 199f
Colon
ascending, 308, 309f
descending, 308, 309f
sigmoid, 308, 309f
transverse, 308, 309f
Commissural fibers, 204, 204f
Common carotid artery, 158, 159f
left, 162, 163f, 260, 261f, 271, 272f
right, 162, 163f
Common facial vein, 160, 161f
Common hepatic duct, 310, 311f, 312, 313f
Common iliac artery, 368, 368f
Common iliac vein, 129, 130f
Communicating artery, posterior, 188, 189f
Communicating vein, anterior, 160, 161f
Condyle(s), 14
femoral, 87, 88f, 95, 96f
lateral, 87, 88f
medial, 87, 88f
occipital, 168
tibial, 88, 89f, 95, 96f
lateral, 88, 89f
medial, 88, 89f
Conjunctiva, 226, 227f, 229, 230f
Constrictor muscles, 156, 157f
Conus elasticus, 136, 137f
Conus medullaris, 31, 32f
Cooper's ligaments, 249, 250f
Coracoacromial ligament, 43, 44f
Coracobrachialis, 57, 61, 63f
Coracoclavicular ligament, 43, 44f
Coracoid process, 43, 44f
Cords, of brachial plexus, 56
Cornea, 226, 227f, 235, 236f
Corniculates, 134, 135f
Coronal plane, 2, 3f
Coronary artery
left, 277, 278f
right, 278, 278f
Coronary sinus, 278, 278f
Coronoid process, 170
Corpora cavernosum
of clitoris, 361, 362f
of penis, 356, 357f
Corpus albicans, 365, 366f
Corpus callosum, 194, 202
Corpus cerebelli, 196
Corpus luteum, 365, 366f
Corpus spongiosum, 356, 357f, 361, 362f
Corpus striatum, 205, 206f
Corti, spiral organ of, 245, 246f
Costal cartilages, 251, 252f, 293, 293f
Costal facets, of thoracic vertebrae, 27, 27f
Costocervical trunk, 158
Costoclavicular ligament, 43, 44f
Costoscapular muscles, 254, 255f

Cowper's glands, 354, 355*f*
Cranial, 2, 3*f*
Cranial fossae, 168, 169
Cranial nerve(s), 209–224
    I (olfactory), 210, 210*f*
    II (optic), 211, 212*f*
    III (oculomotor), 211, 212*f*
    IV (trochlear), 211, 212*f*
    V (trigeminal), 213, 214*f*
    VI (abducens), 211, 212*f*
    VII (facial), 215, 216*f*
    VIII (vestibulocochlear), 217, 217*f*, 237, 238*f*
    IX (glossopharyngeal), 218, 219*f*
    X (vagus), 220, 221*f*, 329
    XI (accessory), 222, 223*f*
    XII (hypoglossal), 223, 224*f*
    in neck, 140, 141*f*
Cranial root, of accessory nerve, 222, 223*f*
Cranium, 168–170, 171f, 172f
Cremaster muscle, 350, 351*f*
Cribriform plate, 169
Cricoarytenoid
    lateral, 154, 155*f*
    posterior, 154, 155*f*
Cricoid cartilage, 133, 135*f*
Cricothyroid, 154, 155*f*
Crista galli, 169
Crown, of tooth, 177, 177*f*
Cruciate ligament
    anterior, 95, 96*f*
    posterior, 95, 96*f*
Crus
    of clitoris, 361, 362*f*
    of penis, 356, 357*f*
Crus cerebri, 198, 199*f*
Cubital vein, median, 80, 81*f*
Cuboid, 90, 91*f*, 97, 98*f*
Cuneiforms, 90, 91*f*, 134
Cystic artery, 315, 315*f*
Cystic duct, 312, 313*f*
Cystic vein, 318, 319*f*

Dartos muscle, 350, 351*f*
Deep, 4, 5*f*
Deep artery, 356, 357*f*
Deep inguinal ring, 350, 351*f*
Deep palmar arch, 78, 79*f*
Deep transverse perineus, 342
Deltoid, 61, 63*f*
Deltoid ligament, 97, 98*f*
Deltoid tuberosity, 45, 46*f*
Dens (odontoid process), 25, 26*f*
Dentine, 177, 177*f*
Dentition, 175–177, 175*f*, 176*f*, 177*f*
    deciduous, 175, 175*f*
    development of, 176, 176*f*
    longitudinal section of, 177, 177*f*
    permanent, 175, 175*f*
Dermis, 22, 23*f*
Descending aorta, 261, 261*f*
Descending colon, 308, 309*f*
Detrusor muscle, 347, 348*f*
Diaphragm
    abdominal, 257, 258*f*, 298, 299*f*
    pelvic, 337–339, 337*f*, 339*f*
Diaphysis, 10, 11*f*, 14
Diencephalon, 194, 200, 201*f*
Digastric, 150, 151*f*
Digestive canal, 302–313, 303*f*, 305*f*, 307*f*, 309*f*, 311*f*, 313*f*
Diploë, 190, 190*f*
Direction, 2–4, 3*f*, 5*f*
Distal, 2, 3*f*, 4, 5*f*
Distal convoluted tubule, 326, 327*f*
Distal phalanx
    of finger, 48, 48*f*
    of toe, 90, 91*f*
Dorsal, 2, 3*f*, 4, 5*f*
Dorsal interossei, 72, 73*f*
Dorsal ramus, 33, 34*f*
Dorsal root ganglion, 33, 34*f*
Dorsalis pedis artery, 127, 128*f*
Dorsiflexion, 6, 9*f*
Duct(s)
    alveolar, 262, 263*f*
    bile, 306, 307*f*, 312, 313*f*
    cochlear, 242, 243*f*, 245, 246*f*
    collecting, of nephron, 326, 327*f*
    cystic, 312, 313*f*
    efferent, 352, 353*f*
    ejaculatory, 349*f*, 354, 355*f*
    hepatic
        common, 310, 311*f*, 312, 313*f*
        left, 312, 313*f*
        right, 312, 313*f*
    lactiferous, 249, 250*f*
    lymphatic, right, 279, 280*f*
    nasolacrimal, 231, 231*f*
    pancreatic
        accessory, 306, 307*f*, 312, 313*f*
        main, 306, 307*f*, 312, 313*f*
    semicircular, 242, 243*f*, 244, 245*f*
    thoracic, 279, 280*f*
Ductus deferens, 349*f*, 352, 353*f*
Duodenum, 304, 305*f*, 306, 307*f*
Dura mater, 31, 32*f*, 190, 190*f*, 194
Dural folds, 191

Ear, 237–246
    external, 237, 238*f*, 239, 240*f*
    inner, 237, 238*f*, 242–246, 243*f*, 245*f*, 246*f*
    middle, 237, 238*f*, 241, 242*f*
Ear drum, 239, 240*f*
Efferent duct, 352, 353*f*
Ejaculation, 356
Ejaculatory duct, 349*f*, 354, 355*f*
Elbow, 51, 52*f*
    extensors of, 65, 66*f*
    flexors of, 65, 66*f*
Enamel cap, of tooth, 177, 177*f*
Endocranium, 190, 190*f*
Endolymphatic sac, 244, 245*f*
Endometrium, 362, 363*f*, 364*f*
Epicanthic fold, 226, 227*f*
Epicondyle(s), 14
    femoral, 87, 88*f*
        lateral, 87, 88*f*
    humeral
        lateral, 45, 46*f*
        medial, 45, 46*f*
Epidermis, 22, 23*f*
Epididymis, 349*f*, 352, 353*f*
Epidural space, 31, 32*f*
Epiglottis, 133, 135*f*, 156, 157*f*
Epiphyseal (growth) plate, 10, 11*f*, 14, 15*f*
Epiphysis (epiphyses), 10, 11*f*, 14
    fusion of, 14, 15*f*
Equilibrium, 244, 245*f*
Erection
    of clitoris, 361
    of penis, 356
Erector spinae, 38, 39*f*, 145
Esophageal hiatus, 257, 258*f*
Esophagus, 156, 157*f*, 162, 163*f*, 261, 261*f*, 302, 303*f*, 304, 305*f*
Ethmoid bone, 169, 171*f*, 172*f*, 173*f*, 186, 187*f*, 228, 228*f*
Eustachian tube, 237, 238*f*, 241, 242*f*
Eversion, 6, 8*f*
Extension, 6, 7*f*
Extensor carpi radialis brevis, 68, 69*f*
Extensor carpi radialis longus, 68, 69*f*
Extensor carpi ulnaris, 68, 69*f*
Extensor digiti minimi, 70, 71*f*
Extensor digitorum brevis, 125, 126*f*
Extensor digitorum communis, 70, 71*f*
Extensor digitorum longus, 123, 124*f*
Extensor hallucis brevis, 125, 126*f*
Extensor hallucis longus, 123, 124*f*
Extensor indicis, 70, 71*f*
Extensor pollicis brevis, 74, 75*f*
Extensor pollicis longus, 74, 75*f*
External, 4, 5*f*
External acoustic meatus, 239, 240*f*
External anal sphincter, 345, 345*f*
External auditory (acoustic) canal, 239, 240*f*
External carotid artery, 158, 159*f*, 164, 165*f*, 188
External ear, 237, 238*f*, 239, 240*f*
External iliac artery, 368, 368*f*
External iliac vein, 129, 130*f*
External intercostal muscle, 256, 256*f*
External jugular vein, 160, 161*f*
External laryngeal nerve, 140, 141*f*
External oblique muscle, 297, 298*f*
Extraocular muscles, 232, 233*f*–234*f*
Eye, 226–236
    bony orbit of, 228, 228*f*
    external features of, 226, 227*f*
    extrinsic muscles of, 232, 233*f*–234*f*
    lacrimal apparatus of, 231, 231*f*
Eyeball, 235, 236*f*
Eyelids, 226, 227*f*, 229, 230*f*

Face, bones of, 169–170, 173*f*, 174*f*
Facet, 14
Facial artery, 158, 159*f*
Facial expression, muscles of, 149, 149*f*, 178, 179*f*
Facial nerve, 140, 141*f*, 142, 215, 216*f*
Facial vein, 160, 161*f*, 191, 192*f*
    common, 160, 161*f*
Falciform ligament, 310, 311*f*
Fallopian tubes, 358, 358*f*, 363, 363*f*, 364*f*
False pelvis, 85
False vocal cords, 136, 137*f*
Falx cerebri, 191
Fascia, superficial, 22, 23*f*
Feet. *See* Foot (feet)
Femoral artery, 116, 127, 128*f*
Femoral neck, 87, 88*f*
Femoral nerve, 102, 103, 104*f*, 105, 106*f*, 116
Femoral triangle, 105, 116
Femoral vein, 116, 129, 130*f*
Femur, 87, 88*f*
    head of, ligament of, 93, 94*f*
Fibrous joints, 16, 17*f*
Fibrous pericardium, 267, 268*f*
Fibula, 88, 89*f*
Fibular artery, 127, 128*f*
Fibular nerve
    common, 102, 103, 104*f*, 105, 106*f*
    deep, 103, 104, 104*f*, 105, 106*f*
    superficial, 103, 104*f*, 105, 106*f*
Fibular vein, 129, 130*f*
Filum terminale, 31, 32*f*
Fimbriae, 363, 363*f*

Fingers
bones of, 48, 48f
muscles of, 70–77, 71f, 73f, 75f
nerves of, 57, 58f
Fissure
oblique, of lung, 264, 265f
orbital
inferior, 228, 228f
superior, 169, 171f, 172f, 213, 228, 228f
Flexion, 6, 7f
Flexor carpi radialis, 68, 69f
Flexor carpi ulnaris, 68, 69f
Flexor digiti minimi, 72, 73f
Flexor digitorum brevis, 125, 126f
Flexor digitorum longus, 123, 124f
Flexor digitorum profundus, 70, 71f
Flexor digitorum superficialis, 70, 71f
Flexor hallucis brevis, 125, 126f
Flexor hallucis longus, 123, 124f
Flexor pollicis brevis, 76, 77f
Flexor pollicis longus, 74, 75f
Flexor retinaculum, 53, 54f, 68
Flocculonodular lobes, of cerebellum, 196
Folds
aryepiglottic, 136, 137f
dural, 191
palatoglossal, 186, 187f
palatopharyngeal, 186, 187f
ventricular, 136, 137f
Folia cerebelli, 196
Follicle
mature (Graafian), 365, 366f
primary, 365, 366f
primordial, 365, 366f
secondary, 365, 366f
Foot (feet)
arch of, 91–92, 92f
bones of, 90–92, 91f
directional terms for, 4, 5f
muscles of, 120–125, 121f–122f
Foramen (foramina), 14
carotid, 168, 172f
incisive, 169, 172f
infraorbital, 169, 171f, 213
interventricular, 207, 208f
intervertebral, 24, 25f
jugular, 168, 172f, 173f, 220
mandibular, 170, 174f, 213
mental, 170, 174f, 213
obturator, 85, 86f, 333, 334f
palatine, 170
sacral, 28, 29f
sciatic, greater, 105
sphenopalatine, 170
stylomastoid, 168, 172f, 215, 216f
supraorbital, 213
vena caval, 257, 258f
vertebral, 24, 25f
Foramen lacerum, 168, 172f, 215, 216f
Foramen magnum, 168, 172f
Foramen ovale, 169, 172f, 213, 218
Foramen rotundum, 169, 172f, 213
Foramen spinosum, 169, 172f
Foramen transversarium, 25, 26f
Forearm
bones of, 45, 47f
muscles of, 65–67, 66f
nerves of, 57, 58f
Foregut, 314
Foreskin, 356, 357f
Fornix, 362, 362f, 363f, 364f
Fossa, 14
lacrimal, 228, 228f
Fossa ovalis, 273, 274f
Fovea capitis, 93, 94f
Fovea centralis, 235, 236f
Frenulum, 186
Frontal bone, 168, 171f, 172f, 173f, 228, 228f
Frontal lobe, 202, 203f
Frontalis, 178, 179f
Fundus
of stomach, 304, 305f
of uterus, 362, 363f

Gag reflex, 218
Gallbladder, 302, 303f, 310, 311f, 312, 313f
Ganglion
celiac, 220
cervical, 142, 144f
middle, 142, 144f
superior, 142, 144f
dorsal root, 33, 34f
otic, 218, 219f
paravertebral, 373, 374f
preaortic, 373, 374f
stellate, 142, 144f
sympathetic, paravertebral, 286, 287f
Gastric artery
left, 314, 315f
right, 315, 315f
Gastric vein
left, 318, 319f
right, 318, 319f
Gastrocnemius, 120, 121f
Gastroduodenal artery, 315, 315f
Gastroepiploic artery
left, 315, 315f
right, 315, 315f
Gastro-omental artery, 315, 315f
Gemellus inferior, 110, 111f
Gemellus superior, 110, 111f
Genial spines, 170
Genioglossus, 184, 185f
Geniohyoid, 150, 151f
Genitalia
female, 358–367, 360f, 362f, 363f, 364f, 366f, 367f
male, 349–357, 349f, 351f, 353f, 355f, 357f
Germinal epithelium, 365, 366f
Gingiva, 177, 177f
Gland(s)
adrenal, 294, 294f
Bartholin's, 359, 360f
bulbourethral, 349f, 354, 355f
lacrimal, 231, 231f
parathyroid, 164, 165f
prostate, 349f, 354, 355f
sebaceous, 22, 23f
Skene's, 359, 360f
sweat, 22, 23f
thyroid, 164, 165f
vestibular, 359, 360f
Glans clitoris, 361, 362f
Glans penis, 356, 357f
Glaucoma, 234
Glenohumeral ligaments, 49, 50f
Glenoid cavity, 43, 44f
Glenoid fossa, 168
Globus pallidus, 205, 206f
Glomerular (Bowman's) capsule, 326, 327f
Glomerulus, 326, 327f
Glossopharyngeal nerve, 140, 141f, 218, 219f
Gluteal nerve
inferior, 102, 103, 104f
superior, 102, 103, 104f
Gluteal tuberosity, 87, 88f
Gluteus maximus, 108, 109f
Gluteus medius, 109, 109f
Gluteus minimus, 109, 109f
Gomphosis, 16, 17f
Gonadal arteries, 328, 328f
Gonadal veins, 328, 328f
Gracilis, 112, 113f, 116, 117f
Gray matter, 31, 32f, 194, 202
Gray ramus communicans, 142, 143, 144f, 286, 287f, 288f, 289f
Great auricular nerve, 138, 139f
Great cardiac vein, 278, 278f
Great cerebral vein, 191, 192f
Great saphenous vein, 129, 130f
Greater curvature, of stomach, 304, 305f
Greater omentum, 300, 301f
Greater petrosal nerve, 215, 216f
Greater sciatic foramen, 105
Greater sciatic notch, 85, 86f, 333, 334f, 335, 336f
Greater trochanter, femoral, 87, 88f
Greater tubercle, humeral, 45, 46f
Gyrus
parahippocampal, 202, 203f
postcentral, 202, 203f
precentral, 202, 203f

Hair follicles, 22, 23f
Hallux, 90, 91f
Hamate, 48, 48f
Hamstring muscles, 112, 113f, 114, 115f
Hand. *See also* Fingers; Wrist
bones of, 48, 48f
directional terms for, 4, 5f
Haustra, 308, 309f
Haversian canal, 10, 11f
Head
blood vessels of, 188–192, 189f, 190f, 192f
bones of, 168–178, 171f, 172f, 173f, 174f
muscles of, 178–185, 179f, 181f, 183f, 185f
Head (of bone), 14
femoral, 87, 88f
fibular, 88, 89f
humeral, 45, 46f
radial, 45, 47f
Hearing, 245, 246f
Heart, 266–278
auscultation of, 274
blood supply of, 271, 272f, 277–278, 278f
chambers of, 269, 270f, 273–275, 274f
conducting system of, 276, 277f
innervation of, 276
location of, 266, 267f
lower margin of, 266, 267f
pericardium of, 267, 268f
pump action of, 269, 270f
upper margin of, 266, 267f
valves of, 273–275, 274f, 275f
Helicotrema, 245, 246f
Helix, 239, 240f
Hemiazygos vein, 283, 284f
Henle, loop of, 326, 327f
Hepatic artery, 310, 311f, 315, 315f
Hepatic duct
common, 310, 311f, 312, 313f
left, 312, 313f
right, 312, 313f
Hepatic veins, 318, 319f
Hiatus
aortic, 257, 258f
esophageal, 257, 258f
urogenital, 338, 339f

Hilum, 320, 321f
Hindbrain, 194, 195f
Hindgut, 314
Hip joint, 93, 94f
Horizontal oblique fissure, of lung, 264, 265f
Humerus, 45, 46f
Hymen, 359
Hyoglossus, 184, 185f
Hyoid, 133, 134f, 135f
Hyolaryngeal muscles, 150–153, 151f, 153f
Hypogastric nerve, 373, 374f
Hypogastric plexus, 373, 374f
Hypoglossal canal, 168, 172f, 173f
Hypoglossal nerve, 140, 141f, 223, 224f
Hypophyseal fossa, 169
Hypophysis, 200, 201f
Hypothalamus, 200, 201f
Hypothenar eminence, 72, 73f

Ileal branches, of superior mesenteric artery, 316, 316f
Ileocolic artery, 316, 316f
Ileocolic vein, 318, 319f
Ileum, 306, 307f
Iliac artery
  common, 368, 368f
  external, 127, 128f, 368, 368f
  internal, 127, 128f, 368, 368f
Iliac crest, 85, 86f
Iliac spine, superior, anterior, 85, 86f
Iliac vein
  common, 129, 130f
  external, 129, 130f
  internal, 129, 130f
Iliacus, 107, 108f
Iliococcygeus, 338, 339f
Iliocostalis, 38, 39f
Iliofemoral ligament, 93, 94f
Iliopsoas, 107, 108f
Iliotibial tract, 108
Ilium, 85, 86f, 333, 334f
Incisive foramen, 169, 172f
Incisor, 175, 175f
Incus, 241, 242f
Inferior, 2, 3f
Inferior alveolar nerve, 213, 214f
Inferior articular processes, vertebral, 24, 25f
Inferior cerebellar artery
  anterior, 188, 189f
  posterior, 188, 189f
Inferior cerebellar peduncle, 196, 197f
Inferior colliculi, 198, 199f
Inferior constrictor, 156, 157f
Inferior gluteal nerve, 102, 103, 104f
Inferior hypogastric plexus, 373, 374f
Inferior lip, of cervical vertebrae, 25, 26f
Inferior lobe, of lung, 264, 265f
Inferior mesenteric artery, 314, 314f, 317, 317f
Inferior mesenteric vein, 318, 319f
Inferior nasal concha, 169, 171f, 173f, 186, 187f
Inferior oblique muscles, of eye, 232, 233f–234f
Inferior orbital fissure, 228, 228f
Inferior rectal artery, 369, 370f
Inferior rectal nerve, 371, 372f
Inferior rectus muscle, 232, 233f–234f
Inferior sagittal sinus, 191, 192f
Inferior suprarenal artery, 322, 323f
Inferior thyroid artery, 158, 159f, 164, 165f
Inferior thyroid vein, 160, 161f, 162, 163f, 164, 165f
Inferior vena cava, 129, 130f, 310, 311f, 328, 328f
Inferior vesical artery, 369, 370f
Infraorbital foramen, 169, 171f, 213
Infraspinatus, 49, 50f, 61, 64f
Infundibulum, 200, 201f, 363, 363f
Inguinal ligament, 297, 298f, 350, 351f
Inguinal ring
  deep, 350, 351f
  superficial, 297, 298f, 350, 351f
Inner ear, 237, 238f, 242–246, 243f, 245f, 246f
Innermost intercostal muscle, 256, 256f
Insula, 202
Integument, 22, 23f
Interclavicular ligament, 43, 44f
Intercondylar eminence, 88, 89f
Intercostal muscles, 256, 256f
Internal, 4, 5f
Internal auditory (acoustic) meatus, 168, 172f, 173f
Internal carotid artery, 158, 159f, 188, 189f
Internal iliac artery, 368, 368f
Internal iliac vein, 129, 130f
Internal intercostal muscle, 256, 256f
Internal jugular vein, 160, 161f, 162, 163f, 164, 165f
Internal laryngeal nerve, 140, 141f
Internal oblique muscles, 297, 298f
Internal pudendal artery, 369, 370f
Internal sphincter urethrae, 347, 348f
Internal thoracic artery, 158, 159f, 281, 282f
Interossei, 72, 73f
Interspinales, 39
Interspinous ligaments, 29, 30f
Intertarsal joint, 97, 98f
Intertransversarii, 39
Intertrochanteric crest, 87, 88f
Interventricular artery, posterior, 278, 278f
Interventricular foramen, 207, 208f
Intervertebral disc, 24, 25f
Intervertebral foramen, 24, 25f
Intestine
  large, 302, 303f, 308, 309f
  small, 302, 303f, 306, 307f
Intrinsic muscles
  of back, 38, 39f, 40f
  of larynx, 154, 155f
Inversion, 6, 8f
Iris, 226, 227f, 235, 236f
Ischial spine, 85, 86f, 333, 334f
Ischial tuberosity, 85, 86f, 333, 334f
Ischiocavernosus, 342, 343f
Ischiococcygeus, 338, 339f
Ischiofemoral ligament, 93, 94f
Ischiopubic ramus, 333, 334f, 335, 336f
Ischiorectal (ischioanal) fossa, 340, 341f
Ischium, 85, 86f, 333, 334f
Isthmus, 202, 203f

Jejunal branches, of superior mesenteric artery, 316, 316f
Jejunum, 306, 307f
Joint(s), 16–19
  acromioclavicular, 43, 44f
  ankle, 90, 91f, 97, 98f
  cartilaginous, 16, 17f
  elbow, 51, 52f
  fibrous, 16, 17f
  hip, 93, 94f
  intertarsal, 97, 98f
  knee, 95, 96f
  of lower limb, 93–98, 94f, 96f, 98f
  radiocarpal, 53, 54f
  shoulder, 49, 50f
  sternoclavicular, 43, 44f
  subtalar, 97, 98f
  synovial, 18, 19f
  talocrural, 97, 98f
  tarsal, transverse, 97, 98f
  wrist, 53, 54f
Joint capsule, 18, 19f
  of elbow, 51, 52f
  of hip, 93, 94f
  of knee, 95, 96f
  of shoulder, 49, 50f
Jugular foramen, 168, 172f, 173f, 220
Jugular notch, 251
Jugular vein
  anterior, 160, 161f
  external, 160, 161f
  internal, 160, 161f, 162, 163f, 164, 165f

Kidneys, 294, 294f, 322–327, 323f, 325f, 327f
  structure of, 324, 325f
Knee joint, 95, 96f
Kyphosis
  coccygeal, 29, 30f
  sacral, 29, 30f
  thoracic, 29, 30f

Labia majora, 358–359, 358f, 360f
Labia minora, 358–359, 358f, 360f
Labial, 4, 5f
Labrum, of glenoid, 49, 50f
Labyrinth, 237, 238f
  bony, 242, 243f, 245, 246f
  membranous, 242, 243f, 245, 246f
Lacrimal apparatus, 231, 231f
Lacrimal bones, 169, 171f, 173f, 228, 228f
Lacrimal canaliculus, 231, 231f
Lacrimal caruncle, 226, 227f
Lacrimal duct, 169
Lacrimal fossa, 228, 228f
Lacrimal gland, 231, 231f
Lacrimal punctum, 226, 227f, 231, 231f
Lacrimal sac, 231, 231f
Lactation, 249
Lactiferous ducts, 249, 250f
Lactiferous sinuses, 249, 250f
Lambdoid suture, 168
Lamina, vertebral, 24, 25f
Large intestine, 302, 303f, 308, 309f
Laryngeal membranes, 136, 137f
Laryngeal muscles, intrinsic, 154, 155f
Laryngeal nerve
  external, 140, 141f
  internal, 140, 141f
  recurrent, 140, 141f, 220, 221f, 285, 285f
  superior, 140, 141f, 220, 221f
Laryngopharynx, 156, 157f
Larynx, 133, 135f
Lateral, 2, 3f
Lateral aperatures, 207, 208f
Lateral canthus, 226, 227f
Lateral collateral ligament, of ankle, 97, 98f
Lateral condyles
  femoral, 87, 88f
  tibial, 88, 89f
Lateral cord, of brachial plexus, 56
Lateral cricoarytenoid, 154, 155f
Lateral epicondyles
  femoral, 87, 88f
  humeral, 45, 46f
Lateral lobes, of cerebellum, 196
Lateral malleolus
  of ankle, 97, 98f
  of knee, 88, 89f

Lateral mammary branches, of lateral thoracic artery, 282, 282f
Lateral mass, of sacral vertebrae, 28, 29f
Lateral pectoral nerve, 56, 57
Lateral pterygoid, 180, 181f
Lateral rectus muscle, 232, 233f–234f
Lateral sacral artery, 369, 370f
Lateral (Sylvian) sulcus, 202
Lateral thoracic artery, 282, 282f
Lateral ventricles, 207, 208f
Lateral/collateral ligament, 95, 96f
Latissimus dorsi, 36, 37f
Left brachiocephalic vein, 279, 280f
Left colic artery, 317, 317f
Left colic (hepatic) flexure, 308, 309f
Left colic vein, 318, 319f
Left common carotid artery, 260, 261f, 271, 272f
Left coronary artery, 277, 278f
Left gastric artery, 314, 315f
Left gastric vein, 318, 319f
Left gastroepiploic artery, 315, 315f
Left gonadal artery, 328, 328f
Left gonadal vein, 328, 328f
Left hepatic duct, 312, 313f
Left lobe, of liver, 310, 311f
Left lung, 264, 265f
Left renal veins, 328, 328f
Left subclavian artery, 260, 261f, 271, 272f
Leg
  bones of, 88, 89f
  muscles of, 114–119, 115f, 117f, 119f
  nerves of, 99–106, 99f, 103f, 104f, 106f
Lens, 235, 236f
Lentiform nucleus, 205, 206f
Lesser curvature, of stomach, 304, 305f
Lesser occipital nerve, 138, 139f
Lesser omentum, 300, 301f
Lesser petrosal nerve, 218, 219f
Lesser sciatic notch, 85, 86f, 333, 334f
Lesser trochanter, femoral, 87, 88f
Lesser tubercle, humeral, 45, 46f
Levator ani, 338, 339f
Levator palpebrae superioris, 229, 230f, 232, 233f
Levator scapulae, 36, 37f, 145
Levator (veli) palatini, 182, 183f
Levatores costarum, 39
Ligament(s), 18, 19f
  acetabular, transverse, 93, 94f
  annular, of elbow, 51, 52f
  broad, 366, 367f
  calcaneofibular, 97, 98f
  capsular, 18, 19f
  collateral
    lateral, 97, 98f
    medial (tibial), 95, 96f
    radial, 51, 52f, 53, 54f
    ulnar, 51, 52f, 53, 54f
  coracoacromial, 43, 44f
  coracoclavicular, 43, 44f
  costoclavicular, 43, 44f
  cruciate
    anterior, 95, 96f
    posterior, 95, 96f
  deltoid, 97, 98f
  falciform, 310, 311f
  glenohumeral, 49, 50f
  iliofemoral, 93, 94f
  inguinal, 297, 298f, 350, 351f
  interclavicular, 43, 44f
  interspinous, 29, 30f
  ischiofemoral, 93, 94f
  lateral/collateral, 95, 96f
  longitudinal, 29, 30f
    anterior, 29, 30f
    posterior, 29, 30f
  ovarian, 366, 367f
  patellar, 95, 96f
  pelvic, 333, 334f
  periodontal, 177, 177f
  pubofemoral, 93, 94f
  radiocarpal, 53, 54f
  round, 366, 367f
  sacroiliac, 85
    anterior, 333, 334f
    posterior, 333, 334f
  sacrospinous, 333, 334f
  sacrotuberous, 333, 334f
  spring, 92, 92f, 97
  stylohyoid, 133
  supraspinous, 29, 30f
  suspensory
    of breast, 249, 250f
    of eyeball, 235, 236f
    of uterus, 366, 367f
  talofibular, 97, 98f
    anterior, 97, 98f
    posterior, 97, 98f
  ulnocarpal, 53, 54f
Ligamentum flavum, 29, 30f
Ligamentum nuchae, 36, 38
Ligamentum teres, 93, 94f, 310, 311f
Limbic lobe, 202
Limbic system, 202
Linea alba, 295, 296f
Linea aspera, 87, 88f
Lingual, 4, 5f
Lingual artery, 158, 159f
Lingual nerve, 140, 141f, 213, 214f
Lingula, lung, 264, 265f
Lip, of cervical vertebrae, 25, 26f
Liver, 294, 294f, 302, 303f, 310, 311f
Lobe(s)
  caudate, of liver, 310, 311f
  left, of liver, 310, 311f
  of lung, 264, 265f
  quadrate, of liver, 310, 311f
  renal, 324, 325f
  right, of liver, 310, 311f
Lobules, of testis, 352, 353f
Longissimus, 38, 39f
Longitudinal fibers, of tongue, 184, 185f
Longitudinal layer, of gastric smooth muscle, 304, 305f
Longitudinal ligament
  anterior, 29, 30f
  posterior, 29, 30f
Longitudinal plane, 2
Longus capitis, 145, 146f
Longus colli, 146, 148f
Loop of Henle, 326, 327f
Lordosis
  cervical, 29, 30f
  lumbar, 29, 30f
Lower limb
  blood vessels of, 127–130, 128f, 130f
  bones of, 85–92, 86f, 88f, 89f, 91f, 92f
  dorsal compartment of, 99, 99f
  joints of, 93–98, 94f, 96f, 98f
  muscles of, 107–126, 108f, 109f, 111f, 113f, 115f, 117f, 119f, 121f–122f, 124f, 126f
    dorsal, 100–101
    ventral, 100–101
  nerves of, 99–106, 99f, 103f, 104f, 106f
  ventral compartment of, 99, 99f
Lumbar lordosis, 29, 30f
Lumbar nerves, 33, 34f
Lumbar plexus, 33, 34f, 102, 371, 372f
Lumbar vertebrae, 27, 28f, 293, 293f
Lumbosacral plexus, 33, 102, 103f
Lumbosacral trunk, 33, 102
Lumbricals, 72, 73f
Lunate, 48, 48f, 53, 54f
Lungs, 264, 265f
Lymph nodes, inguinal, superficial, 116
Lymphatic duct, right, 162, 163f, 279, 280f
Lymphatic vessels
  of abdomen, 320, 321f
  of dermis, 22, 23f
  of thorax, 279, 280f

Macula lutea, 235, 236f
Main pancreatic duct, 312, 313f
Major calyces, of renal pelvis, 324, 325f
Malleolus
  lateral
    of ankle, 97, 98f
    of knee, 88, 89f
  medial
    of ankle, 97, 98f
    of knee, 88, 89f
Malleus, 241, 242f
Mammary gland, 249, 250f
Mammillary bodies, 200, 201f
Mandible, 170, 174f
Mandibular division, of trigeminal nerve, 213, 214f
Mandibular foramen, 170, 174f, 213
Manubrium, 43, 44f, 251, 252f
Marginal branch, of right coronary artery, 278, 278f
Masseter, 180, 181f
Mastication, muscles of, 180, 181f
Mastoid process, 168
Mature (Graafian) follicle, 365, 366f
Maxilla, 169, 171f, 173f, 182f, 186, 187f, 228, 228f
Maxillary artery, 158, 159f
Maxillary division, of trigeminal nerve, 213, 214f
Maxillary sinus, 186, 187f
Medial, 2, 3f
Medial canthus, 226, 227f
Medial condyles
  femoral, 87, 88f
  tibial, 88, 89f
Medial cord, of brachial plexus, 56
Medial epicondyles
  femoral, 87, 88f
  humeral, 45, 46f
Medial malleolus
  of ankle, 97, 98f
  of knee, 88, 89f
Medial mammary branches, of internal thoracic artery, 281, 282f
Medial pectoral nerve, 56, 57
Medial pterygoid, 180, 181f
Medial rectus muscle, 232, 233f–234f
Median antebrachial vein, 80, 81f
Median aperture, 207, 208f
Median cubital vein, 80, 81f
Median nerve, 56, 57, 58f, 59, 60f
Median raphé, 156
Median sagittal plane, 2, 3f
Mediastinum, 260–261, 261f
Medulla oblongata, 194, 196, 197f
Medullary branch, of accessory nerve, 222, 223f
Medullary (marrow) cavity, 10, 11f
Meissner's corpuscles, 22, 23f

Membrane(s)
basilar, 245, 246*f*
laryngeal, 136, 137*f*
otolithic, 244
perineal, 340, 341*f*
quadrangular, 136, 137*f*
synovial, 18, 19*f*
of elbow, 51, 52*f*
of hip, 93
of knee, 95, 96*f*
of shoulder, 49, 50*f*
tectorial, 245, 246*f*
thyrohyoid, 136, 137*f*
tympanic, 237, 238*f*, 239, 240*f*, 241, 242*f*
vestibular, 245, 246*f*
Meninges, 31, 32*f*, 190, 190*f*
Meniscus
lateral, 95, 96*f*
medial, 95, 96*f*
Mental foramen, 170, 174*f*, 213
Mental protuberance, 170
Mesencephalon, 194, 195*f*, 199*f*
Mesenteric artery
inferior, 314, 314*f*, 317, 317*f*
superior, 306, 307*f*, 314, 314*f*, 316, 316*f*
Mesenteric vein
inferior, 318, 319*f*
superior, 318, 319*f*
Mesentery, 300, 301*f*
Mesial, 4, 5*f*
Mesocolon, transverse, 300, 301*f*
Mesovarium, 366, 367*f*
Metacarpals, 48, 48*f*
Metaphysis, 10, 11*f*
Metatarsals, 90, 91*f*
Midbrain, 194, 195*f*, 199*f*
Middle cardiac vein, 278, 278*f*
Middle cerebellar peduncle, 196, 197*f*
Middle cerebral artery, 188, 189*f*
Middle cervical ganglion, 142, 144*f*
Middle colic artery, 316, 316*f*
Middle constrictor, 156, 157*f*
Middle ear, 237, 238*f*, 241, 242*f*
Middle lobe, of lung, 264, 265*f*
Middle nasal concha, 186, 187*f*
Middle phalanx
of finger, 48, 48*f*
of toe, 90, 91*f*
Middle rectal artery, 369, 370*f*
Middle suprarenal artery, 322, 323*f*
Middle thyroid vein, 160, 161*f*, 164, 165*f*
Midgut, 314
Minor calyces, of renal pelvis, 324, 325*f*
Mitral bicuspid valve, 273, 274*f*
Mitral valve, 275, 275*f*
Molar, 175, 175*f*
longitudinal section of, 177, 177*f*
Motor neurons, 33, 35*f*
Movement, terminology for, 6, 7*f*–9*f*
Multifidus, 39, 40*f*
Muscle(s)
of abdomen, 295–299, 296*f*, 298*f*, 299*f*
abdominal diaphragm, 257, 258*f*, 298, 299*f*
abductor digiti minimi, 72, 73*f*
abductor hallucis, 92, 125, 126*f*
abductor pollicis brevis, 76, 77*f*
abductor pollicis longus, 74, 75*f*
adductor brevis, 112, 113*f*
adductor longus, 112, 113*f*
adductor magnus, 112, 113*f*
adductor pollicis, 76, 77*f*
anconeus, 65, 66*f*
ansa cervicalis, 138, 139*f*
ansa subclavia, 143, 144*f*
of arm, 60–64, 63*f*, 64*f*
arytenoideus, 154, 155*f*
back, 36–40
intrinsic, 38, 39*f*, 40*f*
superficial, 36, 37*f*
biceps brachii, 65, 66*f*, 67
long head of, 65, 66*f*
tendon of, 49, 50*f*
short head of, 65, 66*f*
biceps femoris, 114, 115*f*
brachialis, 65, 66*f*
brachioradialis, 65, 66*f*
buccinator, 178, 179*f*
bulbospongiosus, 342, 343*f*
coccygeus, 338, 339*f*
constrictor, 156, 157*f*
inferior, 156, 157*f*
middle, 156, 157*f*
superior, 156, 157*f*
coracobrachialis, 57, 61, 63*f*
costoscapular, 254, 255*f*
cremaster, 350, 351*f*
Dartos, 350, 351*f*
deltoid, 61, 63*f*
detrusor, 347, 348*f*
digastric, 150, 151*f*
dorsal interossei, 72, 73*f*
erector spinae, 38, 39*f*, 145
extensor carpi radialis brevis, 68, 69*f*
extensor carpi radialis longus, 68, 69*f*
extensor carpi ulnaris, 68, 69*f*
extensor digiti minimi, 70, 71*f*
extensor digitorum brevis, 125, 126*f*
extensor digitorum communis, 70, 71*f*
extensor digitorum longus, 123, 124*f*
extensor hallucis brevis, 125, 126*f*
extensor hallucis longus, 123, 124*f*
extensor indicis, 70, 71*f*
extensor pollicis brevis, 74, 75*f*
extensor pollicis longus, 74, 75*f*
external anal sphincter, 345, 345*f*
external intercostal, 256, 256*f*
external oblique, 297, 298*f*
extraocular, 232, 233*f*–234*f*
of facial expression, 149, 149*f*, 178, 179*f*
finger, 70–77, 71*f*, 73*f*, 75*f*
flexor carpi radialis, 68, 69*f*
flexor carpi ulnaris, 68, 69*f*
flexor digiti minimi, 72, 73*f*
flexor digitorum brevis, 125, 126*f*
flexor digitorum longus, 123, 124*f*
flexor digitorum profundus, 70, 71*f*
flexor digitorum superficialis, 70, 71*f*
flexor hallucis brevis, 125, 126*f*
flexor hallucis longus, 123, 124*f*
flexor pollicis brevis, 76, 77*f*
flexor pollicis longus, 74, 75*f*
flexor retinaculum, 53, 54*f*, 68
foot, 120–125, 121*f*–122*f*
forearm, 65–67, 66*f*
frontalis, 178, 179*f*
gastrocnemius, 120, 121*f*
gemellus inferior, 110, 111*f*
gemellus superior, 110, 111*f*
genioglossus, 184, 185*f*
geniohyoid, 150, 151*f*
gluteus maximus, 108, 109*f*
gluteus medius, 109, 109*f*
gluteus minimus, 109, 109*f*
gracilis, 112, 113*f*, 116, 117*f*
hamstring, 112, 113*f*, 114, 115*f*
head, 178–185, 179*f*, 181*f*, 183*f*, 185*f*
hyoglossus, 184, 185*f*
hyolaryngeal, 150–153, 151*f*, 153*f*
iliacus, 107, 108*f*
iliococcygeus, 338, 339*f*
iliocostalis, 38, 39*f*
iliopsoas, 107, 108*f*
inferior oblique, 232, 233*f*–234*f*
inferior rectus, 232, 233*f*–234*f*
infraspinatus, 49, 50*f*, 61, 64*f*
innermost intercostal, 256, 256*f*
intercostal, 256, 256*f*
internal intercostal, 256, 256*f*
internal oblique, 297, 298*f*
interossei, 72, 73*f*
interspinales, 39
intertransversarii, 39
ischiocavernosus, 342, 343*f*
laryngeal, intrinsic, 154, 155*f*
lateral pterygoid, 180, 181*f*
lateral rectus muscle, 232, 233*f*–234*f*
latissimus dorsi, 36, 37*f*
leg, 114–119, 115*f*, 117*f*, 119*f*
levator, 36, 37*f*, 145
levator ani, 338, 339*f*
levator palatini, 182, 183*f*
levator palpebrae superioris, 229, 230*f*, 232, 233*f*
levatores costarum, 39
longissimus, 38, 39*f*
longus capitis, 145, 146*f*
longus colli, 146, 148*f*
of lower limb, 107–126, 108*f*, 109*f*, 111*f*, 113*f*, 115*f*, 117*f*, 119*f*, 121*f*–122*f*, 124*f*, 126*f*
lumbricals, 72, 73*f*
masseter, 180, 181*f*
of mastication, 180, 181*f*
medial pterygoid, 180, 181*f*
medial rectus, 232, 233*f*–234*f*
multifidus, 39, 40*f*
of neck, 145–157, 146*f*, 148*f*, 149*f*, 151*f*, 153*f*, 155*f*, 157*f*
obliquus capitis inferior, 39
obliquus capitis superior, 39
obturator externus, 110, 111*f*
obturator internus, 110, 111*f*
opponens digiti minimi, 72, 73*f*
opponens pollicis, 76, 77*f*
orbicularis oculi, 178, 179*f*, 229, 230*f*
orbicularis oris, 178, 179*f*
palatoglossus, 184, 185*f*
palmar interossei, 72, 73*f*
palmaris longus, 68, 69*f*
papillary, 273, 274*f*
pectineus, 112, 113*f*
pectoralis major, 57, 60, 63*f*, 249, 250*f*
pectoralis minor, 254, 255*f*
pelvic diaphragm, 337–339, 337*f*, 339*f*
of pelvis, 337–345, 339*f*, 341*f*, 343*f*, 344*f*, 345*f*
of perineum, 340–344, 341*f*, 343*f*, 344*f*, 345*f*
peroneus brevis, 120, 121*f*
peroneus longus, 120, 121*f*
peroneus tertius, 122, 122*f*
piriformis, 110, 111*f*
plantaris, 120, 121*f*
platysma, 149, 149*f*
popliteus, 116, 117*f*
pronator quadratus, 67, 67*f*
pronator teres, 67, 67*f*
psoas major, 107, 108*f*, 298, 299*f*
psoas minor, 299, 299*f*
pubococcygeus, 338, 339*f*
puborectalis, 338, 339*f*

quadratus femoris, 110, 111*f*
quadratus lumborum, 36, 37*f*, 298, 299*f*
quadriceps femoris, 118, 119*f*
rectus abdominis, 295, 296*f*
rectus capitis anterior, 145
rectus capitis lateralis, 145
rectus capitis posterior major, 39
rectus capitis posterior minor, 39
rectus cervicis, 152, 153*f*
rectus femoris, 118, 119*f*
rhomboideus, 145
rhomboideus major, 36, 37*f*
rhomboideus minor, 36, 37*f*
rotator cuff, 49, 50*f*
rotatores, 39
sartorius, 116, 117*f*
scalenus anterior, 146, 148*f*
scalenus medius, 146, 148*f*
scalenus posterior, 147, 148*f*
scapulohumeral, 61, 63*f*
semimembranosus, 114, 115*f*
semispinalis, 39, 40*f*
semispinalis capitis, 39, 40*f*, 145
semispinalis cervicis, 146
semitendinosus, 114, 115*f*
serratus anterior, 249, 250*f*, 254, 255*f*
small segmental of back, 39
smooth, of stomach, 304, 305*f*
of soft palate, 182, 183*f*
soleus, 120, 121*f*
sphincter urethrae, 342, 343*f*
spinalis, 38, 39*f*
splenius, 38, 39*f*
splenius capitis, 38, 39*f*, 145
splenius cervicis, 38, 39*f*, 146
stapedius, 241, 242*f*
sternocleidomastoid, 145, 146*f*
sternocostal, 255, 255*f*
sternohyoid, 152, 153*f*
sternothyroid, 152, 153*f*
styloglossus, 184, 185*f*
stylohyoid, 150, 151*f*
suboccipital, 40
subscapularis, 49, 50*f*, 61, 64*f*
superior oblique, 232, 233*f*–234*f*
superior rectus, 232, 233*f*–234*f*
supraspinatus, 49, 50*f*, 61, 64*f*
temporalis, 180, 181*f*
tensor palatini, 182, 183*f*
tensor tympani, 241, 242*f*
teres major, 61, 63*f*
teres minor, 49, 50*f*, 61, 64*f*
thigh, 107–113, 108*f*, 109*f*, 111*f*, 113*f*
thoracohumeral, 60, 63*f*
thumb, 74–77, 75*f*, 77*f*
thyrohyoid, 152, 153*f*
tibialis anterior, 122, 122*f*
tibialis posterior, 120, 121*f*
toes, 123–126, 124*f*, 126*f*
of tongue, 184, 185*f*
transversus abdominis, 297, 298*f*
transversus thoracis, 255, 255*f*
trapezius, 36, 37*f*, 145
triceps brachii, 65, 66*f*
of urogenital triangle, 342, 343*f*, 344*f*
vastus intermedius, 118, 119*f*
vastus lateralis, 118, 119*f*
vastus medialis, 118, 119*f*
wrist, 68, 69*f*
zygomaticus major, 178, 179*f*
Musculocutaneous nerve, 56, 57, 58*f*, 59, 60*f*
Mylohyoid, 150, 151*f*
Mylohyoid nerve, 140, 141*f*
Nasal aperture, 169
Nasal bones, 169, 171*f*, 173*f*
Nasal cavity, 156, 157*f*, 186, 187*f*
Nasal concha
inferior, 169, 171*f*, 173*f*, 186, 187*f*
middle, 169, 186, 187*f*
superior, 169, 186, 187*f*
Nasal septum, 169
Nasal spine, anterior, 169
Nasal visual field, 211, 212*f*
Nasolacrimal duct, 169, 231, 231*f*
Nasopharynx, 156, 157*f*
Navicular, 90, 91, 91*f*, 92*f*, 97, 98*f*
Neck
autonomic nerves in, 142–143, 144*f*
blood vessels of, 158–163, 159*f*, 161*f*, 163*f*
bones of, 133, 134*f*
cartilages of, 133–137, 134*f*, 135*f*, 137*f*
muscles of, 145–157, 146*f*, 148*f*, 149*f*, 151*f*, 153*f*, 155*f*, 157*f*
nerves of, 138–144, 139*f*, 141*f*, 144*f*
root of, 162, 163*f*
Neck (of bone), 14
Nephron, 326, 327*f*
Nerve(s)
abducens, 211, 212*f*
accessory, 140, 141*f*, 222, 223*f*
alveolar, 177, 177*f*
inferior, 213, 214*f*
auricular, great, 138, 139*f*
auriculotemporal, 213, 214*f*
axillary, 56, 57, 59, 60*f*
brachial plexus, 33, 34*f*, 55, 56, 57*f*
cervical, 33, 34*f*
transverse, 138, 139*f*
cervical plexus, 33, 34*f*, 138, 139*f*
coccygeal, 33, 34*f*
cranial, 209–224
I (olfactory), 210, 210*f*
II (optic), 211, 212*f*
III (oculomotor), 211, 212*f*
IV (trochlear), 211, 212*f*
V (trigeminal), 213, 214*f*
VI (abducens), 211, 212*f*
VII (facial), 215, 216*f*
VIII (vestibulocochlear), 217, 217*f*, 237, 238*f*
IX (glossopharyngeal), 218, 219*f*
X (vagus), 220, 221*f*, 329
XI (accessory), 222, 223*f*
XII (hypoglossal), 223, 224*f*
in neck, 140, 141*f*
of dermis, 22, 23*f*
dorsal, of penis (clitoris), 371, 372*f*
facial, 140, 141*f*, 142, 215, 216*f*
femoral, 102, 103, 104*f*, 105, 106*f*, 116
fibular
common, 102, 103, 104*f*, 105, 106*f*
deep, 103, 104, 104*f*, 105, 106*f*
superficial, 103, 104*f*, 105, 106*f*
forearm, 57, 58*f*
glossopharyngeal, 140, 141*f*, 218, 219*f*
carotid branch of, 218, 219*f*
gluteal
inferior, 102, 103, 104*f*
superior, 102, 103, 104*f*
hypogastric, 373, 374*f*
hypoglossal, 140, 141*f*, 223, 224*f*
laryngeal
external, 140, 141*f*
internal, 140, 141*f*
recurrent, 140, 141*f*, 220, 221*f*, 285, 285*f*
superior, 140, 141*f*, 220, 221*f*
lingual, 140, 141*f*, 213, 214*f*
of lower limb, 99–106, 99*f*, 103*f*, 104*f*, 106*f*
lumbar, 33, 34*f*
lumbar plexus, 33, 34*f*, 102
lumbosacral plexus, 33, 102, 103*f*
median, 56, 57, 58*f*, 59, 60*f*
musculocutaneous, 56, 57, 58*f*, 59, 60*f*
mylohyoid, 140, 141*f*
of neck, 138–144, 139*f*, 141*f*, 144*f*
obturator, 102, 103, 104*f*, 105, 106*f*, 371, 372*f*
occipital, lesser, 138, 139*f*
oculomotor, 211, 212*f*
olfactory, 210, 210*f*
optic, 211, 212*f*, 235
pectoral
lateral, 56, 57
medial, 56, 57
of pelvis, 371–374, 372*f*, 374*f*
perineal, 371, 372*f*
petrosal
greater, 215, 216*f*
lesser, 218, 219*f*
pharyngeal, 220, 221*f*
phrenic, 138, 139*f*, 162, 163*f*, 284, 285*f*
pudendal, 371, 372*f*
radial, 56, 57, 58*f*, 59, 60*f*
rectal, inferior, 371, 372*f*
sacral, 33, 34*f*
sciatic, 102, 105, 106*f*, 371, 372*f*
spinal, 33, 34*f*, 35*f*
of abdomen, 329, 330*f*
of neck, 138, 139*f*
of pelvis, 371, 372*f*
thoracic, 284, 285*f*
splanchnic, pelvic, 373, 374*f*
subscapular, 56, 57
supraclavicular, 138, 139*f*
suprascapular, 56, 57, 59, 60*f*
thoracic, 33, 34*f*
thyroid, 143, 144*f*
tibial, 102, 103, 104, 104*f*, 105, 106*f*
trigeminal, 140, 141*f*, 213, 214*f*
trochlear, 211, 212*f*
tympanic, 218, 219*f*
ulnar, 56, 57, 58*f*, 59, 60*f*
of upper limb, 55–60, 57*f*, 58*f*, 60*f*
vagus, 140, 141*f*, 162, 163*f*, 220, 221*f*, 285, 285*f*
pharyngeal branch of, 140, 141*f*
vestibulocochlear, 217, 217*f*
cochlear branch of, 217, 217*f*
Nerve plexus. *See also* Plexus
of hair follicle, 22, 23*f*
Neural arch, 24, 25*f*
Neurons
motor, 33, 35*f*
sensory, 33, 35*f*
Nipple, 249, 250*f*
Node
atrioventricular, 276, 277*f*
sinoatrial, 276, 277*f*
Notch(s)
cardiac, 264, 265*f*
jugular, 251
radial, 45, 47*f*
of ulna, 51, 52*f*
sciatic, 85, 86*f*
greater, 85, 86*f*, 333, 334*f*, 335, 336*f*
lesser, 85, 86*f*, 333, 334*f*
trochlear, 45, 47*f*
of ulna, 51, 52*f*

Nuchal lines, 168
Nuclei, basal, 205, 206f
Nutrient artery, 10, 11f

Oblique fissure, of lung, 264, 265f
Oblique layer, of gastric smooth muscle, 304, 305f
Oblique muscles
 external, of abdomen, 297, 298f
 inferior, of eye, 232, 233f–234f
 internal, of abdomen, 297, 298f
 superior, of eye, 232, 233f–234f
Obliquus capitis inferior, 39
Obliquus capitis superior, 39
Obstetrical pelvis, 85
Obturator externus, 110, 111f
Obturator foramen, 85, 86f, 333, 334f
Obturator internus, 110, 111f
Obturator nerve, 102, 103, 104f, 105, 106f, 371, 372f
Occipital artery, 158, 159f
Occipital bone, 168, 171f, 172f, 173f
Occipital condyles, 168
Occipital lobe, 202, 203f
Occipital nerve, lesser, 138, 139f
Occipital vein, 160, 161f
Occlusal, 4, 5f
Oculomotor nerve, 211, 212f
Olecranon, 45, 47f
Olecranon bursa, 51, 52f
Olecranon fossa, 45, 46f
Olfactory bulb, 210, 210f
Olfactory nerve, 210, 210f
Olives, of medulla oblongata, 196, 197f
Omentum
 greater, 300, 301f
 lesser, 300, 301f
Omohyoid, 152, 153f
Oocyte, 365, 366f
Ophthalmic artery, 188, 189f
Ophthalmic division, of trigeminal nerve, 213, 214f
Ophthalmic vein, 191, 192f
Opponens digiti minimi, 72, 73f
Opponens pollicis, 76, 77f
Opposition, 6, 9f
Optic canal, 169, 171f, 172f, 228, 228f
Optic chiasm, 211, 212f
Optic disc, 235, 236f
Optic nerve, 211, 212f, 235, 236f
Optic tract, 211, 212f
Oral cavity, 186, 187f
Orbicularis oculi, 178, 179f, 229, 230f
Orbicularis oris, 178, 179f
Orbit, 228, 228f
Orbital fissure
 inferior, 228, 228f
 superior, 169, 171f, 172f, 213, 228, 228f
Orbital septum, 229, 230f
Organ nerve, 286, 287f
Orgasm, 361
Oropharynx, 156, 157f
Os coxae, 85, 86f, 333, 334f
Os, external, 362, 363f, 364f
Ossicles, auditory, 237, 238f
Ossification, 14, 15f
Osteon, 10, 11f
Otic ganglion, 218, 219f
Otolithic membrane, 244
Oval window, 241, 242, 242f, 243f, 244, 245f
Ovarian artery, 368, 368f
Ovarian ligament, 366, 367f
Ovary, 358, 358f, 365, 366f

Pacinian corpuscles, 22, 23f
Palate
 hard, 156, 157f
 soft, 186, 187f
  muscles of, 182, 183f
Palatine bones, 170, 172f, 173f, 186, 187f
Palatine foramen, 170
Palatine tonsils, 156, 157f, 186, 187f
Palatoglossal folds, 186, 187f
Palatoglossus, 184, 185f, 186
Palatopharyngeal folds, 186, 187f
Palm, bones of, 48, 48f
Palmar, 4, 5f
Palmar aponeurosis, 68
Palmar arch
 deep, 78, 79f
 superficial, 78, 79f
Palmar interossei, 72, 73f
Palmaris longus, 68, 69f
Palmaris longus tendon, 68
Palpebral conjunctiva, 229, 230f
Palpebral fissure, 226, 227f
Pancreas, 302, 303f, 312, 313f
Pancreatic duct
 accessory, 306, 307f
 main, 306, 307f, 312, 313f
Papilla, of renal collecting ducts, 324, 325f
Papillary layer, 22, 23f
Papillary muscles, 273, 274f
Parahippocampal gyrus, 202, 203f
Parasympathetic nervous system, of pelvis, 373, 374f
Parathyroid glands, 164, 165f
Paravertebral ganglion, 286, 287f, 373, 374f
Parietal bone, 168, 171f, 172f, 173f
Parietal lobe, 202, 203f
Parietal peritoneum, 300, 301f
Parietal serous pericardium, 267, 268f
Patella, 87, 88f, 95, 96f
Patellar ligament, 95, 96f
Pecten pubis, 85
Pectineus, 112, 113f
Pectoral girdle, 43, 44f
Pectoral nerve
 lateral, 56, 57
 medial, 56, 57
Pectoralis major, 57, 60, 63f, 249, 250f
Pectoralis minor, 254, 255f
Pedicles, vertebral, 24, 25f
Peduncles, cerebral, 198
Pelvic brim, 333, 334f
Pelvic cavity, 337, 337f
Pelvic diaphragm, 337–339, 337f, 339f
Pelvic girdle, bones of, 85, 86f
Pelvic splanchnic nerves, 373, 374f
Pelvis
 blood vessels of, 368–370, 368f, 370f
 bones of, 85, 86f, 333, 334f
 false, 85
 ligaments of, 333, 334f
 muscles of, 337–345, 339f, 341f, 343f, 344f, 345f
 nerves of, 371–374, 372f, 374f
 obstetrical, 85
 sex differences in, 335, 336f
Penis, 349f, 356, 357f
 artery of, 369, 370f
 dorsal nerve of, 371, 372f
Pericardial cavity, 259, 260f, 267, 268f
Pericardium, 267, 268f
Pericranium, 190, 190f
Perineal artery, 369, 370f
Perineal body, 338, 339f
Perineal membrane, 340, 341f
Perineal nerve, 371, 372f
Perineum, 337, 337f
 muscles of, 340–344, 341f, 343f, 344f, 345f
Perineus, transverse
 deep, 342
 superficial, 342
Periodontal ligaments, 177, 177f
Periosteum, 10, 11f
Peritoneal cavity, 300, 301f
Peritoneum, 300, 301f, 346, 346f
Peritubular capillary, 326, 327f
Peroneal nerve. *See* Fibular nerve
Peroneus brevis, 120, 121f
Peroneus longus, 120, 121f
Peroneus tertius, 122, 122f
Petrosal nerve
 greater, 215, 216f
 lesser, 218, 219f
Petrosal sinus, superior, 191, 192f
Phalanges
 of foot, 90, 91f
 of hand, 48, 48f
Pharyngeal artery, ascending, 158, 159f
Pharyngeal branches, of glossopharyngeal nerve, 218, 219f
Pharyngeal nerve, 220, 221f
Pharynx, 156, 157f
Phrenic nerve, 138, 139f, 162, 163f, 284, 285f
Pia mater, 31, 32f, 190, 190f, 194
Pineal gland, 200, 201f
Piriformis, 110, 111f
Pisiform, 48, 48f
Pituitary gland (hypophysis), 200, 201f
Planes, 2, 3f
Plantar, 4, 5f
Plantar aponeurosis, 92, 92f
Plantar arteries, 127, 128f
Plantarflexion, 6, 9f
Plantaris, 120, 121f
Platysma, 149, 149f
Pleural cavity, 259, 260f
Pleural dome, 162, 163f
Plexus
 brachial, 33, 34f, 55, 56, 57f
 cervical, 33, 34f, 138, 139f
 hypogastric, 373, 374f
 lumbar, 33, 34f, 102, 371, 372f
 lumbosacral, 33, 102, 103f
 sacral, 33, 34f, 102, 371, 372f
 spinal, 33, 34f
Plica semilunaris, 226, 227f
Plicae circulares, 306, 307f
Pons, 194, 196, 197f
Popliteal artery, 127, 128f
Popliteal vein, 129, 130f
Popliteus, 116, 117f
Porta hepatis, 310, 311f
Portal vein, 310, 311f, 318, 319f, 320, 321f
Postcentral gyrus, 202, 203f
Posterior, 2, 3f
Posterior auricular artery, 158, 159f
Posterior auricular vein, 160, 161f
Posterior cerebral artery, 188, 189f
Posterior chamber, 235, 236f
Posterior communicating artery, 188, 189f
Posterior cord, of brachial plexus, 56
Posterior cricoarytenoid, 154, 155f
Posterior cruciate ligament, 95, 96f
Posterior inferior cerebellar artery, 188, 189f
Posterior interventricular artery, 278, 278f
Posterior longitudinal ligament, 29, 30f

Posterior talofibular ligament, 97, 98f
Posterior tibial artery, 127, 128f
Posterior tibial vein, 129, 130f
Posterior vagal trunk, 220, 221f
Preaortic ganglion, 373, 374f
Precentral gyrus, 202, 203f
Premolar, 175, 175f
Prepuce, 356, 357f
Primary follicle, 365, 366f
Primary oocyte, 365, 366f
Primordial follicle, 365, 366f
Process, 14
Profunda (deep) femoris artery, 127, 128f
Projection fibers, 204, 204f
Pronation, 6, 8f
Pronator quadratus, 67, 67f
Pronator teres, 67, 67f
Prosencephalon, 194, 195f, 200, 201f
Prostate gland, 349f, 354, 355f
Proximal, 2, 3f
Proximal convoluted tubule, 326, 327f
Proximal phalanx
    of finger, 48, 48f
    of toe, 90, 91f
Psoas major, 107, 108f, 298, 299f
Psoas minor, 299, 299f
Pterygoid
    lateral, 180, 181f
    medial, 180, 181f
Pterygoid plates, 169
Pterygomandibular raphé, 156
Pterygopalatine ganglion, 215, 216f
Pubic symphysis, 85, 86f
Pubis, 85, 86f, 333, 334f, 335, 336f
    superior ramus of, 333, 334f
Pubococcygeus, 338, 339f
Pubofemoral ligament, 93, 94f
Puborectalis, 338, 339f
Pudendal artery, internal, 369, 370f
Pudendal nerve, 371, 372f
Pulmonary arteries, 261, 261f, 269, 270f, 271, 272f
Pulmonary circulatory system, 269, 270f
Pulmonary semilunar valve, 273, 274f, 275, 275f
Pulmonary trunk, 271, 272f, 273, 274f
Pulmonary veins, 269, 270f, 271, 272f
Pulp chamber, 177, 177f
Pupil, 226, 227f, 235, 236f
Purkinje fibers, 276, 277f
Putamen, 205, 206f
Pyloric orifice, 304, 305f
Pyloric sphincter, 304, 305f
Pylorus, 304, 305f
Pyramidal decussation, 196, 197f
Pyramidalis, 295
Pyramids, of medulla oblongata, 196, 197f
Pyriform aperture, 169

Quadrangular membrane, 136, 137f
Quadrate lobe, of liver, 310, 311f
Quadratus femoris, 110, 111f
Quadratus lumborum, 36, 37f, 298, 299f
Quadriceps femoris, 118, 119f
    tendon of, 95, 96f

Radial artery, 78, 79f
Radial (bicipital) tuberosity, 45, 47f
Radial collateral ligament, 51, 52f, 53, 54f
Radial nerve, 56, 57, 58f, 59, 60f
Radial notch, 45, 47f, 51, 52f
Radiocarpal joint, 53, 54f
Radiocarpal ligament, 53, 54f
Radius, 45, 47f
    head of, 51, 52f
Rectal artery
    inferior, 369, 370f
    middle, 369, 370f
    superior, 317, 317f, 368, 368f
Rectal nerve, inferior, 371, 372f
Rectal vein, superior, 318, 319f
Rectouterine pouch, 346, 346f
Rectovesical pouch, 346, 346f
Rectum, 308, 309f, 340, 341f
Rectus abdominis, 295, 296f
Rectus capitis anterior, 145
Rectus capitis lateralis, 145
Rectus capitis posterior major, 39
Rectus capitis posterior minor, 39
Rectus cervicis, 152, 153f
Rectus femoris, 118, 119f
Rectus muscle
    inferior, 232, 233f–234f
    lateral, 232, 233f–234f
    medial, 232, 233f–234f
    superior, 232, 233f–234f
Recurrent laryngeal nerve, 140, 141f, 220, 221f, 285, 285f
Renal artery, 322, 323f, 324, 325f, 328, 328f
Renal column, 324, 325f
Renal cortex, 324, 325f
Renal lobe, 324, 325f
Renal pelvis, 324, 325f
Renal pyramids, 324, 325f
Renal tubules, 326, 327f
Renal vein, 322, 323f, 324, 325f, 328, 328f
Respiration, 265
    thoracic movement during, 251–259
Respiratory apparatus, 262–265, 263f, 265f
Respiratory terminal bronchioles, 262, 263f
Rete testis, 352, 353f
Reticular layer, 22, 23f
Retina, 235, 236f
Retinal artery, central, 235, 236f
Retromandibular vein, 160, 161f
Rhombencephalon, 194, 195f, 196, 197f
Rhomboideus, 145
Rhomboideus major, 36, 37f
Rhomboideus minor, 36, 37f
Rib(s), 251, 252f, 293, 293f
    bucket handle movement of, 253f
    costovertebral articulation of, 252f
    first, 43, 44f
Right atrium, 273, 274f
Right brachiocephalic vein, 279, 280f
Right colic artery, 316, 316f
Right colic (hepatic) flexure, 308, 309f
Right coronary artery, 278, 278f
Right gastric artery, 315, 315f
Right gastric vein, 318, 319f
Right gastropiploic artery, 315, 315f
Right gonadal artery, 328, 328f
Right gonadal vein, 328, 328f
Right hepatic duct, 312, 313f
Right lobe, of liver, 310, 311f
Right lung, 264, 265f
Right lymphatic duct, 279, 280f
Right ventricle, 273, 274f
Rima glottidis, 136
Root, of tooth, 177, 177f
Rotation, 6, 8f
Rotator cuff, 49, 50f, 61, 63f, 64f
Rotatores, 39
Round ligament, 366, 367f
Round window, 241, 242, 242f, 243f, 245, 246f
Rugae
    of bladder, 347, 348f
    of stomach, 304, 305f

Saccule, 242, 243f, 244, 245f
Sacral artery, 369, 370f
Sacral canal, 28, 29f
Sacral kyphosis, 29, 30f
Sacral nerves, 33, 34f
Sacral plexus, 33, 34f, 102, 371, 372f
Sacral vertebrae, 28, 29f
Sacroiliac joint, 333, 334f
Sacroiliac ligaments, 85
    anterior, 333, 334f
    posterior, 333, 334f
Sacrospinous ligament, 333, 334f
Sacrotuberous ligament, 333, 334f
Sacrum, 85, 86f, 333, 334f
    articular surface of, 28, 29f, 85
Sacs, alveolar, 262, 263f
Sagittal plane, 2, 3f
Sagittal sinus
    inferior, 191, 192f
    superior, 191, 192f, 207, 208
Sagittal suture, 168
Saphenous vein, 129, 130f
Sartorius, 116, 117f
Scala tympani, 245, 246f
Scala vestibuli, 245, 246f
Scalenus anterior, 146, 148f
Scalenus medius, 146, 148f
Scalenus posterior, 147, 148f
Scaphoid, 48, 48f, 53, 54f
Scapula, 43, 44f
Scapulohumeral muscles, 61, 63f
Schlemm, canal of, 235, 236f
Sciatic nerve, 102, 105, 106f, 371, 372f
Sciatic notch, 85, 86f
    greater, 333, 334f, 335, 336f
    lesser, 333, 334f
Sclera, 226, 227f, 235, 236f
Scrotum, 349–350, 349f, 351f
Sebaceous (oil) glands, 22, 23f
Secondary follicle, 365, 366f
Secondary oocyte, 365, 366f
Sella turcica, 169
Semicircular canals, 242, 243f, 244, 245f
Semicircular duct, 242, 243f, 244, 245f
Semilunar valve
    aortic, 273, 274f, 275, 275f
    pulmonary, 273, 274f, 275, 275f
Semimembranosus, 114, 115f
Seminal vesicle, 349f, 354, 355f
Seminiferous tubules, 352, 353f
Semispinalis, 39, 40f
Semispinalis capitis, 39, 40f, 145
Semispinalis cervicis, 146
Semitendinosus, 114, 115f
Sensory neurons, 33, 35f
Septum, orbital, 229, 230f
Septum pellucidum, 207
Serous pericardium, 267, 268f
Serratus anterior, 249, 250f, 254, 255f
Shoulder, 49, 50f
Sigmoid branches, of inferior mesenteric artery, 317, 317f
Sigmoid colon, 308, 309f
Sigmoid sinus, 168
Sigmoid vein, 318, 319f
Sinoatrial node, 276, 277f
Sinus(es)
    cavernous, 191, 192f
    coronary, 278, 278f
    lactiferous, 249, 250f

Sinus(es)—*Cont.*
maxillary, 186, 187*f*
paranasal, 168
petrosal, superior, 191, 192*f*
sagittal
inferior, 191, 192*f*
superior, 168, 191, 192*f*, 207, 208
sigmoid, 168
sphenoparietal, 191, 192*f*
straight, 191, 192*f*
transverse, 168, 191, 192*f*
Skeleton, 12*f*. *See also* Bone(s)
appendicular, 13, 13*f*
axial, 13, 13*f*
Skene's glands, 359, 360*f*
Skin, 22, 23*f*
Skull, 168–178, 171*f*, 172*f*, 173*f*, 174*f*
Small cardiac vein, 278, 278*f*
Small intestine, 302, 303*f*, 306, 307*f*
Small saphenous vein, 129, 130*f*
Small segmental muscles, of back, 39
Snuff box, anatomical, 74
Soft palate, 186, 187*f*
muscles of, 182, 183*f*
Soleus, 120, 121*f*
Spermatic cord, 350, 351*f*
Spermatic fascia, 350, 351*f*
Sphenoid bone, 169, 171*f*, 172*f*, 173*f*, 186, 187*f*, 228, 228*f*
Sphenopalatine foramen, 170
Sphenoparietal sinus, 191, 192*f*
Sphincter
anal, 345, 345*f*
cardiac, 304, 305*f*
pyloric, 304, 305*f*
Sphincter urethrae, 342, 343*f*
internal, 347, 348*f*
Spinal branch, of accessory nerve, 222, 223*f*
Spinal cord, 31, 32*f*
Spinal nerves, 33, 34*f*, 35*f*
of abdomen, 329, 330*f*
of neck, 138, 139*f*
of pelvis, 371, 372*f*
of thorax, 284, 285*f*
Spinal root, of accessory nerve, 222, 223*f*
Spinalis, 38, 39*f*
Spinous process, vertebral, 24, 25*f*
of axis, 25, 26*f*
lumbar, 27, 28*f*
thoracic, 27, 27*f*
Spiral organ of Corti, 245, 246*f*
Splanchnic nerves, pelvic, 373, 374*f*
Spleen, 294, 294*f*, 320, 321*f*
Splenic artery, 315, 315*f*, 320, 321*f*
Splenic vein, 320, 321*f*
Splenius, 38, 39*f*
Splenius capitis, 38, 39*f*, 145
Splenius cervicis, 38, 39*f*, 146
Sprained ankle, 97
Spring ligament, 92, 92*f*, 97
Squamosal suture, 168
Stapedius, 241, 242*f*
Stapes, 241, 242*f*
Stellate ganglion, 142, 144*f*
Sternal angle, 251
Sternal body, 251, 252*f*
Sternoclavicular joint, 43, 44*f*
Sternocleidomastoid, 145, 146*f*
Sternohyoid, 152, 153*f*
Sternothyroid, 152, 153*f*
Sternum, 251, 252*f*
Stomach, 294, 294*f*, 302, 303*f*, 304, 305*f*
Straight sinus, 191, 192*f*
Stratum corneum, 22, 23*f*
Stratum germinativum, 22, 23*f*
Stratum granulosum, 22, 23*f*
Stratum lucidum, 22, 23*f*
Stratum, superficial, 22, 23*f*
Stroma, ovarian, 365, 366*f*
Styloglossus, 184, 185*f*
Stylohyoid, 150, 151*f*
Stylohyoid ligament, 133
Styloid process, 45, 47*f*, 168
Stylomastoid foramen, 168, 172*f*, 215, 216*f*
Subacromial bursa, 49
Subarachnoid space, 31, 32*f*, 207, 208*f*
Subclavian artery, 78, 79*f*, 158, 159*f*, 164, 165*f*
left, 162, 163*f*, 260, 261*f*, 271, 272*f*
right, 162, 163*f*
Subclavian vein, 80, 81*f*, 160, 161*f*, 162, 163*f*
Subcutaneous layer, 22, 23*f*
Submandibular ganglion, 215, 216*f*
Suboccipital muscles, 40
Subpubic angle, 335, 336*f*
Subscapular bursa, 49
Subscapular nerve, 56, 57
Subscapularis, 49, 50*f*, 61, 64*f*
Subtalar joint, 97, 98*f*
Subtendinous olecranon bursa, 51, 52*f*
Subthalamus, 200, 201*f*
Sulcus, 14
central, 202
lateral (Sylvian), 202
Superficial, 4, 5*f*
Superficial fascia, 22, 23*f*
Superficial inguinal ring, 297, 298*f*, 350, 351*f*
Superficial palmar arch, 78, 79*f*
Superficial stratum, 22, 23*f*
Superficial temporal artery, 158, 159*f*
Superficial transverse perineus, 342
Superior, 2, 3*f*
Superior articular facet, of atlas, 25, 26*f*
Superior articular processes, vertebral, 24, 25*f*
Superior cerebellar artery, 188, 189*f*
Superior cerebellar peduncle, 196, 197*f*
Superior cervical ganglion, 142, 144*f*
Superior colliculi, 198, 199*f*
Superior constrictor, 156, 157*f*
Superior gluteal nerve, 102, 103, 104*f*
Superior hypogastric plexus, 373, 374*f*
Superior laryngeal nerve, 140, 141*f*, 220, 221*f*
Superior lip, of cervical vertebrae, 25, 26*f*
Superior lobe, of lung, 264, 265*f*
Superior mesenteric artery, 306, 307*f*, 314, 314*f*, 316, 316*f*
Superior mesenteric vein, 318, 319*f*
Superior nasal concha, 186, 187*f*
Superior oblique muscles, of eye, 232, 233*f*–234*f*
Superior orbital fissure, 169, 171*f*, 172*f*, 213, 228, 228*f*
Superior petrosal sinus, 191, 192*f*
Superior ramus of pubis, 333, 334*f*
Superior rectal artery, 317, 317*f*, 368, 368*f*
Superior rectal vein, 318, 319*f*
Superior rectus muscle, 232, 233*f*–234*f*
Superior sagittal sinus, 191, 192*f*, 207, 208
Superior suprarenal artery, 322, 323*f*
Superior thyroid artery, 158, 159*f*
Superior thyroid vein, 160, 161*f*, 164, 165*f*
Superior vena cava, 80, 81*f*, 162, 163*f*, 260, 261*f*, 271, 272*f*, 283, 284*f*
Superior vesical artery, 369, 370*f*
Supination, 6, 8*f*
Supinator, 67, 67*f*
Supraclavicular nerve, 138, 139*f*
Supraorbital foramen, 213
Suprapatellar bursa, 95, 96*f*
Suprarenal artery
inferior, 322, 323*f*
middle, 322, 323*f*
superior, 322, 323*f*
Suprarenal vein, 322, 323*f*
Suprascapular artery, 158, 159*f*
Suprascapular nerve, 56, 57, 59, 60*f*
Supraspinatus, 49, 50*f*, 61, 64*f*
Supraspinous ligament, 29, 30*f*
Suspensory ligaments, 235, 236*f*, 366, 367*f*
of breast, 249, 250*f*
of eyeball, 235, 236*f*
of uterus, 366, 367*f*
Sustentaculum tali, 92
Suture
coronal, 168
fibrous, 16, 17*f*
lambdoid, 168
sagittal, 168
squamosal, 168
Swallowing, 182
Sweat (sudoriferous) glands, 22, 23*f*
Sympathetic nervous system
in abdomen, 329, 330*f*
in neck, 142–143, 144*f*
of pelvis, 373, 374*f*
in thorax, 286–289, 287*f*, 288*f*, 289*f*
Sympathetic trunk, 286, 287*f*
Symphysis, 16, 17*f*, 333, 334*f*
Synchondrosis, 16, 17*f*
Syndesmosis, 16, 17*f*
Synovial joints, 18, 19*f*
Synovial membrane, 18, 19*f*
of elbow, 51, 52*f*
of hip, 93
of knee, 95, 96*f*
of shoulder, 49, 50*f*

Taeniae coli, 308, 309*f*
Tail, of epididymis, 352, 353*f*
Talocrural joint, 97, 98*f*
Talofibular ligaments, 97, 98*f*
Talus, 90, 91, 91*f*, 92*f*, 97, 98*f*
Tarsal bones, 90, 91*f*
Tarsal joint, transverse, 97, 98*f*
Tarsal plate, 229, 230*f*
Tectorial membrane, 245, 246*f*
Tectum, 198
Teeth. *See* Tooth (teeth)
Telencephalon, 194, 202–206, 203*f*, 204*f*, 206*f*
Temporal artery, superficial, 158, 159*f*
Temporal bone, 168, 171*f*, 172*f*, 239, 240*f*
Temporal lobe, 202, 203*f*
Temporal visual field, 211, 212*f*
Temporalis, 180, 181*f*
Tendinous intersections, of rectus abdominis, 295, 296*f*
Tendon
Achilles, 90, 120
of long head of biceps Brachii, 49, 50*f*
of quadriceps femoris, 95, 96*f*
Tendon sheath, 18
of shoulder, 49, 50*f*
Tensor fascia latae, 108, 109*f*
Tensor tympani, 241, 242*f*
Tensor (veli) palatini, 182, 183*f*
Tentorium cerebelli, 191, 198
Teres major, 61, 63*f*

Teres minor, 49, 50f, 61, 64f
Terminology
for anatomical position, 2, 2f, 3f
for direction, 2–4, 3f, 5f
for movement, 6, 7f–9f
osteological, 14
Testicular artery, 368, 368f
Testis, 349f, 352, 353f
Thalamus, 194, 200, 201f
Thenar eminence, 76, 77f
Thigh
bone of, 87, 88f
muscles of, 107–113, 108f, 109f, 111f, 113f
Thoracic artery
internal, 158, 159f, 281, 282f
lateral, 282, 282f
Thoracic cavity, 259, 260f
Thoracic duct, 162, 163f, 279, 280f
Thoracic kyphosis, 29, 30f
Thoracic nerves, 33, 34f
Thoracic spinal nerves, 284, 285f
Thoracic vertebrae, 27, 27f
Thoracohumeral muscles, 60, 63f
Thorax
blood vessels of, 281–284, 282f, 284f
bones of, 251–259, 252f
cavity of, 259, 260f
lymphatics of, 279, 280f
muscles of, 254–259, 255f, 256f, 258f
nerves of, 284–289, 285f, 287f, 288f, 289f
Thumb, muscles of, 74–77, 75f, 77f
Thyroarytenoid, 154, 155f
Thyrocervical trunk, 158, 159f
Thyrohyoid, 152, 153f
Thyrohyoid membrane, 136, 137f
Thyroid artery
inferior, 158, 159f, 164, 165f
superior, 158, 159f, 164, 165f
Thyroid cartilage, 133, 135f
Thyroid gland, 164, 165f
Thyroid nerves, 143, 144f
Thyroid vein
inferior, 160, 161f, 162, 163f, 164, 165f
middle, 160, 161f, 164, 165f
superior, 160, 161f, 164, 165f
Tibia, 88, 89f
Tibial artery
anterior, 127, 128f
posterior, 127, 128f
Tibial nerve, 102, 103, 104, 104f, 105, 106f
Tibial vein
anterior, 129, 130f
posterior, 129, 130f
Tibialis anterior, 122, 122f
Tibialis posterior, 120, 121f
Toes, muscles of, 123–126, 124f, 126f
Tongue, 156, 157f
muscles of, 184, 185f
Tonsils, palatine, 156, 157f, 186, 187f
Tooth (teeth), 175–177, 175f, 176f, 177f
development of, 176, 176f
directional terms for, 4, 5f
longitudinal section of, 177, 177f
permanent, 175, 175f
Torus tubarius, 186
Trabeculae, 10, 11f
Trachea, 133, 134f, 162, 163f, 261, 261f, 262, 263f
Tracheal cartilages, 262, 263f
Tragus, 239, 240f
Transverse acetabular ligament, 93, 94f
Transverse cervical nerve, 138, 139f
Transverse colon, 308, 309f
Transverse fibers, of tongue, 184, 185f
Transverse mesocolon, 300, 301f
Transverse perineus
deep, 342
superficial, 342
Transverse plane, 2, 3f
Transverse process, vertebral, 24, 25f
of atlas, 25, 26f
Transverse sinus, 191, 192f
Transverse tarsal joint, 97, 98f
Transversus abdominis, 297, 298f
Transversus thoracis, 255, 255f
Trapezium, 48, 48f
Trapezius, 36, 37f, 145
Trapezoid, 48, 48f
Triangular disc, of wrist, 53, 54f
Triceps brachii, 65, 66f
lateral head of, 65, 66f
long head of, 65, 66f
medial head of, 65, 66f
Tricuspid valve, 273, 274f, 275, 275f
Trigeminal nerve, 140, 141f, 213, 214f
Trigone, 347, 348f
Triquetrum, 48, 48f, 53, 54f
Triradiate cartilage, 85
Trochanter, 14
femoral, 87, 88f
greater, 87, 88f
lesser, 87, 88f
Trochlea, 45, 46f
humeral, 51, 52f
Trochlear nerve, 211, 212f
Trochlear notch, 45, 47f, 51, 52f
Tubercle(s), 14
humeral, 45, 46f
greater, 45, 46f
lesser, 45, 46f
Tuberosity, 14
deltoid, 45, 46f
gluteal, 87, 88f
ischial, 85, 86f
radial (bicipital), 45, 47f
Tunica albuginea, 352, 353f, 365, 366f
Tympanic bone, 239, 240f
Tympanic cavity, 237, 238f, 241, 242f
Tympanic membrane, 237, 238f, 239, 240f, 241, 242f
Tympanic nerve, 218, 219f

Ulna, 45, 47f
radial notch of, 51, 52f
trochlear notch of, 51, 52f
Ulnar artery, 78, 79f
Ulnar collateral ligament, 51, 52f
of wrist, 53, 54f
Ulnar nerve, 56, 57, 58f, 59, 60f
Ulnocarpal ligament, 53, 54f
Umbilical artery, 369, 370f
Uncus, 202, 203f
Upper limb. *See also specific bones, joints, muscles, and nerves*
bones of, 43–48, 44f, 46f, 47f, 48f
dorsal compartment of, 55, 55f
innervation of, 55–60, 57f, 58f, 60f
joints of, 49–54, 50f, 52f, 54f
muscles of, 60–77, 63f, 64f, 66f, 67f, 69f, 71f, 73f, 75f, 77f
ventral compartment of, 55, 55f
Ureteric orifice, 347, 348f
Ureters, 347, 348f
Urethra, 349f, 356, 357f
Urethral meatus, 359, 360f
Urethral orifice, 347, 348f
Urogenital hiatus, 338, 339f
Urogenital triangle, 340, 341f
muscles of, 342, 343f, 344f
Uterine artery, 369, 370f
Uterine tubes, 358, 358f, 363, 363f, 364f
Uterovesical pouch, 346, 346f
Uterus, 358, 358f, 362, 363f, 364f
Utricle, 242, 243f, 244, 245f
Uvula, 156, 157f, 186, 187f

Vagal trunk
anterior, 220, 221f
posterior, 220, 221f
Vagina, 358, 358f, 362, 362f, 363f, 364f
Vaginal orifice, 359, 360f, 362f
Vagus nerve, 140, 141f, 162, 163f, 220, 221f, 285, 285f, 329
pharyngeal branch of, 140, 141f
Vallate papillae, 186, 187f
Valve
mitral, 273, 274f, 275, 275f
semilunar
aortic, 273, 274f, 275, 275f
pulmonary, 273, 274f, 275, 275f
tricuspid, 273, 274f, 275, 275f
Vastus intermedius, 118, 119f
Vastus lateralis, 118, 119f
Vastus medialis, 118, 119f
Vein(s)
antebrachial, median, 80, 81f
anterior communicating, 160, 161f
auricular, posterior, 160, 161f
axillary, 80, 81f
azygos, 271, 272f, 283, 284f
basilic, 80, 81f
brachial, 80, 81f
brachiocephalic, 160, 161f, 162, 163f, 260, 261f, 271, 272f
left, 164, 165f, 279, 280f
right, 279, 280f
cardiac
great, 278, 278f
middle, 278, 278f
small, 278, 278f
cephalic, 80, 81f
cerebral, great, 191, 192f
colic
left, 318, 319f
middle, 318, 319f
right, 318, 319f
communicating, anterior, 160, 161f
cubital, median, 80, 81f
cystic, 318, 319f
facial, 160, 161f, 191, 192f
common, 160, 161f
femoral, 116, 129, 130f
fibular, 129, 130f
gastric
left, 318, 319f
right, 318, 319f
gonadal, 328, 328f
hemiazygos, 283, 284f
accessory, 283, 284f
hepatic, 318, 319f
ileocolic, 318, 319f
iliac
common, 129, 130f
external, 129, 130f
internal, 129, 130f
jugular
anterior, 160, 161f
external, 160, 161f
internal, 160, 161f, 162, 163f, 164, 165f

Vein(s)—*Cont.*
of lower limb, 129, 130*f*
mesenteric
inferior, 318, 319*f*
superior, 318, 319*f*
of neck, 160, 161*f*
occipital, 160, 161*f*
ophthalmic, 191, 192*f*
popliteal, 129, 130*f*
portal, 310, 311*f*, 318, 319*f*, 320, 321*f*
pulmonary, 269, 270*f*, 271, 272*f*
rectal, superior, 318, 319*f*
renal, 322, 323*f*, 324, 325*f*, 328, 328*f*
left, 328, 328*f*
retromandibular, 160, 161*f*
saphenous, 129, 130*f*
great, 129, 130*f*
small, 129, 130*f*
sigmoid, 318, 319*f*
splenic, 318, 320, 321*f*
subclavian, 80, 81*f*, 160, 161*f*, 162, 163*f*
suprarenal, 322, 323*f*
thyroid
inferior, 160, 161*f*, 162, 163*f*, 164, 165*f*
middle, 160, 161*f*, 164, 165*f*
superior, 160, 161*f*, 164, 165*f*
tibial
anterior, 129, 130*f*
posterior, 129, 130*f*
of upper limb, 80, 81*f*
vertebral, 160, 161*f*
Vena cava, 269, 270*f*
inferior, 129, 130*f*, 310, 311*f*, 328, 328*f*
superior, 80, 81*f*, 162, 163*f*, 260, 261*f*, 271, 272*f*, 283, 284*f*
Vena caval foramen, 257, 258*f*
Venae comitantes
of lower limb, 129, 130*f*
of upper limb, 80, 81*f*
Venous sinus, 190, 190*f*
drainage of, 191, 192*f*
Ventral, 2, 3*f*
Ventral ramus, 33, 34*f*
of spinal nerve, 286, 287*f*
Ventral root, 33, 34*f*
of spinal nerve, 286, 287*f*
Ventricle(s), 207
fourth, 207, 208*f*
lateral, 208*f*
left, 269, 270*f*, 271, 272*f*, 273, 274*f*
right, 269, 270*f*, 271, 272*f*, 273, 274*f*
third, 194, 207, 208*f*
Ventricular fold, 136, 137*f*
Vermiform appendix, 308, 309*f*
Vermis, 196
Vertebra prominens, 36
Vertebrae
cervical, 25, 26*f*, 133, 134*f*
coccygeal, 28, 29*f*
lumbar, 27, 28*f*, 293, 293*f*
sacral, 28, 29*f*
structure of, 24, 25*f*
thoracic, 27, 27*f*, 251, 252*f*
Vertebral arteries, 158, 159*f*, 188, 189*f*
Vertebral body, 24, 25*f*
cervical, 25, 26*f*
lumbar, 27, 28*f*
sacral, 28, 29*f*
thoracic, 27, 27*f*
Vertebral column, 24–30, 25*f*
cervical, 25, 26*f*
coccygeal, 28, 29*f*
curvatures of, 29, 30*f*
lumbar, 27, 28*f*
sacral, 28, 29*f*
stabilizing ligaments, 29, 30*f*
thoracic, 27, 27*f*
Vertebral foramina, 24, 25*f*
Vertebral vein, 160, 161*f*
Vesical artery
inferior, 369, 370*f*
superior, 369, 370*f*
Vestibular branch, of vestibulocochlear nerve, 217, 217*f*
Vestibular glands
greater, 359, 360*f*
lesser, 359, 360*f*
Vestibular membrane, 245, 246*f*
Vestibule, 242, 243*f*, 244, 245*f*
Vestibulocochlear nerve, 217, 217*f*, 237, 238*f*
Visceral peritoneum, 300, 301*f*
Visceral serous pericardium, 267, 268*f*
Visual field
nasal, 211, 212*f*
temporal, 211, 212*f*
Vitreous chamber, 235, 236*f*
Vocal cords, 136, 137*f*
Vomer, 170, 171*f*, 172*f*, 173*f*, 186, 187*f*
Vulva, 358–359, 358*f*, 360*f*

White matter, 194, 202, 204, 204*f*
of spinal cord, 32*f*, 33
White ramus communicans, 286, 287*f*, 288*f*, 289*f*
Wrist, 53, 54*f*
bones of, 48, 48*f*
muscles of, 68, 69*f*

Xiphoid process, 251, 252*f*

Zygomatic arch, 168, 169, 171*f*, 172*f*
Zygomatic bone, 169, 171*f*, 172*f*, 228, 228*f*
Zygomaticus major, 178, 179*f*